Developmental Biology

A COMPREHENSIVE SYNTHESIS

Volume 5

The Molecular Biology of Cell Determination and Cell Differentiation

Developmental Biology
A COMPREHENSIVE SYNTHESIS

Editor
LEON W. BROWDER
University of Calgary
Calgary, Alberta, Canada

Developmental Biology

A COMPREHENSIVE SYNTHESIS

Volume 5

The Molecular Biology of Cell Determination and Cell Differentiation

Edited by

LEON W. BROWDER

University of Calgary
Calgary, Alberta, Canada

PLENUM PRESS • NEW YORK AND LONDON

Library of Congress Cataloging in Publication Data

(Revised for vol. 5)

Developmental biology.

Vol. 4 edited by Ralph B. L. Gwatkin.
Includes bibliographies and indexes.
Contents: v. 1. Oogenesis—v. 2. The cellular basis of morphogenesis—[etc.]—v. 5.
The molecular biology of cell determination and cell differentiation.
1. Developmental biology—Collected works. I. Browder, Leon W.
QH491.D426 1985   574.3   85-3406
ISBN 0-306-42735-4

Cover illustration: Drosophila embryo that is homozygous for a strong mutation at the
Krüppel locus. The embryo has been stained with an antibody to the fushi tarazu (ftz)
segmentation protein. The Krüppel mutation causes an altered pattern of ftz stripes.
(Micrograph courtesy of Dr. Sean Carroll, University of Wisconsin, Madison.)

Grandma Kröte is enjoying the balmy summer night at the shores of her beloved pond. Unconcerned about science and scientists, she stubbornly keeps her secrets to herself. (Print by Dr. Johannes F. Holtfreter.)

Contributors

T. Bisseling Department of Molecular Biology, Agricultural University, De Dreijen 11/6703 BC, Wageningen, The Netherlands

Lois A. Chandler Urological Cancer Research Laboratory, Comprehensive Cancer Center and Department of Biochemistry, University of Southern California, Los Angeles, California 90033

Caren Chang Division of Biology, California Institute of Technology, Pasadena, California 91125

Arthur Chovnick Molecular and Cell Biology Department, The University of Connecticut, Storrs, Connecticut 06268

William R. Crain, Jr. Cell Biology Group, Worcester Foundation for Experimental Biology, Shrewsbury, Massachusetts 01545

E. Bryan Crenshaw III Howard Hughes Medical Institute, Eukaryotic Regulatory Biology Program, School of Medicine, University of California, San Diego, La Jolla, California 92093

Martha L. Crouch Department of Biology, Indiana University, Bloomington, Indiana 47405

F. Lee Dutton, Jr. Molecular and Cell Biology Department, The University of Connecticut, Storrs, Connecticut 06268

Ronald Emeson Howard Hughes Medical Institute, Eukaryotic Regulatory Biology Program, School of Medicine, University of California, San Diego, La Jolla, California 92093

Laurence D. Etkin Department of Molecular Genetics, The University of Texas Cancer Systems Center, M.D. Anderson Hospital and Tumor Institute, Houston, Texas 77030

Jeffrey Guise Howard Hughes Medical Institute, Eukaryotic Regulatory Biology Program, School of Medicine, University of California, San Diego, La Jolla, California 92093

Johannes F. Holtfreter   Department of Biology, University of Rochester, Rochester, New York 14627

William R. Jeffery   Center for Developmental Biology, Department of Zoology, University of Texas, Austin, Texas 78712

Peter A. Jones   Urological Cancer Research Laboratory, Comprehensive Cancer Center and Department of Biochemistry, University of Southern California, Los Angeles, California 90033

Stuart Leff   Howard Hughes Medical Institute, Eukaryotic Regulatory Biology Program, School of Medicine, University of California, San Diego, La Jolla, California 92093

Sergio Lira   Howard Hughes Medical Institute, Eukaryotic Regulatory Biology Program, School of Medicine, University of California, San Diego, La Jolla, California 92093

Elliot M. Meyerowitz   Division of Biology, California Institute of Technology, Pasadena, California 91125

N. A. Morrison   Department of Biology, Centre for Plant Molecular Biology, Montreal, Quebec H3A 1B1, Canada

Charles Nelson   Howard Hughes Medical Institute, Eukaryotic Regulatory Biology Program, School of Medicine, University of California, San Diego, La Jolla, California 92093

Christian Nelson   Howard Hughes Medical Institute, Eukaryotic Regulatory Biology Program, School of Medicine, University of California, San Diego, La Jolla, California 92093

Michael G. Rosenfeld   Howard Hughes Medical Institute, Eukaryotic Regulatory Biology Program, School of Medicine, University of California, San Diego, La Jolla, California 92093

Andrew Russo   Howard Hughes Medical Institute, Eukaryotic Regulatory Biology Program, School of Medicine, University of California, San Diego, La Jolla, California 92093

Matthew P. Scott   Department of Molecular, Cellular, and Developmental Biology, University of Colorado, Boulder, Colorado 80309-0347

J. C. Smith   Laboratory of Embryogenesis, National Institute for Medical Research, The Ridgeway, Mill Hill, London NW7 1AA, England

Jamshed R. Tata   Laboratory of Developmental Biochemistry, National Institute for Medical Research, Mill Hill, London NW7 1AA, England

D. P. S. Verma   Department of Biology, Centre for Plant Molecular Biology, Montreal, Quebec H3A 1B1, Canada

William B. Wood   Department of Molecular, Cellular, and Developmental Biology, University of Colorado, Boulder, Colorado 80309-0347

W. Michael Wormington   Department of Biochemistry, Rosenstiel Basic Medical Sciences Research Center, Brandeis University, Waltham, Massachusetts 02254

Preface

This series was established to create comprehensive treatises on specific topics in developmental biology. Such volumes serve a useful role in developmental biology, which is a very diverse field that receives contributions from a wide variety of disciplines. This series is a meeting ground for the various practitioners of this science, facilitating an integration of heterogeneous information on specific topics.

Each volume is comprised of chapters selected to provide the conceptual basis for a comprehensive understanding of its topic as well as an analysis of the key experiments upon which that understanding is based. The specialist in any aspect of developmental biology should understand the experimental background of the specialty and be able to place that body of information in context, in order to ascertain where additional research would be fruitful. The creative process then generates new experiments. This series is intended to be a vital link in that ongoing process of learning and discovery.

The application of the techniques and principles of molecular biology to problems of development has resulted in the most intense period of investigation and discovery in the history of developmental biology. Today, the major thrust of molecular biology is to understand the controls over differential gene expression. The identification of *cis*-acting regulatory sequences and demonstration of their potentiation by *trans*-acting factors have provided us with the framework for understanding how selective transcription of individual genes is mediated. Much remains to be learned, however, about the molecular mechanisms involved in activation and maintenance of transcriptional activity.

Developmental biologists have a broader mission, which is to explain how the complex program of differential gene expression is established and utilized during embryonic development. We recognize two general mechanisms for establishment of cellular domains having unique patterns of gene expression: segregation of cytoplasmic determinants into discrete mitotic derivatives and exposure of cells to extrinsic signals (i.e., embryonic induction). The roles of cytoplasmic determinants are discussed in Chapters 1 and 2. Embryonic induction has been studied most intensively in amphibians, beginning with Spemann's pioneering work. Johannes Holtfreter, Spemann's student, presents a personal perspective on Spemann's organizer in Chapter 4. Contemporary research on embryonic induction is discussed in Chapter 3.

Study of the expression of specific genes during development promotes the understanding of differential gene expression in a broader context, and the developmental systems highlighted in this book have been chosen with that aim in mind. Actin genes, discussed in Chapter 6, are expressed in lineage-specific patterns during sea urchin development. The coincidence of expression of specific actin genes and the establishment of lineages provide the opportunity to study cell determination at the molecular level with a precision never before possible. *Xenopus* ribosomal protein synthesis, discussed in Chapter 8, is subject to both transcriptional and posttranscriptional regulation during early development. Since selective transcription and delayed utilization of messenger RNA are both important strategies in animal development, this system provides an opportunity to learn how the embryo exploits these two regulatory modes to control protein synthesis.

Regulation of protein synthesis by steroids provides a means for ensuring that physiologically significant proteins will be rapidly produced by specific cells when needed by the organism. This strategy is particularly important for regulation of the reproductive cycle. This mode of regulation also provides the investigator with an important experimental advantage: the ability to modulate gene expression either *in vivo* or *in vitro* by the administration of exogenous hormone. The vitellogenin genes of *Xenopus*, discussed in Chapter 9, provide one of the best-studied examples of genes under steroid regulation.

There are two general strategies for studying gene control of development. One is to begin with the phenomenon and use molecular techniques to isolate the genes that are involved in its control. The cloned genes can then be experimentally modified and the effects of these modifications assessed. The other approach is to begin with a mutant gene that influences a developmental event. By comparison of the wild-type and mutant genes, the genetic basis for the defect and the regulation of the normal process can be determined. The latter approach has been particularly well exploited by investigators of *Drosophila* development. Genes affecting pattern formation of this insect have yielded a massive amount of information, which is masterfully discussed by Scott in Chapter 5. One of the best-studied gene loci in eukaryotes is the *rosy* locus of *Drosophila*, which encodes xanthine dehydrogenase (XDH). Mutant genes that have tissue-specific effects on XDH synthesis have been cloned, and the nucleotide sequences that exert this effect have been pinpointed. This process is discussed in Chapter 10. Further exploitation of this system should yield valuable information regarding the molecular mechanisms involved in tissue-specific gene expression.

In *Xenopus* development, the zygote is initially transcriptionally inert. Transcription is first detected at the mid-blastula stage. As discussed in Chapter 7, the rate of cell division influences the onset of transcription during *Xenopus* development. The relationship between mitosis and cell differentiation has fascinated developmental biologists for many years and is now under intense investigation. We shall soon learn a great deal about this relationship as it pertains to both normal development and malignant transformation.

One of the most surprising recent findings of molecular biology is that individual gene sequences may contain information that can be utilized to produce different proteins in different cells. The mammalian neuroendocrine system provides a remarkable example of differential processing of messengers to produce tissue-specific products, as is discussed in Chapter 11.

An important characteristic of cell determination is the stability of the determined state. Once determined to differentiate, cells normally retain their specificity, which they can transmit to their progeny at cell division. Any hypothesis of differential gene regulation must account for this fact. One means of achieving stability might involve heritable modifications to the nucleotides themselves. A correlation between hypomethylation of cytosine residues and transcriptional activity exists in higher animals and appears to play a role in determination and stability of differential transcription, as is discussed in Chapter 12.

Plant molecular biology has not yet yielded as much information as animal molecular biology, but it is emerging as an extremely important and active area of investigation, particularly due to the potential economic and social impact of the manipulation of plant gene expression. Mankind relies on plant proteins for food, clothing, and shelter. The ability to manipulate production of plant proteins has the potential to revolutionize the economies of both the developed and underdeveloped nations of the world. Part III of this book presents three of the finest systems for the study of differential gene expression in plants: Chapter 13 describes *Arabidopsis*, which has emerged as an outstanding organism for the conduct of genetic and molecular biological research; Chapter 14 discusses gene expression during seed development in flowering plants; and Chapter 15 deals with the synergistic relationship between plants and bacteria, which is necessary for nitrogen fixation. Not only is nitrogen fixation vital for nutrient enrichment, but it also provides an excellent experimental system for studying induced gene expression.

The publication of a book during a period of rapid progress, in which scientific journals report spectacular discoveries with incredible frequency, is risky. However, the authors have not attempted to present definitive answers to the problems of cell determination and differentiation. Rather, they bring us to the present level of understanding and lead us to the future, which is indeed going to be very exciting. I think they have succeeded admirably. I am proud of their achievements and I am confident that this book will serve both to inform its readers and to stimulate investigators to supplement our rapidly growing body of knowledge.

Leon Browder

Calgary, Alberta

Contents

I. Specification of Regional Patterns of Cell Fate in Animal Development

II. Developmental Regulation of Gene Expression in Animals

Chapter 6 • Regulation of Actin Gene Expression during Sea Urchin Development

William R. Crain, Jr.

Chapter 7 • Regulation of the Mid-Blastula Transition in Amphibians

Laurence D. Etkin

Chapter 8 • Expression of Ribosomal Protein Genes during *Xenopus* Development

W. Michael Wormington

III. The Molecular Biology of Plant Growth and Development

Chapter 15 • Development and Differentiation of the Root Nodule:
                    Involvement of Plant and Bacterial Genes

    N. A. Morrison, T. Bisseling, and D. P. S. Verma

I

Specification of Regional Patterns of Cell Fate in Animal Development

Chapter 1

The Role of Cytoplasmic Determinants in Embryonic Development

WILLIAM R. JEFFERY

1. Introduction

Many different types of cells arise from the totipotent egg during embryonic development. Two distinct concepts have been formulated to explain how this diversity of cell types is generated. The first is that the egg already contains regionalized components known as **cytoplasmic determinants** that influence cell fate as a consequence of their differential segregation into various blastomeres during cleavage. The second is that certain embryonic cells, whose unique properties were established earlier by the action of localized determinants, influence the development of adjacent cells by **induction.** Thus, the diversification of embryonic cells is thought to be the result of differential gene expression mediated by the combined activity of cytoplasmic determinants and inductive processes.

The purpose of this chapter is to review critically the concept of cytoplasmic determinants and their role in metazoan embryogenesis. First, the properties of cytoplasmic determinants are examined, and the classic evidence for the existence of such factors is reviewed. Then, the observations and experiments that provide information on the identity, segregation, mechanism of action, and manner of localization of cytoplasmic determinants are discussed and analyzed. Finally, a summary and evaluation of the concept of cytoplasmic determinants is presented, based on integration of both classic and modern data.

1.1. The Concept of Cytoplasmic Determinants

Although cytoplasmic determinants have been the focus of considerable research and a number of reviews (Wilson, 1925; Beams and Kessel, 1974;

WILLIAM R. JEFFERY • Center for Developmental Biology, Department of Zoology, University of Texas, Austin, Texas 78712.

Davidson, 1986; Raff, 1977; Subtelny and Konigsberg, 1979; Jeffery, 1983; Jeffery and Raff, 1983), seldom has the concept been defined precisely (for exceptions, however, see Whittaker, 1979; Slack, 1984; Gerhart *et al.,* 1984). For our purposes, cytoplasmic determinants are defined as entities localized in specific cytoplasmic regions of the egg or embryonic cell that bias the cells containing them to assume particular properties or fates during embryogenesis. Entities considered to be determinants range from those that specify some of the more global features of embryos, such as the formation of the body axis and germ layers, to those that govern the phenotypes of individual differentiated cells, such as the expression of histospecific enzymes. Determinants are described as entities, rather than as particular kinds of molecules or organelles, because they have yet to be isolated and characterized. Unfortunately, until this step has been achieved, determinants can be studied only by indirect means.

1.2. Egg Cytoplasmic Localizations

The visible regionalization of the egg cytoplasm, known as **cytoplasmic localization,** has been described for a large number of animal eggs, particularly invertebrates. For many years, cytoplasmic localization was thought to signify the presence of determinants. Some examples of localization are reviewed here, which also provide most of the available information on cytoplasmic determinants. Descriptions of many other cytoplasmic localizations can be found in the comprehensive reviews of Wilson (1925) and Davidson (1986).

Possibly the most dramatic example of localization occurs in ascidian eggs. Near the turn of the century, Conklin (1905) described five differently colored cytoplasmic regions in the unfertilized *Styela* egg. The three most obvious of these regions are the transparent **ectoplasm,** residing in the animal hemisphere; the gray **endoplasm,** located mainly in the vegetal hemisphere; and the yellow **myoplasm,** distributed uniformly in the egg periphery. The different regions reflect an uneven distribution of variously pigmented granules (e.g., yellow lipid granules complexed with mitochondria in the myoplasm) and yolk particles in the egg cytoplasm (Berg and Humphreys, 1960). Between fertilization and first cleavage, these cytoplasmic regions are rearranged by a process known as **ooplasmic segregation** (Conklin, 1905; Jeffery, 1984a). During ooplasmic segregation, the myoplasm and ectoplasm stream vegetally, while the endoplasm flows into the animal hemisphere. The ectoplasm subsequently moves back into the animal hemisphere, the endoplasm returns to its original position in the vegetal hemisphere, and the myoplasm shifts upward from the vegetal pole, extending into a yellow crescent in the future vegetal-posterior region of the zygote (Fig. 1). The plane of the first cleavage bisects the yellow myoplasmic crescent, distributing each of the different regions equally into the first two blastomeres. During subsequent cleavages, however, the cytoplasmic regions are segregated into different blastomeres (Fig. 1), so that by the 110-cell stage each region is located in a different cell lineage. The ectoplasm, endoplasm, and myoplasm enter the epidermal and neural cells, endodermal and

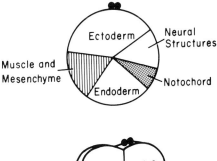

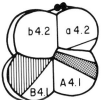

Figure 1. A schematic illustration of the fate maps of 1- and 8-cell *Styela* embryos indicating the embryonic derivatives of various cytoplasmic regions. The terminology for identifying the blastomeres of 8-cell embryos is that of Conklin (1905).

notochord cells, and mesenchyme and muscle cells, respectively. These localizations, which provide a natural fate map of the egg, are also considered to reflect regional differences in the distribution of determinants.

Another classic example of cytoplasmic localization is the **polar lobe,** a cytoplasmic protrusion that forms transiently at the vegetal pole of spiralian eggs (Verdonk and Cather, 1983). Depending on the species, the polar lobe varies in size (Fig. 2) and can form during the maturation divisions, the early cleavages, or both. In some small polar lobes, unique organelles, such as the vegetal body of the mollusc *Bithynia*, have been identified (Fig. 2B). During first cleavage, cytoplasm from the polar lobe enters only one of the first two blastomeres. This blastomere is known as the CD cell and is usually larger than its counterpart that receives no polar lobe cytoplasm, the AB cell. During second cleavage, a smaller polar lobe forms from the CD blastomere, and its cytoplasmic contents are shunted into only one cell of the 4-cell embryo, the D cell. A, B, and C blastomeres are usually smaller than D and are devoid of polar lobe materials. Dorsal features of the embryo and mesodermal derivatives arise from the D cell. Thus, the polar lobe appears to be a mechanism used to shunt cytoplasmic determinants responsible for subsequent germ layer and embryonic axis formation into specific blastomeres during cleavage.

The eggs of some holometabolous insects also contain distinctive localizations at their anterior and posterior poles (Fig. 3). The cytoplasmic region at the anterior pole is thought to contain determinants responsible for the development of head segments (Kalthoff, 1979). A cytoplasmic region at the posterior pole, the oosome, contains the **polar granules,** a localized group of electron-dense organelles believed to be inherited by the germ cells (Mahowald, 1968). After fertilization and oviposition, the insect egg develops as a syncytium; the zygotic nucleus undergoes a series of divisions within the internal yolk-filled cytoplasm. Subsequently, many nuclei migrate to the periphery of the egg; those that enter the oosome obtain special qualities: They cease division pre-

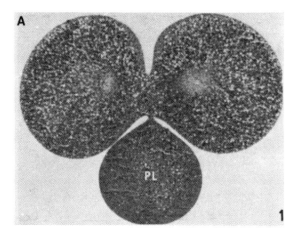

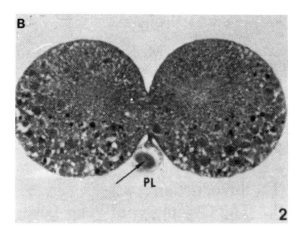

Figure 2. Cross-sections through the animal–vegetal axis of cleaving *Dentalium* (**A**) and *Bithynia* (**B**) embryos illustrating the position and size of the polar lobe (PL). The vegetal body in the polar lobe of *Bithynia* is indicated by an arrow. (From Dohmen and Verdonk, 1979.)

cociously and become the nuclei of pole cells, the first cells formed in the embryo. Pole cells eventually enter the interior of the embryo during gastrulation; many of them migrate to the gonad and form germ cells. In insect species that exhibit chromosome elimination during early development, pole cell nuclei are protected from loss of chromosomes by their localization in the oosome (Beams and Kessel, 1974). Therefore, the oosome is considered a localization containing germ cell determinants.

Although many embryonic localizations are more subtle than those described above, they also may be correlated with the presence of determinants. For instance, organelles known as vegetal (or P) granules are localized in the vegetal (posterior) hemisphere of *Caenorhabditis elegans* (Strome and Wood, 1982; Wolf *et al.*, 1983; see Chapter 2) and *Ascaris* (Beams and Kessel, 1974) eggs and are segregated into the germ cell lineage during early development. The first cleavage of nematode eggs is equatorial, so that most of the vegetal cytoplasm is distributed to the vegetal (P1) blastomere (Fig. 4). The animal (AB)

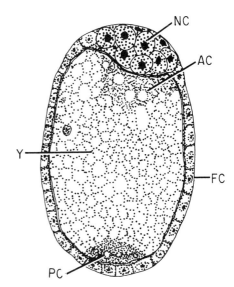

Figure 3. The anterior and posterior (oosome) cytoplasmic regions in an oocyte of the dipteran insect, *Miastor americana*. NC, nurse cell complex; FC, follicle cells; AC, anterior cytoplasm; PC, posterior cytoplasm; Y, yolk. (Redrawn from Hegner, 1914.)

blastomere gives rise to only somatic cells, while P1 gives rise to both somatic and germ cells. *Ascaris* (but not *C. elegans*) undergoes chromatin diminution, a process involving the selective destruction of portions of each chromosome, which begins in the somatic progenitor cells during the early cleavages (Boveri, 1887). P1 subsequently divides into EMSt, a somatic progenitor cell whose chromatin is diminished, and P2, a germline progenitor cell whose chromatin remains intact. This process is repeated after every cleavage in the descendants of P2 until the fifth cleavage, when P5 (which forms the germline) becomes entirely segregated from the somatic line, and chromatin diminution ceases. The vegetal granules, which are segregated from the egg into P5 through P1, P2, P3, and P4 (Strome and Wood, 1982), are thought to represent localized germ cell determinants.

Localizations also occur in vertebrate eggs (Gerhart et al., 1983). Unfertilized amphibian eggs contain at least four distinct cytoplasmic regions: a rim of pigment granules in the animal cortex; a region of transparent cytoplasm (derived from the oocyte GV) mixed with small yolk platelets in the animal hemisphere; a middle zone containing a mixture of medium-sized yolk

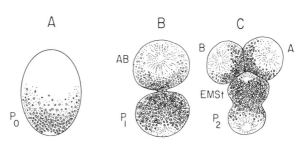

Figure 4. Early cleavage in *Ascaris megalocephala* embryos. (**A**) An uncleaved egg. (**B**) A 2-cell embryo. (**C**) The transition from a 3- to a 4-cell embryo. (Redrawn in part from Boveri, 1910a.)

A

B

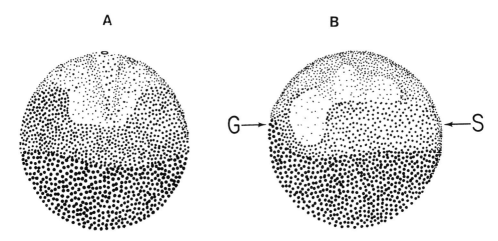

Figure 5. A schematic illustration of a section through the animal–vegetal axis of unfertilized (**A**) and fertilized (**B**) eggs of the frog *Discoglossus pictus* showing the organization of cytoplasmic regions. The animal (upper), middle, and vegetal (lower) cytoplasmic regions are shown as well as the position of the sperm entry point (S) and gray cresent (G). (Redrawn from Klag and Ubbels, 1975.)

platelets, mitochondria, and ribosomes; and a vegetal hemisphere region comprised of densely packed, large yolk platelets (Fig. 5). Sperm penetration in the animal hemisphere results in a reorganization of the egg cytoplasm; the pigmented animal cortex and the contents of the middle zone move toward the point of sperm entry, which becomes the ventral side of the embryo. Simultaneously, the animal and vegetal cytoplasmic regions are displaced in the opposite direction, forming a gray crescent 180° from the sperm entry point, which becomes the dorsal side of the embryo. During embryogenesis, ectoderm is derived from the animal region, mesoderm forms from the middle zone, and endoderm originates from the vegetal region. Thus, each germ layer receives a different type of localized cytoplasmic inclusion.

2. Evidence for the Existence of Cytoplasmic Determinants

Although the localizations described earlier may signify the position of determinants in the egg, they do not, by themselves, provide evidence for the existence of determinants. Evidence for the existence of cytoplasmic determinants includes (1) the progressive restriction of the fates of isolated blastomeres; (2) the invariant segregation of developmental potential between specific blastomeres; (3) the alteration of developmental fate after the removal, destruction, or reorientation of egg cytoplasmic regions; and (4) the assumption of new developmental fates after a cytoplasmic region is transplanted from one region of the egg or embryo to another.

2.1. Progressive Restriction of Blastomere Fates

Embryos characterized by the so-called mosaic type of development frequently show restricted blastomere fates as early as the 2- or 4-cell stage (Wilson, 1925). For instance, blastomeres isolated from 8-cell ascidian embryos, when the pigmented cytoplasmic regions have begun to be segregated into separate cells, each form only the tissue that would be expected according to the fate map (Reverberi, 1971). Likewise, AB and CD blastomeres isolated from polar lobe-containing embryos exhibit different fates in culture (Verdonk and Cather, 1983). In the gastropod *Ilyanassa* (Fig. 6), partial embryos derived from the CD cell form foot, shell, and eye, whereas partial embryos derived

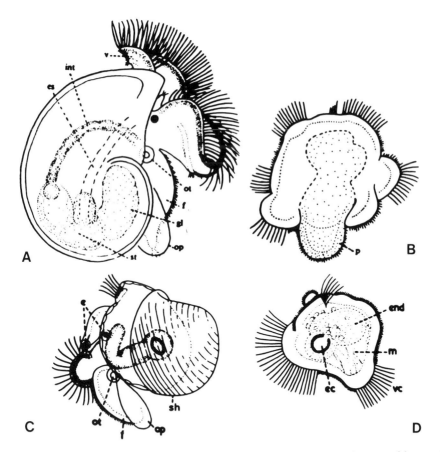

Figure 6. A comparison of normal, lobeless, and partial *Ilyanassa* larvae. (**A**) A normal larva with shell, velum (v), intestine (int), esophagus (es), otocyst (ot), foot (f), digestive gland (gl), stomach (st), eye (dark spot), and operculum (op). (**B**) A lobeless larva lacking most of the normal structures. (**C**) A partial larva that developed from an isolated CD cell showing many obvious structures, including shell(s), operulum, foot, otocyst, and eyes (e). (**D**) A partial larva that developed from an isolated AB cell showing muscle (m), velar cells (vc), endoderm (end), and enteric cavity (ec). (From Verdonk and Cather, 1983.)

from the AB cell form the velum, muscle, gut, and pigment cells (Clement, 1956). Early restriction of blastomere fate in these and other embryos suggests that developmental potential is unevenly segregated during the early cleavages.

2.2. Invariant Segregation of Developmental Potential

Invariant segregation of developmental potential is illustrated by cleavage-arrest experiments. In these experiments, the spatial expression of a cell lineage or tissue-specific marker is examined in embryos arrested at one or more of the early cleavage stages by treatment with cytochalasin B. This treatment blocks cytokinesis, but cleavage-arrested cells continue to undergo mitotic cycles and DNA replication and eventually express tissue-specific markers only in cells expected according to the fate map.

A clear example of this phenomenon occurs in ascidian embryos, in which the spatial distribution of several histospecific enzymes (Whittaker, 1973; 1977) and morphological markers (Crowther and Whittaker, 1984) has been thoroughly investigated. Acetylcholinesterase (AChE), an enzyme expressed in the larval muscle cell lineage, appears in a maximum of two cells of cleavage-arrested 2-, 4-, or 8-cell embryos, four cells of cleavage-arrested 16-cell embryos, six cells of cleavage-arrested 32-cell embryos, and eight cells of cleavage-arrested 64-cell embryos. The cleavage-arrested cells that express AChE at each developmental stage precisely match the position and maximum number of cells belonging to the muscle lineage (Fig. 7). Other histospecific markers in cleavage-arrested ascidian embryos, including alkaline phosphatase in gut cells, tyrosinase in brain melanocytes, a colloidal substance in notochord cells, and extracellular secretions in epidermal cells, also exhibit patterns of expression that match their respective cell lineages.

Cleavage-arrest experiments have also been conducted in nematode embryos, in which convenient markers exist for differentiated gut, muscle, and hypodermal cells. The gut cell lineage of *C. elegans* embryos is derived from the E cell, which arises at third cleavage as one of the progeny of EMSt. Autofluorescent rhabditin granules, which normally appear in the gut precursor cells at the 200–300-cell stage, are expressed exclusively in the endodermal lineage and its antecedents (EMSt and P1) in cleavage-arrested embryos (Laufer *et al.*, 1980). Similarly, in cleavage-arrested embryos, paramyosin, a muscle cell protein, develops only in the muscle progenitor cells, and a cuticle protein, produced by the hypodermis, appears only in the hypodermal progenitor cells (Cowan and McIntosh, 1985).

Although early restriction in blastomere fate and invariant segregation of developmental potential are often cited as evidence for the existence of cytoplasmic determinants, they suggest only that the fates of isolated blastomeres or cells of cleavage-arrested embryos are programmed by segregated factors. These experiments alone do not indicate whether such factors reside in the nucleus or cytoplasm.

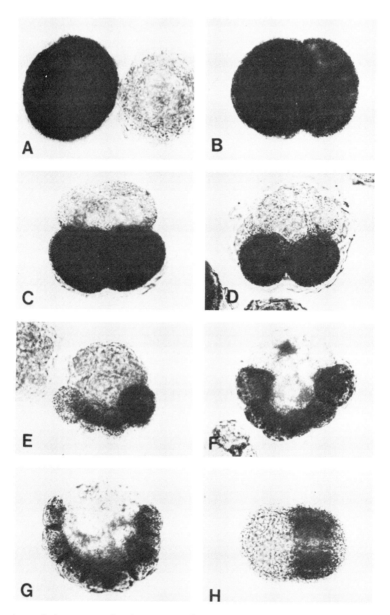

Figure 7. Acetycholinesterase development in cleavage-arrested ascidian embryos. (**A**) 1-cell; (**B**) 2-cell; (**C**) 4-cell; (**D**) 8-cell; (**E**) 16-cell; (**F**) 32-cell; (**G**) 64-cell; and (**H**) a normal tailbud stage embryo showing enzyme activity in the tail muscle cells. (From Whittaker, 1979.)

2.3. Alteration of Developmental Fate after Removal, Destruction, or Reorientation of Egg Cytoplasmic Regions

The issue of whether nuclear or cytoplasmic determinants are responsible for the early embryonic restriction and segregation of developmental potential has been examined by experiments in which the developmental fate of cells is followed after specific regions of the egg cytoplasm are removed, destroyed, or reoriented.

When fertilized ascidian eggs are cut into fragments perpendicular to the animal–vegetal axis, only those fragments that contain a part of the vegetal cytoplasm develop into complete larvae (Ortolani, 1958). Thus, determinants important for larval development appear to be localized in the vegetal cytoplasm of ascidian eggs.

Detailed cytoplasmic deletion experiments have been conducted with polar lobe-forming embryos. The polar lobe can be separated by mechanical agitation because it is usually attached to the rest of the embryo by only a thin isthmus of cytoplasm. When the polar lobe is removed during the first cleavage, the lobeless embryo exhibits changes in the pattern and chronology of cell division, and, later, significant defects appear in larval morphology (Verdonk and Cather, 1983). In the scaphopod *Dentalium*, larvae that develop from lobeless embryos are missing the apical tuft (a ciliary organ), the mesodermal bands, and the post-trochal region, from which parts of the mantle, foot, and shell are formed (Wilson, 1904). In the gastropod *Ilyanassa* (Fig. 6), larvae that develop from lobeless embryos lack the shell, foot, heart, eyes, operculum, and tentacles (Crampton, 1896; Clement, 1952). In general, it appears that determinants specifying some mesodermal structures are located within the anucleate polar lobe in spiralian embryos.

In insect embryos, the oosome can be eliminated by ligature, cauterization, or ultraviolet (UV) irradiation without affecting the nucleus (Hegner, 1911, Geigy, 1931; Okada *et al.*, 1974). Embryos lacking the oosome never develop pole cells or functional germ cells, and the nuclei that enter the posterior region of these embryos are subsequently incorporated into general blastoderm cells. In the dipteran *Wachtliella*, the oosome apparently protects future germ line nuclei from chromosome elimination (Geyer-Duszynska, 1959, 1961), since, after ligation, cauterization, or UV irradiation of the oosome, chromosomal elimination occurs in each of the embryonic nuclei and no germ cells are formed.

Cytoplasmic deletion experiments, similar to those conducted with mollusc and insect embryos, have recently been extended to vertebrate embryos. When the nucleate animal hemisphere region of *Xenopus laevis* eggs is separated from the anucleate vegetal hemisphere region by ligation, the animal hemisphere develops into a ciliated structure lacking a dorsal–ventral axis (Gurdon *et al.*, 1985). By contrast, animal hemisphere fragments ligated so that they include a part of the vegetal hemisphere cytoplasm form ciliated embryos with partial axial structures. Nucleate vegetal or equatorial egg fragments also form axial structures but do not develop cilia, unless they include a part of the animal hemisphere cytoplasm. These experiments suggest that determinants specifying ectodermal ciliation are located in the animal cytoplasm of *Xenopus* eggs, whereas determinants required for dorsal–ventral axis specification are concentrated in the vegetal cytoplasm.

The egg-ligation studies were then extended to the molecular level by demonstrating that information localized in the vegetal region of *Xenopus* eggs is required for subsequent activation of the muscle actin genes during neurulation (Gurdon *et al.*, 1985). In these experiments, various parts of the egg cytoplasm were separated by ligation, and the ability of the nucleate fragment to express the muscle actin genes was determined by S1 nuclease protection analysis using a cloned DNA probe specific for the 3′ noncoding region of a muscle actin transcript. The results showed that a given nucleate egg fragment requires at least a portion of the vegetal hemisphere cytoplasm to activate the muscle actin genes. Moreover, cytoplasm near the future dorsal side of the vegetal hemisphere was shown to be more active in promoting muscle actin gene expression than cytoplasm near the presumptive ventral side. These experiments suggest that determinants controlling muscle actin gene expression are concentrated in the vegetal hemisphere cytoplasm near the future dorsal side of uncleaved *Xenopus* zygotes.

Thus, in a variety of species, the deletion of an anucleate egg cytoplasmic region severely affects embryonic development. These experiments indicate that determinants responsible for fundamental embryonic processes, such as dorsal–ventral axis formation, mesoderm development, germ line divergence, and the differentiation of specific cell types, reside in the egg cytoplasm.

More information on the location of determinants has been obtained by reorientation of cytoplasmic regions in either the egg or embryo. The quality of egg cytoplasm received by different blastomeres during cleavage has been altered in two ways: by a change in the position of the spindle apparatus and cleavage furrows or by centrifugation.

The yellow myoplasm of *Styela* embryos is partitioned to only two cells at the third cleavage. These two cells, which are muscle cell progenitors, are the only cells that express AChE when 8-cell embryos are cleavage arrested. When 4-cell *Styela* embryos are flattened by compression (Fig. 8), the spindle apparatus is reoriented, causing the third cleavage to be meridional rather than equatorial (Whittaker, 1980). This causes the myoplasm to enter four instead of two cells. Some of the cells that obtain myoplasm in this way now become competent to express AChE. In a similar experiment, when the position of the equatorial cleavage furrow between the muscle and ectodermal lineage cells of 8-cell embryos (Fig. 1) is modified by micromanipulation, so that cytoplasm earmarked for the muscle cells may instead enter the epiderimal lineage, the latter sometimes synthesizes AChE (Whittaker, 1982). The nuclear lineages remain unaltered in these experiments, providing strong evidence that ascidian muscle cell determinants are present in the cytoplasm. The results suggest further that the presence of muscle cell determinants can influence the fate of nonmuscle cells.

In spiralian embryos, polar lobe formation can be prevented by treatment

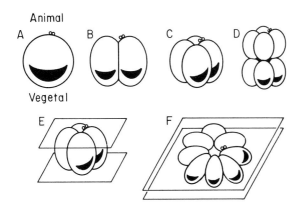

Animal

Vegetal

Figure 8. A schematic illustration of the location of the myoplasm in normal (**A–D**) and compressed (**E–F**) *Styela* embryos. The myoplasm (filled crescents) is distributed to 2 cells of the normal 8-cell embryo (**D**), but 4 cells of the compressed 8-cell embryo (**F**). (Redrawn from Whittaker, 1980.)

with cytochalasin B, sodium dodecyl sulfate (SDS), or by compression immediately before the first cleavage. Under these conditions, the first cleavage is equalized, and the cytoplasm of the polar lobe is evenly distributed between the first two blastomeres. Subsequently, each cell that receives part of the polar lobe contents forms dorsal structures. This results in a double embryo, replete with two shells (Guerrier *et al.*, 1978). Similar results were obtained by equalizing the first cleavage of spiralian eggs that do not exhibit polar lobes but still divide into unequal AB and CD cells (Tyler, 1930).

Normal cytoplasmic distribution can also be modified by increasing the number of sperm that enter the egg. In dispermic *Ascaris* eggs, the vegetal region containing the potential for germ cell determination is sometimes segregated into two or three cells instead of into P1 and P2 as occurs in normal eggs (see Fig. 4) (Boveri, 1910*a*). This inclusion of vegetal cytoplasm in these blastomeres results in protection of their nuclei from chromatin diminution. These experiments suggest that cytoplasmic determinants normally protect the germ line nuclei from diminution and have recently been extended by testing the pattern of diminution in cleavage-arrested 1- and 2-cell *Ascaris* embryos (Oliver and Shen, 1986). The cleavage-arrested embryos continue to undergo nuclear division and show either all diminished or all undiminished nuclei in a given cell. These results confirm earlier studies suggesting that both protectant and diminisher determinants exist in *Ascaris* eggs (Beams and Kessel, 1974) and provide strong evidence that the germ cell determinants are located in the vegetal egg cytoplasm.

Cytoplasmic reorientation has also been achieved by subjecting eggs to centrifugal force. Centrifugation of uncleaved *Ascaris* eggs drives the vegetal granules to the centrifugal pole (Boveri, 1910*b*). The first cleavage furrow then forms at right angles to the plane of vegetal granule stratification, thereby separating the centrifugal zone of compacted granules into a spherical mass completely separate from each of the first two blastomeres. Consequently, the nucleus in each blastomere is diminished, providing further indication that the vegetal granules may contain the protectants of chromatin diminution. Likewise, when eggs of the dipteran *Wachtliella* are centrifuged, the contents of the oosome are dispersed throughout the syncytial egg before pole cell formation, and chromosomal elimination occurs in all nuclei. Occasionally, however, when particulate material from the dispersed oosome contacts a nucleus in another part of the egg, that nucleus is protected from chromosome elimination (Greyer-Duszynska, 1959). These experiments suggest that cytoplasmic determinants can protect nuclei from chromatin diminution or chromosome elimination.

Centrifugation has also been used to rearrange the cytoplasm of amphibian eggs. In amphibian embryos, dorsal–ventral axis determination occurs during a critical period about midway between fertilization and the first cleavage (Gerhart et al., 1983). When fertilized *Rana fusca* eggs are centrifuged before this critical period, such that the fluid portions of the animal and vegetal cytoplasms are reversed in position, development is essentially normal, but ciliated ectodermal structures are now derived from the vegetal hemisphere and endodermal structures develop from the pigmented animal hemisphere (Pasteels, 1941). In contrast to these results, centrifugation of uncleaved *Xenopus* eggs during or after the critical period for dorsal–ventral axis determination yields a high proportion of embryos with double axes (Gerhart et al., 1983). These experiments show that the early arrangement of the egg cytoplasm affects the pattern of amphibian development. As indicated in these examples, reorientations of egg cytoplasm that do not affect nuclei may still alter developmental fate, providing convincing evidence for the existence of cytoplasmic determinants.

2.4. Alteration of Developmental Fate by Transplantation of Egg Cytoplasmic Regions

The strongest evidence for the existence of cytoplasmic determinants is obtained when a portion of the egg cytoplasm can be shown to alter developmental fate after it is removed from one part of an egg or embryo and transferred into another part of a different egg or embryo. This experiment has been successful in only a few instances. Unsuccessful published attempts to transfer determinants, as well as the few successful ones, are discussed below.

In the ascidian *Halocynthia*, cytoplasm was removed from the muscle progenitor cells and injected into one of the other progenitor cells of 8-cell embryos (Deno and Satoh, 1984). After cleavage arrest with cytochalasin B, the injected 8-cell embryos were tested for AChE expression. Cleavage-arrested ascidian embryos express tissue-specific enzymes in the maximum number of blastomeres expected for the cell lineage from which the relevant tissue is derived. For instance, AChE is expressed in a maximum of two muscle progenitor cells in cleavage-arrested 8-cell embryos (Whittaker, 1973). If muscle (or AChE) determinants were transferred to other progenitor cells by cytoplasmic injection and were to alter the fate of the recipients, a maximum of three cells (one AChE-positive cell would be derived from an injected progenitor cell) would be expected to express AChE. Unfortunately, three cells displayed AChE activity in only 2% of injected embryos (Deno and Satoh, 1984). These results may represent an underestimate of the number of injected cells that expressed AChE, however. The typical position of cells becomes distorted in cleavage-arrested 8-cell embryos, whereas the polar bodies, the sole markers of polarity in the unpigmented ascidian species used in this experiment, are not visible at the time of assaying for enzyme development. Therefore, in embryos that expressed AChE activity in only one or two cells, one of these cells could have been an injected cell. Another explanation for the low yields may be that myoplasm amounting to only 3% of the total volume of the recipient cell cytoplasm was injected. If a threshold of determinant activity is required for a response, this level may not have been attained. Clearly, no final conclusions can be made concerning the transplantability of ascidian muscle AChE determinants from these experiments alone.

Attempts to transplant cytoplasmic determinants in polar lobe-containing embryos have also been reported. At second cleavage, cytoplasm was removed from the polar lobe of a *Dentalium* embryo with a micropipette and injected into the B blastomere of another embryo at the 4-cell stage (van den Biggelaar, cited in Verdonk and Cather, 1983). The injected B cell, however, failed to obtain the dorsalizing qualities of the D cell; the injected embryo showed normal development without any apparent dorsal duplications. In similar experiments, a large proportion of the polar lobe cytoplasm was removed from a *Dentalium* embryo at the first cleavage (van den Biggelaar, cited in Verdonk and Cather, 1983). The cytoplasm removed from the lobe was replaced by cytoplasm that flowed into the lobe from the animal hemisphere, but embryonic development was normal. One explanation for these results may be that the polar lobe determinants are not located in fluid egg cytoplasm. Thus, they would not necessarily be transplantable by the methods used in this study. Further discussion of the absence of determinants in the fluid egg cytoplasm is presented in Section 5.1.

Illmensee and Mahowald (1974, 1976) reported the first successful transplantation of cytoplasmic determinants in *Drosophila*. The basic scheme for these transplantations is illustrated in Fig. 9. Cytoplasm from the oosome of eggs or syncytial embryos was transplanted to the anterior or mid-ventral region of a precellular blastoderm embryo. The presence of the transplanted cytoplasm led to the formation of histologically identifiable pole cells, which formed precociously at ectopic sites in the recipient. In a similar experiment, normal oosome cytoplasm was injected into the anterior-lateral region of UV-irradiated *Drosophila* eggs whose own oosome was inactivated. This transplant also resulted in the formation of ectopic anterior pole cells (Okada et al., 1974). The ability of the ectopic pole cells produced in the experiments of Illmensee and Mahowald (1974, 1976) to form germ cells was determined by a second transplantation experiment. The ectopically formed pole cells were transplanted to the posterior region of host embryos of the same age, but of different genotype (Illmensee and Mahowald, 1974, 1976) (Fig. 9). Some of the transplanted pole cells migrated to the gonads and became germ cells, and the hosts

DONOR   RECIPIENT   **HOST**

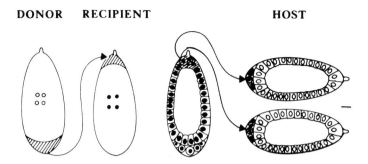

Figure 9. A schematic illustration of the transplantation of cytoplasm from the oosome and pole cells in *Drosophila* embryos. Cytoplasm from the posterior pole of a syncytial embryo is injected in the anterior pole of a recipient syncytial embryo. After ectopic pole cell development, pole cells from the anterior pole of the recipient embryo are transplanted to the posterior pole region of a host embryo at the cellular blastoderm stage. (From Mahowald *et al.*, 1979.)

produced progeny showing the genotype of the eggs from which the oosome cytoplasm was originally derived.

The reason that these elegant experiments, which have now been repeated in other laboratories (cited in Boswell, 1985), were successful, whereas other determinant transplantation experiments failed or achieved minimal success, may be due to the nature of insect germ cell determinants. If the polar granules contain germ cell determinants, it can be argued that these relatively large structures can be transferred to other regions by a micropipette more successfully than smaller determinants. It is also possible that insect germ cell determinants are a special class that exist in the fluid cytoplasm or are less strongly attached to rigid cell structures than other determinants. This viewpoint is supported by experiments presented in the previous section indicating that determinants located in the oosome can be displaced by low centrifugal forces.

A remarkable series of experiments has been reported recently concerning the establishment of dorsal–ventral pattern in *Drosophila* embryos that are relevant to a discussion of determinant transplantation (Anderson *et al.*, 1985b). The establishment of the dorsal–ventral pattern in *Drosophila* larvae is controlled by a set of about ten maternal-effect genes (see Section 3.2.1.d) (Anderson and Nüsslein-Volhard, 1984). Genetic studies suggest that, within this group of genes, the product of the *Toll* locus plays a central role in axis formation (Anderson *et al.*, 1985a). *Drosophila* females that lack *Toll* gene activity produce dorsalized larvae. Injection of wild-type cytoplasm or cytoplasmic RNA into young *Toll* embryos restores their ability to develop a normal dorsal–ventral pattern. Most significantly, however, the position of the injected cytoplasm or RNA defines the ventral portion of the embryo. These studies show that cytoplasmic determinants involved in *Drosophila* axis formation can be transplanted to new positions in a mutant embryo, where they act to set up a dorsal–ventral axis.

2.5. The Reality of Cytoplasmic Determinants

The many experiments described in this section show that determinants, either by themselves or in combination with other developmental mechanisms, can control at least four features of early animal development: (1) dorsal–ventral axis determination; (2) germ layer formation; (3) germ cell determination; and, to some extent, (4) the diversification of somatic cell lineages. Moreover, cytoplasmic deletion, reorientation, and transplantation experiments leave little doubt that these entities are located in the egg cytoplasm. The following sections examine the identities, manner of localization and segregation, and mechanism of action of cytoplasmic determinants.

3. Identity of Cytoplasmic Determinants

The identity of cytoplasmic determinants has received much attention since their discovery in the early part of the twentieth century. Unfortunately, the precise nature of determinants has not been elucidated. Although speculations have ranged widely, the most likely candidates now appear to be macromolecules or specialized organelles. Some of the important advances in this area are discussed in this section.

3.1. Organelles as Candidates for Determinants

Soon after the discovery of cytoplasmic determinants, attention was focused on the possibility that localized organelles might be determinants. In most cases, experiments designed to test this possibility were negative. The particulate inclusions of the ascidian myoplasm, which constitute a complex of pigment granules and mitochondria (Berg and Humphreys, 1960), are an appropriate example. Two observations suggest that this complex does not contain determinants. First, centrifugation of eggs immediately before the first cleavage sometimes causes the inclusions to be dispersed from the myoplasm and stratify at one end of the cell, where they may subsequently enter only one blastomere of the 2-cell embryo (Conklin, 1931). Some of the centrifuged embryos that show this phenomenon continue to develop normally and form larval muscle cells. Moreover, nonmuscle cells that acquire a high concentration of the dispersed complex after centrifugation do not subsequently become muscle cells. Second, direct-developing ascidian species, which form vestigial larval tail muscle cells, do not display mitochondrial localizations in their myoplasm (Whittaker, 1979b,c). Clearly, the muscle cell determinants are neither pigment granule–mitochondria complexes nor substances that are tightly associated with these organelles.

Despite the negative findings for the ascidian myoplasmic complexes, there are some instances in which positive correlations do exist between lo-

calized egg organelles and determinants. The polar lobe of the mollusc *Bithynia* has been shown by extirpation experiments to contain determinants that specify the development of the dorsal features of the embryo (Cather and Verdonk, 1974). When the uncleaved *Bithynia* egg is centrifuged, the vegetal body (see Fig. 2B) can sometimes be displaced intact from the vegetal region of the egg without dispersing its contents (van Dam *et al.*, 1982). The polar lobe still forms in centrifuged eggs but sometimes without the vegetal body; in these embryos, polar lobe removal does not cause subsequent developmental defects. These results suggest that the vegetal body or other materials removed from the vegetal region of the egg by centrifugation, rather than the polar lobe *per se*, contain the dorsal determinants.

There is evidence that the polar granules may be germ cell determinants in insect eggs (see Sections 2.3 and 2.4). This correlation has been extended by analysis of a class of maternal effect mutations in *Drosophila*, known as *grand-childless* (gs). In gs mutants, the F1 progeny of homozygous females are sterile (Fielding, 1967). A number of gs mutants have been isolated and characterized. The F1 progeny of females homozygous for strong alleles of one of these mutations, *tudor*, produce eggs that lack polar granules, do not form pole cells, and die during late embryogenesis (Boswell and Mahowald, 1985). In weaker *tudor* alleles, however, there is a reduction in polar granule number (and size) and a corresponding decrease in the proportion of functional pole cells formed during embryogenesis. This correlation is consistent with the possibility that the polar granules contain germ cell determinants. The possibility that maternal messenger RNA (mRNA), which is transiently associated with the polar granules, may be the actual germ cell determinant is taken up in the following section.

Thus, in some instances, specialized organelles localized in the egg cytoplasm may be determinants. It is also possible that the organelles themselves are not the determinants but are a matrix with which determinants are closely associated.

3.2. Macromolecules as Candidates for Determinants

In recent years, attention has focused on maternal proteins or RNA as candidates for cytoplasmic determinants. Proteins can be spatially localized in cells and can interact with a variety of different cellular inclusions. The only evidence implicating maternal proteins as potential determinants, however, is derived from studies of the *o* maternal-effect mutant of the Mexican axolotl (Brothers, 1979). Females homozygous for *o* produce eggs that arrest at the gastrula stage, regardless of the sperm genotype. The eggs of homozygous *o* females can be rescued, however, by microinjection of nucleoplasm from wild-type oocyte GVs or cytoplasm from mature wild-type eggs. Preliminary characterization of the *o* corrective activity localized in the GV indicates that it is a macromolecule sensitive to high temperature and trypsin. These characteristics suggest that the *o* corrective factor is a protein.

3.2.1. RNA as a Determinant

Evidence implicating maternal mRNA as a cytoplasmic determinant has accumulated steadily in recent years. Four observations, which will be discussed in the following sections, support this idea: (1) the presence of RNA in organelles that appear to contain cytoplasmic determinants; (2) the UV sensitivity and photoreactivation exhibited by some determinants; (3) the RNase sensitivity of some determinants; and (4) the ability of purified RNA fractions to rescue UV-irradiated embryos, and embryos affected by maternal-effect mutations that alter early embryonic events controlled by determinants.

3.2.1a. RNA in Localized Organelles.   Many localized egg organelles that may be determinants also have been shown to contain RNA. Only a few examples are considered in this chapter; the detailed review of Jeffery (1983) may be consulted for other examples.

One example of RNA localization is found in the polar lobe of *Bithynia* embryos (see Fig. 2B) (Dohmen and Verdonk, 1974). This organelle first appears at the vegetal pole of the oocyte and remains in this position until just before first cleavage, when it is enclosed in the polar lobe. When the polar lobe and its contents are resorbed by the CD blastomere after first cleavage, the vegetal body disintegrates, presumably releasing its contents into the cytoplasm. The vegetal body stains intensely with the basophilic dyes pyronin, acridine orange, and Hoechst 33258, which suggests that it contains large amounts of RNA. The structural elements of the vegetal body consist of groups of membrane-enclosed vesicles with adjacent electron-dense granules, mitochondria, and ribosomes. At present, it is not certain which of the vegetal body elements contain RNA, since cytochemical tests have been done only with material prepared for light microscopy. The mitochondria and ribosomes present, however, are probably not sufficiently abundant to account entirely for the intensity of RNA staining registered by the vegetal body.

Polar granules in *Drosophila* may be storage sites for localized maternal RNA molecules that code for proteins involved in germ cell determination (Mahowald, 1971; Mahowald *et al.*, 1979). According to this hypothesis, these RNA molecules are synthesized in the nurse cells of *Drosophila* ovaries during oogenesis and are transported to the growing oocyte, where they become localized in the developing oosome by their associations with the polar granules. They would be sequestered in an inactive form until fertilization, when the polar granules disperse and develop polyribosomes at their surface, presumably due to the translational activation of stored mRNA.

Cytochemical tests coupled with electron microscopy suggest that the polar granules of *Drosophila* do contain RNA (Mahowald, 1971). Indium trichloride, a substance that binds phosphate groups in nucleic acids, stains the polar granule matrix and surrounding ribosomes. Other observations are consistent with the possibility that the polar granules contain mRNA. First, the electron density exhibited by these structures after indium trichloride staining is homogeneous, rather than granular in appearance, as might be expected if the

polar granules consist of tightly packaged ribosomes. By contrast, the ribosomes surrounding the polar granules show a granular type of staining. Second, polyribosomes appear at the borders of the polar granules after fertilization. Third, ribosomal proteins are not found in isolated polar granules (Waring *et al.*, 1978). Although none of these data are proof for the existence of mRNA in the polar granules, they support the possibility that the staining of polar granules for RNA is not based merely on the presence of ribosomes.

The presence of RNA in the polar granules appears to be a transient phenomenon (Mahowald, 1971). In *Drosophila*, RNA staining is most intense in the mature oocyte, declines gradually during intravitelline cleavage and blastoderm formation, and finally disappears after pole cell formation. Since no RNA component can be detected by biochemical means in crude polar granule preparations derived from isolated pole cells, it is likely that the RNA is either dissociated or degraded before pole cell formation (Allis *et al.*, 1977). Isolated polar granules do contain a basic 93,000-M_r polypeptide, however (Waring *et al.*, 1978). It is possible that this polypeptide is responsible for RNA localization in the polar granules before pole cell formation.

There are many reported cases of RNA localization in unique egg organelles, especially organelles that enter primordial germ cells. These organelles also contain protein, however, and perhaps other constituents as well. Thus, the designation of RNA as their active component is quite tentative.

3.2.1b. UV Irradiation and Photoreactivation Studies. Further evidence that nucleic acids function as cytoplasmic determinants can be obtained from UV irradiation studies coupled with photoreactivation by visible light. UV irradiation causes sterility (Okada *et al.*, 1974) and allows chromosome elimination (Bantock, 1970) or chromatin diminution (Moritz, 1967) to occur in all the nuclei of certain insect and nematode eggs, respectively. The most effective use of UV irradiation to probe the molecular nature of determinants, however, has been made using the eggs of the chironomid *Smittia* (Kalthoff, 1979). UV irradiation focused on the anterior pole region of this egg prevents the subsequent development of normal head segments. Instead, the anterior portion of the embryo develops a set of mirror-image abdominal segments, including all normal posterior constituents except pole cells. The embryos that show this abnormality are known as **double abdomens.** Double abdomen formation is thought to involve an inactivation of determinants that specify anterior segment identities. A detailed action spectrum has been constructed for double abdomen formation by UV irradiation. The construction of an action spectrum is a necessary step toward the identification of a UV target, because it is correlated precisely with the absorption spectrum of the inactivated molecular component(s). The action spectrum obtained for the production of double abdomens is a bimodal curve with peaks at about 265 and 285 nm, which are near the absorption maxima of nucleic acids and proteins, respectively. Thus, both nucleic acids and proteins are potential candidates for anterior determinants in *Smittia* eggs.

Photoreactivation experiments have provided evidence that the anterior determinants are nucleic acids. When pyrimidine dimers are formed in DNA or RNA by UV irradiation they can often be enzymatically excised and repaired in the presence of visible light, a process known as photoreactivation. Reversal of UV-induced double abdomens is obtained in *Smittia* embryos by a subsequent exposure of the UV-irradiated eggs to visible light at wavelengths of 310–460 nm (Jackle and Kalthoff, 1978). Although this result suggests that the factors responsible for normal head segmentation are nucleic acids, it should be recalled that the UV action spectrum also includes a peak in the region of maximal protein absorption. A second peak in this region of the spectrum would be expected if the anterior determinants were nucleoproteins. This possibility has been subjected to an experimental test (Kalthoff, 1979). It was reasoned that since photoreactivation is unknown for UV-damaged proteins, visible light should be more effective as a rescue agent following UV irradiation at 265 nm than after irradiation at 285 nm, provided that two independent UV targets exist. If the determinant is a nucleoprotein, however, photoreactivation should be as effective after UV irradiation at 285 nm as it is at 265 nm due to the transfer of excitation energy through the macromolecular complex. Visible light was found to be equally effective in rescuing double abdomens produced by UV irradiation at either 265 or 285 nm, suggesting that the target is a macromolecular complex, possibly a nucleoprotein. The information discussed in the next section shows that the anterior determinants of *Smittia* eggs are specifically inactivated by RNase, and therefore the actual determinant may be RNA.

Recently, UV irradiation and photoreactivation have also been used to study the nature of germ cell determinants in the oosome of *Smittia* (Brown and Kalthoff, 1983). Pole cell formation can be inhibited by UV irradiation; the action spectrum for this effect shows a distinct peak at 260 nm (Fig. 10), suggesting that the UV target is a nucleic acid. This interpretation is supported by the fact that the effect of UV on pole cell formation is photoreversible. Thus, nucleic acids might be determinants at both the anterior and posterior poles of

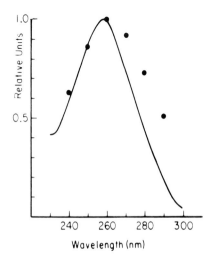

Figure 10. The action spectrum for UV inhibition of pole cell formation in *Smittia* eggs. Solid line, absorbancy of purified *Smittia* RNA; dots, quantum efficiency of pole cell inhibition. (From Brown and Kalthoff, 1983.)

insect eggs. Neither the UV action spectrum nor the photoreactivation results alone, however, conclude whether RNA or DNA is the active component. It seems likely, however, that the germ cell determinants are RNA because DNA is not a major component of the oosome, and injected RNA can rescue the effects of UV irradiation on pole cell formation in *Drosophila*.

3.2.1c. Local Exposure of Egg Cytoplasmic Regions to Enzymes. The UV irradiation and photoreactivation studies on anterior determinants in *Smittia* eggs have been complemented by local application of enzymes (Kandler-Singer and Kalthoff, 1976). In these experiments, the anterior egg cytoplasm was exposed to RNase, DNase, or protease, either by microinjection or by puncture of the anterior pole while the egg was immersed in aqueous solutions containing the enzymes. About one half the eggs treated with RNases, but none of those exposed to DNase or protease, became double abdomens. To prove that the RNase activity rather than the presence of the enzyme molecules *per se* was responsible for double abdomen formation, the eggs were treated with fragments of RNase S, either individually or after being mixed together. RNase S molecules can be proteolytically cleaved into two fragments, which are inactive alone but become enzymatically active again when mixed. The mixture of enzyme fragments, but not the individual fragments, formed double abdomens. Thus, the determinant(s) involved in the specification of normal head segments in *Smittia* embryos is likely to be composed of maternal RNA.

3.2.1d. RNA Rescue Studies. One way to test whether determinants are composed of RNA would be to rescue an embryonic defect imposed by the modified determinants with purified RNA. Two instances in which this has been accomplished are discussed below.

It has been shown that *Drosophila* embryos sterilized by UV irradiation can form pole cells and functional germ cells after microinjection with cytoplasm from the oosome of unirradiated eggs (Okada *et al.*, 1974; Warn, 1975). Recently, the capacity to form pole cells in UV-irradiated embryos has also been restored by microinjection of a subcellular fraction containing ribosomes and mRNA (Ueda and Okada, 1982), or purified RNA isolated from this subcellular fraction (Togashi and Okada, 1982). These results suggest that the cytoplasmic component responsible for pole cell restoration in *Drosophila* is an RNA molecule. Pole cells that were formed when the subcellular fraction or the purified RNA were injected did not become functional germ cells. Despite the rescue, sterile individuals developed from the injected embryos. This could mean that at least two different determinants, one regulating pole cell development and another controlling the conversion of pole cells to germ cells, are involved in specifying the germ line in *Drosophila*.

Rescue experiments using isolated cytoplasmic components have also provided evidence that maternal mRNA molecules are involved in the process of dorsal–ventral axis determination in *Drosophila* (Anderson and Nüsslein-Volhard, 1984). The establishment of the dorsal–ventral axis is directed by a group of maternal-effect genes, defined in a saturation screen for female sterile muta-

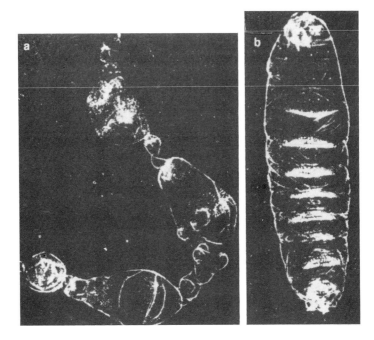

Figure 11. Rescue of the maternal effect mutation *snake* by injection of poly(A)⁺ RNA. (**A**) The cuticle of an uninjected *snake* embryo showing differentiation of only the dorsal structures. (**B**) The cuticle of a *snake* embryo injected with poly(A)⁺ RNA from wild-type embryos showing a complete dorsal–ventral pattern, including the ventral denticle belts. (From Anderson and Nüsslein-Volhard, 1984. Reprinted by permission from *Nature*, Vol. 311, No. 5983, pp. 223–227. Copyright © 1984 Macmillan Journals Limited.)

tions. For each gene, lack of function in the mother results in F1 larvae lacking ventral pattern elements. Dorsalized larvae exhibit a malformed cuticle consisting only of a tubelike structure with fine hairs. These larvae lack the normal lateral and ventral cuticular elements, including filzkörper structures and belts of denticles (Fig. 11a). In the case of one of these mutants, *dorsal*, it was already known that injection of wild-type cytoplasm into preblastoderm embryos mediates a partial rescue of the ventral pattern elements (Santamaria and Nüsslein-Volhard, 1983). Recently, it was found that wild-type cytoplasm is also capable of rescuing ventral structures to various degrees in six more of these mutants (*tube*, *snake*, *easter*, *Toll*, *spätzle*, and *pelle*) (Anderson and Nüsslein-Volhard, 1984). The most dramatic effect is obtained with *snake*, which is the only mutant of this group in which the dorsal–ventral pattern can be rescued completely by injection of wild-type cytoplasm. When cytoplasmic RNA, purified from wild-type, preblastoderm embryos was injected into *snake* embryos, the normal dorsal–ventral pattern was completely restored (Fig. 11b). The isolated RNA was further fractionated into poly(A)⁻ and poly(A)⁺ RNA, and, after microinjection into *snake* embryos, a rescue was obtained only with poly(A)⁺ RNA. This suggests that the active factor is mRNA. The other five maternal effect mutations can also be rescued to some extent by the injection of wild-type cytoplasm or poly(A)$^+$ RNA (Anderson and Nüsslein-Volhard, 1984; Muller-Holtkamp *et al.*, 1985). These results show that mRNA is a normal storage form of the morphogenetic information from at least six of the genes that specify the dorsal–ventral axis in *Drosophila*. Presumably, these mRNAs are synthesized during oogenesis, stored in the egg, and regionally translated during early development.

3.3. Current Status of Cytoplasmic Determinants

The studies reviewed in this section suggest that maternal RNA, possibly mRNA, is currently the strongest candidate for a cytoplasmic determinant, especially in insect eggs. It should be kept in mind, however, that there may be many different kinds of cytoplasmic determinants, and that these entities could consist of more than one kind of molecule or cellular structure. Clearly, much more research is necessary before a generalization can be made concerning the identities of cytoplasmic determinants.

4. Localization and Segregation of Cytoplasmic Determinants

Since the identity of cytoplasmic determinants is still unresolved, research on their origin and distribution is naturally somewhat speculative. It is common to assume, however, that there are several critical stages in the life history of a determinant (Freeman, 1979). These stages include **synthesis, localization, segregation during cleavage,** and **eventual function in a restricted portion of the embryo.**

4.1. Origin of Cytoplasmic Determinants

What little is known about the origin of cytoplasmic determinants suggests that they are produced during oogenesis. Visible localizations that may be associated with cytoplasmic determinants appear during oogenesis. The oosome and polar granules of *Drosophila* are already present and positioned in mid-vitellogenic oocytes (Mahowald *et al.*, 1979). However, the oosome of immature oocytes does not promote pole cell development when transplanted to the anterior pole region of a preblastoderm stage embryo (Illmensee *et al.*, 1976). Since oosome transplantations from mature eggs reportedly promote the formation of ectopic pole cells in pre-blastoderm embryos, the pole cell determinants are likely positioned or activated in the oosome sometime between the end of oogenesis and egg maturation. It is possible that determinants or activation factors are made in the nurse cells and liberated into the egg cytoplasm at the end of oogenesis when the nurse cells are destroyed and taken up by the oocyte.

Like the insect egg oosome, the vegetal body of the *Bithynia* polar lobe is visible in early vitellogenic oocytes as a small cluster of electron-dense vesicles in the vegetal cytoplasm (Dohmen and Verdonk, 1974). During vitellogenesis it grows by the addition of more vesicles, which are probably obtained from the endoplasmic reticulum. Presumably, cytoplasmic determinants other than those discussed here also appear during oogenesis, but definitive evidence awaits their isolation and characterization.

4.2. Localization of Cytoplasmic Determinants

After their synthesis, cytoplasmic determinants may be localized in specific cytoplasmic regions of the egg, presumably by their association with a localization matrix. The pattern of specific determinant localizations is unknown, but several possibilities are: (1) restriction to certain regions of the egg; (2) distribution throughout the egg, but along a gradient; or (3) uniform distribution throughout the egg, but function only in particular cytoplasmic regions.

The timing of determinant localization has been examined by removing portions of egg cytoplasm or blastomeres at different times during early development and assessing the effect on the development of particular embryonic features. Experiments of this kind have shown that determinants can be localized either during oogenesis, during the cytoplasmic rearrangements that occur after fertilization, or during early cleavage (Freeman, 1979). Examples of determinants localized during each of these intervals are considered below.

4.2.1. Prelocalized Determinants

Cytoplasmic determinants that are localized during oogenesis are often termed prelocalized because their positions are fixed in the unfertilized egg. Determinants in the oosome and vegetal body clearly fall within this category. The polar lobe determinants are an experimentally verified example of prelocalized determinants. Determinants that specify the formation of the larval apical tuft and post-trochal region appear to be pre-localized in the polar lobe of *Dentalium* eggs (see Section 2.3). There are several lines of evidence suggesting that the determinants for these two structures may exhibit different sublocalizations within the polar lobe. Wilson (1904) found that when the polar lobe was removed at first cleavage, lobeless embryos developed into larvae lacking both the apical tuft and post-trochal region, but when the smaller polar lobe was removed at second cleavage, lobeless embryos developed into larvae still lacking post-trochal regions but exhibiting apical tufts. In another experiment, sections of the polar lobe, from its lower-most to upper-most regions, were excised at first cleavage (Geilenkirchen *et al.*, 1970). These results showed that when the lower 60% of the polar lobe was removed, embryos developed into larvae lacking the post-trochal region, but containing an apical tuft. In contrast, when a section containing the lower 80% of the polar lobe was removed, embryos developed into larvae lacking both the post-trochal region and the apical tuft. These experiments suggest that there is a layering of apical tuft and post-trochal determinants in the polar lobe, with the former localized more extensively toward the animal pole of the egg than the latter.

Evidence that the apical tuft and post-trochal determinants are pre-localized in *Dentalium* eggs is derived from experiments similar to those discussed above in which sections of the vegetal hemisphere region (up to 70% of the volume of the first polar lobe) were removed from the uncleaved egg (Verdonk *et al.*, 1971). This resulted in the prevention of the development of the post-trochal region, but not the apical tuft. Recently, Render and Guerrier (1984) confirmed and extended these results in an examination of the development of size-regulated polar lobes. Wilson (1904) reported that vegetal fragments of unfertilized *Dentalium* eggs, which were subsequently fertilized, formed polar lobes in proportion to the size of the vegetal fragment. Render and Guerrier (1984) found that these smaller polar lobes are necessary for the development of the larval post-trochal region, but not the apical tuft, suggesting that the position of the determinants is fixed before fertilization and is not significantly altered thereafter.

4.2.2. Localization of Determinants during Ooplasmic Segregation

The extensive cytoplasmic rearrangements that accompany some egg maturation divisions or may occur as a response to fertilization are thought to concentrate and/or localize determinants so they can be properly partitioned to the blastomeres during cleavage. The most dramatic example of this phenomenon occurs in ascidian eggs, in which the colored cytoplasmic regions undergo extensive rearrangements after fertilization. Fragments cut from any part of an unfertilized ascidian egg have the potential to develop into complete larvae after fertilization (Ortolani, 1958). If similar fragments are prepared from fertilized eggs that have completed ooplasmic segregation, however, only the fragments containing part of the vegetal hemisphere cytoplasm retain the capacity to form complete larvae. More recent microsurgical experiments, in which 15% or less of the cytoplasm removed from the vegetal pole region of fertilized ascidian eggs caused the formation of radialized larvae, suggest that determinants involved in specifying the embryonic axis may be localized at the vegetal pole (Bates and Jeffery, 1987b). Since the vegetal pole cytoplasm enters cells that are the first to invaginate during gastrulation, these experiments also suggest that determinants segregated to the vegetal pole region of ascidian eggs after fertilization may define the site of gastrulation.

The vegetal granules, presumed germ cell determinants in nematode eggs, are also localized during cytoplasmic rearrangements that occur after fertilization (Strome and Wood, 1982) (see Chapter 2). Immunofluorescence, using a monoclonal antibody to the vegetal granules, shows that they are uniformly distributed in the unfertilized egg. After fertilization, they are gradually segregated into the vegetal pole region, where they are enclosed in the vegetal (P1)

blastomere at first cleavage. These results suggest that some cytoplasmic determinants in ascidian and nematode eggs are evenly distributed before fertilization and become localized during ooplasmic segregation.

4.2.3. Localization of Determinants during Cleavage

In ctenophores (Freeman, 1976) and nemertines (Freeman, 1978),
cytoplasmic determinants are fixed in their final positions during the early
cleavages. The apical tuft of the larvae of the nemertine *Cerebratulus* develops
from the four animal hemisphere cells of 8-cell embryos, while the gut develops from the four vegetal hemisphere cells of 8-cell embryos. If the four animal
blastomeres are removed at the 8-cell stage, the partial vegetal embryos develop
into larvae without apical tufts (Horstadius, 1937). Likewise, if the four vegetal
blastomeres are removed from 8-cell embryos, the partial animal embryos develop into larvae without guts. This experiment shows that the position of
apical tuft and gut determinants are localized by the 8-cell stage. To investigate
the timing of localization of apical tuft and gut determinants at earlier stages,
unfertilized eggs, uncleaved zygotes, 2-cell embryos, and 4-cell embryos were
cut into animal and vegetal hemisphere fragments. The fragments were then
tested for their ability to form larvae with apical tufts and guts (Hörstadius,
1971; Freeman, 1978). Apical tuft determinants in *Cerebratulus* did not appear
to be localized in the animal hemisphere portions of the embryo until during
the interval between the first and second cleavages (Fig. 12). The gut determinants began to be concentrated in the vegetal hemisphere of uncleaved eggs
after the second maturation division, but this process was not completed until
the interval between the first two cleavages (Fig. 12). There is additional evidence that these determinants continue to undergo localization during subsequent cleavage cycles (Freeman, 1979), presumably until they enter the larval
apical tuft and gut cell lineages.

The ctenophore *Mnemiopsis* contains several distinct cell types, including
ciliated comb plate cells, positioned in rows along the surface of the larva, and
bioluminescent photocytes, found within the larval radial canals. In

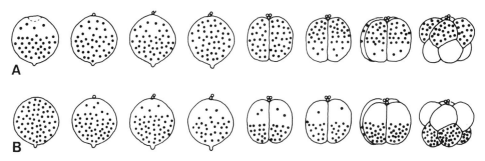

Figure 12. A schematic diagram showing the progressive distribution of apical tuft (**A**) and gut (**B**)
determinants during early *Cerebratulus* development. (From Freeman, 1978.)

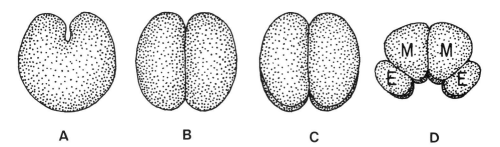

Figure 13. Early cleavage of a ctenophore embryo. (**A**) 1-cell stage in the process of cleavage; (**B**) 2-cell stage; (**C**) 4-cell stage; (**D**) 8-cell stage showing the E and M macromeres.

ctenophore embryos, the third cleavage is oblique, producing four internal M blastomeres and four smaller external E blastomeres (Fig. 13). Blastomere isolation studies show that at the 8-cell stage, the ability to form photocytes is confined to the four M blastomeres, whereas the ability to develop comb plate cilia is restricted to the four E blastomeres. Experiments similar to those described above for *Cerebratulus* embryos have been conducted with early cleavage-stage *Mnemiopsis* embryos. At the 2-cell stage, comb plate cell determinants appear to be intermingled with photocyte determinants in each blastomere. At the 4-cell stage, however, comb plate cell determinants are already localized in the presumptive E cell (external) region of each blastomere, while photocyte determinants are still evenly distributed. It is concluded that *Mnemiopsis* comb plate determinants become localized during second cleavage, whereas photocyte determinants are localized during third cleavage.

4.2.4. Regional Activation of Cytoplasmic Determinants

In addition to those determinants that are localized during development, it is also possible that some cytoplasmic determinants are uniformly distributed but functional only in particular regions of the egg or specific blastomeres of the early embryo because of their localized elimination or activation. Gerhart *et al.* (1981, 1984) have obtained duplications of the dorsal–ventral axis in *Xenopus* by forced reorientation of the egg cytoplasm. In their experiments, tilting or centrifugation of the egg during the critical period for axis determination abolishes the orientation of the original axis and generates a new axis corresponding to the direction of gravity or centrifugal force. When centrifuged eggs are centrifuged again, but in the opposite direction, embryos develop two complete body axes, each oriented according to one direction of centrifugal force (Fig. 14). A duplication of the dorsal–ventral axis in this way is difficult to explain by the localization of axial determinants in only one portion of the egg. If strict localization were involved, only one half the determinants would have had to be driven to the opposite side of the egg during the second centrifugation to specify the supernumerary axis. These results can be explained,

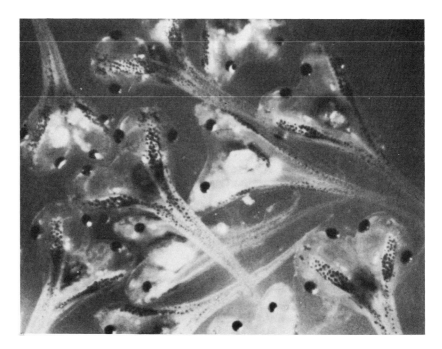

Figure 14. *Xenopus laevis* tadpoles with double axes produced by centrifugation of the uncleaved egg. (From Gerhart *et al.*, 1983.)

however, by assuming that the first centrifugation causes the activation of uniformly distributed axial determinants in one region of the egg, while the second centrifugation activates these determinants in another region of the egg. Thus, an activator substance, rather than the axial determinants themselves, is thought to be localized and reoriented during centrifugation. Although further research will be necessary to determine whether uniformly distributed axial determinants and localized activators actually exist in *Xenopus* eggs, this hypothesis is supported by the fact that *Toll*$^+$ cytoplasm obtained from any region (including dorsal areas) of a wild-type *Drosophila* egg can establish a ventral region at the site of injection in a mutant *Toll* embryo (Anderson *et al.*, 1985b). Thus, the *Toll* gene product may be another example of a determinant that is uniformly distributed, but activated normally in only one region of the egg.

4.3. Distribution of Cytoplasmic Determinants during Development

For localized determinants to ultimately function, mechanisms must exist to ensure their precise distribution to particular cells during embryogenesis. There are at least two ways to achieve this. First, determinants may be actively segregated into particular cells either before or during each cleavage. The progressive localization of apical tuft determinants in *Cerebratulus* embryos and comb plate cilia and photocyte determinants in *Mnemiopsis* embryos during the early cleavages are examples of this situation. Second, determinants may be fixed in their final positions before cleavage, either by physical localization or regional activation, and be distributed to the correct embryonic cells by simple partitioning, which is regulated by the position of the cleavage planes or by cytoplasmic asymmetries such as the polar lobe. Prelocalized determinants and determinants fixed in their positions during ooplasmic segregation apparently employ the second strategy.

4.4. Localization and Segregation of Macromolecules

The possibility that macromolecules serve as determinants has initiated a number of studies on the localization and segregation of maternal proteins and mRNA during early development. Unfortunately, it has been difficult to obtain information on the spatial distribution of macromolecules because it is usually impossible to fractionate eggs into separate cytoplasmic regions for bio-chemical analysis. This problem has been approached by allowing the egg to cleave and biochemically analyzing differences between separated embryonic cells, by studying large eggs that can be divided manually into sections for biochemical analysis, or by utilizing *in situ* methods in which sectioned eggs or embryos are exposed to antibodies or nucleic acid probes.

4.4.1. Spatial Distribution of Maternal Proteins

Since proteins are unevenly distributed in somatic cells, uneven protein distributions were expected in eggs as well. However, if proteins are to be considered as possible cytoplasmic determinants, it is important to verify whether unique protein distributions occur in different egg cytoplasmic re-gions and whether localized proteins segregate during early development. The first experiments addressing these questions were done on amphibian eggs. These eggs are large enough to be frozen in mass and sectioned perpendicular to the animal–vegetal axis. The protein composition of pooled sections can then be examined by gel electrophoresis. Using this approach, it was shown that about 25% of the abundant proteins of *Xenopus* eggs exist only along a portion of the animal–vegetal axis (Moen and Namenwirth, 1977). Subsequent studies, however, showed that the majority of these differences in *Xenopus* eggs were quantitative rather than qualitative, and that in axolotl eggs, only a few quantitative differences in proteins could be observed along the animal–vegetal axis (Jäckle and Eagleson, 1980). Although these experiments suggest that some proteins may be distributed unevenly along the animal–vegetal axis, they do not provide precise information on the spatial distribution of proteins between the various cytoplasmic regions of amphibian eggs.

The spatial distribution of proteins in *Styela* eggs was recently examined by comparing the proteins of mass-isolated yellow crescents to those present in the remaining cytoplasmic regions (Jeffery, 1985). Yellow crescents can be fractionated in mass because the myoplasm contains a unique cytoskeletal domain, which is more resistant to disruption by homogenization than the other egg cytoplasmic regions. Gel electrophoretic analysis of isolated yellow crescents showed that 32% of the egg proteins are enriched by several fold in the yellow crescent, whereas 11% of the total egg proteins are present exclusively in the myoplasm. These studies show that cytoplasmic regions of distinct developmental fate may contain a subset of egg proteins.

In studies on protein localization in axolotl oocytes, a few proteins were found exclusively in the GV (Jäckle and Eagleson, 1980). These proteins are of particular interest because the activity that rescues the maternal effect mutation *o* is located in the GV and appears to be a protein. Using GV proteins as antigens, monoclonal antibodies were prepared and used to examine the fates of amphibian GV proteins during early development (Dreyer *et al.*, 1982). After GV breakdown, the antigens become localized primarily in the animal hemisphere cytoplasm. During cleavage, the antigens enter the animal hemisphere blastomeres and are found later in ectodermally derived tissues. At specific times during embryogenesis, the antigens shift from the cytoplasm to the nuclei. Thus, these GV-localized proteins in axolotl eggs behave as would be expected of determinants that affect gene activity. A number of protein antigens are also present in sea urchin eggs that are localized in germ layers or specific areas of the embryo, but little else is known about these interesting molecules (McClay *et al.*, 1983).

4.4.2. Spatial Distribution of Maternal mRNA

Because of the accumulating evidence that determinants may be composed of maternal mRNA, the identification of localized maternal mRNAs has been attempted for more than a decade. Most of the early studies on maternal RNA distribution were either negative or ambiguous and will not be considered here (for a detailed review, see Jeffery, 1983). Recent studies with amphibian, sea urchin, ascidian, and *Drosophila* embryos, however, suggest that some maternal mRNAs are localized in the egg and differentially distributed between the embryonic cells.

The most extensive evidence for maternal mRNA localization exists in *Xenopus laevis* oocytes, eggs, and early embryos. Evidence for mRNA localization along the animal-vegetal axis has required the ability to align and section quantities of frozen early developmental stages perpendicular to the animal-vegetal axis. Three different experiments have been carried out to identify localized mRNAs. First, eggs or gastrulae were divided into animal, middle, and vegetal sections and the poly(A)$^+$ RNA from each region was compared by hybridization with cDNAs made from animal and vegetal poly(A)$^+$ RNAs. This experiment demonstrated that 3 to 5% of the total poly(A)$^+$ RNA sequences were enriched 2- to 20-fold in the vegetal region of uncleaved eggs and gastrulae (Carpenter and Klein, 1982). Second, fully grown oocytes were divided into the same 3 regions, and poly(A)$^+$ RNA extracted from each region was compared by *in vitro* translation (King and Barklis, 1985). This experiment showed that

17 out of about 600 detectable mRNAs were enriched by 10- to 100-fold in the vegetal region of oocytes (King and Barklis, 1985). Similar experiments suggest that a small number of maternal mRNA species are also localized in the vegetal hemisphere of eggs and cleavage stage embryos (Smith, 1986). Third, a cDNA library prepared from RNA of unfertilized eggs was screened with cDNAs made from animal and vegetal region poly(A)$^+$ RNAs (Rebagliati et al., 1985). In this experiment, 0.2% of the clones were specific to the vegetal region, while 1.2% of the clones were specific to the animal region. These studies suggest that a small proportion of Xenopus egg mRNAs may be differentially localized, although most of these molecules are uniformly distributed along the animal-vegetal axis. Recent sequencing of a cDNA clone corresponding to one of the maternal mRNAs localized in the animal pole region of Xenopus eggs codes for the α-chain of a mitochondrial ATPase (Weeks and Melton, 1987). Although the developmental purpose of localizing a mitochondrial ATPase mRNA is unknown, it is possible that the protein product of this message may affect processes such as respiration, ion flow, or metabolism, and regionally alter the physiological state of the embryonic cells that it enters.

According to two of the studies described above (Carpenter and Klein, 1982; King and Barklis, 1985), the localized mRNAs are mainly present in the vegetal region, while the other suggests that localized mRNAs are found in either vegetal or animal regions (Rebagliati et al., 1985), with a greater proportion present in the latter. It is interesting that some localized maternal mRNAs exist in the vegetal region of amphibian eggs because of the possibility that axial determinants or their activators may also be concentrated in this region.

The sea urchin embryo has also been used to study RNA localizations. The fourth cleavage of sea urchin embryos is unequal, resulting in the formation of a 16-cell embryo composed of 8 animal mesomeres, 4 subequatorial macromeres, and 4 vegetal micromeres. There is also a restriction in developmental potential associated with this cleavage, since the micromeres are fated to become the primary mesenchyme cells. The small size of the micromeres relative to the other blastomeres allows them to be separated from mixtures of mesomeres and macromeres on density gradients. Consequently, a direct comparison of the sequence complexity of RNA derived from the three cell types is possible. The sequence complexity of unique RNA molecules derived from the micromeres and from mixtures of mesomeres and macromeres was compared by hybridization with single copy DNA probes (Rodgers and Gross, 1978). Using DNA tracers made from total egg RNA or mesomere–macromere RNA as probes, the micromere RNA formed 20 to 30% fewer hybrids than did RNA molecules derived from the other blastomeres (Fig. 15). This result suggests that the macromeres and/or mesomeres contain a significant number of single-copy RNA sequences that are absent from the micromeres. This difference corresponds to a total coding capacity for about 10,000 average-sized polypeptides, assuming all these sequences represent mRNAs. However, the overall pattern of protein synthesis in populations of micromeres and mesomere–macromere mixtures is qualitatively identical in two-dimensional gels in which as many as 3000 different polypeptides are compared (Tufaro and Brandhorst, 1979). Thus, if the

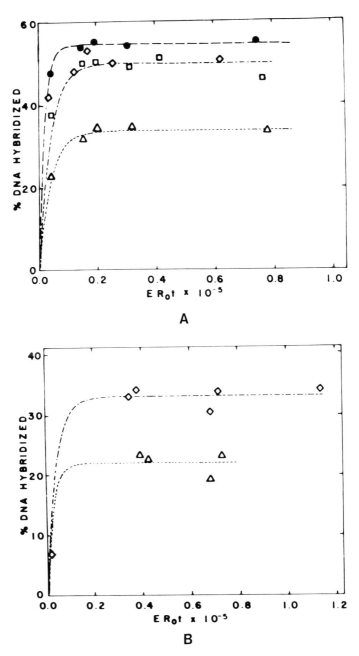

Figure 15. A demonstration of differential maternal RNA localization in the blastomeres of 16-cell sea urchin embryos by molecular hybridization. (**A**) Hybridization of a total egg single-copy DNA tracer with total egg RNA (●), total 16-cell embryo RNA (□), mesomere–macromere RNA (◇), and micromere RNA (△) drivers. (**B**) Hybridization of a total mesomere–macromere, single-copy DNA tracer with mesomere–macromere RNA (◇) and micromere RNA (△) drivers. (From Rodgers and Gross, 1978, copyright © M.I.T.)

single-copy RNA sequences localized in the mesomeres and/or macromeres are mRNAs, they are either of low abundance, are translationally inactive at this stage of embryogenesis, or their translation products are not resolved in two-dimensional gels.

Examination of the spatial distribution of mRNA in ascidian eggs not only provides information on mRNA distribution; but, because of the presence of the colored cytoplasmic regions, it also permits mRNAs to be mapped with respect to the embryonic fate map. When RNA from isolated yellow crescents was translated in a cell free system, quantitative (but not qualitative differences) could be observed in the abundant messages of the myoplasm relative to the other cytoplasmic regions of the egg (Jeffery, 1985). This result suggests that some messages may be enriched in the myoplasm. RNA localization in ascidian eggs has also been investigated by in situ hybridization with cloned DNA probes (Jeffery et al., 1983). Hybridization of sectioned eggs and early embryos with a histone DNA probe showed that histone mRNA was evenly distributed between the various cytoplasmic regions, a result that was expected because histone synthesis is presumably required in each of the cell lineages. On the other hand, actin mRNA was shown to be concentrated in the ectoplasm and myoplasm (Fig. 16). The localization of actin mRNA in the ectoplasm was expected since earlier in situ hybridization studies using a poly (U) probe indicated that total poly(A)$^+$ RNA was highly concentrated in the ectoplasm (Jeffery and Capco, 1978). These earlier studies also showed a paucity of total poly(A)$^+$ RNA in the myoplasm, as did data obtained from isolated yellow crescents (Jeffery, 1985). Thus, it appears that actin mRNA is specifically enriched in the myoplasm. Recent studies have shown that ascidian eggs contain maternal mRNA coding for the muscle form of actin (Jeffery et al., 1986; Tomlinson et al., 1987). Thus, it is possible that localized maternal mRNA serves as a determinant for myofilaments in ascidians.

An advantage of using in situ hybridization to investigate the spatial distribution of maternal mRNA is that the origin, fate, and any changes in the distribution of these molecules can be traced. For instance, one can ask: When does mRNA of ascidian eggs become localized in the ectoplasm and myoplasm? What is the behavior of mRNA molecules during ooplasmic segregation? To which cells is mRNA partitioned during the early cleavages? In vitellogenic oocytes, actin mRNA is already localized in the GV plasm and the peripheral egg cytoplasm from which the ectoplasm and myoplasm arise, respectively (Jeffery et al., 1983). After GV breakdown, the actin mRNA in the GV becomes the ectoplasmic actin mRNA of the mature egg. When the various cytoplasmic regions stream through the egg after fertilization, the actin mRNA maintains its association with the ectoplasm and myoplasm. Thus, the position of actin mRNA molecules, like cytoplasmic regions and determinants, appears to be fixed during ooplasmic segregation. During the early cleavages, the actin mRNA in the ectoplasm is distributed mainly to the ectodermal cell lineages, while the actin mRNA in the myoplasm is distributed to the muscle and mesenchyme cell lineages. The maintenance of mRNA associations with specific cytoplasmic regions during ooplasmic segregation and cleavage suggests that

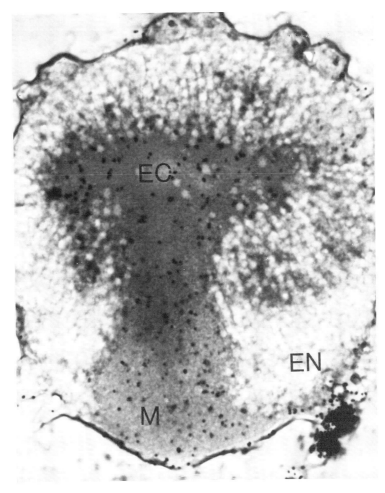

Figure 16. The spatial distribution of actin mRNA in a blastomere of a 2-cell ascidian embryo as determined by *in situ* hybridization with a cloned actin DNA probe. EN, endoplasm; EC, ectoplasm; M, myoplasm. (From Jeffery *et al.*, 1983.)

these molecules are tenaciously associated with a structural framework in the egg, a topic that will be considered in the next section.

In *situ* hybridization has also been used to map the distribution of a maternal transcript in oocytes, eggs, and early embryos of *Drosophila* (Mlodzik *et al.*, 1985). In *Drosophila*, there is a gene known as *caudal*, whose function is unknown, but has been identified and cloned because it contains a sequence that is highly homologous to the homeo box, a conserved DNA sequence shared by several homeotic genes. The *caudal* mRNA accumulates in the nurse cells and the oocyte during oogenesis, and the maternal transcripts then become distributed in a concentration gradient along the anterior–posterior axis of syncytial blastoderm embryos.

The spatial distribution during oogenesis of one of the maternal mRNAs

localized in the vegetal pole region of *Xenopus* eggs has recently been examined by *in situ* hybridization (Melton, 1987). Surprisingly, this mRNA is uniformly distributed in the cytoplasm of immature oocytes. During vitellogenesis, however, it becomes concentrated as a crescent in the vegetal pole region of the oocyte and is retained in this location throughout the remainder of oogenesis. At present, the mechanisms that cause the vegetally-localized mRNA to enter the vegetal pole region are unknown, however, this behavior does not appear to be due to changes in the structure of the mRNA molecule itself. Northern blots of RNA isolated before, during, or after the localization process show that this mRNA is present as a mature (spliced) polyadenylated molecule throughout oogenesis (Melton, 1987).

There is also evidence from *in situ* hybridization (Venezsky *et al.*, 1981) and egg fractionation (Showman *et al.*, 1982) experiments that maternal histone mRNA may be localized within the maternal pronucleus of sea urchin eggs. The histone mRNA is intranuclear until the time of nuclear envelope breakdown at first cleavage, when histone transcripts appear in the cytoplasm and load onto polyribosomes (Wells *et al.*, 1981). Although the function of histone mRNA sequestration in sea urchin eggs is unknown, this observation does demonstrate that eggs possess mechanisms to localize specific maternal mRNAs.

Evidence from these investigations allows the tentative conclusion that a small number of maternal mRNA species may be localized in eggs and early embryos.

5. Mechanism of Cytoplasmic Determinant Localization

Since cytoplasmic determinants are probably responsible for regulation of a number of different activities in developing embryos, and since these entities become localized at different times during early development, it is possible that they use different mechanisms for their localization. One common feature that determinants may share, at least some time during their life history, is an attachment to a structural matrix in the egg. This hypothetical structural matrix has been called a **promorphological scaffold** (Freeman, 1979).

5.1. Evidence That Most Cytoplasmic Determinants Are Absent from the Fluid Egg Cytoplasm

Attempts to transfer cytoplasmic determinants from one region of the egg or embryo to another were discussed in Section 2.4. The problems investigators experienced in those studies, and other results considered below, support the possibility that most determinants are absent from the fluid egg cytoplasm. Early evidence for this possibility came from centrifugation studies with embryos containing polar lobes (Morgan, 1935; Verdonk, 1968; Clement, 1968). In *Ilyanassa*, the polar lobe is normally filled with vegetal yolk granules, whereas yolk-free cytoplasm is located in the animal hemisphere. When these eggs are centrifuged with the animal pole oriented in the direction of the centrifugal force, the yolk moves into the animal hemisphere, displacing the animal hemisphere cytoplasm into the vegetal hemisphere. Thus, when the polar lobe forms, it contains cytoplasm originally present in the animal hemisphere. If the polar lobes of centrifuged eggs are removed at first cleavage, the same set of larval defects is obtained as in control embryos. This suggests that the polar lobe determinants are not readily displaced by centrifugation, indicating that they are not present in the fluid cytoplasm. A similar conclusion was reached in experiments in which about half of the polar lobe cytoplasm was removed from *Dentalium* embryos with a micropipette. After this operation, most of the original polar lobe cytoplasm was replaced by cytoplasm that originated in the animal hemisphere, but the embryo still developed into a normal larva (Verdonk and Cather, 1983).

Cytoplasmic determinants in ascidian eggs are also resistant to displacement by centrifugal force. As mentioned in Section 3.1, when uncleaved *Styela* eggs are centrifuged, cytoplasmic regions may become rearranged within the cell (Conklin, 1931). When the egg divides, the two blastomeres often contain visibly different cytoplasmic regions (Conklin, 1931), but despite this cytoplasmic reorientation the centrifuged embryos develop into normal larvae.

When the shell of *C. elegans* eggs or embryos is punctured by a laser microbeam, a bleb of cytoplasm is often extruded (Laufer and von Ehrenstein, 1981). Surprisingly, complete larvae can be obtained from eggs that lose 20% of their cytoplasm (from either pole) or 60% of the cytoplasm from one or several different cells during the early cleavage stages. These results suggest that cytoplasmic determinants are anchored to rigid components and therefore are not extruded outside the eggshell with the fluid cytoplasm.

What are the structural components in eggs that anchor determinants? Several possibilities exist: First, they may be anchored to the plasma membrane. Plasma membrane association would explain each of the experiments discussed above, with the exception of the laser experiment, in which the plasma membrane is blebbed out of the egg shell with the extruded cytoplasm. Second, they may be associated with a gelled cytoplasm in the egg cortex. Many eggs, however, seem to lack an extensive cortical region. Third, it is possible that determinants are associated with cytoskeletal domains. The cytoskeleton has been recognized as a ubiquitous structural component of all eukaryotic cells, including eggs (Moon *et al.*, 1983; Jeffery and Meier, 1983; Jeffery, 1985*c*; Wylie *et al.*, 1985). Evidence for each of these possibilities will be discussed separately in the following sections.

5.2. Egg Plasma Membrane Domains as Potential Localization Matrices

Studies on the plasma membrane of somatic cells have shown that, in some cases, this structure does not behave like an ideal fluid; that is, membrane components may not be freely diffusible throughout the bilayer. This phe-

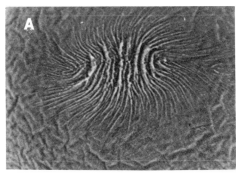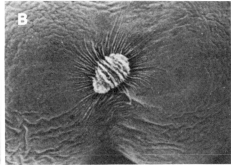

Figure 17. Scanning electron micrographs showing folds in the plasma membrane at the vegetal pole and in the polar lobe of embryos of the mollusc, *Crepidula fornicata*. (**A**) Uncleaved egg. (**B**) First cleavage and polar lobe. (From Dohmen, 1983.)

nomenon, which may be due to the presence of different cytoskeletal domains beneath the bilayer, causes unique membrane domains to be established. A number of different studies suggest that plasma membrane domains exist in eggs and early embryos.

The first indication that membrane domains may exist in eggs came from morphological studies with mollusc eggs (Dohmen and van der Mey, 1977). Scanning electron microscopy indicated that local surface differentiations, including microvilli and/or extensive surface folds, are present in the vegetal pole region (Fig. 17). This membrane domain is extruded with the polar lobes during the early cleavages and eventually enters the D cell lineage. These results were subsequently extended by freeze-fracture studies, which indicated that the vegetal pole region of *Nassarius* eggs is characterized by a much higher density of intra-membrane particles than the remainder of the egg (Speksnijder *et al.*, 1985a). In addition, fluorescence photobleaching analysis showed there were differences in lipid lateral mobility between the animal and vegetal hemisphere regions of *Nassarius* eggs (Speksnijder *et al.*, 1985b). Electrophysiological studies have shown that the polar lobe membrane is more electrically excitable than membrane in other regions of the *Dentalium* embryo (Jaffe and Guerrier, 1981). This increased excitability was attributed to a regional localization of different ion channels. These studies on mollusc eggs, which are supported by parallel investigations on *Xenopus* eggs (Bluemink and Tertoolen, 1978; Dictus *et al.*, 1984), suggest that membrane domains are potential candidates for determinant localization matrices.

5.3. Egg Cortical Regions as Potential Localization Matrices

The potential role of the egg cortex in localizing cytoplasmic determinants has been accepted for years, yet there are few published experiments that directly approach this question. Transplantation experiments of the gray crescent cortex of uncleaved *Xenopus* eggs resulted in double axes in host embryos,

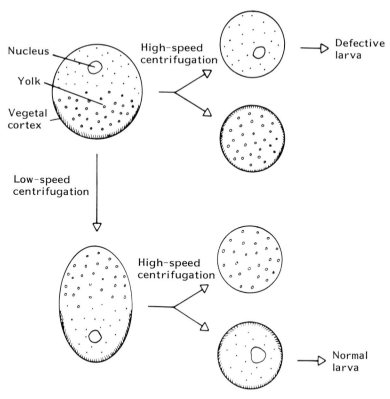

Figure 18. A schematic illustration of the centrifugation experiment of Clement (1968), which suggests that a vegetal cortex is required for normal *Ilyanassa* development. (From Slack, 1983.)

indicating that this region contains axial determinants (Curtis, 1960; 1962); however, the double axes now appear to be the result of cytoplasmic displacement due to tilting eggs, rather than the result of the cortical transplantations (Gerhart et al., 1981).

The role of the vegetal–cortical region in development was tested in *Ilyanassa* eggs by the two-step centrifugation experiment illustrated in Fig. 18 (Clement, 1968). When uncleaved eggs were centrifuged at high speed, they fragmented into nucleate animal halves and anucleate vegetal halves. In this experiment, the vegetal cortex remained in the vegetal halves. The nucleate animal halves, which were missing vegetal cortex, developed into defective larvae that lacked many of the structures normally specified by the polar lobe determinants. When uncleaved eggs were stratified by low speed centrifugation, so that the nucleus entered the vegetal region, and then fragmented by high speed centrifugation, nucleate vegetal halves could be obtained that contained the vegetal cortex. The nucleate vegetal halves with the vegetal cortex developed into normal larvae. Although these experiments are consistent with the possibility that the polar lobe determinants are localized in the vegetal cortex, they do not exclude vegetal plasma membrane or, as discussed below, cortical cytoskeletal domains as potential localization matrices.

5.4. Egg Cytoskeletal Domains as Potential Localization Matrices

The most extensive work on the role of the egg cytoskeleton in localization has been done with ascidian and annelid eggs. The first evidence that the cytoskeleton may be important in localizing determinants was provided when *Styela* eggs, extracted with nonionic detergent, were shown to contain at least three different cytoskeletal domains that reflected precisely the cytoplasmic territories of the ectoplasm, endoplasm, and myoplasm (Jeffery and Meier, 1983). The myoplasmic cytoskeletal domain is the best characterized of these domains. It consists of two parts: an actin filamentous network, called the plasma membrane lamina (PML) (Fig. 19A), which is closely associated with the plasma membrane, and a more internal filamentous lattice, which contains suspended myoplasmic pigment granules (Fig. 19B) and links the PML to the other cytoskeletal domains. The distribution of the myoplasmic cytoskeletal domain follows that of the myoplasm at each developmental stage. It is initially localized in the periphery of the unfertilized egg, becomes attenuated into a crescent after ooplasmic segregation, and is partitioned specifically to the muscle cell lineage during cleavage.

The myoplasmic cytoskeletal domain appears to play an important role in ooplasmic segregation. During this process, the PML is condensed into a tighter network and recedes into the vegetal hemisphere. As the PML recedes, it drags with it the remainder of the myoplasmic cytoskeletal domain including the myoplasmic pigment granules and, presumably, any bound determinants that might also be present in this matrix. The recession of the PML is sensitive to the microfilament inhibitor cytochalasin, but not to the microtubule-depolymerizing drug colchicine, suggesting that actin microfilaments are important for its movement. The behavior of the PML is correlated with that of ascidian muscle cell determinants in centrifugation experiments. Moderate centrifugation, which does not markedly affect subsequent development (Conklin, 1931), causes the deep filamentous lattice and its associated components, but not the PML, to be dispersed (Jeffery and Meier, 1984). Stronger centrifugation, which does affect development, also disperses or destroys the PML. These results suggest that the PML may be a matrix that localizes cytoplasmic determinants in ascidian eggs.

The role of cytoskeletal elements in localizing the vegetal granules of *C. elegans* eggs has recently been tested using drugs and mutants that exhibit abnormal spindle orientations (Strome and Wood, 1983). In the unfertilized egg, the vegetal granules are dispersed randomly throughout the cytoplasm. After fertilization, the vegetal granules gradually become localized in the vegetal hemisphere during ooplasmic segregation, when they are partitioned into the P1 cell during first cleavage. In mutant zygotes with abnormal spindle orientations, and in wild-type zygotes treated with drugs that depolymerize microtubules,

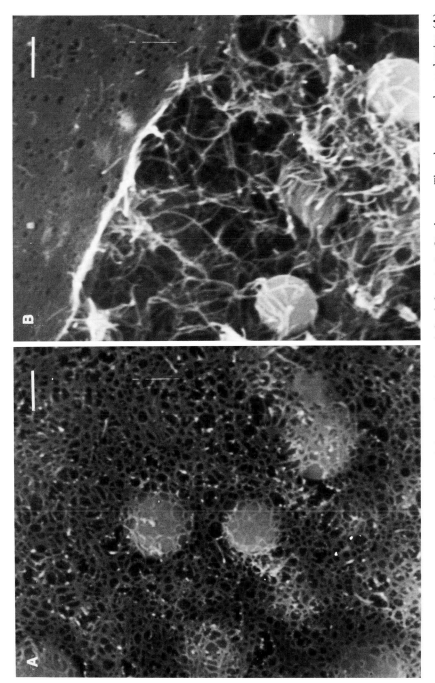

Figure 19. Scanning electron micrographs of the myoplasmic cytoskeletal domain in *Styela* eggs. The plasma membrane lamina (**A**) overlies a deep filamentous system (**B**) containing pigment granules. Scale bars: 1 μm. (From Jeffery and Meier, 1983.)

vegetal-granule segregation occurs normally. In contrast, treatment of wild-type zygotes with cytochalasin inhibits vegetal-granule segregation.

The results suggest that cytoskeletal elements may be involved in fixing the position of cytoplasmic determinants localized during ooplasmic segregation, and that the important components of the egg cytoskeleton may be actin microfilaments.

5.5. Role of the Egg Cytoskeleton in mRNA Localization

The tenacious association of maternal mRNA with specific egg cytoplasmic regions during ooplasmic segregation suggests that these molecules may be associated with a structural framework in the egg cytoplasm. This possibility has been investigated by extracting eggs or early embryos with non-ionic detergents and determining whether mRNA is retained in the insoluble cytoskeletal residue. If so, its pattern of spatial localization is then studied to see if it is similar to that of the intact egg or embryo. Extraction of the eggs and early embryos of *Styela* or the annelid *Chaetopterus* results in the retention of most of the total poly(A)$^+$ RNA in the cytoskeletal residue (Jeffery, 1984b; Jeffery, 1985c). Moreover, *in situ* hybridization of sections of cytoskeletal residues indicates that the pattern of mRNA localization observed in intact eggs and embryos is reflected in their cytoskeletons. In *Styela* eggs, most total poly(A)$^+$ RNA sequences are localized in the ectoplasmic cytoskeletal domain, while most actin mRNA is localized in both the ectoplasmic and myoplasmic cytoskeletal domains (Jeffery and Capco, 1978; Jeffery, 1984b). In *Chaetopterus* eggs, a large proportion of the poly(A)$^+$ RNA and actin mRNA is located in a cortical region that is partly shunted to the polar lobe and the CD blastomere at first cleavage (Jeffery and Wilson, 1983). In detergent-extracted *Chaetopterus* eggs, these RNAs retain their cortical localization in a specialized cytoskeletal domain (Jeffery, 1985c). These results suggest that mRNAs are localized by virtue of their association with specific egg cytoskeletal domains and that their final destinations in the embryo may be governed by the regionalization and partitioning of the egg cytoskeleton during cleavage.

5.6. Localization of Cytoplasmic Determinants in a Cortical Complex

In the preceding sections, evidence was presented implicating the egg plasma membrane, cortex, and cytoskeleton in the localization of cytoplasmic determinants. One can now ask the question: Are these egg components individually involved in determinant localization, or do they somehow function together? Based on evidence from ascidian and *Chaetopterus* eggs, it was proposed recently that the egg plasma membrane, cortical cytoskeleton, and perhaps what is envisioned as the cortical gel, are intimately associated as a supra-organellar complex (Jeffery, 1984a). A model for the cortical complex of ascidian eggs is shown in Fig. 20.

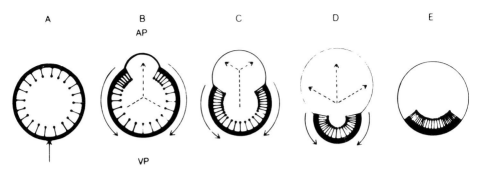

Figure 20. A schematic illustration of the movements of the hypothetical cortical complex in uncleaved ascidian eggs. (**A**) An unfertilized egg with an arrow indicating the future focal point of ooplasmic segregation. (**B–E**) Fertilized eggs in the process of ooplasmic segregation. In each diagram, the thick egg boundaries represent parts of the plasma membrane with integral proteins connected to the cortical (myoplasmic) cytoskeleton. The lines attached to the inner surface of the membrane represent the cortical cytoskeleton with associated organelles and mRNA (filled circles). In **B–E**, the continuous arrows represent the direction of movement of the cortical complex, while the broken arrows represent the direction of movement of the fluid internal cytoplasm. AP = animal pole; VP = vegetal pole. (From Jeffery and Meier, 1983.)

Evidence for the existence of a cortical complex in ascidian eggs stems from the observation that membrane components, the myoplasmic (cortical) cytoskeletal domain, cytoskeletal associated mRNA, and cortical organelles move in concert into the vegetal hemisphere during ooplasmic segregation. Connections can sometimes be seen between these structures by electron microscopy and after *in situ* hybridization (Jeffery and Meier, 1983; Jeffery, 1984b). A cortical cytoskeletal domain containing localized mRNA and specific cortical organelles is also present in *Chaetopterus* eggs (Jeffery, 1985c). In a series of experiments similar to the centrifugation studies on *Ilyanassa* eggs (Clement, 1968) (see Section 5.3) high speed centrifugation was used to fragment *Chaetopterus* eggs such that the nucleus is present in the animal fragment; the cortical domain, with localized cytoskeletal elements, organelles, and mRNA molecules, is present in the vegetal fragment (Swalla *et al.*, 1985). The results indicated that the nucleate animal halves were unable to form normal larvae unless they contained part of the cortical domain. These studies suggest that a supraorganellar complex located in the egg cortex may be the ultimate localization matrix for cytoplasmic determinants. It remains to be decided, however, where the determinants would be found in this complex, and whether it also exists in other eggs.

5.7. Role of Asters in Segregating Cytoplasmic Determinants

The attachment of determinants to localized plasma membrane or cytoskeletal domains would provide a mechanism for partitioning determinants that are pre-localized in the unfertilized egg, or whose localization in the zygote is fixed during ooplasmic segregation; however, segregation of determinants during cleavage is more difficult to explain by these mechanisms. There are several lines of evidence which suggest that asters formed during the maturation divisions and early cleavages may play an instrumental role in segregating the apical tuft and gut determinants of *Cerebratulus* eggs (Freeman, 1978). First, an exposure of eggs to ethyl carbamate at the time of the second maturation division inhibits aster formation while allowing development to proceed, but the apical tuft and gut determinants do not become localized in the animal and vegetal blastomeres by the 8-cell stage. Second, when the meiotic apparatus is cut out of eggs, the eggs do not form maturation asters or polar bodies after fertilization, although they undergo cleavages with astral cycles and develop into normal embryos. In these embryos, no progress is made in localizing gut determinants in the uncleaved egg (see Fig. 12). Normal determinant localization is observed, however, beginning in 2-cell embryos, presumably during the astral cycle associated with first cleavage. Third, when supernumerary asters are induced by treating eggs with hypertonic sea water, there is a precocious localization of apical tuft and gut determinants in the uncleaved egg. These results suggest that the apical tuft and gut determinants in *Cerebratulus* may use the astral cycle to cue their localization.

6. Mechanism of Cytoplasmic Determinant Action

Until cytoplasmic determinants are identified, it is difficult to present a refined account of their mechanism(s) of action. In the meantime, however, there are a few generalizations that can be made based on recent experiments and our current understanding of the nature of determinants.

At least five important events would be expected to occur during the life history of a cytoplasmic determinant. The origin, localization, and partitioning of cytoplasmic determinants have already been considered. The remaining events involve activation and, ultimately, function. How a determinant is activated depends on its identity. Proteins may be activated by any number of amino acid modifications, proteolytic cleavages, or higher order structural changes, whereas mRNAs may be activated by translation. The translational activation of mRNA determinants is part of the unresolved issue of how maternal mRNAs are activated in general during early development. Morgan (1934) first proposed that cytoplasmic determinants may function by promoting differential gene expression in specific cell lineages, and this idea has been a theme in subsequent hypotheses of how determinants function (Davidson, 1986). It must be appreciated, however, that a number of developmental events that involve the function of determinants are completed before the beginning of embryonic transcription (see Chapters 6 and 7). Thus, some determinants do not require embryonic gene activity for at least their initial developmental functions.

6.1. Cytoplasmic Determinants That Function Independently of Immediate Gene Transcription

The evidence that some cytoplasmic determinants function without imme-diate embryonic gene transcription is that (1) they function in situations where nuclear transcription is impossible or blocked by drugs; (2) they produce a response before the time in development that embryonic transcription is initi-ated; or (3) they are defined by strict maternal-effect genes. For example, the dorsal–ventral axis of amphibians is set up by a series of cytoplasmic rear-rangements that occur in the uncleaved egg. It is thought that these movements position determinants in the proper place to enter vegetal cells, and that these cells subsequently attain the capacity to induce mesoderm development (Nieuwkoop, 1977; Gerhart et al., 1983; Gimlich and Gerhart, 1984). The induc-tion of mesoderm at the 256–512-cell stage (Asashima, 1980) is unlikely to require embryonic gene transcription because RNA synthesis begins much later at the mid-blastula transition (Newport and Kirschner, 1982) (see Chapter 7), when the embryo consists of ~4000 cells. Thus, it is unlikely that the vegetal determinants of amphibian embryos function initially by affecting differential gene activity. The same situation is true for pole cell development and dorsal–ventral pattern formation in Drosophila embryos. The pole cells are formed before the cellular blastoderm stage, when new transcription is initiated (McKnight and Miller, 1976), and, as discussed earlier, the dorsal–ventral pat-tern is determined by the activity of at least ten maternal effect genes (Anderson and Nüsslein-Volhard, 1984; Anderson et al., 1985a,b). Since the pole cell determinants behave as nucleic acids in their response to UV irradiation and photoreactivation (Brown and Kalthoff, 1983), and many of the mutations that affect the dorsal–ventral axis can be rescued by the injection of poly(A)$^+$ RNA (Anderson and Nüsslein-Volhard, 1984), it is reasonable to propose that the determinants that control these events are composed of maternal mRNA mole-cules. This may also be true for the vegetal determinants of amphibian eggs, although this possibility remains to be demonstrated. The activation of a mRNA determinant may simply involve its temporally- and/or spatially con-trolled translation. It is also possible, especially in cases where an inductive function is specified, that determinants may be localized organelles or struc-tures that render special properties to the cell(s) they eventually occupy. For example, Gerhart et al. (1983) discuss the Golgi apparatus, which is already localized at one pole of the Xenopus oocyte before vitellogenesis. Localized Golgi activities, such as exocytosis, endocytosis, and membrane turnover, may provide a special internal environment for the mesoderm-inducing, vegetal cells. A similar case could be made for other cytoplasmic organelles and mem-brane components, including those membrane receptors and channels that may be involved in intercellular communication.

Ascidian embryos may be a special situation in which maternal factors—possibly mRNA—may specify the phenotypic features of differentiated cells without requiring new gene transcription. This possibility was originally sug-gested by the morphology exhibited by interspecific andromerogons (Minganti,

1959). When anucleate merogons (fragments of eggs) of one ascidian species are fertilized with the sperm of another species, zygotes often develop to the larval stage as haploid andromerogons. The andromerogon larvae frequently exhibit some of the morphological features characteristic of the maternal rather than the paternal parent. This suggests that egg cytoplasmic factors, rather than the zygotic nucleus, may specify the development of some of the larval features. Further indication of a role for maternal information in later developmental events was obtained in studies of alkaline phosphatase expression (Whittaker, 1977). The synthesis of this enzyme in the endodermal cell lineage is insensitive to actinomycin D, suggesting that it does not require new transcription (Whittaker, 1977). Thus, the alkaline phosphatase determinants are postulated to be localized maternal mRNA molecules coding for the enzyme itself. If so, the function of these determinants is more complicated than translational activation of the alkaline phosphatase mRNA *per se,* because it has recently been demonstrated that alkaline phosphatase is not produced in anucleate merogons of ascidian eggs (Bates and Jeffery, 1987a). It is possible that nuclear factors may also be involved in alkaline phosphatase development.

6.2. Cytoplasmic Determinants That Require Gene Transcription for Their Function

Although the idea that cytoplasmic determinants may act by causing differential gene activity is more than fifty years old (Morgan, 1934), there are surprisingly few investigations that have attempted to verify it. The most pertinent information on this issue comes from ascidian embryos, in which the expression of muscle cell AchE and myofibrils, brain cell tyrosinase, epidermal cell secretions, and notochordal cell inclusions—which appear to be specified by localized cytoplasmic determinants—have been found to be highly sensitive to actinomycin D (Whittaker, 1973; Crowther and Whittaker, 1984). This sensitivity suggests that the expression of these markers requires embryonic RNA synthesis. If so, one would expect that mRNAs encoding these tissue-specific markers are synthesized exclusively in the progenitor cells of each tissue during early development. This has only been demonstrated for AChE, whose mRNA first appears in the muscle cell lineage (Meedel and Whittaker, 1984) of late gastrulae (Perry and Melton, 1983; Meedel and Whittaker, 1983).

Assuming some cytoplasmic determinants function by mediating differential gene activity, does an individual determinant activate genes required to specify a given cell lineage or tissue, suppress genes required to specify another cell lineage or tissue, or both? Answers to this question may be obtained by examining the developmental potential of embryonic cells containing mixtures of different determinants. Cells of mixed lineage have been obtained by arresting embryos in cleavage before the determinants for tissue-specific markers have segregated. When these cells were examined for the expression of tissue-specific markers, one of two results has been obtained. In cleavage-arrested 4-cell ascidian embryos, the posterior blastomeres, which contain a mixture of determinants normally segregated into epidermal and muscle cells, develop simultaneously myofibrils and epidermal cell secretions (Crowther and Whittaker, 1984). Cleavage-arrested 1-cell zygotes of ascidians also express simultaneously muscle AChE and endodermal alkaline phosphatase (Whittaker, 1975), although if the alkaline phosphatase determinants are maternal mRNAs, this observation may not be as significant as the codistributed myofibrils and epidermal substances. These results suggest that some determinants in ascidians may be noninterfering activators of transcription.

Different results have been obtained for nematode embryos assayed in the same way (Cowan and McIntosh, 1985). In cleavage-arrested 2-cell embryos, individual P1 blastomeres (see Fig. 4), which are antecedants of the muscle, endodermal, and hypodermal cell lineages, can express the markers for only one cell lineage at a time. This result suggests that different determinants of nematode eggs may exhibit a mutually exclusive function. One explanation for the behavior of mutually exclusive determinants is that they may suppress genes required for specification of other cell lineages.

Independent evidence suggesting that determinants may act as suppressors has been obtained in recent studies on the apical tuft and post-trochal bristle determinants in the annelid *Sabellaria* (Render, 1983). *Sabellaria* embryos form a polar lobe at the first and second cleavages. As discussed earlier for *Dentalium* embryos, when the first polar lobe is removed, the larval apical tuft and post-trochal bristles do not develop. When the smaller second polar lobe is removed, however, the larvae develop apical tufts, but not post-trochal bristles. Actinomycin D studies suggest that both apical tuft and post-trochal bristle determinants require new transcription for their function (Guerrier, 1971). The second polar lobe has been postulated to contain a determinant that acts as a suppressor of apical tuft development based on results of the following experiments. First, whereas partial embryos containing the C but not the D cell form apical tufts, those containing the D cell but not the C cell do not form this structure. Second, when the second polar lobe is removed, cultured C and D cells from lobeless embryos can both form apical tufts. Third, when the second cleavage is equalized, such that C and D cells both receive second polar lobe material, no apical tuft is formed by either cell. The proposed localization and distribution of the post-trochal bristle, apical tuft, and apical tuft suppressor determinants are shown in Fig. 21. Although extrapolation of these results to the other systems would be premature, the *Sabellaria* polar lobe studies suggest that cytoplasmic determinants might function both as regional activators and suppressors of gene activity.

7. Cytoplasmic Determinants Revisited

There is strong evidence, based on cytoplasmic extripation, reorientation, and transplantation experiments, that cytoplasmic determinants exist in a variety of different animal eggs. Such determinants appear to affect certain embryonic events, including dorsal–ventral axis formation, germ layer determina-

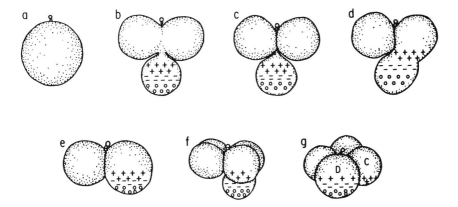

Figure 21. A schematic illustration of the distribution of post-trochal bristle (○), apical tuft (+), and apical tuft suppressor (−) determinants during early development of *Sabellaria*. (From Render, 1983.)

tion, and germ and somatic cell specification. Although cytoplasmic determinants have yet to be isolated and identified, evidence based on UV action spectrum data, photoreactivation, RNase sensitivity, and rescue experiments with purified RNA indicates that at least some determinants are likely to be maternal RNA molecules. It should be emphasized, however, that other determinants may be proteins, organelle systems, or regulatory ions, such as Ca^{2+} (Jaffe, 1983). The lack of knowledge concerning the molecular identities of determinants has hampered studies on their life history and function. It seems likely, however, that determinants can be localized during oogenesis, during ooplasmic segregation in the uncleaved egg, or during initial cleavages. They also appear to be bound to cellular structures, such as the plasma membrane, the egg cortex, the cytoskeleton, or a complex containing some or all of these structures. There appear to be at least two types of determinants based on their mode of action: those that do not require embryonic gene expression and those that promote specific gene activation in different regions of the early embryo.

Determinants have been broadly defined as cytoplasmic entities localized in the egg or early embryo that bias the properties or fates of the cells they enter during early embryogenesis. In general, both the classic and modern experiments discussed here support this definition. Determinants are cytoplasmic, at least in the uncleaved egg, localized or locally activated, and a number of different ways in which they bias embryonic cells have been examined. A possible criticism, however, is that this definition does not reflect the number of determinant-requiring steps in the biasing process. For example, a minimum of three different entities, a regulatory factor, a localization matrix, and an activator of the regulatory factor, seem to be required even for the simplest embryonic event triggered by a cytoplasmic determinant. In practice, many components are probably required. Recall that dorsal–ventral axis determination in *Drosophila* appears to require the products of at least ten maternal-effect genes (Anderson and Nüsslein-Volhard, 1985*a,b*). Thus, the biasing event in the definition may actually be a sequence of events approaching the complexity of a metabolic pathway. If so, which components of the pathway are to be considered determinants? An argument could be made for designating either all the components of the hypothetical pathway, only the regulatory factors in the pathway, or only the active final product as determinants. Before this question can be answered, it will be necessary to understand thoroughly the sequence of events and the identity of the components involved in cytoplasmic determinant function.

ACKNOWLEDGMENTS.   Financial support was provided by grants HD-13970 and GM-25119 from the National Institutes of Health, grant DCB–84116763 from the National Science Foundation, and by the Muscular Dystrophy Association and the University of Texas Research Institute during the preparation of this manuscript. I am also grateful to B. Brodeur and S. Daugherty for photographic assistance, to J. E. Speksnijder, B. Swalla, J. Venuti, and M. White for critically reviewing parts of the manuscript, to J. Escobar for typing the manuscript, and to J. Young for artwork.

References

Allis, C. D., Waring, G. L., and Mahowald, A. P., 1977, Mass isolation of pole cells from *Drosophila melanogaster, Dev. Biol.* **56:**372–381.

Anderson, K. V., and Nüsslein-Volhard, C., 1984, Information for the dorsal–ventral pattern of the *Drosophila* embryo is stored as maternal mRNA, *Nature (Lond.)* **311:**223–227.

Anderson, K. V., Bokla, L., and Nüsslein-Volhard, C., 1985*a*, Establishment of dorsal–ventral polarity in the *Drosophila* embryo: The induction of polarity by the *Toll* gene product, *Cell* **42:**791–798.

Anderson, K. V., Jürgens, G., and Nüsslein-Volhard, C., 1985*b*, The establishment of dorsal–ventral polarity in the *Drosophila* embryo: Genetic studies on the role of the *Toll* gene product, *Cell* **42:**779–789.

Asashima, M., 1980, Inducing effects of the grey crescent region of early developmental stages of *Ambystoma mexicanum, Wilhelm Roux Arch.* **188:**123–126.

Bantock, C. F., 1970, Experiments on chromosome elimination in the gall midge, *Mayetiola destructor, J. Embryol. Exp. Morphol.* **24:**257–286.

Bates, W. R., and Jeffery, W. R., 1987*a*, Alkaline phosphatase expression in ascidian egg fragments and andromerogons, *Dev. Biol.* **119:**382–389.

Bates, W. R., and Jeffery, W. R., 1987*b*, Localization of axial determinants in the vegetal pole region of ascidian eggs, *Dev. Biol.* **124:**65–76.

Beams, H. W., and Kessel, R. G., 1974, The problem of germ cell determinants, *Int. Rev. Cytol.* **39:**413–479.

Berg, W. E., and Humphreys, W. J., 1960, Electron microscopy of four-cell stages of the ascidians *Ciona* and *Styela, Dev. Biol.* **2:**42–60.

Bluemink, J. E., and Tertoolen, L. G. J., 1978, The plasma membrane IMP pattern as related to animal/vegetal polarity in the amphibian egg, *Dev. Biol.* **62:**334–343.

Boswell, R. E., 1985, A short review of germ cell determination in *Drosophila melanogaster*, in: *The Cellular and Molecular Biology of Invertebrate Development* (R. H. Sawyer and R. M. Showman, eds.), pp. 187–195, University of South Carolina Press, Columbia, South Carolina.

Boswell, R. E., and Mahowald, A. P., 1985, *Tudor*, a gene required for assembly of the germ plasm in *Drosophila melanogaster, Cell* **43**:97–104.

Boveri, T., 1887, Uber Differenzierung der Zellkerne wahrend der Furchung des Eies von *Ascaris megalocephala, Anat. Anz.* **2**:688–693.

Boveri, T., 1910*a*, Die Potenzen der *Ascaris*–blastomeren bei abgeanderter Furchung, *Festschr. Z. Geburt. R. Hertwig* **3**:131–214.

Boveri, T., 1910*b*, Uber die Teilung centrifugierter Eier *Ascaris megalocephala, Arch. Anat. Entwicklungsmech.* **30**:101–125.

Brothers, A. J., 1979, A specific case of genetic control of early development: The o maternal effect mutation of the Mexican axolotl, in: *Determinants of Spatial Organization* (S. Subtelny and I. R. Konigberg, eds.), pp. 167–183, Academic, New York.

Brown, P. M., and Kalthoff, K., 1983, Inhibition by ultra-violet light of pole cell formation in *Smittia* sp. (Chiromidae, Diptera): Action spectrum and photoreversibility, *Dev. Biol.* **97**:113–122.

Carpenter, C. D., and Klein, W. H., 1982, A gradient of poly(A)$^+$ RNA sequences in *Xenopus laevis* eggs and embryos, *Dev. Biol.* **99**:408–417.

Cather, J. N., and Verdonk, N. H., 1974, The development of *Bithynia tentaculata* (Prosobranchia, Gastropoda) after removal of the polar lobe, *J. Embryol. Exp. Morphol.* **31**:415–422.

Clement, A. C., 1952, Experimental studies on germinal localization in *Ilyanassa*. l. The role of the polar lobe in determination of the cleavage pattern and its influence in later development, *J. Exp. Zool.* **121**:593–626.

Clement, A. C., 1956, Experimental studies on germinal localization in *Ilyanassa*. II. The development of isolated blastomeres, *J. Exp. Zool.* **132**:427–446.

Clement, A. C., 1968, Development of the vegetal half of the *Ilyanassa* egg after removal of most of the yolk by centrifugal force, compared with the development of animal halves of similar visible composition, *Dev. Biol.* **17**:165–186.

Conklin, E. G., 1905, The organization and cell lineage of the ascidian egg, *J. Acad. Nat. Sci. (Philadelphia)* **13**:1–119.

Conklin, E. G., 1931, The development of centrifuged eggs of ascidians, *J. Exp. Zool.* **60**:1–119.

Cowan, A. E., and McIntosh, J. R., 1985, Mapping the distribution of differentiation potential for intestine, muscle, and hypodermis during early development in *Caenorhabitis elegans, Cell* **4**:923–932.

Crampton, H. E., 1896, Experimental studies on gastropod development, *Arch. Entwicklungsmech. Org.* **3**:1–19.

Crowther, R., and Whittaker, J. R., 1984, Differentiation of histospecific ultrastructural features in cells of cleavage-arrested early ascidian embryos, *Wilhelm Roux Arch.* **194**:87–98.

Curtis, A. S. G., 1960, Cortical grafting in *Xenopus laevis, J. Embryol. Exp. Morphol.* **8**:163–173.

Curtis, A. S. G., 1962, Morphogenetic interactions before gastrulation in the amphibian, *Xenopus laevis*—The cortical field, *J. Embryol. Exp. Morphol.* **10**:410–422.

Davidson, E. H., 1986, *Gene Activity in Early Development*, 3rd ed., Academic, Orlando, Florida.

Deno, T., and Satoh, N., 1984, Studies on the cytoplasmic determinant for muscle cell differentiations in ascidian embryos, an attempt at transplantation of the myoplasm, *Dev. Growth Diff.* **26**:43–48.

Dictus, W. J. A. G., van Zoelen, E. J. J., Tetteroo, P. A. T., Tertoolen, L. G. J., De Laat, S. W., and Bluemink, J. G., 1984, Lateral mobility of plasma membrane lipids in *Xenopus* eggs: Regional differences related to animal/vegetal polarity become extreme upon fertilization, *Dev. Biol.* **101**:201–211.

Dohmen, M. R., 1983, The polar lobe in eggs of molluscs and annelids: Structure, composition, and function, in: *Time, Space, and Pattern in Embryonic Development* (W. R. Jeffery and R. A. Raff, eds.), pp. 197–220, A. R. Liss, New York.

Dohmen, M. R., and Verdonk, N. H., 1974, The structure of a morphogenetic cytoplasm present in the polar lobe of *Bithynia tentaculata* (Gastropoda, Prosobranchia), *J. Embryol. Exp. Morphol.* **31**:423–433.

Dohmen, M. R., and Verdonk, N. H., 1979, The ultrastructure and role of the polar lobe in development of molluscs, in: *Determinants of Spatial Organization* (S. Subtelny and I. R. Konigsberg, eds.), pp. 3–27, Academic, Orlando, Florida.

Dohmen, M. R., and van der Mey, J. C. A., 1977, Local surface differentiations at the vegetal pole of the eggs of *Nassarius reticulatus, Buccinium undatum,* and *Crepidula fornicata* (Gastropoda, Prosobranchia), *Dev. Biol.* **61:**104–113.

Dreyer, C., Scholz, E., and Hausen, P., 1982, The fate of oocyte nuclear proteins during early development of *Xenopus laevis, Roux Arch.* **191:**228–233.

Fielding, C. F., 1967, Developmental genetics of the mutant *grandchildless* of *Drosophila subobscura, J. Embryol. Exp. Morphol.* **17:**375–384.

Freeman, G., 1976, The role of cleavage in the localization of developmental potential in the ctenophore *Mnemiopsis leidyi, Dev. Biol.* **49:**143–177.

Freeman, G., 1978, The role of asters in the localization of the factors that specify the apical tuft and the gut of the nemertine *Cerebratulus lacteus, J. Exp. Zool.* **206:**81–107.

Freeman, G., 1979, The multiple roles which cell division can play in the localization of developmental potential, in: *Determinants of Spatial Organization* (S. Subtelny and I. R. Konigsberg, eds.), pp. 53–76, Academic, Orlando, Florida.

Geigy, R., 1931, Action de l'ultraviolet sur le pole germinal dans l'oeuf de *Drosophila melanogaster* (caustration et mutabilité), *Rev. Suisse Zool.* **38:**187–288.

Geilenkirchen, W. L. M., Verdonk, N. H., and Timmermans, L. P. M., 1970, Experimental studies on the morphogenetic factors localized in the first and the second polar lobe of *Dentalium* eggs, *J. Embryol. Exp. Morphol.* **23:**237–243.

Gerhart, J., Ubbels, G., Black, S., Hara, K., and Kirschner, M., 1981, A reinvestigation of the role of the grey crescent in axis formation in *Xenopus laevis, Nature (Lond.)* **292:**511–516.

Gerhart, J., Black, S., Gimlich, R., and Scharf, S., 1984, Control of polarity in the amphibian egg, in: *Time, Space, and Pattern in Embryonic Development* (W. R. Jeffery and R. A. Raff, eds.), pp. 261–286, A. R. Liss, New York.

Geyer-Duszynska, I., 1959, Experimental research on chromosome elimination in cecidomyidae (Diptera), *J. Exp. Zool.* **141:**341–448.

Geyer-Duszynska, I., 1961, Spindle disappearance and chromosome behaviour after partial-embryo irradiation in Cecidomyidae (Diptera), *Chromosoma* **12:**233–247.

Gimlich, R. L., and Gerhart, J. C., 1984, Early cellular interactions promote embryonic axis formation in *Xenopus laevis, Dev. Biol.* **104:**117–130.

Guerrier, P., 1971, A possible mechanism for control of morphogenesis in the embryo of *Sabellaria alveolata* (Annelide, polychaete), *Exp. Cell. Res.* **67:**215–218.

Guerrier, P., van den Biggelaar, J. A. M., van Dongen, C. A. M., and Verdonk, N. H., 1978, Significance of the polar lobe for the determination of dorsoventral polarity in *Dentalium vulgare* (da Costa), *Dev. Biol.* **63:**233–242.

Gurdon, J. B., Mohun, T. J., Fairman, S., and Brennan, S., 1985, All components required for the eventual activation of muscle-specific actin genes are localized in the sub-equatorial region of an uncleaved amphibian egg, *Proc. Natl. Acad. Sci. USA* **82:**139–143.

Hegner, R. W., 1911, Experiments with chrysomelid beetles. III. The effects of killing parts of the egg of *Leptinotarsa decemlineata, Biol. Bull.* **20:**237–251.

Hegner, R. W., 1914, Studies on germ cells. I. The history of the germ cells in insects with special reference to the "Keinbahn" determinants, *J. Morphol.* **25:**375–509.

Hörstadius, S., 1937, Experiments on the early development of *Cerebratulus lacteus, Biol. Bull.* **73:**317–342.

Hörstadius, S., 1971, Nemertinae in: *Experimental Embryology of Marine and Freshwater Invertebrates* (G. Reverberi, ed.), pp. 164–174, North-Holland, Amsterdam.

Illmensee, K., and Mahowald, A. P., 1974, Transplantation of polar plasm in *Drosophila:* Induction of germ cells at the anterior pole of the egg, *Proc. Natl. Acad. Sci. USA* **71:**1016–1020.

Illmensee, K., and Mahowald, A. P., 1976, Autonomous function of germ plasm in a somatic region of the *Drosophila* egg, *Exp. Cell. Res.* **97:**127–140.

Illmensee, K., Mahowald, A. P., and Loomis, M. R., 1976, The ontogeny of germ plasm during oogenesis in *Drosophila, Dev. Biol.* **49:**40–65.

Jäckle, H., and Kalthoff, K., 1978, Photoreactivation of RNA in UV-irradiated insect eggs (*Smittia* sp., Chironomidae, Diptera). I. Photosensitive production and light-dependent disappearance of pyrimidine dimers, *Photochem. Photobiol.* **27:**309–315.

Jäckle, H., and Eagleson, G. W., 1980, Spatial distribution of abundant proteins in oocytes and fertilized eggs of the Mexican axolotl (*Ambystoma mexicanum*), *Dev. Biol.* **75**:492–499.

Jaffe, L. F., 1983, Sources of calcium in egg activation: A review and hypothesis, *Dev. Biol.* **99**:265–276.

Jaffe, L. A., and Guerrier, P., 1981, Localization of electrical excitability in the early embryo of *Dentalium*, *Dev. Biol.* **83**:370–373.

Jeffery, W. R., 1983, Maternal RNA and the embryonic localization problem, in: *Control of Embryonic Gene Expression* (M. A. Q. Siddiqui, ed.), pp. 73–114, CRC Press, Boca Raton, Florida.

Jeffery, W. R., 1984a, Pattern formation by ooplasmic segregation in ascidian eggs, *Biol. Bull.* **166**:277–298.

Jeffery, W. R., 1984b, Spatial distribution of messenger RNA in the cytoskeletal framework of ascidian eggs, *Dev. Biol.* **103**:482–492.

Jeffery, W. R., 1985a, Identification of proteins and mRNAs in isolated yellow crescents of ascidian eggs, *J. Embryol. Exp. Morphol.* **89**:275–287.

Jeffery, W. R., 1985b, Patterns of maternal mRNA distribution and their role in early development, in: *The Cellular and Molecular Biology of Invertebrate Development* (R. H. Sawyer and R. M. Showman, eds.), pp. 125–151, University of South Carolina Press, Columbia, South Carolina.

Jeffery, W. R., 1985c, The spatial distribution of maternal mRNA is determined by a cortical cytoskeletal domain in *Chaetopterus* eggs., *Dev. Biol.* **110**:217–229.

Jeffery, W. R., and Capco, D. G., 1978, Differential accumulation and localization of maternal poly (A)-containing RNA during early development of the ascidian, *Styela*, *Dev. Biol.* **67**:152–166.

Jeffery, W. R., and Meier, S., 1983, A yellow crescent cytoskeletal domain in ascidian eggs and its role in early development, *Dev. Biol.* **96**:125–143.

Jeffery, W. R., and Meier, S., 1984, Ooplasmic segregation of the myoplasmic actin network in stratified ascidian eggs, *Wilhelm Roux Arch.* **193**:257–262.

Jeffery, W. R., and Raff, R. A. (eds.), 1983, *Time, Space, and Pattern in Embryonic Development*, Alan R. Liss, New York.

Jeffery, W. R., and Wilson, L., 1983, Localization of messenger RNA in the cortex of *Chaetopterus* eggs and early embryos, *J. Embryol. Exp. Morphol.* **75**:225–239.

Jeffery, W. R., Tomlinson, C. R., and Brodeur, R. D., 1983, Localization of actin messenger RNA during early ascidian development, *Dev. Biol.* **99**:408–417.

Jeffery, W. R., Bates, W. R., Beach, R. L., and Tomlinson, C. R., 1986, Is maternal mRNA a determinant of tissue-specific proteins in ascidian embryos? *J. Embryol. Exp. Morphol.* **97**(Suppl.):1–14.

Kalthoff, K., 1979, Analysis of a morphogenetic determinant in an insect embryo (*Smittia* sp., Chironomidae, Diptera), in: *Determinants of Spatial Organization* (S. Subtelny and I. R. Konigsberg, eds.), pp. 97–126, Academic, Orlando, Florida.

Kandler-Singer, I., and Kalthoff, K., 1976, RNase sensitivity of an anterior morphogenetic determinant in an insect egg (*Smittia* sp., Chironomidae, Diptera), *Proc. Natl. Acad. Sci. USA* **73**:3739–3743.

King, M. L., and Barklis, E., 1985, Regional distribution of maternal messenger RNA in the amphibian oocyte, *Dev. Biol.* **112**:203–212.

Klag, J. J., and Ubbels, G. A., 1975, Regional morphological and cytological differentiation of the fertilized egg of *Discoglossus pictus* (Anura), *Differentiation* **3**:15–20.

Laufer, J. S., and von Ehrenstein, G., 1981, Nematode development after removal of egg cytoplasm: Absence of localized unbound determinants, *Science* **211**:402–405.

Laufer, J. S., Bazzicalupo, P., and Wood, W. B., 1980, Segregation of developmental potential in early embryos of *Caenorhabditis elegans*, *Cell* **19**:569–577.

Mahowald, A. P., 1968, Polar granules of *Drosophila*. II. Ultrastructural changes during early embryogenesis, *J. Exp. Zool.* **167**:237–262.

Mahowald, A. P., 1971, Polar granules of *Drosophila*. IV. Cytochemical studies showing the loss of RNA from polar granules during early stage of embryogenesis, *J. Exp. Zool.* **176**:355–352.

Mahowald, A. P., 1977, The germ plasm of *Drosophila*: An experimental system for the analysis of determination, *Am. Zool.* **17**:551–563.

Mahowald, A. P., Allis, C. D., Karrer, K. M., Underwood, E. M., and Waring, G. L., 1979, Germ

plasm and pole cells of *Drosophila*, in: *Determinants of Spatial Organization* (S. Subtelny and I. R. Konigsberg, eds.), pp. 127–146, Academic, Orlando, Florida.

McClay, D. R., Cannon, G. W., Wessel, G. M., Fink, R. D., and Marchase, R. B., 1983, Patterns of antigenic expression in early sea urchin development, in: *Time, Space, and Pattern in Embryonic Development* (W. R. Jeffery and R. A. Raff, eds.), pp. 157–169, Alan R. Liss, New York.

McKnight, S. L., and Miller, O. L., 1976, Ultrastructural patterns of RNA synthesis in early embryogenesis of *Drosophila melanogaster*, *Cell* **8**:305–319.

Meedel, T. H., and Whittaker, J. R., 1983, Development of translationally active mRNA for larval muscle acetylcholinesterase during ascidian embryogenesis, *Proc. Natl. Acad. Sci. USA* **80**:4761–4765.

Meedel, T. H., and Whittaker, J. R., 1984, Lineage segregation and developmental antonomy in expression of functional muscle acetylcholinesterase mRNA in ascidian embryos, *Dev. Biol.* **105**:479–487.

Melton, D. A., 1987, Translocation of a localized maternal mRNA to the vegetal pole of *Xenopus* oocytes, *Nature*, **238**:80–82.

Minganti, A., 1959, Androgenetic hybrids in ascidians, 1. *Ascidia malaca* (♀) × *Phallusia mammillata* (♂), *Acta Embryol. Morph. Exp.* **2**:244–256.

Mlodzik, M., Fjose, A., and Gehring, W. J., 1985, Isolation of *caudal*, a *Drosophila* homeo box-containing gene with maternal expression whose transcripts form a concentration gradient at the pre-blastoderm stage, *EMBO J.* **4**:2961–2969.

Moen, T. L., and Namenwirth, M., 1977, The distribution of soluble proteins along the animal-vegetal axis of frog eggs, *Dev. Biol.* **58**:1–10.

Moon, R. T., Nicosia, R. F., Olsen, C., Hille, M., and Jeffery, W. R., 1983, The cytoskeletal framework of sea urchin eggs and embryos: Developmental changes in the association of messenger RNA, *Dev. Biol.* **95**:447–458.

Morgan, T. H., 1934, *Embryology and Genetics*, Columbia University Press, New York.

Morgan, T. H., 1935, Centrifuging the egg of *Ilyanassa* in reverse, *Biol. Bull.* **68**:268–279.

Moritz, K. B., 1967, Die Blastomerendifferenzierung für Soma and Kleinbahn bei *Parascaris equorum*. II. Untersuchungen mittels UV-Bestrahlung und Zeitrifugierung, *Wilhelm Roux Arch.* **159**:203–266.

Muller-Holtkamp, F., Knippler, D. G., Seigert, E., and Jäckle, H., 1985, An early role of maternal mRNA in establishing the dorsoventral pattern in *pelle* mutant *Drosophila* embryos, *Dev. Biol.* **110**:238–246.

Newport, J., and Kirschner, M., 1982, A major developmental transition in early *Xenopus* embryos. II. Control of the onset of transcription, *Cell* **30**:687–695.

Nieuwkoop, P., 1977, Origin and establishment of embryonic polar axes in amphibian development, *Curr. Top. Dev. Biol.* **11**:115–132.

Okada, M., Kleinman, I. A., and Schneiderman, H. A., 1974, Restoration of fertility in sterilized *Drosophila* eggs by transplantation of polar cytoplasm, *Dev. Biol.* **37**:43–54.

Oliver, B. C., and Shen, S. S., 1986, Cytoplasmic control of chromosome diminution in *Ascaris suum*, *J. Exp. Zool.* **239**:41–55.

Ortolani, G., 1958, Cleavage and development in egg fragments of ascidians, *Acta Embryol. Morph. Exp.* **1**:247–272.

Pasteels, J., 1941, Recherches sur les facteurs initiaux de la morphogenese chez les Amphibiens anoures. V. Les effects de la pesanteur sur l'oeuf de *Rana fusca* maintenu en position anormale en avant la formation du croissant gris, *Arch. Biol.* **52**:321–339.

Perry, H. E., and Melton, D. A., 1983, A rapid increase in acetylcholinesterase mRNA during ascidian embryogenesis as demonstrated by microinjection into *Xenopus laevis* oocytes, *Cell Diff.* **13**:233–238.

Raff, R. A., 1977, The molecular determination of morphogenesis, *Bio. Sci.* **27**:394–401.

Rebagliati, M. R., Weeks, D. L., Harvey, R. P., and Melton, D. A., 1985, Identification and cloning of localized maternal mRNAs from *Xenopus* eggs, *Cell* **42**:769–777.

Render, J. A., 1983, The second polar lobe of the *Sabellaria cementarium* embryo plays an inhibitory role in apical tuft formation, *Wilhelm Roux Arch.* **192**:120–129.

Render, J. A., and Guerrier, P., 1984, Size regulation and morphogenetic localization in the *Dentalium* polar lobe, *J. Exp. Zool.* **232**:79–86.

Reverberi, G., 1971, Ascidians, in: *Experimental Embryology of Marine and Freshwater Invertebrates* (G. Reverberi, ed.), pp. 507–550, North-Holland, Amsterdam.

Rodgers, W. H., and Gross, P. R., 1978, Inhomogeneous distribution of egg RNA sequences in the early embryo, *Cell* **14**:279–288.

Santamaria, P., and Nüsslein-Volhard, C., 1983, Partial rescue of *dorsal*, a maternal effect mutation affecting the dorso-ventral pattern of the *Drosophila* embryo, by the injection of wild-type cytoplasm, *EMBO J.* **2**:1695–1699.

Showman, R. M., Wells, D. E., Anstrom, J., Hursh, D. A., and Raff, R. A., 1982, Message specific sequestration of maternal histone mRNA in the sea urchin egg, *Proc. Natl. Acad. Sci. USA* **79**:5944–5947.

Slack, J. M. W., 1983, *From Egg to Embryo: Determinative Events in Early Development*, Cambridge University Press, London.

Smith, R. C., 1986, Protein synthesis and messenger RNA levels along the animal–vegetal axis during early *Xenopus* development, *J. Embryol. Exp. Morphol.* **95**:15–35.

Strome, S., and Wood, W. B., 1982, Immunofluorescence visualization of germ-line specific cytoplasmic granules in embryos, larvae, and adults of *Caenorhabitis elegans*, *Proc. Natl. Acad. Sci. USA* **79**:1558–1562.

Strome, S., and Wood, W. B., 1983, Generation of asymmetry and segregation of germ-like granules in early *C. elegans* embryos, *Cell* **35**:15–25.

Speksnijder, J. E., Dohmen, M. R., Tertoolen, L. G. J., and DeLaat, S. W., 1985a, Regional differences in the lateral mobility of plasma membrane lipids in a molluscan embryo, *Dev. Biol.* **110**:207–216.

Speksnijder, J. E., Mulder, M. M., Hage. W. J., Dohmen, M. R., and Bluemink, J. G., 1985b, Regional differences in numerical particle distribution in the plasma membrane of a molluscan egg, *Dev. Biol.* **108**:38–48.

Subtelny, S., and Konigsberg, I. R. (eds.), 1979, *Determinants of Spatial Organization*, Academic, Orlando, Florida.

Swalla, B. J., Moon, R. T., and Jeffery, W. R., 1985, Developmental significance of a cortical cytoskeletal domain in *Chaetopterus* eggs, *Dev. Biol.* **111**:434–450.

Togashi, S., and Okada, M., 1982, Restoration of pole-cell forming ability to UV-sterilized *Drosophila* embryos by injection of an RNA fraction extracted from eggs, in: *The Ultrastructure and Functioning of Insect Cells* (H. Akai, ed.), pp. 41–44, Japanese Society for Insect Cells, Tokyo.

Tomlinson, C. R., Bates, W. R., and Jeffery, W. R., 1987, Development of a muscle actin specified by maternal and zygotic mRNA in ascidian embryos, *Dev. Biol.* **123**:470–482.

Tufaro, F., and Brandhorst, B. P., 1979, Similarity of proteins synthesized by isolated blastomeres of early sea urchin embryos, *Dev. Biol.* **72**:390–397.

Tyler, A., 1930, Experimental production of double embryos in annelids and molluscs, *J. Exp. Zool.* **57**:347–407.

Ueda, R., and Okada, M., 1982, Induction of pole cells in sterilized *Drosophila* embryos by injection of subcellular fraction from eggs, *Proc. Natl. Acad. Sci. USA* **67**:6946–6950.

van Dam, W. I., Dohmen, M. R., and Verdonk, N. H., 1982, Localization of morphogenetic determinants in a special cytoplasm present in the polar lobe of *Bithynia tentaculata* (Gastropoda), *Wilhelm Roux Arch.* **191**:371–377.

Verdonk, N. H., 1968, The effect of removing the polar lobe in centrifuged eggs of *Dentalium*, *J. Embryol. Exp. Morphol.* **19**:33–42.

Verdonk, N. H., and Cather, J. N., 1983, Morphogenetic determination and differentiation, in: *The Mollusca*, Vol. 3: *Development* (N. H. Verdonk, J. A. M. van den Biggelaar, and A. S. Tomba, eds.), pp. 215–252, Academic, Orlando, Florida.

Verdonk, N. H., Geilenkirchen, W. L. M., and Timmermans, L. P. M., 1971, The localization of morphogenetic factors in uncleaved eggs of *Dentalium*, *J. Embryol. Exp. Morphol.* **25**:57–63.

Waring, G. L., Allis, C. D., and Mahowald, A. P., 1978, Isolation of polar granules and the identification of polar granule specific protein, *Dev. Biol.* **66**:197–206.

Warn, R., 1975, Restoration of the capacity to form pole cells in UV-irradiated *Drosophila* embryos, *J. Embryol. Exp. Morphol.* **33**:1003–1011.

Weeks, D. L., and Melton, D. A., 1987, A maternal mRNA localized to the animal pole of *Xenopus* eggs encodes a subunit of mitochondrial ATPase, *Proc. Nat. Acad. Sci. USA* **84**:2798–2802.

Wells, D. E., Showman, R. M., Klein, W. H., and Raff, R. A., 1981, Delayed recruitment of maternal mRNA in sea urchin embryos, *Nature (Lond.)* **292**:277–278.

Whittaker, J. R., 1973, Segregation during ascidian embryogenesis of egg cytoplasmic information for tissue-specific enzyme development, *Proc. Natl. Acad. Sci. USA* **70**:2096–2100.

Whittaker, J. R., 1975, Differentiation of two histospecific enzymes in the same cleavage-arrested ascidian egg, *Biol. Bull.* **149**:451.

Whittaker, J. R., 1979a, Cytoplasmic determinants of tissue differentiation in the ascidian egg, in: *Determinants of Spatial Organization* (S. Subtelny and I. R. Konigsberg, eds.), pp. 29–51, Academic, Orlando, Florida.

Whittaker, J. R., 1979b, Development of vestigial tail muscle acetylcholinesterase in embryos of anuran ascidian species, *Biol. Bull.* **156**:393–407.

Whittaker, J. R., 1979c, Development of tail muscle acetylcholinesterase in ascidian embryos lacking mitochondrial localization and segregation, *Biol. Bull.* **157**:344–355.

Whittaker, J. R., 1980, Acetylcholinesterase development in extra cells caused by changing the distribution of myoplasm in ascidian embryos, *J. Embryol. Exp. Morphol.* **55**:343–354.

Whittaker, J. R., 1982, Muscle lineage cytoplasm can change the developmental expression in epidermal lineage cells of ascidian embryos, *Dev. Biol.* **93**:463–470.

Wilson, E. B., 1904, Experimental studies on germinal localization. I. The germ regions in the egg of *Dentalium*, *J. Exp. Zool.* **1**:1–72.

Wilson, E. B., 1925, *The Cell in Development and Heredity*, 3rd ed., Macmillan, New York.

Wolf, N., Priess, J., and Hirsch, D., 1983, Segregation of germline granules in early embryos of *Caenorhabitis elegans*: An electron microscopic analysis, *J. Embryol. Exp. Morphol.* **73**:297–306.

Wylie, C. C., Brown, D., Godsave, S. F., Quarmby, J., and Heasman, J., 1985, The cytoskeleton of *Xenopus* oocytes and its role in development, *J. Embryol. Exp. Morphol.* **89**:1–15.

Chapter 2

Determination of Pattern and Fate in Early Embryos of *Caenorhabditis elegans*

WILLIAM B. WOOD

1. Introduction

The free-living hermaphroditic soil nematode *Caenorhabditis elegans* has become a widely used organism for study of invertebrate development. Its advantages include ease of cultivation in the laboratory and a short life cycle of about 72 hr at 22°C. In addition, it is transparent at all stages, so that development can be followed microscopically in living specimens; it is also cellularly simple, having only 959 somatic cells in the adult hermaphrodite. These latter two features have permitted determination of the complete cell lineage, from the zygote to the adult animal (Sulston and Horvitz, 1977; Kimble and Hirsh, 1979; Sulston *et al.*, 1983). The pattern and timing of cell divisions is almost completely invariant from animal to animal. Perhaps most important, *C. elegans* is well suited to genetic analysis of development; mutational identification and mapping of about 700 loci on its six chromosomal linkage groups has already been accomplished (Wood *et al.*, 1988).

Characteristic of many lower invertebrates, *C. elegans* appears to have a highly mosaic pattern of early embryonic development, in which cell fates are primarily determined by ancestry rather than position. As in other mosaic embryos, the mechanisms that dictate pattern formation, cell fate determination, and timing of embryonic events remain poorly understood, but the experimental advantages of this organism provide opportunities to investigate these mechanisms using a variety of experimental techniques ranging from micromanipulation to molecular genetics. This chapter outlines current knowledge of *C. elegans* embryogenesis and its genetic control, followed by a discussion of possible mechanisms for determination of cell fates in the early embryo.

WILLIAM B. WOOD • Department of Molecular, Cellular, and Developmental Biology, University of Colorado, Boulder, Colorado 80309-0347.

2. Overview of *Caenorhabditis elegans* Embryology

Embryogenesis in *C. elegans,* from fertilization to hatching, takes about 14 hr at 22°C. The process can be considered conveniently in three major stages (Fig. 1). The first, including zygote formation and early cleavage, establishment of the embryonic axes, and determination of the somatic and germline founder

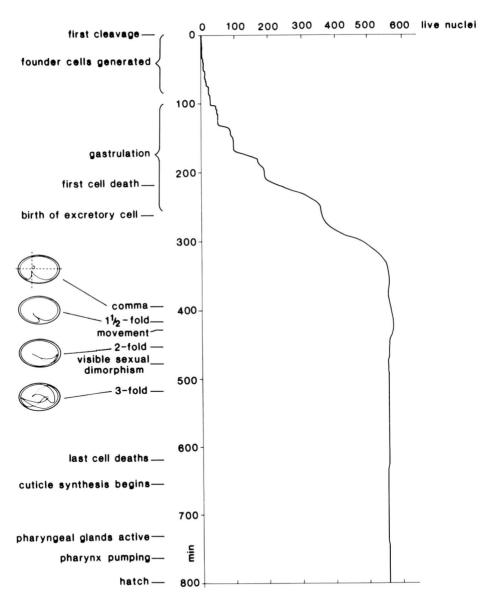

Figure 1. Stages, marker events, and number of nuclei during embryogenesis. Fertilization normally occurs at about −40 min. (From Sulston *et al.*, 1983.)

cell fates, takes place during the first 2 hr postfertilization. The second, including gastrulation, completion of most cell proliferation, and the beginning of cell differentiation and organogenesis, continues until about halfway through embryogenesis. The third stage, including morphogenesis as well as completion of embryonic cell differentiation and organogenesis, occupies the remainder of embryogenesis and concludes with hatching.

3. Description of Cellular Events

3.1. Stage 1: Zygote Formation and Establishment of Embryonic Axes

The *C. elegans* hermaphrodite, shown as a young adult in Fig. 2, is self-fertilizing. During the fourth larval stage about 150 ameboid sperm are produced in each of the two arms of the symmetric gonad and stored in the spermathecae. At the final molt, the germline switches to exclusive production of oocytes. Mature oocytes pass from the oviduct into the spermatheca and become fertilized by engulfment of sperm as they exit the spermatheca into the uterus. At this time, the anterior–posterior polarity of the embryo is evident, with the sperm pronucleus at the posterior pole (Albertson, 1984). It is not known whether sperm entry establishes the posterior pole, or whether a functional anterior–posterior polarity is already present in the oocyte before fertilization.

Figure 3 shows Nomarski photomicrographs of an egg in successive stages between fertilization and the end of first cleavage. Following the standard convention for *C. elegans*, eggs are oriented with the anterior pole to the left. Shortly after fertilization, the maternal pronucleus, which was arrested in first meiotic metaphase, resumes meiosis and, within 20 min, the first and second polar bodies have been extruded at the anterior pole. Meanwhile, the egg has formed a tough shell around itself, consisting of an inner vitelline envelope impermeable to most solutes, a middle chitinous layer, and an outer layer consisting of lipids and probably collagenous, crosslinked proteins (Chitwood and Chitwood, 1974). Also during this time, turbulent cytoplasmic movements are accompanied by contractions of the anterior cell membrane and a pronounced constriction, termed **pseudocleavage,** near the equator, involving formation and then regression of a cleavage furrow (Fig. 3a–c). Concurrently with pseudocleavage, the egg pronucleus migrates posteriorly toward the sperm pronucleus, which moves a short distance toward the center (Fig. 3b), until the two meet in the posterior half of the egg (Fig. 3c). On the basis of staining of *C. elegans* eggs with anti-tubulin antibodies, the centripetal movement of the sperm pronucleus appears to be mediated by growth of astral microtubules from the centriolar regions adjacent to the sperm pronucleus (Albertson, 1984). Further growth of the asters after encounter of the two pronuclei moves them to the center of the egg, where the growing spindle rotates 90° and moves to its final position along the anterior–posterior axis, slightly posterior to the center

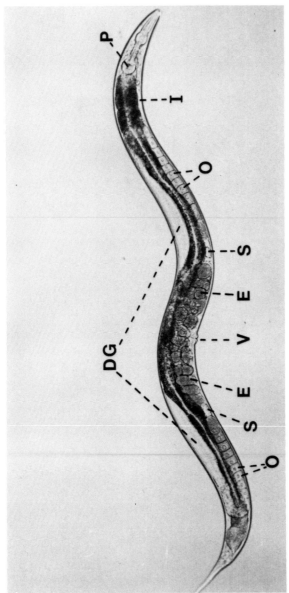

Figure 2. Nomarski photomicrograph of an adult hermaphrodite *Caenorhabditis elegans*. Note transparency of entire animal, the dark intestine running along its length, and the symmetric gonad, with one reflexed lobe extending anteriorly and the other posteriorly. The distal arm of each lobe is a syncytium of germline nuclei; the proximal arm contains oocytes. The spermatheca lies in each lobe between the oocytes and the adjacent uterus. P, pharnyx; I, intestine; O, oocytes; S, spermatheca; E, developing embryos in uterus; V, vulva; DG, distal gonad. Actual length of the animal is ~1 mm. (Micrograph courtesy of Carol Trent.)

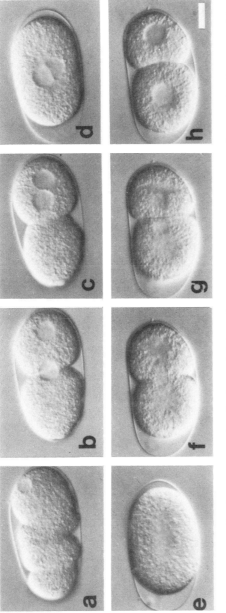

Figure 3. Nomarski photomicrographs of a *Caenorhabditis elegans* embryo at successive stages between fertilization and first cleavage. Embryos in this and subsequent figures are oriented with posterior pole to the right. (**a**) Formation of pronuclei and contractions of anterior membrane (~20 min after fertilization at 25°C). Egg pronucleus at anterior pole is not visible in this focal plane. (**b**) Pseudocleavage and pronuclear migration. (**c**) Meeting of the two pronuclei in posterior half of zygote. (**d**) Movement to center and rotation of pronuclei. (**e**) Formation of the first mitotic spindle. (**f**) Anaphase and beginning of first cleavage. Centrosomes are visible as round, granule-free regions; note the asymmetric location of the spindle along the anterior–posterior axis. (**g**) Telophase; note smaller disc-shaped centrosome in posterior (P1) cell, as compared with larger spherical centrosome in anterior (AB) cell. (**h**) A 2-cell embryo. For additional explanation, see text. Scale bar: 10 μm. (From Strome and Wood, 1983, copyright © M.I.T.)

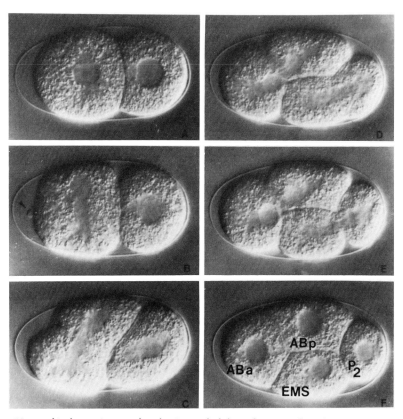

Figure 4. Nomarski photomicrographs of a *Caenorhabditis elegans* embryo between the 2-cell and 4-cell stages. Note that the larger anterior AB cell begins and completes cleavage somewhat before the smaller P1 cell. Arrowhead designates polar body. For further explanation, see text. (Micrographs courtesy of Ann Cowan.)

of the egg (Fig. 3d,e). The pronuclei then fuse, and first cleavage begins immediately, at about 35 min postfertilization (Fig. 3f,g). The first cleavage is asymmetric, giving rise to the larger anterior ectodermal founder cell AB and the smaller posterior germline cell P1 (Figs. 3h and 4a).

At about 10 min after completion of the first cleavage, the AB cell begins second cleavage (Fig. 4b), followed 3 min later by cleavage of P1 to produce EMS and P2 (Fig. 4d,e). The polarity of the dorsal–ventral axis first becomes apparent during AB cleavage (Fig. 4c), which places the posterior daughter (AB·p) dorsal to EMS, leaving AB.a at the anterior and P2 at the posterior pole of the embryo (Fig. 4f).

3.2. Stage 1: Generation of the Founder Cells

Like first cleavage, the next three P-cell cleavages are also asymmetric, following a stem-cell pattern in which each division gives rise to a larger somatic founder cell and a smaller germline (P-cell) daughter (Fig. 5). First cleavage produces the AB founder cell; the second somatic daughter, EMS divides asymmetrically to give the smaller E and the larger MS founder cells. Shortly thereafter, P2 divides to produce the C founder cell and the germline cell P3, which subsequently divides to produce the last somatic founder cell D and the germline founder cell P4.

The cleavages that produce the founder cells are asynchronous as well as asymmetric (Fig. 6). Following its birth, each founder cell and its progeny exhibit a characteristic rate of cleavage that is roughly proportional to founder cell size and follows the order of origin, from the fastest, AB, to the slowest, D (P4 is omitted from the comparison, because it cleaves only once more during embryogenesis) (see Fig. 5).

The tissues and cell types that derive from each of the founder cells (Fig. 5) are known from the determination by Sulston *et al.* (1983) of the complete cell lineage during embryogenesis. The E cell is exclusively an endodermal precursor, giving rise to the entire gut; the D cell is strictly mesodermal, producing only body wall muscle; and the P4 cell gives rise only to the germ line. The other three founder cells, AB, MS, and C, however, produce both ectodermal

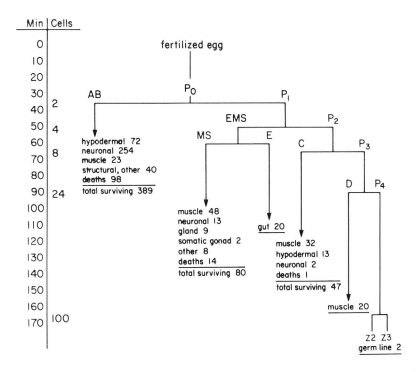

Figure 5. Lineage pattern of early cleavages in the *C. elegans* embryo, showing derivation of the six founder cells AB, MS, E, C, D, and P4. Left-hand scale indicates min after fertilization at 25°C; horizontal lines connecting sister cells indicate approximate times of cleavages. Second scale at left shows number of cells in the embryo with time. Below each of the founder cells are indicated the types and numbers of cells (at hatching) that derive from it. (Adapted from Sulston *et al.*, 1983.)

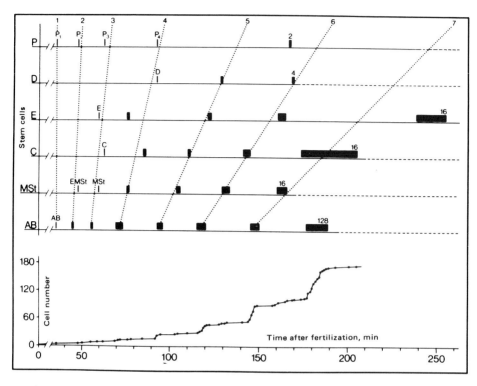

Figure 6. Timing of early cleavages in *Caenorhabditis elegans*, showing differences in cell cycle times for the six founder cell lineages. Stem cells refer to founder cells; EMSt and MSt are previously used designations for EMS and MS, respectively. In the top half of the diagram, a horizontal line for each founder cell is marked to show the cell division events of that lineage. Order from top to bottom is by length of cell cycle. Time of origin of each founder cell is indicated by a labeled vertical line. Black boxes indicate the time, in the indicated lineage, from division of the first cell to division of the last cell to divide in a given round of division. The number over the last box for each lineage indicates the number of cells of that lineage present after that division. (···) Rounds of cell divisions. (- - -) Lineages not followed further in these experiments. The bottom half of the diagram shows the increase in total nuclei (cell number) with time. (From Deppe *et al.*, 1978.)

and mesodermal derivatives, although descendants of AB are primarily ecto-dermal and those of MS primarily mesodermal (Fig. 5).

3.3. Stage 2: Gastrulation

Just before gastrulation, which begins at 100 min after first cleavage, the 16 AB descendants lie anteriorly and laterally, and the four C descendants lie posteriorly and dorsally. The four MS derivatives lie ventrally near the middle of the embryo, and the two E derivatives, D and P4 lie ventrally and posteriorly. At the start of gastrulation (Fig. 7a), the two E cells sink inward, followed by P4 and the MS cells. Later, as the entry zone widens and lengthens (Fig. 7b), most

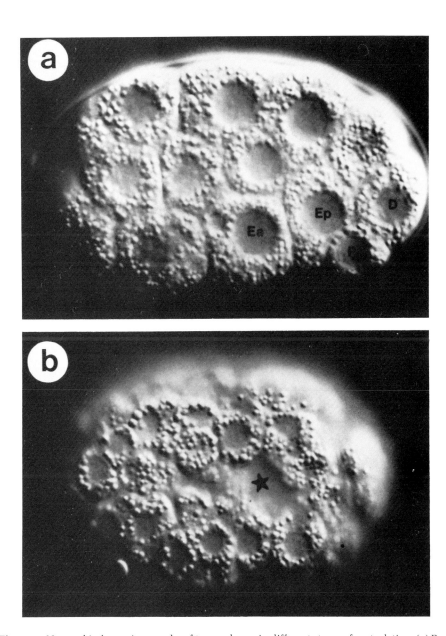

Figure 7. Nomarski photomicrographs of two embryos in different stages of gastrulation. (**a**) Beginning of gastrulation (~100 min after first cleavage); left lateral view focused on mid-plane of embryo. Ea and Ep are daughters of the E founder cell, giving rise to the intestine; D and P_4 are founder cells that give rise to body muscle and to the germline, respectively (see Fig. 5). (**b**) Late gastrulation (~210 min); ventral view, focused on ventral surface of embryo. Star (*) indicates cleft through which the MS cells have just entered. (From Sulston *et al.*, 1983.)

of the remaining myoblasts from the C and D lineages and then the AB-derived pharyngeal precursors sink into the interior. The ventral cleft is closed by 290 min. As gastrulation proceeds, further divisions of the E cells and the pharyngeal precursors form a central cylinder, and the body myoblasts become positioned between it and the outer cell layer. Cell division and organogenesis continue until about 350 min.

3.4. Stage 3: Morphogenesis

At about 350 min after first cleavage, cell proliferation largely ceases, and morphogenesis begins (Fig. 1). The first twitching movements, indicative of muscle function, are observed at about 400 min. Sexual dimorphism can first be observed at about 470 min; a set of four neurons undergoes programmed cell death in the hermaphrodite, whereas two other cells undergo cell death in the male (Sulston et al., 1983).

During the period of ~350–650 min, the embryo becomes transformed from a spheroid of ~550 cells into a cylindrical worm of the same volume, still inside the egg shell, with an increase in length of about four fold. To accomplish this transformation, cells of the animal change shape coordinately in a process that appears to be controlled by the surface hypodermal cells derived from the AB and C lineages (Priess and Hirsh, 1986). These cells are connected by belt desmosomes and contain bands of both actin filaments and microtubules that run circumferentially in alternating bundles around the animal. Lengthening is probably accomplished by tightening of the actin filaments against the internal hydrostatic pressure of the embryo, with the microtubular bundles serving as structural supports to maintain shape. At ~650 min, a collagenous cuticle begins to be secreted by the hypodermal cells, and the circumferential bundles begin to disperse. From this point on, shape is maintained by the cuticle, as evidenced by cuticle-defective mutants, which elongate normally but then revert to a spheroidal shape late in embryogenesis (Priess and Hirsh, 1986). The pretzel-stage embryo, now moving actively and about four times the length of the original zygote, begins pharyngeal pumping at ~760 min and hatches at ~800 min, probably with the aid of one or more hatching enzymes that catalyze breakdown of the eggshell from within.

3.5. General Features of Embryogenesis

Contrary to earlier assumptions based on partial lineages (reviewed in Chitwood and Chitwood, 1974), there is no strict correlation between founder cells and either germ layers or tissues. Although three of the founder cells (E, D, and P4) give rise to cells of single tissues, the other three, (AB, MS, and C) each contribute to two germ layers, ectoderm and mesoderm, and to several tissues (Fig. 5). Conversely, most tissues of the adult, with the clear exceptions of the intestine and the germline, are of mixed lineage, deriving from more than one founder cell.

Patterns of division during embryogenesis put most cells near their final locations; of the 558 cells present at hatching in the hermaphrodite, only 12 undergo long-range migrations during embryogenesis. A common cell fate is programmed cell death; 113 of the 671 cells generated during hermaphrodite embryogenesis die shortly after their birth. Like most other cell fates, these deaths are specific, cell-autonomous, and invariant from one embryo to the next. The bilateral symmetry of the animal at the cellular level does not always arise from bilaterally symmetric lineage patterns during embryogenesis, especially in the nervous system. In fact, the embryo is bilaterally asymmetric from the 6-cell stage onward. The lineage patterns later in embryogenesis appear to have been modified during the course of evolution to produce bilateral symmetry; thus, lineally nonhomologous cells take on appropriate fates near the positions where they are needed rather than lineal homologues migrating to these positions (Sulston *et al.*, 1983).

3.6. Macromolecular Synthesis during Embryogenesis

Little is known about the details of macromolecular synthesis in the embryo. Polyadenylated messenger RNAs (mRNAs) are first detected in nuclei at about the 100-cell stage (~150 min) by *in situ* hybridization to squashed embryos using labeled polyuridylate as a probe (Hecht *et al.*, 1981). Collagen mRNA and an increase over maternal levels of actin mRNA are also first detected at ~150 min by similar methods using cloned fragments of the corresponding genes as probes (Edwards and Wood, 1983). One tissue-specific enzyme, a gut esterase, is first detected at ~200 min (Edgar and McGhee, 1986). Several tissue-specific antigens become detectable by immunofluorescence staining at about this time as well (S. Strome and W. B. Wood, unpublished data).

4. Determination of Cell Fates and Patterns in the Embryo

4.1. Evidence for Cell-Autonomous Determinants

Several lines of evidence indicate that founder cell fates in the early embryo are lineally, rather than positionally, determined, apparently by asymmetric segregation of cell-autonomous determinants into the appropriate daughter cell at determinative cleavages. As an assay for determination, a convenient tissue-specific differentiation marker is provided by autofluorescent granules containing tryptophan catabolites that accumulate exclusively in gut cells beginning at ~200 min (Babu, 1974; Laufer *et al.*, 1980). Recently, the gut-specific esterase mentioned in the preceding section has been used for the same purpose. Permeabilized embryos treated at early stages with the drugs cytochalasin and colchicine to block further cleavage were shown to express gut granules (Laufer et al., 1980) or gut esterase (Edgar and McGhee, 1986) at about the normal time after first cleavage, and only in gut precursor cells present at the time of cleavage arrest. No expression was observed in arrested 1-cell embryos. For subsequent stages, expression was observed in only the P1 cell in arrested 2-cell embryos, the EMS cell in 4-cell embryos, and the E cell in 8-cell embryos. This result indicates that expression of gut-specific differentiation markers in E cells and their precursors does not require the normal environment of the developing embryo, but rather appears to be dictated in a cell-autonomous manner by a determinant that segregates internally from P1 to EMS to E during the first three cleavages.

Results with partial embryos and isolated blastomeres, derived by mechanical or enzymatic removal of the eggshell and lysis of some blastomeres support these conclusions. Partial embryos derived from P1 alone show twitching, indicative of muscle function, as well as expression of gut markers, and isolated E cells give rise to progeny that express gut markers in the absence of other cells (Laufer et al., 1980; Edgar and McGhee, 1986).

Rudimentary cytoplasmic transfer experiments carried out using laser microsurgery are consistent with the view that the presumed gut determinant is cytoplasmic (Wood et al., 1984). The P1 nucleus can be removed from a 2-cell embryo by extrusion through a small hole burned with a laser microbeam into the posterior pole of the egg shell, which will often reseal subsequently, leaving a P1 cytoplast adjacent to the intact AB cell. The AB cell will then proliferate through several divisions, but gut granules are never expressed. However, if the P1 cytoplast is fused to one of the AB cell progeny by laser-induced disruption of the membrane between them, gut granules are subsequently expressed in about half of the treated embryos.

Ablation experiments using a laser microbeam to kill single embryonic cells appear to support the view that all cells in the early embryo, and most cells throughout embryogenesis, are autonomously determined by internally segregating factors, independent of influences from neighboring cells. In general, ablation of an embryonic cell causes the cell types comprising the normal progeny of that cell to be absent from the embryo and has no effect on the fates of neighboring cells or their progeny. Except for a few cases in late embryogenesis, there is no evidence for regulation to compensate for removal of a cell by ablation (Sulston et al., 1983).

However, interpretation of these findings may be complicated by the fact that the remains of an ablated cell are still present in the embryo. More recent results suggest that within a founder cell lineage (e.g., the AB lineage) fates may be positionally influenced, even at early cleavage stages (Priess and Thomson, 1987). Moreover, experiments with isolated blastomeres suggest that cell–cell interactions may be important in maintaining correct polarity of early cleavages, which may in turn be important for fate determinations (Laufer et al., 1980; Schierenberg, 1987; L. Edgar and T. Hyman, personal communications). Additional work is needed to ascertain the extent of cell autonomy in determination of embryonic cell fates.

4.2. Localization of Germline Granules in the Embryo

In attempts to observe localization of possible macromolecular lineage-specific determinants in early cleavage, Strome and Wood (1982; 1983) and subsequently Strome (personal communication) injected mice with homogenates of early embryos, prepared hybridoma cell lines from spleen lymphocytes, and screened them by immunofluorescence microscopy for production of monoclonal antibodies that recognize lineage-specific antigenic components. Only one such component has been identified. It is present in early embryos as cytoplasmic granules, termed **P granules,** that are observed only in the P cells and subsequently in all germline cells except mature sperm throughout the life cycle. Seventeen independently derived hybridomas have been found to produce antibody that reacts with P granules, out of a total of about 1500 lines tested. Two apparently similar monoclonal antibodies were found in an independent screen by Yamaguchi *et al.* (1983).

At fertilization (Fig. 8a), P granules are observed by immunofluorescence microscopy to be uniformly distributed as small particles throughout the ooplasm (Strome and Wood, 1983). As the pronuclei approach each other, the P granules become localized around the posterior periphery of the embryo (Fig. 8b), so that after first cleavage they are present exclusively in the P1 cell (Fig. 8c). Similar prelocalization and segregation occur at each of the subsequent P-cell cleavages, always placing the P granules into the germline daughter. The granules appear to increase in size and decrease in number during the first three cell cycles. Before the cleavage of P3, they become associated with the nuclear envelope, where they are found during the remainder of the life cycle. Occasionally, a few granules are found in the D cell immediately after P3 cleavage (Fig. 8d), but they disappear before the subsequent cleavage of D, as if the granules are unstable in somatic cytoplasm.

4.3. Evidence for the Mechanism of P-Granule Localization

Drug experiments with early embryos indicate that P-granule localization is dependent on the function of an actin-based motility system (Strome and Wood, 1983). In permeabilized embryos treated immediately after fertilization with any one of several microtubule inhibitors, such as colcemid, pronuclear migration is blocked, but the P granules nevertheless become localized at the posterior pole. By contrast, in embryos treated with either of the actin microfilament inhibitors cytochalasin B or D, the pronuclei migrate together, but P granules coalesce near the center of the embryo, rather than becoming localized posteriorly. Moreover, none of the other normal manifestations of zygotic anterior–posterior asymmetry is observed, such as the early contractions of the anterior membrane and the characteristic disc-shaped morphology of the posterior centrosome at first cleavage (see Fig. 3g). The behavior of actin-containing components in the zygote as observed by fluorescence microscopy is consistent with their proposed role in P-granule localization (Strome, 1986). These

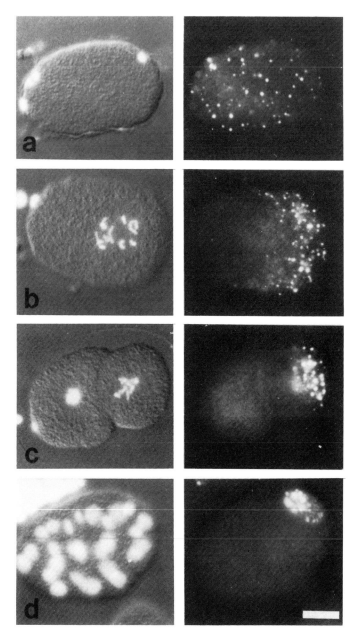

Figure 8. Localization of P granules after fertilization. Embryos removed from hermaphrodites were fixed and stained with diamidinophenylindole (DAPI) to visualize chromosomes and with an appropriate monoclonal antibody and fluoresceinated secondary antibody to visualize P granules. (Left panels) Nomarski images simultaneously epiilluminated with an appropriate wavelength to show DAPI-stained chromosomes. (Right panels) Same set of embryos epiilluminated with a different wavelength to visualize antibodies bound to P granules. (**a**) Zygote after completion of meiosis before pronuclear migration. In the Nomarski–DAPI image, the egg pronucleus and both polar bodies can be seen at the anterior pole (left), and the sperm pronucleus is near the posterior pole. P

results suggest that at fertilization, two independent systems of cytoskeletal machinery are set in motion: a microtubule-based system necessary for moving the pronuclei and subsequently mediating mitosis and an actin-based system responsible for posterior localization of the P granules and for generation of several other asymmetries in the zygote.

4.4. Origin, Nature, and Possible Function of P Granules

Although immunofluorescence signals from early embryos have not been quantitated, the P granules observed throughout embryogenesis appear to be maternal in origin, based on lack of an obvious increase in total fluorescence. Consistent with this view are experiments with a fertile mutant (Wood *et al.*, 1984) in which the P granules do not stain with one of the anti-P-granule monoclonal antibodies (K76). When mutant hermaphrodites are mated to wild-type males, the outcross progeny embryos, which are heterozygous for the mutation, show no antigen detectable by immunofluorescence using K76 antibody until the start of gonad proliferation midway through the first larval stage.

Although the segregation behavior of P granules is as expected for a lineage-specific determinant, there is no evidence that they are determinative for the germline, and their function remains unknown. However, they are probably homologous to the granular inclusions seen by electron microscopy in the germline cytoplasm of many organisms and called **germinal plasm, nuage,** or **polar granules** (see Chapter 1 and Eddy, 1975, for additional information). These inclusions have also been observed in *C. elegans*, where their size distribution and localization in the different P cells of early embryos correlate well with the distributions observed for P granules by immunofluorescence (Krieg *et al.*, 1978; Wolf *et al.*, 1983).

4.5. Control of Lineage-specific Cell Cycle Periods

In the early embryo, the progeny of a given somatic founder cell divide synchronously, with a cell-cycle period that is different for each lineage (Deppe *et al.*, 1978) (see Fig. 6). The lineage-specific differences in cell-cycle timing appear to be important in establishing the correct spatial patterns of the cleavage-stage blastomeres (Schierenberg *et al.*, 1980).

Using laser microsurgery on early embryos, Schierenberg and Wood (1985) carried out blastomere fusion and cytoplasmic transfer experiments to investi-

---

granules are dispersed throughout the cytoplasm. (**b**) Zygote at the time of pronuclear conjunction, showing localization of P granules around the posterior periphery. (**c**) A 2-cell embryo in which the P1 cell (posterior) is in prophase and P granules are prelocalized in the region of cytoplasm destined for the next P cell daughter. (**d**) A 26-cell embryo in which most of the P granules are localized in the P4 cell, but some are detected in its sister cell, D, below and to the right. Scale bar: 10 μm. (From Strome and Wood, 1983, copyright © M.I.T.)

gate the basis for the timing differences between lineages. The results indicate that control of cell-cycle period is cytoplasmic: The period of a cell from one lineage can be substantially altered by introduction of cytoplasm from a cell of another lineage with a different period. Short term effects of foreign cytoplasm on the timing of the subsequent mitoses differ depending on the position of the donor cell in the cell cycle. The observations are consistent with the presence of a cytoplasmic oscillator for which the period depends on the concentration of some component that is partitioned to the different founder cells in different concentrations and that regulates the activity of mitosis-inducing or mitosis-inhibiting factors, or both, and thus controls the period of the cell cycle.

5. Mutations That Affect Embryogenesis

5.1. Isolation of Embryonic Lethal Mutants

One of the attractions of *C. elegans* as an experimental organism is its suitability for genetic analysis. Treatment of fourth-stage hermaphrodite larvae with chemical mutagens such as EMS results in a high mutation rate of approximately 5×10^{-4} mutations per gene per gamete. Mutations can also be easily generated by ionizing radiation. Self-fertilization of the mutagenized animals produces F1 progeny that are heterozygous for a newly induced mutation. Such a heterozygote, by self-fertilization, will produce one fourth F2 progeny that are homozygous for the mutation. Mapping and complementation testing of mutations can be done in the usual manner by crosses of hermaphrodites with males, using appropriate markers to distinguish self-progeny from outcross-progeny. A variety of morphological and behavioral mutants defining genes on each of the six linkage groups serve as convenient markers for mapping of new mutations.

Genes essential for embryogenesis were first identified by screening mutagenized animals for temperature-sensitive (*ts*) lethal mutations (Hirsh and Vanderslice, 1976). Such mutations result in an altered gene product that is functional at a permissive temperature (generally 16°C) but nonfunctional at a restrictive temperature (generally 25°C), in contrast to the normal gene product, which functions at both temperatures. Animals carrying such a mutation can be propagated at the permissive temperature but exhibit embryonic lethality at the restrictive temperature. In this and other such screens, carried out subsequently in several laboratories (Wood *et al.*, 1980; Miwa *et al.*, 1980; Cassada *et al.*, 1981), 10–28% of the lethals recovered have shown phenotypes of embryonic lethality; i.e., homozygous mutant animals shifted to nonpermissive temperature as third- or fourth-stage larvae self fertilize to produce embryos that fail to hatch.

The screens cited above identified a total of about 55 genes required for embryogenesis. More recent screens for *ts* mutations have identified additional genes (e.g. Kemphues, 1987; Priess *et al.*, 1987; Kemphues *et al.*, 1988; H. Schnabel and R. Schnabel, personal communication), and several more such genes, so far less well characterized, have been defined by nonconditional (absolute) lethals isolated in screens employing partial chromosome duplications or balancer chromosomes that permit convenient propagation of heterozygotes (Meneely and Herman, 1981; Rogalski *et al.*, 1982; Sigurdson *et al.*, 1984). The frequency distribution of multiple mutations within single genes defined by the *ts* mutants has been used to estimate the total number of such genes at about 200 (Cassada *et al.*, 1981); however, this could be an overestimate of the genes required exclusively for embryogenesis, as discussed in Section 5.2.2.

5.2. Characterization of Genes Defined by Embryonic Lethal Mutants

5.2.1. Time of Action

Information on when these genes act in development has been obtained from parental-effect tests on the *ts* mutations that define them (Wood *et al.*, 1980; Miwa *et al.*, 1980; Cassada *et al.*, 1981). Among the 47 genes for which these tests are complete, four classes can be distinguished. Mutations in the first class (28 genes) show a *strict* parental effect; i.e., the homozygous mutant (*m*/*m*) self-progeny of a heterozygous mutant (*m*/+) hermaphrodite escape embryonic lethality, and the *m*/+ outcross-progeny of an *m*/*m* hermaphrodite mated to a wild-type (+/+) male die as embryos. Expression of these genes in the maternal parent therefore appears to be both necessary and sufficient for embryonic survival.

Mutations in the second class (11 genes) show a *partial* parental effect; *m*/*m* self-progeny of *m*/+ hermaphrodites survive, but so do *m*/+ outcross-progeny of *m*/*m* hermaphrodites. In 10 of these cases, survival of outcross-progeny depends on introduction of the + allele from the male parent; therefore, expression of these genes in either the maternal parent or the embryo is sufficient for embryonic survival. In the remaining case, which is somewhat unusual, the *m*/*m* as well as the *m*/+ outcross-progeny of *m*/*m* hermaphrodites mated to *m*/+ males survive, indicating that expression of this gene in the paternal parent is sufficient for embryonic survival.

Mutations in the third class (three genes) behave in a strictly nonmaternal fashion; i.e., *m*/*m* self-progeny of *m*/+ hermaphrodites die as embryos, and *m*/+ outcross-progeny of *m*/*m* hermaphrodites survive. Expression of these genes in the embryo is therefore both necessary and sufficient for embryonic survival.

For mutations in the fourth class (five genes), neither the *m*/*m* self-progeny embryos of *m*/+ hermaphrodites nor the *m*/+ outcross-progeny embryos of *m*/*m* hermaphrodites survive, indicating that expression of this gene is required *both* maternally *and* embryonically for embryonic survival.

The preponderance of maternal effects among the mutationally identified genes required for embryonic survival is consistent with the view that rapid rates of cell division in early embryogenesis necessitate maternal synthesis and storage in the oocyte of most of the gene products required during this period.

More detailed analysis of when these genes act has been carried out for the *ts* mutations by temperature-shift experiments, to determine time and duration of the temperature-sensitive period (TSP) leading to embryonic lethality. Interpretation of such experiments is complicated by the uncertainty of whether the TSP defines the time of synthesis or the time of function of the mutant gene product (for discussion, see Wood *et al.,* 1980). Nevertheless, the results in general are consistent with expectations from the parental-effect tests: Strict maternal mutants tend to show the earliest TSPs, either before or close to the time of fertilization, and partial maternal mutants show later TSPs, generally during the first half of embryogenesis. Only three strict nonmaternal complementation groups include *ts* mutants, which have TSPs during the second half of embryogenesis.

5.2.2. Possible Gene Functions

An inherent problem with genetic analysis of embryogenesis is the difficulty of distinguishing genes that control pattern formation and cell fates (the developmentally most interesting ones) from genes that control general metabolic and cellular functions also required in embryogenesis. Presumably, null (complete loss-of-function) mutations that result in a specific embryonic defect only are more likely to define genes that fall into the interesting category than mutations that cause defects at several different developmental stages. For classifying essential genes in this regard, *ts* mutations have the advantage that their effects on different stages can be tested by shifting homozygous mutant animals to the nonpermissive temperature at different times in the life cycle. They have the disadvantage that they often do not result in a null phenotype, so that inferences of gene function from *ts* mutant phenotypes, particularly when based on analysis of only one *ts* mutant allele, must be viewed with great caution. Nevertheless, analysis of *ts* mutants has so far provided the most useful available information on the nature of genes required for embryogenesis.

In the set of 55 genes discussed above identified by *ts* embryonic lethal mutants, only about 14 are possible candidates for exclusively embryonic function, and only five of these are defined by more than one allele. Among this set of 14, 11 show strict maternal, three show partial maternal, and only one shows strict nonmaternal, requirements for function. Most of the strict maternal mutants show defects in meiosis or first cleavage (Wood *et al.,* 1980; Miwa *et al.,* 1980; Denich *et al.,* 1984; Kemphues *et al.,* 1986), so that subsequent anomalies may be the result of aberrant chromosome distribution rather than lack of specific factors required for determination of cleavage patterns or cell fates. Mutants defective in the single nonmaternal gene (*let-2*) arrest near the end of embryogenesis and may have a defect in basement membrane function. In summary, it seems possible that few of the interesting genes for control of embryogenesis are included in this set of 55.

More recently, Kemphues *et al.* (1988) have isolated potentially interesting strict maternal effect mutations at new loci, which may be important in the asymmetric partitioning of cytoplasmic components in the early embryo. These mutations show normal meiosis, affect only embryogenesis, and are characterized by synchronous and usually symmetrical, rather than the normal asynchronous and asymmetric, early cleavages. In several of these mutants, for example, P granules do not show asymmetric localization or segregation; gut differentiation markers do not appear, and morphogenesis does not take place.

Perrimon et al. (1986) showed that in Drosophila, maternal-effect embryonic lethality can result from rare hypomorphic alleles of essential genes whose null phenotypes are general lethality rather than pure embryonic defects. Inclusion of such mutations in frequency calculations can lead to overestimates of the number of genes mutable to maternal-effect embryonic lethality. Analysis of 29 maternal-effect-lethal loci from a C. elegans saturation screen of a region of chromosome II indicates that the same phenomenon occurs in this organism, and has led to a more precise estimate of the number of pure maternal-effect genes required for embryogenesis in C. elegans as between 25 and 60 (K. Kemphues, personal communication).

6. Possible Mechanisms of Determination

The mechanism by which the fates of blastomeres are determined in mosaic embryos represents a longstanding problem. Fates appear to be dictated by internally segregating factors, based on the kinds of evidence cited above for C. elegans. There is considerable evidence in several organisms, including C. elegans, that these factors are cytoplasmic, rather than asymmetrically distributed chromosomal components. DNA strands derived from C. elegans sperm have been shown to be randomly distributed to embryonic cells during development, thereby ruling out segregation of paternally derived DNA to specific cells as a determination mechanism (Ito and McGhee, 1987). However, a round of critically timed DNA replication may be required for expression of lineage-specific differentiation markers in ascidians (Satoh and Ikegami, 1981) as well as in C. elegans (Edgar and McGhee, 1988).

Although the cytoplasmic location of fate-determining factors seems very likely, their nature remains unclear. Classic observations of cytoplasmic localization and segregation of visible cytoplasmic components, for example, in the ascidian Styela (Conklin, 1905) (see Chapter 1), led to the notion of maternally derived cytoplasmic determinants, informational macromolecules (we would assume today) that are segregated specifically to different lineages and later somehow promote the appropriate differential patterns of gene expression. An alternative possibility is that maternally derived components are responsible for producing asymmetric distributions in the early embryo of small molecules or ions. These gradients could confer polarity and positional information, somehow dictating early differential synthesis of embryonic gene products that determine cell fates.

There is currently little basis for deciding between these two possibilities or eliminating them in favor of another, in C. elegans or any other mosaic embryo. There is convincing experimental evidence for presence in the polar plasm of Drosophila oocytes and early embryos of determinative factors that dictate functional germline development (Illmensee and Mahowald, 1974; Niki, 1986). Similar evidence for germline determinants based on morphological criteria has been obtained in amphibians (Wakahara, 1978) and in nematodes, in which polar plasm appears to prevent chromosome diminution in *Ascaris* (Boveri, 1910) and may be involved in preserving P granules in *C. elegans* (Fig. 8d). However, the evidence is less clear for determinants of somatic cell fates, raising the possibility that the germ line represents a special case.

P-granule experiments show that *C. elegans* embryos have the capability of segregating a cytoplasmic component to a specific lineage, but there is no evidence that the granules are determinative factors. Moreover, our failure to find other antigenic components segregating to specific somatic lineages again suggests that such all-or-none segregation could be a process unique to the germline.

Ultimately, insights into embryonic determination mechanisms should come from analysis of mutations that affect embryogenesis. So far, however, the studies of embryonic lethal mutants reviewed above have just begun to be helpful in this regard. Analysis of homoeotic mutations that affect post-embryonic cell lineages has identified genes that appear to be directly involved in the specification of different daughter cell fates in determinative cell divisions (reviewed in Sternberg and Horvitz, 1984). However, no such mutations have been identified among the embryonic mutants, with one recent, surprising exception (Priess *et al.*, 1987). The **determinant model,** postulating maternally derived informational macromolecules required for specification of founder-cell fates would predict the possibility of maternal effect mutants that lack a particular lineage but are otherwise embryonically normal; no such mutants have been found. The **gradient model** would predict that, as in *Drosophila* (Scott, 1987), genes expressed in the early embryo, defined by nonmaternal mutations may be important for early patterning and cell fate determination. So far, too few nonmaternal mutations have been analyzed in *C. elegans* to assess this prediction. In general, although mutant analysis provides a promising approach to elucidating embryonic determination mechanisms, definitive answers must await the results of more extensive screens for embryonic mutants and more detailed characterizations of mutant phenotypes than have so far been reported.

ACKNOWLEDGMENTS.   The author is grateful to D. Albertson, L. Edgar, K. Kemphues, J. Priess, E. Schierenberg, S. Strome, and J. Sulston for comments on the manuscript, and to those mentioned in the text for communication of unpublished results. Research from the author's laboratory was supported by grants HD-11762 and HD-14958 from the National Institutes of Health.

References

Albertson, D., 1984, Formation of the first cleavage spindle in nematode embryos, *Dev. Biol.* **101**:61−72.

Babu, P., 1974, Biochemical genetics of *Caenorhabditis elegans*, *Mol. Gen. Genet.* **135**:39−44.

Boveri, T., 1910, Ueber die Teilung centrifugierte Eier von *Ascaris megalocephala*, *Wilhelm Roux. Arch. Entwicklungsmech. Org.* **30**:101−125.

Cassada, R., Isnenghi, E., Culotti, M., and von Ehrenstein, G., 1981, Genetic analysis of temperature-sensitive embryogenesis mutants in Caenorhabditis elegans, Dev. Biol. **84**:193–205.

Chitwood, B. G. and Chitwood, M. B., 1974, Introduction to Nematology, University Park Press, Baltimore.

Conklin, E. G., 1905, The organization and cell-lineage of the ascidian egg, J. Acad. Natl. Sci. Phila. **13**:1–119.

Denich, K. T. R., Schierenberg, E., Isnenghi, E., and Cassada, R., 1984, Cell lineage and developmental defects of temperature-sensitive embryonic arrest mutants of the nematode Caenorhabditis elegans, Wilhelm Roux Arch. Dev. Biol. **193**:164–179.

Deppe, U., Schierenberg, E., Cole T., Krieg, C., Schmitt, D., Yoder, B., and von Ehrenstein, G., 1978, Cell lineages of the embryo of the nematode Caenorhabditis elegans, Proc. Natl. Acad. Sci. USA **75**:376–380.

Eddy, E. M., 1975, Germ plasm and the differentiation of the germ cell line. Int. Rev. Cytol. **43**:229–280.

Edgar, L. G. and McGhee, J. D., 1986, Embryonic expression of a gut-specific esterase in Caenorhabditis elegans, Dev. Biol. **114**:109–118.

Edgar, L. G. and McGhee, J. D., 1988, DNA synthesis and the control of embryonic gene expression in Caenorhabditis elegans, Cell, in press.

Edwards, M. K., and Wood, W. B., 1983, Location of specific messenger RNAs in Caenorhabditis elegans by cytological hybridization, Dev. Biol. **97**:375–390.

Hirsh, D., Kemphues, K. J., Stinchcomb, D. T., and Jefferson R., 1985, Genes affecting early development in Caenorhabditis elegans, Cold Spring Harbor Symp. Quant. Biol. **50**:69–78.

Hirsh, D. and Vanderslice, R., 1976, Temperature-sensitive developmental mutants of Caenorhabditis elegans, Dev. Biol. **49**:200–235.

Hecht, R. M., Gossett, L. A., and Jeffery, W. R., 1981, Ontogeny of maternal and newly transcribed mRNA analyzed by in situ hybridization during development of Caenorhabditis elegans, Dev. Biol. **83**:374–379.

Illmensee, K., and Mahowald, A. P., 1974, Transplantation of posterior polar plasm in Drosophila. Induction of germ cells at the anterior pole of the egg, Proc. Natl. Acad. Sci. USA **71**:1016–1020.

Ito, K. and McGhee, J. D., 1987, Parental DNA strands segregate randomly during embryonic development of Caenorhabditis elegans, Cell **49**:329–336.

Kemphues, K. J., 1987, Genetic analysis of embryogenesis in Caenorhabditis elegans, in: Developmental Genetics of Higher Organisms (G. M. Malacinski, ed.), Macmillan, New York.

Kemphues, K. J., Priess, J. R., Morton, D. G., and Cheng, N., 1988, Identification of genes required for cytoplasmic localization in early C. elegans embryos, Cell, in press.

Kemphues, K. J., Wolf, N., Wood, W. B., and Hirsh, D., 1986, Two loci required for cytoplasmic organization in early embryos of Caenorhabditis elegans, Dev. Biol. **113**:449–460.

Kimble, J., and Hirsh, D., 1979, The embryonic cell lineages of the hermaphrodite and male gonads in Caenorhabditis elegans, Dev. Biol. **70**:396–417.

Krieg, C., Cole, T., Deppe, U., Schierenberg, E., Schmitt, D., Yoder, B., and von Ehrenstein, G., 1978, The cellular anatomy of embryos of the nematode Caenorhabditis elegans, Dev. Biol. **65**:193–215.

Laufer, J. S., Bazzicalupo, P., and Wood, W. B., 1980, Segregation of developmental potential in early embryos of Caenorhabditis elegans, Cell **19**:569–577.

Miwa, J., Schierenberg, E., Miwa, S., and von Ehrenstein, G., 1980, Genetics and mode of expression of temperature sensitive mutations arresting embryonic development in Caenorhabditis elegans, Dev. Biol. **76**:160–174.

Meneely, P. M. and Herman, R. K., 1981, Suppression and function of X-linked lethal and sterile mutations in Caenorhabditis elegans, Genetics **97**:65–84.

Niki, Y., 1986, Germline-autonomous sterility of P-M dysgenic hybrids and their application to germline transfers in Drosophila melanogaster, Dev. Biol. **113**:255–258.

Perrimon, N., Mohler, D., Engstrom, L., and Mahowald, A. P., 1986, X-linked female sterile loci in Drosophila melanogaster, Genetics **113**:695–712.

Priess, J. and Hirsh, D., 1986, C. elegans morphogenesis: The role of the cytoskeleton in elongation of the embryo, Dev. Biol. **117**:156–173.

Priess, J. R. and Thomson, J. N., 1987, Cellular interactions in early *Caenorhabditis elegans* embryos, *Cell* **48:**241–250.

Priess, J. R., Schnabel, H., and Schnabel, R., 1987, The *glp-1* locus and cellular interactions in early *C. elegans* embryos, *Cell* **51:**601–611.

Rogalski, T. M., Moerman, D. G., and Baillie, D. L., 1982, Essential genes and deficiencies in the *unc-22 IV* region of *Caenorhabditis elegans*, *Genetics* **102:**725–736.

Satoh, N. and Ikegami, S., 1981, A definite number of aphidicolin-sensitive cell-cyclic events are required for acetylcholinesterase development in the presumptive muscle cells of the Ascidian embryos, *J. Embryol. Exp. Morphol.* **61:**1–13.

early embryos of the nematode *C. elegans, J. Cell Biol.,* **103:**2241–2252.

Strome, S., and Wood, W. B., 1982, Immunofluorescence visualization of germ-line-specific cytoplasmic granules in embryos, larvae, and adults of *Caenorhabditis elegans, Proc. Natl. Acad. Sci. USA* **79:**1558–1562.

Strome, S., and Wood, W. B., 1983, Generation of asymmetry and segregation of germ-line-granules in early *C. elegans* embryos, *Cell* **35:**15–25.

Sulston, J. E., and Horvitz, H. R., 1977, Post-embryonic cell lineages of the nematode *Caenorhabditis elegans, Dev. Biol.* **56:**110–156.

Sulston, J. E., Schierenberg, E., White, J. G., and Thomson, J. M., 1983, The embryonic cell lineage of the nematode *Caenorhabditis elegans, Dev. Biol.* **100:**64–119.

Wakahara, M., 1978, Introduction of supernumerary primordial germ cells by injecting vegetal pole cytoplasm into *Xenopus* eggs, *J. Exp. Zool.* **203:**159–164.

Wood, W. B. (ed.), 1988, *The nematode Caenorhabditis elegans*, Cold Spring Harbor Laboratory, Cold Spring Harbor, New York.

Wood, W. B., Schierenberg, E., and Strome, S., 1984, Localization and determination in early embryos of *Caenorhabditis elegans*, in: *Molecular Biology of Development* (E. H. Davidson and R. Firtel, eds.) pp. 37–49, Alan R. Liss, New York.

Wood, W. B., Hecht, R., Carr, S., Vanderslice, R., Wolf, N., and Hirsh, D., 1980, Parental effects of genes essential for early development in *C. elegans, Dev. Biol.* **74:**446–469.

Wolf, N., Priess, J., and Hirsh, D., 1983, Segregation of germline granules in early embryos of *Caenorhabditis elegans*: An electron microscopic analysis, *J. Embryol. Exp. Morphol.* **73:**297–306.

Yamaguchi, Y., Murakami, K., Furusawa, J., and Miwa, J., 1983, Germ line-specific antigens identified by monoclonal antibodies in the nematode *Caenorhabditis elegans. Dev. Growth Diff.* **25:**121–131.

Chapter 3

Cellular Interactions in Establishment of Regional Patterns of Cell Fate during Development

J. C. SMITH

1. Introduction

The central problem in developmental biology remains: How does the fertilized egg develop into an embryo with the correct spatial pattern of cellular differentiation? Two general solutions to this problem have been proposed. The first involves **cytoplasmic localization,** wherein specific determinants are physically localized within the egg cell and differentially inherited by the cleavage products. Thus, cells that inherit a muscle cell determinant would form muscle, whereas those that inherit the neuroplasm would form nervous tissue. Organisms that use cytoplasmic localization as the major mechanism for specification of cell fate are relatively few, with ascidians being perhaps the best example (see Chapter 1). Most species therefore make use of a mechanism involving a sequence of **inductive interactions,** in which the generation of new cell types results from interactions between two existing cell types. This process requires that an initial polarity, or asymmetry, be present in the fertilized egg or imposed on the embryo by its environment. This asymmetry might occur as a cytoplasmic localization, but it need not be as specific as the ascidian system, which specifies individual cell types; it would be sufficient, for example, to imagine a gradient of a substance the high point of which defines the prospective dorsal side of the egg and whose minimum value defines the ventral side.

In birds and mammals, the origins and identities of the initial polarities are little understood, although for birds there is good evidence that polarity is imposed by gravity (see Eyal-Giladi, 1984); in the mouse, L. J. Smith (1980, 1985) suggested that embryonic axis determination depends on the orientation of the embryo with respect to the axes of the uterine horn. By contrast, in

J. C. SMITH • Laboratory of Embryogenesis, National Institute for Medical Research, The Ridgeway, Mill Hill, London NW7 1AA, England.

amphibia the origin of embryonic polarity—if not the identities of the molecules involved—is well understood and can even be experimentally manipulated. This fact alone would make the amphibian embryo a good starting point for the study of inductive interactions, but it offers several further advantages: (1) The amphibian embryo is large and therefore amenable to microdissection; (2) the embryos develop away from the mother and so are accessible to experimentation at all developmental stages; and (3) the embryos themselves are available in quantity, thus providing material for biochemical analysis. Finally,

rican clawed frog *Xenopus laevis*; I shall begin by describing normal development in *Xenopus*.

2. Normal Development of *Xenopus laevis*

Xenopus is an anuran amphibian—it lacks a tail. The development of *Xenopus* (Fig. 1; see Figs. 2 and 3) is broadly similar to that of other amphibian species, but it does differ in details, especially from the tailed amphibia, the urodeles (Keller, 1976; Smith and Malacinski, 1983; Brun and Garson, 1984). For present purposes, however, these differences do not concern us.

2.1. Fertilization

The egg of *Xenopus laevis* is spherical, with a diameter of ~1.4 mm (Fig. 1a). Before fertilization, the egg is radially symmetric with a distinct animal–vegetal axis, in which the so-called animal hemisphere is heavily pigmented and the vegetal hemisphere is pale (Fig. 1b). The egg is contained within a vitelline envelope and a three-layered jelly coat, and when it is spawned, the animal–vegetal axis is held in position by the vitelline envelope randomly with respect to gravity.

Unlike the urodeles, *Xenopus* is monospermic, and, at fertilization, sperm entry may occur anywhere around the animal hemisphere. Sperm entry is followed by a complex series of cortical and cytoplasmic rearrangements, one of which involves the release of the contents of the so-called cortical granules into the perivitelline space, which thus fills with liquid. This frees the egg to orientate itself with respect to gravity, so that the animal pole faces upwards and the vegetal pole downwards. Another visible manifestation of fertilization is the appearance of a heavily pigmented **sperm entry point** (**SEP**) on one side

of the animal hemisphere, while the side opposite the SEP becomes lighter in pigmentation. The sperm entry half of the egg eventually forms the posterior and ventral structures of the larva, with the other half forming head and dorsal structures.

2.2. Cleavage

~~cleavage is above the equator of the embryo so that the cells~~
are smaller than the vegetal pole cells (Fig. 1c,d).

By the 8-cell stage, a cavity has appeared in the embryo, whose floor is formed by the four vegetal pole cells and whose walls and roof are made by the animal pole cells (Fig. 1d). This cavity is called the **blastocoel;** it enlarges throughout development due to a Na^+-K^+-ATPase present in the internal membranes of the embryo (Slack and Warner, 1973). This ion pump increases the solute concentration in the blastocoel, which then swells due to osmotic uptake of water. The presence of the blastocoel is important in the forthcoming gastrulation movements, and it may also play a role in limiting the extent of mesoderm induction at the blastula stage.

2.3. The Blastula Stage

The rapid series of synchronous cell divisions continues until cleavage 12, when there are 4096 cells. The cells at the animal pole are smaller than those at the vegetal pole (Fig. 1e,f), but this reflects a tendency for cells to cleave away from the side with more yolky cytoplasm; it does not indicate a higher rate of cell division in the animal hemisphere. After the twelfth cleavage, however, the cell division cycle slows and becomes asynchronous. At the same time, cell motility (observed in single cells) commences, and the rate of RNA synthesis increases dramatically, such that all 6 hr of transcription before this point are equivalent to less than 10 sec of transcription after it (Newport and Kirschner, 1982; Kimelman et al., 1987). This coordinated change in the embryo's behavior has been termed the **mid-blastula transition** (Gerhart, 1980; see Chapter 7).

The mid-blastula transition occurs about 6–7 hr after fertilization. For the next 2–3 hr, cell numbers increase more slowly, while the blastomeres undergo subtle movements that precede the overt and dramatic movements of gastrulation. The **pregastrulation movements** include a thinning and increase in sur-

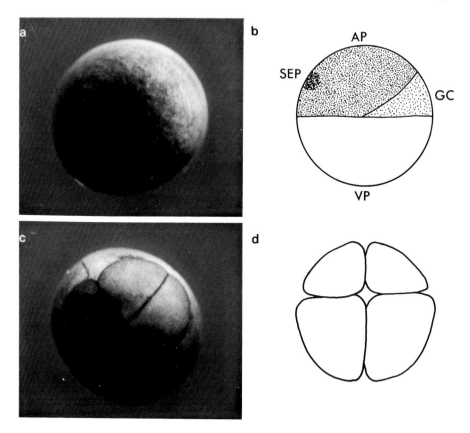

Figure 1. Early development of *Xenopus laevis*. (**a,b**) Fertilized egg. (**c,d**) The 8-cell stage. (**e,f**) Blastula. (**g,h**) The beginning of gastrulation. In (**a**), (**c**), and (**e**) the embryo is viewed from above, so that the animal pole is visible. In (**g**) the vegetal hemisphere of the embryo is visible. (**b**), (**d**), (**f**), and (**h**) are line drawings viewed from the side, with the dorsal half of the embryo to the right. AP, animal pole; VP, vegetal pole; SEP, sperm entry point; GC, gray crescent; BLC, blastocoel; DL, dorsal lip of the blastopore (arrowed in **h**). The diameter of the *Xenopus* embryo is ~1.4 mm.

face area of the animal region, known as **epiboly,** together with a decrease in the area of the vegetal region. The pregastrulation movements may result from the motile behavior of individual cells first observed at the mid-blastula transition. Gastrulation proper occurs about 10 hr after fertilization, when there are about 20,000 cells.

2.4. Gastrulation

The movements of gastrulation convert a radially symmetric ball of cells into a three-layered structure with bilateral symmetry and a head-to-tail axis. These movements have been described in great detail for *Xenopus* by Keller (1986) in Vol. 2 of this series, so they are discussed only briefly here.

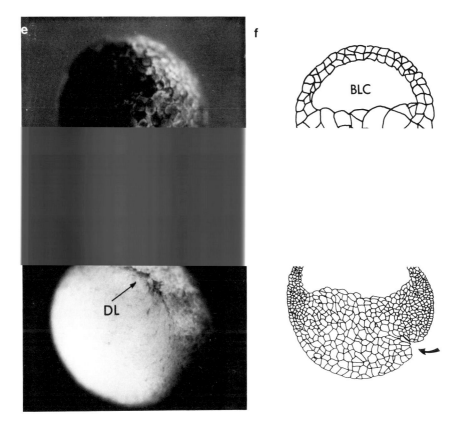

Figure 1. (*continued*)

The first sign of gastrulation is an infolding of cells near the vegetal pole of the embryo, on the dorsal side (Fig. 1 g,h). This indentation, called the **blastopore,** spreads laterally and then ventrally, toward the sperm-entry side of the embryo. Thus, the slit becomes an arc, then a semicircle, and finally a complete circle (Fig. 2). The formation of the blastopore reflects the movements of an equatorial annulus of cells inside the embryo, which—beginning on the dorsal side—turns back on itself, or involutes, and moves toward the animal pole. These are the cells that will form the mesoderm, and the involution movements result in the formation of the **mesodermal mantle.**

The superficial cells on the animal side of the blastopore lip are dragged inside the embryo by the deep cells and form the roof of the archenteron, or primitive gut; their place is taken by cells at the animal cap undergoing epiboly. As gastrulation proceeds, the blastopore circle at the vegetal pole becomes smaller, with the cells on the animal pole side of the blastopore covering the superficial vegetal pole cells, which then form the floor of the archenteron. The most dramatic movements of gastrulation are thus complete when the circle at the vegetal pole vanishes to form a slit. By this time, the archenteron has

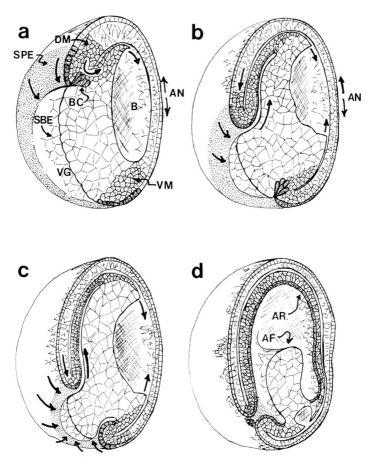

Figure 2. Gastrulation in *Xenopus laevis.* (**a**) Early gastrula. (**b,c**) Mid-gastrula. (**d**) Late gastrula.
The embryos are sectioned mid-sagittally and positioned with the vegetal pole toward the observer
and slightly to the left. The arrows indicate major cell movements. AN, Animal pole; VG, vegetal
pole; AR, archenteron roof; AF, archenteron floor; B, blastocoel; BC, bottle cells; DM, dorsal mesoderm; SBE, subblastoporal endoderm; SPE, suprablastoporal endoderm. (From Keller, 1986.)

expanded at the expense of the blastocoel, and the three germ layers of the
embryo have reached their definitive positions. In reality, however, gastrulation continues beyond this point, because cells are still being added to the
mesoderm germ layer from a thick circumblastoporal collar, which has not yet
involuted at the slit blastopore stage (Keller, 1976; Cooke, 1979). Gastrulation
therefore overlaps with the next clearly definable stage in *Xenopus* development: **neurulation.**

2.5. Neurulation

Shortly after the slit blastopore stage, the dorsal ectoderm becomes thicker
and the cells more closely packed (Tarin, 1971); externally, the neural plate

becomes visible as a slight prominence of the dorsal surface (Fig. 3A,B). These features become more obvious as neurulation proceeds, and then—beginning at the posterior end of the embryo—the neural plate begins to roll itself up to form the neural tube (Fig. 3C,D). The formerly lateral edges of the neural plate meet in the dorsal midline of the embryo, fuse, and are covered by the dorsalmost non-neural ectoderm, which forms epidermis. Finally, some regionalization becomes visible in the neural tube with subdivision of the brain into prosencephalon, mesencephalon, and rhombencephalon.

2.6. Later Development

The earliest visible differentiated structure in the amphibian embryo is the **notochord,** a strip of cells along the dorsal midline of the mesoderm, which becomes separated from the more lateral mesoderm on either side. The function of the notochord and the reason for its precocious behavior are unknown (Malacinski and Youn, 1982; Smith and Watt, 1985). However, it has a characteristic appearance in histological sections and is specifically stained by a monoclonal antibody to keratan sulfate (Smith and Watt, 1985; Dale *et al.,* 1985); thus, it does at least serve as an easily distinguishable marker of the most dorsal mesoderm.

Further regionalization in the mesoderm is marked by other cell types. Just lateral to the notochord are the somites, which in *Xenopus* are almost entirely represented by muscle cells. Ventral to the somites, particularly in the anterior mesoderm, come the cells that form the pronephros, recognizable as a single-cell thick tubular epithelium, whereas ventral to these cells is the lateral plate mesoderm. Finally, the most ventral mesoderm, particularly in the posterior half of the body, forms the blood islands.

Of the remaining two germ layers, the endoderm is slower to differentiate and eventually forms the pharynx, lungs, stomach, liver, and intestine. The ectoderm forms the entire epidermis of the embryo, with the brain and spinal cord being formed by the neural tube. Tissue from the folds of the neural plate that comes to lie between the dorsal part of the neural tube and the dorsal epidermis is the neural crest. The neural crest cells undergo extensive migration and form the autonomic nervous system, melanocytes, and parts of the skull. Figure 3E,F summarizes these patterns of cell differentiation in the early tailbud embryo.

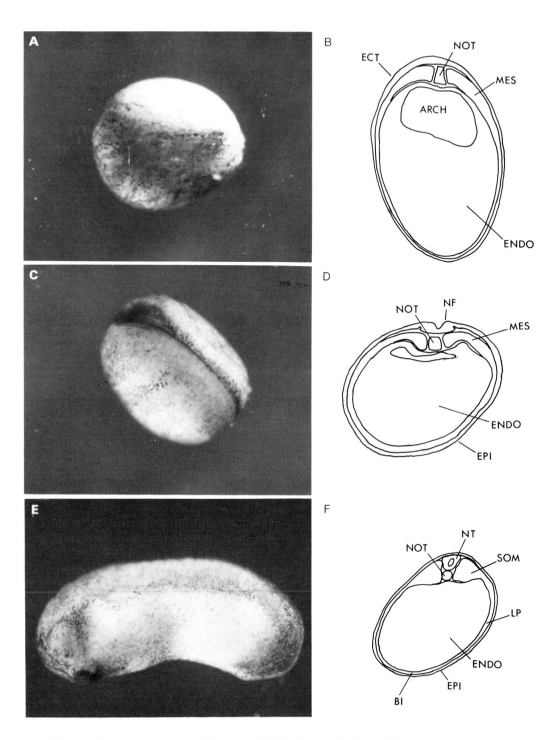

Figure 3. Later development of *Xenopus*. (**A,B**) Early neurula. Note thickened dorsal ectoderm. (**C,D**) Mid-neurula. Note neural folds. (**E,F**) After neural tube closure. In (**A**) and (**C**) the dorsal surface of the embryo is visible, and in (**E**) the embryo is viewed from the side. (**B**), (**D**), and (**F**) are *camera lucida* drawings of transverse sections through embryos. ARCH, archenteron; ECT, ectoderm; ENDO, endoderm; MES, mesoderm; NOT, notochord; NF, neural folds; EPI, epidermis; NT, neural tube; SOM, somite; LP, lateral plate mesoderm; BI, blood islands.

3. Fate Maps, Specification, and Determination

The most powerful tool of the experimental embryologist is the **fate map,** a description of what each part of the embryo at a given early stage will give rise to at a later stage in development. The fate map is important firstly because its very existence for a particular embryo implies a certain regularity and order in the developmental process. However, for our purposes, the main reason for the importance of the fate map is that it enables one to interpret experiments in

by the fate map, one may infer that the cells were removed from the embryo before they received the appropriate inductive signal.

Specification should not be confused with **determination,** which is a measure of the irreversibility of a development decision. Tests for determination involve subjecting cells to influences that may divert them down different developmental pathways, usually by juxtaposing them with cells from another part of the embryo. Cells that are determined will resist this attempt to change their developmental fate. Clearly, cells that are determined are also specified, but specified cells are not necessarily determined.

3.1. Fate Maps

Classic fate maps of amphibian embryos by Vogt (1929) and Pasteels (1942) employed vital dyes (Nile blue sulfate or neutral red). Pieces of agar impregnated with these stains were pressed against the embryo in the desired position until the cells themselves became stained. The subsequent behavior of the cells could then be followed. This technique was applied to *Xenopus* by Pasteels (1949) and Sirlin (1956), but the most comprehensive study is that of Keller (1975, 1976). The vital dyes have the advantages of being easily applied and of being visible while development is taking place (as long as the stained cells remain in the outer surface of the embryo). However, there are several disadvantages to these dyes: (1) There is a slight tendency for color to diffuse from cell to cell; (2) most of the stain is lost during processing for histological sections; and (3) individual stained cells that have migrated away from the main mass are difficult to spot.

All these difficulties have recently been overcome with the advent of strict lineage-linked markers that can be detected unambiguously in single cells in histological sections (Fig. 4). These methods take advantage of the fact that amphibian embryos do not undergo net growth, so that inert markers applied to cells do not become diluted out. The first to be applied to *Xenopus* was the

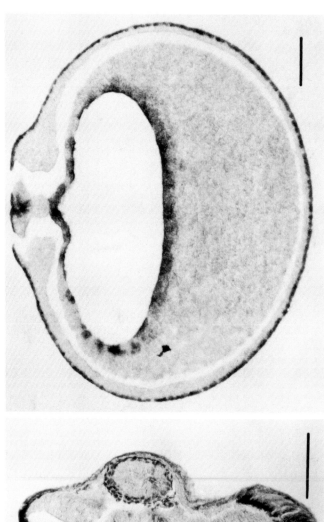

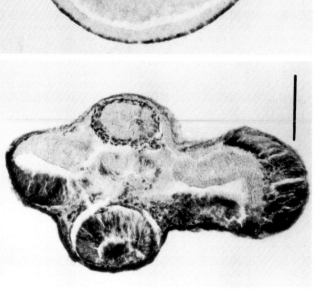

Figure 4. Use of cell-lineage labels to construct fate maps. (**a**) The use of horseradish peroxidase (HRP). One cell of a 2-cell stage embryo was injected with a solution of HRP. The embryo was allowed to develop to the tailbud stage before being fixed and sectioned; peroxidase activity was detected by standard histochemical techniques. This section through the head shows that the left-hand side of the embryo is labeled. (The dark cells at the bottom are within the heavily pigmented sucker.) (**b**) The use of Bolton—Hunter reagent (BHR). Just before the onset of gastrulation, the outer surface of a *Xenopus* embryo was labeled with BHR. The locations of the labeled cells were examined after neural tube closure. A transverse section is shown with the lumen of the neural tube, the epidermis, and the lining of the archenteron derived from the outer surface. Scale bars: 0.2 mm. (From J. C. Smith, 1985.)

enzyme horseradish peroxidase (HRP) (Jacobson and Hirose, 1978). A solution of HRP is microinjected into the cell of interest. HRP is too large to be passed from cell to cell through gap junctions but remains active and is inherited by the progeny of the injected cell. HRP activity can subsequently be detected in frozen sections by standard techniques (Fig. 4a; see Fig. 9). A similar approach is used with the fluorescent lineage tracers fluorescein–dextran–amine (FDA) and rhodamine–dextran–amine (RDA). These are molecules consisting of a fluorochrome linked to dextran linked to lysine (Gimlich and Braun, 1985). The dextran prevents passage of the molecule f   ll   ll

sectioning, labeled cells can be detected by autoradiography. The Bolton–Hunter reagent has a short chemical half-life; it is therefore best used for labeling isolated cells or sheets of tissue (Katz et al., 1982; Smith and Malacinski, 1983) (Fig. 4b).

Another method suited to single cells makes use of the fluorescent dye tetramethyl rhodamine isothiocyanate (TRITC). This has been used to label single isolated blastomeres before transplanting them into intact embryos (Heasman et al., 1984). The cells are labeled simply by washing them in a solution of TRITC; the dye is so strongly absorbed by yolk granules and other protein constituents that it does not pass from cell to cell.

The most complete fate map of *Xenopus* employing one of the new cell markers has been constructed by Dale and Slack (1987a), who used FDA and concentrated on the 32-cell stage. Allowing for epiboly, their results agree well with those of Keller (1975, 1976) (Fig. 5) for the early gastrula stage, although the single-cell resolution possible with FDA emphasizes cell mixing and the statistical nature of the fate map. Other investigators have tended to fate-map

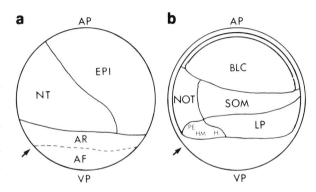

Figure 5. Fate map of the *Xenopus* late blastula–early gastrula. (**a**) Superficial layer of cells. (**b**) Deep layer of cells. AP, animal pole; VP, vegetal pole; BLC, blastocoel; EPI, epidermis; SOM, somite; LP, lateral plate mesoderm; NT, neural tube; NOT, notochord; AR, archenteron roof; AF, archenteron floor; PE, pharyngeal endoderm; HM, head mesenchyme; H, heart. Arrow indicates position of the dorsal lip of the blastopore.

only the stage and region of the embryo they are studying at the time. However, their results supplement and confirm the more complete fate maps and allow one to follow the fates of particular blastomeres from stage to stage. Examples of such fate maps are included in papers by Jacobson and Hirose (1978, 1981), Hirose and Jacobson (1979), Jacobson (1983, 1984), Smith and Slack (1983), Dale *et al.* (1985), Cooke and Webber (1985), Gimlich and Cooke (1983), Gimlich and Gerhart (1984), and Heasman *et al.* (1984).

3.2. Specification

A tissue explant, or cell, is said to be specified to form a particular structure if, after isolation from the embryo and transfer to a neutral environment, that is the structure it forms. In practice the usefulness of this concept depends on the existence of a neutral environment. For cells from mammalian or avian embryos, which frequently require serum supplements for growth and differentiation, a neutral environment is difficult to achieve. The situation is more favorable for explanted amphibian cells, because they can survive on their yolk reserves and divide and differentiate in a simple buffered salts solution (e.g., Holtfreter's solution) whose components are well defined. Holtfreter's solution consists solely of NaCl, 3.5 g/liter; KCl, 0.05 g/liter; $CaCl_2 \cdot 2H_2O$, 0.10 g/liter; and $NaHCO_3$, 0.20 g/liter. Furthermore, there is evidence from axolotl embryos that varying the concentrations of these simple salts does not affect the observed states of specification (Cleine and Slack, 1985). One note of caution to be sounded, however, is that single cells explanted from blastula to gastrula stage embryos usually do not differentiate at all, although they will do so in the company of similar cells. This suggests that cells produce substances that permit the expression of specified states. The alternative explanation, that single cells cannot become specified, is unlikely, because single *Xenopus* cells can become determined (Heasman *et al.*, 1984; see Section 3.3).

The specified states of cells change during development, as implied above; when this occurs, it suggests that the cells have received new developmental information—an inductive signal. For example, if prospective neural tissue is isolated from an embryo at the early gastrula stage and cultured in a simple salts solution it forms epidermis; if the progeny of the same cells are isolated at the late gastrula stage, they form nervous tissue (Spemann, 1938). This suggests that the cells received an inductive signal between these two times.

3.3. Determination

A second type of developmental decision is determination, a measure of the irreversibility of a decision. Tests for determination involve placing specified cells in contact with cells from another part of the embryo and asking whether they can be diverted from their normal fate. Since it is important to distinguish the cells under test from their neighbors, the former are usually identified with one of the cell lineage labels described above. A recent study of determination in amphibian embryos focussed on vegetal pole cells (Heasman *et al.*, 1984). Single vegetal pole cells were isolated from blastula to gastrula stage embryos of *Xenopus*, labeled with TRITC, and transferred to the blastocoels of host embryos. When cells were taken from mid- to late blastula embryos, their progeny populated all three germ layers of the host embryos. However, single vegetal pole cells from early gastrulae contributed progeny only to endoderm—the normal fate of these cells. This experiment shows that vegetal pole cells become determined to form endoderm ~~...~~ ~~...~~

4. Regional Molecular Markers

Analysis of inductive interactions generally involves tissue recombination experiments and the subsequent identification of the different cell types formed. Until recently, it has been necessary to rely on conventional histological techniques. However, markers are now being developed that permit unequivocal identification of cell types, even if the cells in question are in explants or surrounded by an inappropriate tissue. In addition, it is frequently possible to obtain quantitative or at least semiquantitative assessment of the results (Cooke and Smith, 1987).

Almost all the markers in current use identify particular differentiated cell types such as muscle, notochord, or epidermis, many of which are listed in Table I. Work is in progress in many laboratories to discover molecular markers of the determined states that arise before overt differentiation (see Slack *et al.*, 1985; Carrasco and Malacinski, 1987; Kintner and Melton, 1987).

5. Origins of Polarity in the Fertilized Egg

The sequence of inductive interactions that turns the fertilized egg into an embryo is described in this section. For these interactions to occur, there must already exist in the egg regional differences or cytoplasmic localizations that determine whether particular cells in the blastula are going to produce or respond to an inductive signal. The first set of localizations discussed concerns the animal–vegetal axis, which is already established when the egg is laid. We shall then go on to look at the dorsoventral axis.

Table I. Cell Type-Specific Markers Used in Studies on Induction in *Xenopus*

Marker designation	Type[a]	Specificity	Reference
Ectoderm			
PNA	Lectin	Epidermis	Slack (1985)
Antikeratin	pAB	Epidermis	Dale *et al.* (1985)
2F7.C7	mAB	Epidermis	Jones (1985)
Epi 1	mAB	Epidermis	Akers *et al.* (1986)
DG 81	cDNA	Epidermis	Sargent *et al.* (1986)
A4	cDNA	Epidermis	Dworkin-Rastl *et al.* (1986)
4H8	mAB	Neural plate[b]	Slack *et al.* (1985)
N-CAM	pAB	Neural plate	Balak *et al.* (1987)
N-CAM	cDNA	Neural plate	Kintner and Melton (1987)
D8	cDNA	Nervous system	Dworkin-Rastl *et al.* (1986)
Mesoderm			
MZ15	mAB	Notochord	Smith and Watt (1985)
MHC 2	pAB	Muscle	Dale *et al.* (1985)
12/101	mAB	Muscle	Kintner and Brockes (1984)
M2	cDNA	Muscle	Mohun *et al.* (1984)
α-actin	cDNA	Muscle	Sargent *et al.* (1986)
A2,B5,B9,D1,H1	cDNA	Muscle	Dworkin-Rastl *et al.* (1986)
C6	mAB	Mesonephros	Slack *et al.* (1985)
Antiglobin	pAB	Blood	Cooke and Smith (1987)
Endoderm			
AH6	mAB	Gut lumen	Slack *et al.* (1985)
VC1	mAB	Germ cells, Liver, Gut lining	Wylie *et al.* (1985)
DG42	cDNA	Endoderm	Sargent *et al.* (1986)

[a]pAB, polyclonal antibody; mAB, monoclonal antibody.
[b]This antibody is specific for a few cells in the axolotl neural plate.

5.1. Animal–Vegetal Axis

During oogenesis, the egg develops an animal–vegetal polarity. This polarity is reflected in the pigment pattern of the egg: The animal hemisphere is heavily pigmented, whereas the vegetal hemisphere is very pale. The egg cytoplasm studied in histological sections also reflects this polarity, particularly in the distribution of yolk platelets. Yolk proteins constitute at least 50% of the total egg protein, and the platelets range in size from 2 to 15 μm. The vegetal hemisphere contains the larger platelets from 10 to 15 μm, whereas the platelets in the animal hemisphere are 2–4 μm. Between these two regions there is a transitional zone where platelets are 3–8 μm. This graded distribution of yolk platelets is maintained in unfertilized eggs irrespective of their orientation, although after fertilization they are able to shift such that the larger ones face gravity. However, after fertilization, the eggs themselves rotate so that the vegetal hemisphere points downward. Thus, the yolk platelets maintain their vegetal position.

There is no evidence that yolk proteins themselves are cytoplasmic determinants, but the distribution of platelets does serve to illustrate that cytoplasmic constituents can be localized according to the egg axis. This conclusion was confirmed and extended by the elegant work of Moen and Namenwirth (1977), who developed a technique for sectioning hundreds of fertilized *Xenopus* eggs along the animal–vegetal axis before extracting the proteins and analyzing them by polyacrylamide gel electrophoresis.

...maternal mRNA is widely regarded as a likely class of molecule to serve as a cytoplasmic determinant. This has been shown, for example, to be required for expression of the dorsoventral axis in *Drosophila* (Anderson and Nüsslein-Volhard, 1984). Initial studies with *Xenopus* demonstrated that concentrations of both total and poly(A)$^+$ RNA vary along the animal–vegetal axis of oocytes and eggs (Capco, 1982; Capco and Jäckle, 1982; Capco and Jeffrey, 1982; Phillips, 1982). Carpenter and Klein (1982) then found, using Moen and Namenwirth's technique together with cDNA hybridization analysis, that a small number of maternal poly(A)$^+$ RNAs were enriched 2–20-fold in the vegetal third of the *Xenopus* egg and embryo. These workers did not investigate the animal pole, but King and Barklis (1985) included this region in their study, in which they isolated mRNA from different regions of the egg before translating it *in vitro* and analyzing the products by two-dimensional gel electrophoresis. Seventeen mRNAs showing significant qualitative or quantitative regional differences were found, of which 10 were concentrated at the vegetal pole. These results contrast with those of Moen and Namenwirth (1977), who found protein heterogeneity primarily in the animal region.

A different approach to studying the spatial localization of specific maternal mRNAs was taken by Rebagliati *et al.* (1985), who screened an oocyte cDNA library with [^{32}P]-cDNA made from poly(A)$^+$RNA obtained from either the vegetal or the animal region of unfertilized eggs. Most of the clones were equally represented in the animal and vegetal halves, but one vegetal-specific and three animal-specific clones were isolated and shown to remain localized during early development.

It is tempting to speculate that some of the differentially localized proteins or RNA molecules that have been identified are indeed cytoplasmic determinants. Future experiments will no doubt address this point by microinjecting mRNA or protein into particular blastomeres. For present purposes, it is sufficient to make the point that the molecules one would expect to act as cytoplasmic determinants do indeed show spatial heterogeneity.

5.2. Dorsoventral Axis

Before fertilization, histological sections of eggs show a heterogeneous distribution of cytoplasmic components in the animal–vegetal axis, but there is no regional organization to predict which side will form head and dorsal structures and which posterior and ventral: the egg is radially symmetric. The dorsoventral axis is only defined at fertilization, with the side of the egg on which the sperm enters forming posterior and ventral structures. During the 90 min between fertilization and first cleavage, visible markers of polarity appear on the egg surface. The first of these is the sperm entry point, (see Section 2.1). The second visible marker is the so-called **gray crescent,** which appears about halfway through the interval from fertilization to first cleavage on the prospective dorsal side of the embryo (Fig. 1b). In fact, the gray crescent is less clear in *Xenopus* than it is in the eggs of other species, but the displacement of subcortical cytoplasm relative to the egg surface, which causes gray crescent formation, occurs in *Xenopus* in the same way that it does in other species, like *Rana* (Vincent *et al.*, 1986).

These cytoplasmic shifts can be observed in histological sections, where the most obvious events include the displacement of a central region of yolk-free cytoplasm away from the sperm entry side and the movement of yolk granules such that the initial radially symmetric configuration becomes bilaterally symmetric (Gerhart, 1980; Ubbels *et al.*, 1983). However, the clearest study of the movements has been made by Vincent *et al.* (1986), who used Nile blue chloride to mark the subcortical yolk platelets and fluorescent potato lectin to stain the egg surface. The relative contributions of these two layers to gray crescent formation were studied by immobilizing the outer egg surface in gelatin. In immobilized eggs, the outer surface of the egg remains rigid and undeformed, while the entire subcortical cytoplasm rotates about 30°, with that in the animal hemisphere moving toward the prospective dorsal side and that in the vegetal hemisphere moving away from it. Superimposed on this rotation is a convergence of yolk platelets in the animal hemisphere toward the sperm entry point. In freely orientating eggs, there is no apparent movement of the vegetal subcortical cytoplasm, because the dense yolk mass remains oriented in gravity; thus, if the egg surface is marked with potato lectin, it can be seen to rotate such that the vegetal region moves *toward* the prospective dorsal side.

In *Xenopus* the pigment granules of the animal hemisphere are located in the subcortical layer marked by Nile blue, so that rotation does not alter the pigment pattern of freely orientating eggs. By contrast, in *Rana* pigment is present in both subcortical and cortical layers, but with most in the latter. Rotation of the surface thus exposes a region of lightly pigmented subcortical cytoplasm, which becomes visible as the gray crescent.

It is the rotation of subcortical cytoplasm rather than convergence toward the SEP that specifies the dorsoventral axis. One argument in favor of this interpretation is that the future dorsoventral axis of the embryo is much better predicted by the rotation movements than by the position of SEP. Further evidence is that a variety of treatments that inhibit the rotation movements, including ultraviolet (UV) light irradiation of the vegetal surface and cold and pressure shocks, cause the loss of dorsal anterior structures (Scharf and Gerhart, 1983) (Fig. 6), whereas a brief period of oblique orientation, in which the normal vertical axis of the egg is held horizontally, completely rescues the

We can now discuss the three inductive interactions that turn the fertilized amphibian egg into a recognizable vertebrate embryo. The first of these interactions acts along the first axis to be defined, the animal–vegetal axis and results in the formation of an equatorial mesodermal rudiment under the influence of a signal from the vegetal hemisphere. The evidence that such an interaction occurs comes from a comparison of the normal fate and the state of specification of cells in the animal pole at the 8-cell stage and the observation that a change in that state of specification occurs as development proceeds, giving a clue as to when induction takes place. Similar arguments will be put forward to illustrate the existence of the remaining two interactions.

6.1. Fate and Specification Maps of the Mesoderm

Injections of cell lineage markers into the blastomeres of the 8-cell embryo show that the four animal pole cells form virtually all the ectoderm and much of the mesoderm of the tailbud stage (Cooke and Webber, 1985; Dale and Slack, 1987a) (Fig. 7). However, isolated animal pole quartets from embryos that show a well-defined horizontal cleavage form little or no mesoderm, whether assayed histologically or biochemically using a muscle-specific actin probe (Kageura and Yamana, 1983; Gurdon et al., 1985b). Instead, virtually all the cells form epidermis. When regions of prospective dorsal mesoderm are isolated at successively later stages, however, notochord and muscle do differentiate; the results suggest that mesoderm specification occurs around the 128-cell stage (Nakamura et al., 1970) (but see Section 10.1). Thus, the early *Xenopus* embryo can be considered to consist of two cell types: presumptive ectoderm in the animal hemisphere and presumptive endoderm in the vegetal hemisphere (Jones and Woodland, 1986). By the early blastula stage, a third cell type—

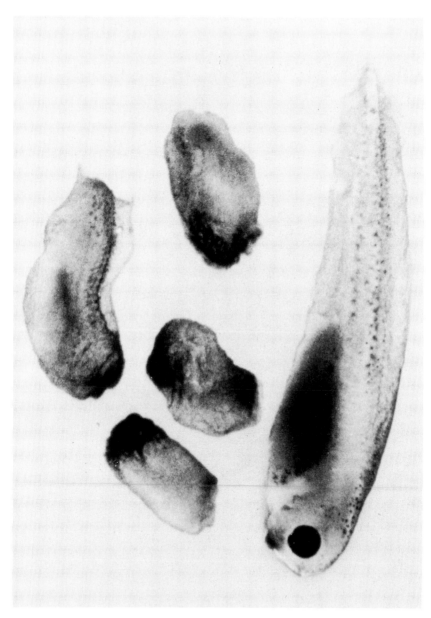

Figure 6. One normal *Xenopus* embryo and three radially symmetric posterior–ventral embryos produced by UV irradiation of the vegetal hemisphere. The embryo at the bottom is only partially ventralized. (Kindly provided by Dr. Jonathan Cooke.)

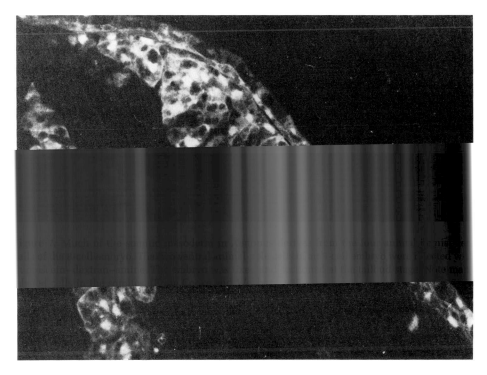

Figure 7. Much of the somitic mesoderm in *Xenopus* derives from the four animal hemisphere cells of the 8-cell embryo. The two ventral animal pole cells of an 8-cell embryo were injected with fluorescein–dextran–amine. The embryo was fixed and sectioned at the tailbud stage. Note many labeled cells in the somites. (From Cooke and Webber, 1985.)

mesoderm—has arisen, and it is possible to show that this is formed as the result of an inductive interaction.

6.2. Mesoderm Formation through an Inductive Interaction

Mesoderm induction was first demonstrated by Nieuwkoop (1969, 1973) in urodele embryos. His experimental design is shown in Fig. 8. Explants of blastula animal pole tissue form epidermis in isolation, but when they are placed adjacent to cells from the vegetal hemisphere, large amounts of mesodermal cell types are formed, which, according to cell marking experiments, are derived from the animal pole component (Nieuwkoop and Ubbels, 1972). Sudarwati and Nieuwkoop (1971) showed that similar experiments could be conducted on *Xenopus*.

A recent study on mesoderm induction in *Xenopus* has made use of cell lineage labels and cell-type specific markers to demonstrate unequivocally that animal pole cells that normally make ectoderm can be converted to mesoderm (Dale *et al.*, 1985). This shows that mesoderm induction is a genuine instruc-

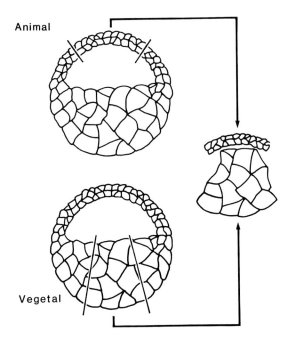

Figure 8. An *in vitro* combination of animal and vegetal pole regions.

tive interaction and not simply a permissive phenomenon permitting selective differentiation of precommitted mesodermal cells.

The normal fate of the animal pole region under study was established by orthotopic grafts between embryos uniformly labeled with a lineage marker and unlabeled embryos (Fig. 9). After being grafted at the early gastrula stage, all the cells from the animal pole region were found in ectodermal derivatives, mostly epidermis (Fig. 9), but also some head mesenchyme. After grafting at the early and mid-blastula stages, most labeled cells were again found in ecto-dermal derivatives, but they also contributed to mesoderm, including notochord, somite, and lateral plate. However, the contribution to mesoderm was signifi-cantly less than that produced by the same regions after mesoderm induction (Fig. 10a,b). For example, at the mid-blastula stage, isolated animal pole regions formed no mesoderm; in normal development, 3% of the tissue gave rise to mesoderm (presumably due to *in vivo* induction), but in animal–vegetal com-binations, in which the animal pole components were labeled with FDA, more than 40% of the tissue formed mesoderm. Identification of mesoderm in this study employed antibodies specific for notochord and muscle to aid identifica-tion of the different cell types.

The results of these animal–vegetal combination experiments depended on the stage of the embryos. In combinations made at the early blastula stage, the animal pole components always formed muscle and about one-half the cases formed notochord. A few contained erythrocytes, mesothelium, and pro-nephros. At the mid-blastula stage, virtually all the mesoderm was present as muscle, but mesoderm induction was poor at the early gastrula stage, with only

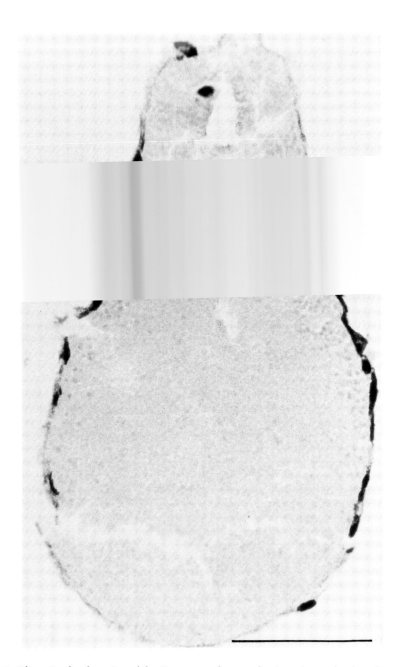

Figure 9. The animal pole region of the *Xenopus* early gastrula gives rise predominantly to epidermis. The animal pole region of a *Xenopus* early gastrula embryo, labeled uniformly with horseradish peroxidase (HRP), was grafted orthotopically to an unlabeled host embryo. The embryo was allowed to develop to the tailbud stage before being fixed and sectioned, and peroxidase activity was detected by standard techniques. Transverse section shows labeled cells in the epidermis and one labeled cell in the neural tube. Scale bar: 0.2 mm.

a few positive cases, each producing small amounts of mesenchyme. The lack of mesoderm induction in these latter cases was found to be due to a loss of the ability of the animal pole component to respond to induction; early blastula animal pole tissue was induced to form mesoderm by gastrula vegetal cells, but the inverse combination gave very few positive results. The loss of responsiveness of the animal pole to mesoderm induction (i.e., loss of **competence**) is followed by the acquisition of competence to respond to a new signal: neural induction.

Finally, it should be pointed out that it is not possible to prove formally that all the mesoderm in the amphibian embryo arises through induction. Indeed, Gurdon *et al.* (1985*a,b*) suggest that some mesoderm, perhaps that derived from the four vegetal pole cells at the 8-cell stage, forms as a result of cytoplasmic localization. The evidence for this rests on egg-ligation experiments, which demonstrate that all the cytoplasmic components required for the eventual activation of muscle-specific actin genes are localized in a region below the equator and predominantly on the gray crescent side. These experiments may indeed indicate the localization of muscle determinants, although another interpretation is that the region mapped represents the border between two regions of cytoplasm, one capable of producing a mesoderm-inducing signal and one capable of responding (see Section 10.1).

6.3. The Regional Character of Mesoderm Induction

Vegetal pole regions of early blastula *Xenopus* embryos usually induce muscle and notochord from animal pole regions, although in a few cases, erythrocytes, mesothelium, and pronephros are formed (Dale *et al.*, 1985). In other words, the typical result is for the animal pole component to form dorsal structures and only occasionally to form ventral cell types. In the axolotl, Broterenbrood and Nieuwkoop (1973) showed that ventral cell types were formed with high frequency when ventral vegetal cells were used as the inducers. In further experiments, Dale *et al.* (1985) found that the same is true in *Xenopus*. Thus, whereas dorsal vegetal cells almost invariably induce muscle and notochord from animal pole regions, ventral vegetal regions induce mesothelium and erythrocytes (Fig. 10).

This result suggests that there are at least two distinct regions in the vegetal hemisphere: one that induces ventral mesodermal structures and another that induces dorsal structures. It is interesting to consider this conclusion within the context of the experiments mentioned in Section 5.2, in which administration of low doses of UV light to the vegetal hemisphere of fertilized eggs results in the formation of ventralized embryos (Fig. 6). These embryos lack dorsal structures but make up for this deficiency by developing an excess of ventral mesodermal structures, particularly red blood cells (RBCs) (Cooke and Smith, 1987). It is as if the dorsal vegetal pole cells had lost the ability to induce dorsal structures, a conclusion elegantly confirmed by Gimlich and Gerhart (1984), who replaced what should have been the dorsal vegetal pole cells of a UV-irradiated embryo with those from a normal embryo (Fig. 11). This resulted in

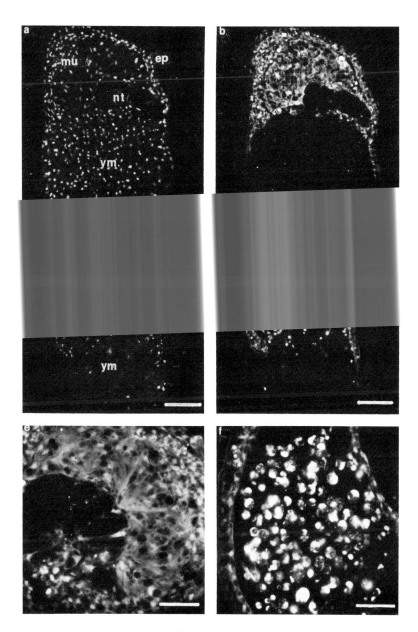

Figure 10. Mesoderm induction by dorsal and ventral regions of the vegetal hemisphere. In both cases, the animal pole component was labeled with fluorescein–dextran–amine (FDA). (**a,b**) An animal pole–dorsal vegetal pole combination allowed to develop for 3 days; (**a**) section is stained with 4, 6-diamidino-2-phenyl indole (DAPI) to show nuclei; (**b**) is the same section viewed through fluorescein filters. Epidermis (ep), muscle (mu), and notochord (nt) are derived from animal pole tissue. The yolk mass (ym) is not labeled with FDA and is therefore derived from the vegetal pole component. (**c,d**) An animal pole–ventral vegetal pole combination. (**c**) is stained by DAPI fluorescence; (**d**) is stained by fluorescein fluorescence. Epidermis (ep), blood cells (bc), and mesothelium (mt) are derived from the animal pole component, whereas the yolk mass is unlabeled and therefore originates from the vegetal pole. (**e**) Higher-power view of muscle and notochord in an animal pole–dorsal vegetal pole combination. (**f**) Higher-power view of labeled blood cells in an animal pole–ventral vegetal pole combination. Scale bars in (**a–d**): 100 μm. Scale bars in (**e**) and (**f**): 50 μm. (From Dale *et al.*, 1985.)

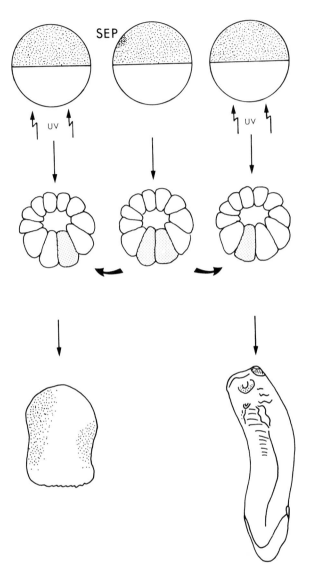

Figure 11. Rescue of UV-irradiated embryos by transplantation of dorsal–vegetal, but not ventral–vegetal, blastomeres. SEP, sperm entry point. The SEP predicts the future ventral half of the embryo.

the rescue of the UV syndrome, the resulting embryos frequently forming a dorsal axis. This rescue was clearly due to dorsal mesoderm induction, and not to self-differentiation of the graft, because when dorsal vegetal blastomeres labeled with FDA were grafted, they only contributed to endoderm.

7. Dorsalization

Dorsal vegetal pole cells induce dorsal mesoderm from overlying animal pole cells. It seems that further regionalization of the mesodermal rudiment results from an interaction between the dorsal organizer region and adjacent ventral mesoderm.

7.1. Fate and Specification Maps of the Dorsoventral Axis of the Mesoderm

After *in vitro* combination, ventral vegetal pole blastomeres are capable of

form only blood and mesenchyme, whereas at later stages they should form large amounts of muscle. Unfortunately, the situation is not so simple.

The earliest stage at which the *Xenopus* embryo can be divided into dorsal and ventral halves is the 4-cell stage. Both Kageura and Yamana (1983) and Cooke and Webber (1985) performed separations at this stage; both groups obtained similar results. The dorsal halves usually formed an embryo with a large head and small tail. The ventral halves, however, gave rise to a range of structures that at one extreme resembled radially symmetrical muscleless UV embryos (see Fig. 6) and at the other resembled what that ventral half would have made in normal development (Cooke and Webber, 1985). Thus, it may be that the state of specification of the ventral half of the embryo varies from individual to individual, although this seems unlikely in view of the consistent results obtained in induction experiments with ventral vegetal pole regions (Dale *et al.*, 1985). Another possibility is that the variability is due to the physical act of separating the two halves of the embryo. This could either cause some ventral blastomeres to lose muscle-forming potential or alternatively bring about partial dorsalization of the ventral half. Such a dorsalization could occur through shifts in the subcortical cytoplasm relative to the cell surface, as the shape of the ventral blastomeres changed from a half-sphere to a smaller complete sphere (see Section 5.2).

My inclination is to regard the latter view as correct. This is because ventral halves isolated at the 4-cell stage give rise to radially symmetric structures in 29% of cases (combined results of Kageura and Yamana, 1983; Cooke and Webber, 1985), whereas those isolated at the 8-cell stage do so 49% of the time (Kageura and Yamana, 1984). It seems most unlikely that the ventral half of the embryo would lose developmental information during development but quite probable that it would become more resistant to artificial dorsalization (Gerhart *et al.*, 1981). Consistent with this is the observation that, at the 32-cell stage, cells isolated from the ventral half of the embryo form significantly less muscle than cells from the dorsal half—the reverse of their normal fates (Nakamura, 1978). Thus, it seems that at early stages the ventral half of the embryo does not contain the information to form somitic muscle.

This view is reinforced by recent experiments of Dale and Slack (1987*b*), who made ventral-half isolates from embryos at the 4-cell stage to the beginning of gastrulation. Most of these differentiated as extreme ventral mesoderm, consisting of blood, mesenchyme, and mesothelium, but some, particularly those made at the 4-cell stage, contained highly organized somite tissue. Dale and Slack presented direct evidence that these types of differentiation, intermediate between extreme dorsal and extreme ventral, arise because ventrovegetal blastomeres can acquire an intermediate-type inductive potency as a result of operative procedures.

7.2. Dorsalization

If the instruction to the ventral half of the embryo to form muscle is not derived from the ventral vegetal pole cells, where does it come from? Experiments by Slack and Forman (1980), Smith and Slack (1983), Smith *et al.* (1985), and Dale and Slack (1987*b*) suggest that it is from the dorsal marginal zone, or **organizer.** Small wedges about 30° wide taken from *Xenopus* early gastrula ventral marginal zones (VMZs) differentiate in isolation as mesenchyme, mesothelium, and RBCs, surrounded by epidermis. Isolated wedges of dorsal marginal zone (DMZ) form predominantly notochord, with some muscle and neural tissue. When FDA-labeled VMZs are juxtaposed with unlabeled DMZs, however, the ventral tissue forms large masses of muscle, while the dorsal tissue still differentiates as notochord (Smith *et al.*, 1985; Dale and Slack, 1987*b*) (Fig. 12). Similar results have been obtained using interspecific grafts of *Xenopus* VMZ with axolotl DMZ (Slack and Forman, 1980) and by grafting HRP-labeled VMZs into the dorsal midline of unlabeled embryos (Smith and Slack, 1983). These results suggest that the further subdivision of the mesoderm into different cell types depends upon an interaction within the mesodermal rudiment itself, in which the most dorsal prospective mesoderm can dorsalize adjacent ventral mesoderm. It is noteworthy that, unlike mesoderm induction, dorsalization can be demonstrated at the early gastrula stage and perhaps even later than this. Dorsalization should therefore be considered as an entirely separate interaction from mesoderm induction.

It has been argued (see Smith *et al.*, 1985) that dorsalization is the main inductive interaction at work in the organizer graft of Spemann and Mangold (1924). In the organizer graft, dorsal marginal zone tissue from one embryo is transplanted to the ventral marginal zone of another. This results in the formation of a double-dorsal mirror-symmetric embryo (Fig. 13a), in which only the notochord, and a few cells in the somites, are derived from the graft (Fig. 13b,c). It seems likely that the somite tissue of the secondary embryo arises through dorsalization of adjacent ventral marginal zone cells, although it is not possible to rule out the possibility that some of it derives from the migration of cells determined to be somite in the host axis (J. Cooke and A. Smallcombe, personal

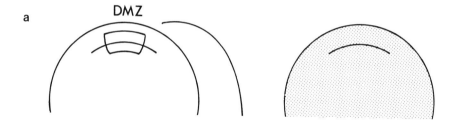

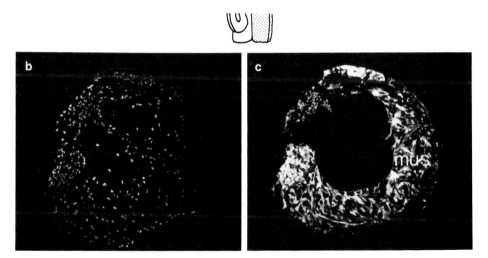

Figure 12. Dorsalization. Isolated ventral marginal zone (VMZ) tissue normally forms red blood cells, mesenchyme, and mesothelium, while dorsal marginal zone (DMZ) cells form notochord and muscle. (**a**) FDA-labeled VMZ tissue is placed next to an unlabeled DMZ explant. The combination was allowed to develop for three days and was then fixed and sectioned. (**b**) Section stained with DAPI to show nuclei. (**c**) The same section viewed through fluorescein filters. The FDA-labeled ventral marginal zone differentiates as muscle (mus). (**b** and **c** kindly provided by Dr. Les Dale.)

communication). The organizer graft also reveals the third inductive interaction involved in the formation of the body plan: neural induction.

8. Neural Induction

The neural tubes of the secondary embryos formed in response to the organizer graft are derived from host tissue (Fig. 13c). This is perhaps the most dramatic demonstration of neural induction, although the phenomenon can also be illustrated by comparing the states of specification and determination of the prospective nervous system at different stages, as already described for mesoderm induction and dorsalization.

The original experiments studying fate, specification, and determination in the nervous system were carried out by Spemann in 1921, using newt embryos (see Spemann, 1938; see also Chapter 4). Spemann exchanged prospective epidermis and prospective neural plate between two embryos at the early gastrula stage and found that each graft behaved according to its new position: Prospective epidermis formed brain, and prospective nervous tissue formed skin. If this experiment was performed at the late gastrula stage, however, a different result was obtained, the grafts behaving according to their origin rather than to their new position. This suggested that the determination, and perhaps specification, of epidermis and neural tissue occur during gastrulation. Furthermore, the result of the organizer graft suggested that a signal inducing neural tube formation is derived from the dorsal marginal zone and that, in the absence of such a signal, gastrula ectoderm would form epidermis. This latter conclusion was confirmed by Holtfreter (1933), who maintained axolotl blastulae in a hypertonic salt solution and found that the presumptive mesoderm and endoderm rolled away from the ectoderm rather than involuting beneath it. This type of movement is called **exogastrulation;** as a result, in the absence of involuting mesoderm, the entire ectoderm developed as epidermis, with no neural differentiation (see Chapter 4). Experiments similar to these have now been carried out in *Xenopus laevis* using cell lineage labels and region-specific markers, and the general conclusions have been confirmed.

8.1. Fate and Specification Maps of Prospective Neural Tissue

Accurate fate maps for the nervous system of *Xenopus* morulae and blastulae have been produced by Jacobson, using HRP injection into individual blastomeres (Jacobson and Hirose, 1978, 1981; Hirose and Jacobson, 1979; Jacobson, 1983). Allowing for epiboly of the animal hemisphere, these maps are similar to, but much more detailed than, the fate map of Keller (1975) for the early gastrula stage, when the prospective nervous system is on the dorsal side of the embryo, extending some way above the equator, about half way to the animal pole (see Fig. 5).

Experiments studying the state of specification of the prospective nervous system have been carried out by Akers *et al.* (1986). Molecular markers for the neural plate were not yet available, so like Slack (1985), these investigators relied on the lack of expression of an epidermal marker to indicate neural differentiation. As would be predicted from the work of Spemann, isolated prospective epidermis invariably expressed the Epi 1 marker, irrespective of the stage at which it was removed from the embryo. Presumptive neural tissue, however, only expressed the epidermal marker if it was isolated before the midgastrula stage; if isolated at later stages this tissue did not express the Epi 1

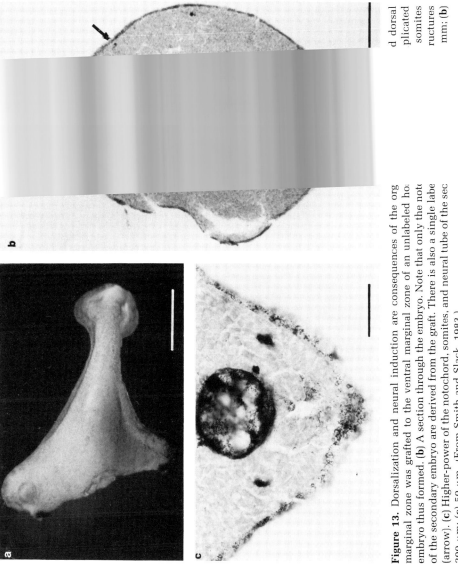

Figure 13. Dorsalization and neural induction are consequences of the org marginal zone was grafted to the ventral marginal zone of an unlabeled ho. embryo thus formed. (**b**) A section through the embryo. Note that only the nott of the secondary embryo are derived from the graft. There is also a single labe (arrow). (**c**) Higher-power of the notochord, somites, and neural tube of the sec 200 μm; (**c**) 50 μm. (From Smith and Slack, 1983.)

d dorsal
plicated
somites
ructures
mm; (**b**)

antigen. Thus, as Spemann himself concluded, neural specification occurs during late gastrulation.

8.2. Neural Induction

Neural induction can be demonstrated in two ways: by *in vitro* combinations of early gastrula ectoderm with the dorsal marginal zone or by the organizer graft.

Combinations of presumptive epidermis with the DMZ have been carried out by Slack (1985) and Akers *et al.* (1986). As an epidermal cell-type marker, Slack used fluorescein isothiocyanate-conjugated peanut lectin (FITC-PNA), whereas Akers *et al.* used their Epi 1 antibody. Slack's experiments furthermore combined the cell-type specific marker with the lineage marker RDA, to distinguish the animal pole component of the combination from the dorsal marginal zone. Both experiments showed clearly that expression of epidermal markers in prospective epidermis is supressed by the presence of the marginal zone, implying that neural induction has occurred.

Cell lineage labels and region-specific markers have also been employed to study the organizer graft. Figure 13 shows that the additional neural tube resulting from an organizer graft is derived from host tissue, consistent with it being derived by induction; similar results have been obtained by Jacobson (1984) and Recanzone and Harris (1985) and were obtained in the original paper by Spemann and Mangold (1924). However, as Jacobson (1982) has pointed out, two objections remained to be satisfied before it could be concluded that neural induction was actually occurring. It had to be shown that (1) the cells in the secondary neural tube were cells that normally would have formed ventral epidermis and that they had not migrated from the primary neural tube, and (2) the cells in the secondary neural tube were genuine neuroepithelial cells and not epidermal cells rolled into a cylinder.

Experiments by Gimlich and Cooke (1983), Jacobson (1984), and Slack *et al.* (1984) have established beyond doubt that the cells in the secondary neural tube would otherwise have formed epidermis. These workers injected cell lineage markers (FDA or HRP) into ventral cells of early *Xenopus* embryos that normally contribute almost exclusively to epidermis and never contribute to neural structures. Then, when these embryos reached the early gastrula stage, unlabeled dorsal marginal zones were grafted into their ventral marginal zones. The secondary neural tubes that resulted were composed largely of labeled cells (Fig. 14); they clearly came from cells that should have made epidermis and not from cells of the primary neural tube that had migrated to the ventral side.

The question of whether the secondary neural tube is composed of genuine neuroepithelial cells was approached by Slack (1985), who again used lack of expression of the FITC-PNA receptor as an indicator of neural differentiation. The same cells injected by Gimlich and Cooke (1983) and Jacobson (1984) were injected with RDA; at the beginning of gastrulation, an unlabeled dorsal mar-

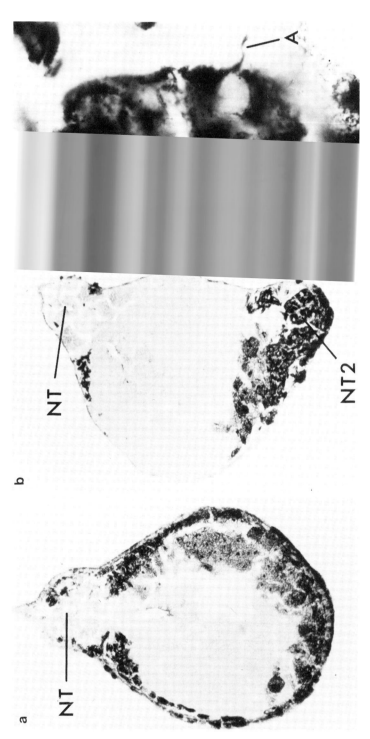

Figure 14. Demonstration of neural induction. The ventral vegetal pair of blastomeres at the 1 of these cells. They contribute to epidermis, lateral mesoderm, and endoderm but not to the ne zone is grafted to the ventral marginal zone of such an embryo at the early gastrula stage, a dc neural tube (NT2) is labeled with HRP. It must therefore be derived from cells that normally t High-power view of the secondary neural tube showing an axon (A) growing into the surro n HRP. (**a**) The normal fate unlabeled dorsal marginal d, in which the secondary central nervous system. (**c**) et al., 1984.)

ginal zone was grafted to the VMZ of these host embryos. Again, the secondary neural tubes contained labeled cells, originally destined to be epidermis; however, they did not stain with FITC-PNA, indicating that they had been diverted from epidermal differentiation, presumably toward neuroepithelium.

This conclusion has recently been confirmed using a new neuroepithelial marker, the neural-cell adhesion molecule (N-CAM). N-CAM can be detected in histological sections using a polyclonal antibody (Balak *et al.*, 1987). In the early neurula the protein is present in the neural plate, notochord, and prospective somite mesoderm, but the mesodermal staining disappears by the time of somite formation, leaving only the neural tube positive. In support of the results of Slack (1985) and Akers *et al.* (1986), Jacobson and Rutishauser (1986) have found that neither animal pole ectoderm nor dorsal marginal zone tissue express N-CAM when cultured separately, whereas combinations of the two regions do. In addition, they showed that the secondary neural tube formed after an organizer graft stains strongly with the anti-N-CAM antibody. Thus, neural induction results in the specification of genuine neuroepithelial cells.

A different approach has been adopted by Kintner and Melton (1987), who have isolated a cDNA clone for *Xenopus* N-CAM. They detected a marked increase in N-CAM RNA during gastrulation, and *in situ* hybridization shows that this is localized exclusively to the neural plate and is not expressed by surrounding epidermis. N-CAM RNA expression does not occur in isolated animal pole ectoderm unless it is cultured in contact with vegetal pole cells, when a portion of the ectoderm is induced to form dorsal mesoderm, which, in turn, induces neural tissue. Interestingly, N-CAM RNA is quite strongly expressed in exogastrulae, but this is primarily at the junction of ectoderm and mesoderm, with the possibility that some spread of neural induction into deep ectoderm is occurring by homeogenetic induction (see below).

Like mesoderm induction, neural induction includes the transmission of regional information. This was first suggested by the results of Spemann (1938), who found that the dorsal lip of the blastopore from an early gastrula would induce head and trunk structures, with the appropriate nervous tissue, whereas the dorsal lip from a late gastrula would induce only trunk and tail. This implied that the craniocaudal character of involuting dorsal mesoderm is acquired during gastrulation and is then transmitted to the overlying neural plate. One might imagine a process in which the longer it is before cells involute, the more caudal their state of specification would become (Slack, 1983; Nieuwkoop *et al.*, 1985). Thus, the first mesoderm cells to involute would acquire an anterior state of specification. This state would be transmitted to all the cells of the overlying dorsal ectoderm during gastrulation but would be superseded, except in the most anterior regions, by a slightly more posterior state due to a signal from the cells involuting immediately behind the first group. By the end of gastrulation, therefore, the craniocaudal specification of the neuroectoderm would coincide with that of the underlying mesoderm (Fig. 15). This conclusion is reinforced by the experiment of Mangold (1933), who implanted regions of the archenteron roof of the early *Triturus* neurula into

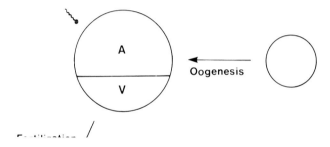

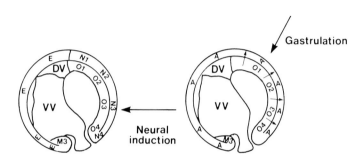

Figure 15. Sequence of cytoplasmic localizations and inductive interactions during early amphibian development. During oogenesis, differences arise between the animal (A) and vegetal (V) halves of the egg. Fertilization results in a subdivision of the vegetal half into dorsovegetal (DV) and ventral vegetal (VV) regions. During mesoderm induction, the dorsovegetal region induces the organizer (O) from the animal half of the embryo, and the ventral region induces a general ventral mesoderm (M), which is subdivided into several regions during dorsalization, under the influence of the organizer. During gastrulation, the organizer acquires craniocaudal positional values (01 to 04), and these are transmitted to the overlying ectoderm as part of the neural induction process. This produces neuroepithelium with homologous craniocaudal positional values (N1 to N4), while the uninduced animal pole material becomes epidermis (E). (From J. C. Smith, 1985, with permission.)

early gastrulae and found that the induced nervous systems corresponded to the original craniocaudal position of the grafted dorsal mesoderm.

Finally, it can be shown that homeogenetic induction occurs in the neural plate; i.e., that newly induced neuroepithelium is itself a neural inductor. This was first demonstrated by Mangold and Spemann (1927) and confirmed and elaborated upon by Nieuwkoop and colleagues (1952a–c). Vertical folds of non-neural gastrula ectoderm were grafted to a medial position in the neural plate of host embryos. The basal regions of the folds differentiated as nervous system, forming structures similar to those in the surrounding tissue. However, neural structures also differentiated toward the apices of the folds, over a range similar to the lateral extent of the neural plate at the craniocaudal position in question. Furthermore, the structures formed within this activated zone were always more anterior than those in the basal region. Thus, within the large activated zone obtained by implanting a fold in the wide anterior neural plate, di- and telencephalic structures might be formed. However, within the smaller activated zone resulting from a graft to the spinal cord level, a whole brain could result.

These results, and others like them, have led to the introduction of various double-gradient models to account for neural induction, which may be studied elsewhere (Saxen and Toivonen, 1962; Yamada, 1958; Slack, 1983; Nieuwkoop et al., 1985). For the present purposes, it is sufficient to note first that homeogenetic induction does occur between the neural plate and surrounding ectoderm and second that this does not result in the entire ectodermal surface becoming neuralized, probably because the spread of neuralization is limited by a loss in competence of the ectoderm (Nieuwkoop et al., 1985).

9. Summary of Inductive Interactions in Early Amphibian Development

The three inductive interactions responsible for turning the fertilized egg into a vertebrate embryo (mesoderm induction, dorsalization, and neural induction) are summarized in Fig. 15. Although these interactions can only be demonstrated by abnormal juxtaposition of different regions of an embryo, there is good evidence from the changes in states of specification we can observe that they do occur in normal development.

After neural induction, the basic vertebrate body plan is complete, and later development builds on this framework through further interactions to form other specified states and cell types. It is not possible to go into these interactions here, but a detailed description may be found in Nieuwkoop et al. (1985).

10. Molecular Mechanisms of Induction

The biochemical mechanism of induction has been studied since Spemann and Mangold's discovery of the organizer in 1924, but progress in this field has been lamentably slow compared with the flood of information in, say, immunology and molecular biology. This failure is due in part to the small number of scientists working on the problem, but it is also attributable to the fact that the molecular questions being posed have been based on an incomplete understanding of the embryology of induction. The use of cell-lineage labels and region-specific markers is leading to the necessary understanding, and we can now pose meaningful questions. Thus, for each inductive interaction, we should ask:

1. When is the signal for induction produced, and how it is transmitted?
2. What is the molecular nature of the inducing signal?
3. What are the early molecular and cellular responses to induction?
4. How is this response translated into cellular activities such as division, locomotion, and gene expression?

(1970) suggested that specification of the mesoderm occurs at ~~~ stage (see Section 6). However, one strong objection to these experiments is that it is not possible to draw an accurate line between the vegetal-inducing cells and the animal-responding cells; isolated presumptive mesoderm may, for example, include some vegetal-inducing cells. Thus, although cases in which no mesoderm is formed may be interpreted to say that mesoderm induction has not occurred, positive cases do not necessarily indicate that it has.

A technique to overcome this difficulty has been introduced by Sargent *et al.* (1986). These investigators used three germ layer-specific gene probes to monitor their results, one encoding a cytokeratin (ectoderm), one encoding α-actin (mesoderm), and one encoding an unidentified endodermal product. When fertilized, *Xenopus* eggs were freed from their vitelline envelope and cultured in a calcium- and magnesium-free solution; they continued to divide, but the cells separated from each other and formed a loose heap. Calcium and magnesium were returned to the disaggregated embryos at the time when controls commenced gastrulation and all three germ-layer specific markers were subsequently expressed. (As also shown by Gurdon *et al.*, 1984, calcium and magnesium ions are required during gastrula and neurula stages for α-actin gene expression.) By contrast, when the cells of dissociated embryos were dispersed from their heap and kept separate throughout the period up to gastrulation, addition of calcium and magnesium did not permit expression of the α-actin gene. Levels of the ectoderm- and endoderm-specific genes, however, were enhanced. Thus, expression of a mesoderm-specific gene during gastrulation seems to require the close proximity of cells at the blastula stage but not firm adhesion, implying perhaps that mesoderm induction can occur without specialized cell–cell contact. The timing of mesoderm induction during normal development could now be studied by disaggregating embryos at different stages and asking at what stage α-actin gene expression becomes independent of cell dispersion. The results so far indicate that induction has not been completed by the 64–128-cell stage, and further work is no doubt in progress.

Thus, Sargent *et al.* suggest that mesoderm induction occurs later than the 128-cell stage and is independent of strong calcium-mediated adhesion. They also point out that since α-actin is not expressed in embryos whose cells were dispersed during blastula stages, this makes it unlikely that the proposed cytoplasmic determinant of Gurdon *et al.* (1985*a,b*) acts in a strictly cell-autonomous fashion. It is more likely, they suggest, that it exercises its effect by cell–cell interactions, as suggested in Section 6.2.

10.2. Mesoderm-Inducing Factors

The nature of the signal produced by the vegetal hemisphere of the amphibian embryo is unknown, but extracts of a variety of tissues and cells, some from the most unlikely sources, have been shown to have mesoderm inducing activity. Until recently the best characterized of these substances had been Tiedemann's "vegetalizing factor," which is isolated from the trunks of 9–13-day chick embryos (Tiedemann and Tiedemann, 1959; Born *et al.*, 1972; Geithe *et al.*, 1981; Schwartz *et al.*, 1981; Born, Davids, and Tiedemann, 1987). This factor is a protein of 28–30,000 M_r, which separates into smaller chains of 13,000–15,000 M_r in formic acid. It binds to heparin. When the vegetalizing factor is applied, in the form of an insoluble pellet, to amphibian blastula ectoderm, it causes the formation of a range of mesodermal cell types, including notochord, muscle, kidney, and blood (Asashima and Grunz, 1983; Grunz, 1983). Other heterologous sources of mesoderm inducing factors include guinea pig bone marrow (Toivonen, 1953), HeLa cells (Saxen and Toivonen, 1958), and carp swimbladder (Kawakami, 1976); some success has even been achieved with extracts of *Xenopus* blastulae and gastrulae (Faulhaber, 1972; Faulhaber and Lyra, 1974), although the limited amount of material available from the latter source may rule it out as a starting material for purification.

Although the chick vegetalizing factor is unlikely to be identical to a natural *Xenopus* mesoderm inducing factor, it is undoubtedly a useful tool for the analysis of mesoderm induction, both in studying the mechanism of action of inducing factors and in looking at the response of cells in the animal hemisphere to induction. Progress along these lines has been rather slow, however, partly because the vegetalizing factor is always assayed as an insoluble pellet. This makes dilution curves, for example, difficult to interpret, and it has been a matter of some importance to obtain a purified mesoderm-inducing factor active in soluble form (Yamada and Takata, 1961). Two exciting discoveries indicate that this aim may soon be achieved. The first is that a *Xenopus* cell line, called XTC (Pudney *et al.*, 1973), secretes a mesoderm-inducing factor (Smith, 1987). When animal pole ectoderm is cultured in XTC conditioned medium it forms muscle and notochord, rather than 'atypical epidermis' (Fig. 16). Medium conditioned by several different cell lines is ineffective. Preliminary characterization of the XTC factor indicates that it is a heat-stable protein of 16,000 M_r. Purification of the factor is in progress, and since it originates

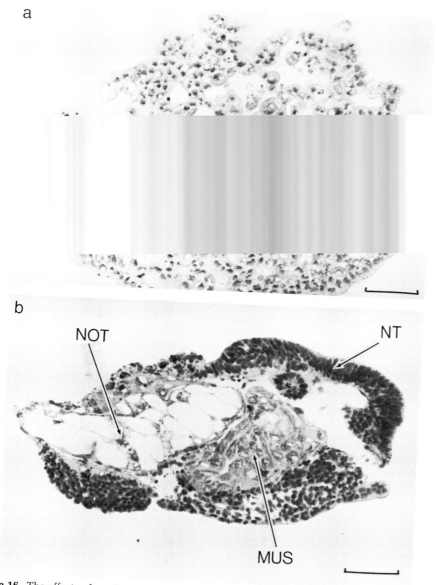

Figure 16. The effects of conditioned medium from the XTC cell line. (a) A piece of animal pole tissue from a *Xenopus* blastula was allowed to develop in a simple salts solution for three days before being fixed and sectioned. It formed atypical epidermis. (b) A similar piece of tissue was cultured in XTC conditioned medium. The cells have formed notochord (NOT), muscle (MUS), and nervous tissue (NT). Scale bars: 200 μm.

from a *Xenopus* source, there is a good chance that it will be related to the natural inducer.

The second discovery, made by Slack *et al.* (1987), is that fibroblast growth factor (FGF) and a variety of other heparin-binding growth factors also induce animal pole ectoderm to form mesoderm. Although there is no evidence yet that *Xenopus* embryos contain FGF, it is noteworthy that heparin, which binds to FGF-like molecules, can inhibit mesoderm induction in animal–vegetal combinations. It is also interesting that FGF tends to induce ventral mesoderm rather than dorsal, so it is possible that this is the ventral mesoderm induction signal (see Fig. 15) while the XTC factor is the dorsal signal.

10.3. Response to Mesoderm Induction

The lack of a mesoderm-inducing factor active in soluble form, only overcome recently, has hampered attempts to study the molecular response to mesoderm induction. Thus, studies on inducing factor receptor phosphorylation, phosphoinositide turnover, intracellular pH, or inducible genes, analogous to those carried out with growth factors on quiescent cells (see Stiles, 1983; Hopkins and Hughes, 1985), have not been carried out.

One attempt has been made, however, to analyze the effects of different concentrations of vegetalizing factor, and of exposure times, on the cell types induced from *Xenopus* early gastrula ectoderm (Grunz, 1983). In this study, purified vegetalizing factor was diluted with γ-globulin (which has no inducing activity), and further variation was introduced by using different sized pellets. Lastly, the animal pole regions were exposed to these pellets for different times. The tendency observed was that short periods of treatment with low concentrations of inducer resulted in the formation of ventral structures (blood cells and heart), whereas more prolonged treatment with higher concentrations of vegetalizing factor induced more dorsal structures, including notochord and somite. Finally, the highest concentrations of inducer produced yolk-rich cells that looked like endoderm, although a molecular marker of endoderm was not available to verify this. Taken at face value, these results suggest—contrary to the model in Fig. 15—that a gradient of a single mesoderm-inducing factor is sufficient to account for the spatial distribution of mesodermal cell types (Weyer *et al.*, 1977). This interpretation is not consistent, however, with the fate map at blastula stages, which shows that heart, blood, and lateral plate tissue arise from cells nearer the vegetal pole than notochord and somite (see Fig. 5). Further experiments with soluble mesoderm-inducing factors, preferably the natural *Xenopus* inducer(s), will be required to settle this issue.

10.4. The Cellular Response to Mesoderm Induction

Unlike the molecular response to soluble mesoderm-inducing factors, the cellular response has been studied. Symes and Smith (1987) have observed that an early response by animal pole explants to mesoderm induction is a change in shape, involving elongation and constriction. By several criteria, these movements resemble the gastrulation movements exhibited by isolated dorsal marginal zone regions (Keller, 1986). Thus, the movements always commence at the same time sibling host embryos begin convergent extension (Keller, 1986); the rate of elongation of induced explants is similar to that of isolated dorsal marginal zone regions (Keller *et al.*, 1985) and like the gastrulation movements of whole embryos (Cooke, 1973), the change in shape of induced explants will occur in the absence of cell division.

10.5. Gene Expression in Response to Mesoderm Induction

One of the most fruitful lines of research into molecular aspects of mesoderm induction concerns the activation of mesoderm-specific genes, particularly the cardiac actin gene, which is strongly expressed in larval somite muscle (Mohun *et al.*, 1984). Using a sensitive SP6 assay (see legend to Fig. 17) activation of this gene can first be detected at the late gastrula stage, making this the earliest molecular marker of mesoderm induction available (Gurdon *et al.*, 1985). The aim of this work by Gurdon and colleagues (see Gurdon, 1987) is to analyze the molecular mechanism of induction by working from the factors involved in gene activation backward toward the initial inductive signal received by the cell (perhaps meeting halfway those who work on inducing factors and who are following the chain of second messengers toward the nucleus). Much of the work so far has concentrated on analyzing the temporal aspects of induction, to identify the conditions that any proposed chain of events must satisfy.

In the absence of a purified *Xenopus* mesoderm-inducing factor Gurdon *et al.* (1985a,b) have continued to use animal–vegetal combinations to study mesoderm induction and actin gene activation. As would be predicted from the results of Nieuwkoop (1969), using histological techniques, and Dale *et al.* (1985), using immunofluorescence, animal–vegetal combinations expressed the cardiac actin gene, whereas either component cultured alone failed to do so. By separating the animal and vegetal regions at different times after the combinations were made, it was confirmed that it was only the animal component that contained cardiac actin transcripts and furthermore that at least 2.5 hr contact with vegetal cells is required for the activation of this gene.

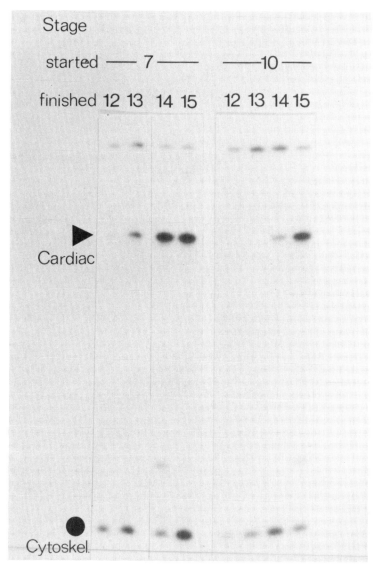

Figure 17. Timing of muscle-specific actin gene activation after making animal–vegetal combinations at different stages. Pieces of animal pole tissue were taken from early blastula (stage 7) or early gastrula (stage 10) embryos, combined with mid-blastula vegetal pole regions, and incubated until the stage indicated. RNA was extracted from the combinations and analyzed with a 380-base SP6-promoted RNA probe. Cardiac actin mRNA protects about 285 bases of this probe from RNAase T1 digestion (▶), whereas the ubiquitous cytoskeletal actin mRNA protects about 135 nucleotides (●). The latter provides a convenient internal control for the absence of cardiac transcripts in explants analyzed at early stages. Combinations made at stage 7 transcribe cardiac actin genes at stage 12–13 (early neurula), whereas conjugates made at stage 10 activate muscle-specific genes at stage 14 (mid-neurula). The interval between stages 7 and 10 is 5 hr, whereas that between 12–13 and 14 is only 2 hr. Thus, the tendency is for muscle-specific genes to be activated at a certain developmental stage, rather than a specific time after making the animal–vegetal combination. (From Gurdon et al., 1985a.)

As with the gastrulation-like movements of induced animal pole regions, the time of actin gene activation is determined more by the developmental stage of the responding tissue than by the time at which induction occurs (Fig. 17). Thus, transcription usually commences at the late gastrula stage, irrespective of when the animal–vegetal combination was made; it is as if the animal pole cells possess an actin gene activation clock in addition to their gastrulation clock. Since, in Gurdon's hands, the animal pole cells are still responsive to induction at the early gastrula stage, the minimum interval between induc-

for tissue-specific transcription (Mohun, *et al.* 1986; see also Wilson, *et al.* 1986). It will be interesting eventually to investigate other genes that may be activated by mesoderm induction, such as those expressed by blood cells (e.g., β-globin) (Banville and Williams, 1985). This may provide an answer to the question of how it is that dorsal vegetal pole regions induce muscle from animal pole cells whereas ventral vegetal poles induce blood (see Section 6.3).

11. Conclusions

A sequence of three inductive interactions involved in amphibian embryogenesis is described. Does a similar sequence operate in other vertebrate species, particularly mammals? The mammalian embryo is unsuited to the kind of experiment that can be carried out so easily on amphibia, but avian descriptive embryology is similar to that of mammals, and among the vertebrates, birds are second only to the amphibia in accessibility to experimental manipulation. In addition, although the cell lineage labels that can be applied to amphibia cannot be applied to chick embryos, because chick embryos grow, tritiated thymidine has been used as a lineage marker (Nicolet, 1970), as has the fact that quail cells can be distinguished from chick in heterospecific combinations by their Feulgen-positive nucleoli (Le Douarin, 1973).

Using grafts of [3H]thymidine-labeled tissue or quail cells, three different inductive interactions have been demonstrated in chick embryos. The first of these, the induction of the primitive streak, is analogous to mesoderm induction in *Xenopus* (Slack, 1983; Nieuwkoop *et al.*, 1985). The primitive streak, in both birds and mammals, represents the site through which the cells of the epiblast pass on their way to form mesoderm and endoderm, and there is evidence that the primitive streak is induced from the epiblast by the underlying hypoblast (Waddington, 1932, 1933; Azar and Eyal-Giladi, 1981; see Nieuwkoop *et al.*, 1985). In the second interaction, analogous to dorsalization, grafts of Hensen's node (the presumptive notochord) can be shown to induce lateral mesoderm to form somite (Hornbruch *et al.*, 1979), and, as in the amphibian organizer graft, this experiment also demonstrates the third interaction: neural induction (Gallera, 1971).

Thus, the interactions we study in amphibian embryos have their counterparts in chick, and probably mouse, development as well. Analysis of inductive interactions in amphibian embryos, which are so much more accessible than many other species, is therefore important in coming to understand the process in all vertebrates. It is to be hoped that progress in the years ahead will be more dramatic than that achieved since 1924.

Note added in proof: Kimelman and Kirschner (1987) have recently confirmed that FGF induces muscle actin gene expression in isolated *Xenopus* animal pole regions (see page 116), and demonstrated that *Xenopus* embryos contain an mRNA encoding a protein homologous to FGF. They also show that transforming growth factor-β (TGFβ) enhances the ability of FGF to induce muscle actin, although TGFβ alone has no effect. In an accompanying paper, Weeks and Melton (1987) report that the vegetal–specific mRNA described on page 93 codes for a factor related to TGFβ.

References

Akers, R. M., Phillips, C. R., and Wessells, N. K., 1986, Expression of an epidermal antigen used to study tissue induction in the early *Xenopus laevis* embryo, *Science* **231**:613–616.

Anderson, K. V., and Nüsslein-Volhard, C., 1984, Information for the dorso-ventral pattern of the *Drosophila* embryo is stored as maternal mRNA, *Nature (Lond.)* **311**:223–227.

Asashima, M., and Grunz, H., 1983, Effects of inducers on inner and outer gastrula and ectoderm layers of *Xenopus laevis*, *Differentiation* **23**:206–212.

Azar, Y., and Eyal-Giladi, H., 1981, Interaction of epiblast and hypoblast in the formation of the primitive streak and the embryonic axis in chick, as revealed by hypoblast-rotation experiments, *J. Embryol. Exp. Morphol.* **61**:133–144.

Balak, K., Jacobson, M., Sunshine, J., and Rutishauser, U., 1987, Neural cell adhesion molecule expression in *Xenopus* embryos. *Dev. Biol.* **119**:540–550.

Banville, D., and Williams, J. G., 1985, Developmental changes in the pattern of larval β-globin gene expression in *Xenopus laevis*. Identification of two early larval β-globin mRNA sequences. *J. Mol. Biol.* **184**:611–620.

Bolton, A. E., and Hunter, W. M., 1973, The labelling of proteins to high specific activities by conjugation to a ^{125}I-containing acylating agent, *Biochem. J.* **133**:529–539.

Born, J., Davids, M., and Tiedemann, H., 1987, Affinity chromatography of embryonic inducing factors on heparin-Sepharose, *Cell Diff.* **21**:131–136.

Born, J., Geithe, H. P., Tiedemann, H., Tiedemann, H., and Kocher-Becker, U., 1972, Isolation of a vegetalizing inducing factor, *Biol. Chem. Hoppe-Seyler* **353**:1075–1084.

Boterenbrood, E. C., and Nieuwkoop, P. D., 1973, The formation of the mesoderm in urodelean amphibians. V. Its regional induction by the endoderm, *Roux Arch.* **173**:319–332.

Brun, R. B., and Garson, J. A., 1984, Notochord formation in the Mexican Salamander (*Ambystoma mexicanum*) is different from notochord formation in *Xenopus laevis*, *J. Exp. Zool.* **229**:235–240.

Capco, D. G., 1982, The spatial pattern of RNA in fully grown oocytes of an amphibian, *Xenopus laevis*, *J. Exp. Zool.* **219**:147–154.

Capco, D. G., and Jäckle, H., 1982, Localized protein synthesis during oogenesis of *Xenopus laevis*—Analysis by *in situ* translation, *Dev. Biol.* **94**:41–50.

Capco, D. G., and Jeffery, W. R., 1982, Transient localizations of messenger RNA in *Xenopus laevis* oocytes, *Dev. Biol.* **89**:1–12.

Carpenter, C. D., and Klein, W. H., 1982, A gradient of poly(A)$^+$ RNA sequences in *Xenopus laevis* eggs and embryos, *Dev. Biol.* **91**:43–49.

Carrasco, A. E., and Malacinski, G. M., 1987, Localization of *Xenopus* homoeo-box gene transcripts during embryogenesis and in the adult nervous system, *Dev. Biol.* **121**:69–81.

Cleine, J. H., and Slack, J. M. W., 1985, Normal fates and states of specification of different regions in the axolotl gastrula, *J. Embryol. Exp. Morphol.* **86**:247–269.

Dale, L., and Slack, J. M. W., 1987*a*, Fate map for the 32-cell stage of *Xenopus laevis*, *Development* **99**:527–551.

Dale, L., and Slack, J. M. W., 1987*b*, Regional specification within the mesoderm of early embryos of *Xenopus laevis*, *Development* **100**:279–295.

Dale, L., Smith, J. C., and Slack, J. M. W., 1985, Mesoderm induction in *Xenopus laevis*: A quantitative study using a cell lineage label and tissue-specific antibodies, *J. Embryol. Exp. Morphol.* **89**:289–312.

Dworkin-Rastl, E., Kelley, D. B., and Dworkin, M. B., 1986, Localization of specific mRNA sequences in *Xenopus laevis* embryos by *in situ* hybridization, *J. Embryol. Exp. Morphol.* **91**:153–168.

Eyal-Giladi, H., 1984, The gradual establishment of cell commitments during the early stages of chick development, *Cell Diff.* **14**:245–255.

Faulhaber, I., 1972, Die Induktionsleistung subcellularer Fraktion aus der Gastrula von *Xenopus laevis*, *Roux Arch.* **171**:87–103.

Faulhaber, I., and Lyra, L., 1974, Ein Vergleich der Induktionsfahigkeit von Hullenmaterial der Dotterplattchen—und der Microsomenfraktion aus Furchungs—sowie Gastrula—und Neurulastadien des Krallenfrosches *Xenopus laevis*, *Roux Arch.* **176**:151–157.

Gallera, J., 1971, Primary induction in birds, *Adv. Morph.* **9**:149–180.

Geithe, H. P., Asashima, M., Asahi, K-I., Born, J., Tiedemann, H., and Tiedemann, H., 1981, A vegetalizing inducing factor. Isolation and chemical properties, *Biochem. Biophys. Acta* **676**:350–356.

Gerhart, J. C., 1980, Mechanisms regulating pattern formation in the amphibian egg and early embryo, in: *Biochemical Regulation and Development*, Vol. 2, (R. F. Goldberger, ed.), pp. 133–293, Plenum, New York.

Gerhart, J., Ubbels, G., Black, S., Hara, K., and Kirschner, M., 1981, A reinvestigation of the role of the grey crescent in axis formation in *Xenopus laevis*, *Nature (Lond.)* **292**:511–516.

Gimlich, R. L., and Braun, J., 1985, Improved fluorescent compounds for tracing cell lineage, *Dev. Biol.* **109**:509–514.

Gimlich, R. L., and Cooke, J., 1983, Cell lineage and the induction of second nervous systems in amphibian development. *Nature (Lond.)* **306**:471–473.

Gimlich, R. L., and Gerhart, J. C., 1984, Early cellular interactions promote embryonic axis formation in *Xenopus laevis*, *Dev. Biol.* **104**:117–130.

Grunz, H., 1983, Changes in the differentiation pattern of *Xenopus laevis* ectoderm by variation of the incubation time and concentration of vegetalizing factor, *Roux Arch.* **192**:130–137.

Gurdon, J. B., 1987, Embryonic induction—molecular prospects, *Development* **99**:285–306.

Gurdon, J. B., Brennan, S., Fairman, S., and Mohun, T. J., 1984, Transcription of muscle-specific actin genes in early Xenopus development: Nuclear transplantation and cell dissociation, *Cell* **38**:691–700.

Gurdon, J. B., Fairman, S., Mohun, T. J., and Brennan, S., 1985*a*, The activation of muscle-specific actin genes in Xenopus development by an induction between animal and vegetal cells of a blastula, *Cell* **41**:913–922.

Gurdon, J. B., and Fairman, S., 1986, Muscle gene activation by induction and the nonrequirement for cell division. *J. Embryol. Exp. Morphol.* **97**(suppl.):75–84.

Gurdon, J. B., Mohun, T. J., Fairman, S., and Brennan, S., 1985*b*, All components required for the eventual activation of muscle-specific actin genes are localized in the sub-equatorial region of an uncleaved amphibian egg, *Proc. Natl. Acad. Sci. USA* **82**:139–143.

Heasman, J., Wylie, C. C., Hausen, P., and Smith, J. C., 1984, Fates and states of determination of single vegetal pole blastomeres of X. laevis, *Cell* **37**:185–194.

Hirose, G., and Jacobson, M., 1979, Clonal organization of the central nervous system of the frog. I. Clones stemming from individual blastomeres of the 16-cell and earlier stages, *Dev. Biol.* **71**:191–202.

Holtfreter, J., 1933, Die totale Exogastrulation, eine Selbstablosung des Ektoderms von Ento-mesoderm, *Roux Arch.* **129**:699–793.

Hopkins, C. R., and Hughes, R. C. (eds.), 1985, *Growth Factors: Structure and Function, J. Cell Sci.* (Suppl. 3).

Hornbruch, A., Summerbell, D., and Wolpert, L., 1979, Somite formation in the early chick embryo following grafts of Hensen's node, *J. Embryol. Exp. Morphol.* **51**:51–62.

Jacobson, M., 1982, Origins of the nervous system in amphibians, in: *Neuronal Development* (N. C. Spitzer, ed.), pp. 45–99, Plenum, New York.

Jacobson, M., 1983, Clonal organization of the central nervous system of the frog. III. clones stemming from individual blastomeres of the 128-, 256-, and 512-cell stages, *J. Neurosci.* **3**:1019–1038.

Jacobson, M., 1984, Cell lineage analysis of neural induction: Origins of cells forming the induced nervous system, *Dev. Biol.* **102**:122–129.

Jacobson, M., and Hirose, G., 1978, Origin of the retina from both sides of the embryonic brain: A contribution to the problem of crossing at the optic chiasma, *Science* **202**:637–639.

Jacobson, M., and Hirose, G., 1981, Clonal organization of the central nervous system of the frog. II. Clones stemming from individual blastomeres of the 32- and 64-cell stages, *J. Neurosci.* **1**:271–284.

Jacobson, M., and Rutishauser, U., 1986, Induction of neural cell adhesion molecule (NCAM) in Xenopus embryos, *Dev. Biol.* **116**:524–531.

Jones, E. A., 1985, Epidermal development in *Xenopus laevis:* The definition of a monoclonal antibody to an epidermal marker, *J. Embryol. Exp. Morphol.* **89**(suppl.):155–166.

Jones, E. A., and Woodland, H. R., 1986, Development of the ectoderm in Xenopus: Tissue specification and the role of cell association and division, *Cell* **44**:345–355.

Kageura, H., and Yamana, K., 1983, Pattern regulation in isolated halves and blastomeres of early Xenopus laevis, *J. Embryol. Exp. Morphol.* **74**:221–234.

Kageura, H., and Yamana, K., 1984, Pattern regulation in defect embryos of *Xenopus laevis, Dev. Biol.* **101**:410–415.

Katz, M. J., Lasek, R. J., Osdoby, P., Whittaker, J. R., and Caplan, A. I., 1982, Bolton-Hunter reagent as a vital stain for developing systems, *Dev. Biol.* **90**:419–429.

Kawakami, I., 1976, Fish swimbladder: An excellent mesodermal inductor in primary embryonic induction, *J. Embryol. Exp. Morphol.* **36**:315–320.

Keller, R. E., 1975, Vital dye mapping of the gastrula and neurula of *Xenopus laevis*. I. Prospective areas and morphogenetic movements of the superficial layer, *Dev. Biol.* **42**:222–241.

Keller, R. E., 1976, Vital dye mapping of the gastrula and neurula of *Xenopus laevis*. II. Prospective areas and morphogenetic movements of the deep layer, *Dev. Biol.* **51**:118–137.

Keller, R. E., 1986, The cellular basis of amphibian gastrulation, in: *Developmental Biology: A Comprehensive Synthesis*, Vol. 2 (L. Browder, ed.), pp. 241–327, Plenum, New York.

Keller, R. E., Danilchik, M., Gimlich, R., and Shih, J., 1985, The function and mechanism of convergent extension during gastrulation of *Xenopus laevis, J. Embryol. Exp. Morph.* **89** (suppl.):185–209.

Kimelman, D. and Kirschner, M., 1987, Synergistic induction of mesoderm by FGP and TGF-β and the identification of an mRNA coding for FGF in the early Xenopus embryo, *Cell* **51**:869–877.

Kimelman, D., Kirschner, M., and Scherson, T., 1987, The events of the midblastula transition in *Xenopus* are regulated by changes in the cell cycle, *Cell* **48**:399–407.

King, M. L., and Barklis, E., 1985, Regional distribution of maternal messenger RNA in the amphibian oocyte, *Dev. Biol.* **112**:203–212.

Kintner, C. R., and Brockes, J. P., 1984, Monoclonal antibodies identify blastemal cells derived from dedifferentiating muscle in newt limb regeneration, *Nature (Lond.)* **308**:67–69.

jungeren Keim, ein Beispiel homoeogenetischer oder assimilatorische Induktion, *Roux Arch.* **111**:341–422.

Moen, T. L., and Namenwirth, M., 1977, The distribution of soluble proteins along the animal–vegetal axis of frog eggs, *Dev. Biol.* **58**:1–10.

Mohun, T. J., Garrett, N., and Gurdon, J. B., 1986, Upstream sequences required for tissue-specific activation of the cardiac actin gene in Xenopus laevis embryos, *EMBO J.* **5**:3185–3193.

Mohun, T. J., Brennan, S., Dathan, N., Fairman, S., and Gurdon, J. B., 1984, Cell type-specific activation of actin genes in the early amphibian embryo, *Nature (Lond.)* **311**:716–721.

Nakamura, O., 1978, Epigenetic formation of the organizer, in: *Organizer—A Milestone of a Half-Century from Spemann* (O. Nakamura and S. Toivonen, eds.), pp. 179–220, Elsevier/North-Holland Biomedical, Amsterdam.

Nakamura, O., Takasaki, H., and Mizohata, T., 1970, Differentiation during cleavage in *Xenopus laevis.* I. Acquisition of self-differentiation capacity of the dorsal marginal zone, *Proc. Jpn Acad.* **46**:694–699.

Newport, J., and Kirschner, M., 1982, A major developmental transition in early Xenopus embryos. I. Characterization and timing of cellular changes at the midblastula stage, *Cell* **30**:675–686.

Nicolet, G., 1970, Analyse autoradiographique de la destination des cellules invaginées an niveau du noeud de Hensen de la ligne primitive achévée de l'embryon de poulet, *J. Embryol. Exp. Morphol.* **23**:79–108.

Nieuwkoop, P. D., 1952c, Activation and organization of the central nervous system in amphibians. III. Synthesis of a new working hypothesis, *J. Exp. Zool.* **120**:83–108.

Nieuwkoop, P. D., 1969, The formation of mesoderm in Urodelean amphibians. I. Induction by the endoderm, *Roux Arch.* **162**:341–373.

Nieuwkoop, P. D., 1973, The "organization centre" of the amphibian embryo, its origin, spatial organization and morphogenetic action, *Adv. Morph.* **10**:1–39.

Nieuwkoop, P. D., and Ubbels, G. A., 1972, The formation of mesoderm in Urodelean amphibians. IV. Quantitative evidence for the purely "ectodermal" origin of the entire mesoderm and of the pharyngeal endoderm, *Roux Arch.* **169**:185–199.

Nieuwkoop, P. D., and others, 1952a, Activation and organization of the central nervous system in amphibians. I. Induction and activation, *J. Exp. Zool.* **120**:1–31.

Nieuwkoop, P. D., and others, 1952b, Activation and organization of the central nervous system in amphibians. II. Differentiation and organization, *J. Exp. Zool.* **120**:33–81.

Nieuwkoop, P. D., Johnen, A. G., and Albers, B., 1985. *The Epigenetic Nature of Early Chordate Development*, Cambridge University Press, Cambridge.

Pasteels, J., 1942, New observations concerning the maps of presumptive areas of the young amphibian gastrula (*Amblystoma* and *Discoglossus*), *J. Exp. Zool.* **89**:255–281.

Pasteels, J., 1949, Observations sur la localisation de la plaque prechordele et l'entoblaste presomptifs au cours de la gastrulation chez *Xenopus laevis*, *Arch. Biol. (Liege)* **60**:235–250.

Phillips, C. R., 1982, The regional distribution of poly(A) and total RNA concentrations during early *Xenopus* development, *J. Exp. Zool.* **223**:265–275.

Pudney, M., Varma, M. G. R., and Leake, C. J., 1973, Establishment of a cell line (XTC-2) from the South African clawed toad, *Xenopus laevis, Experientia* **29**:466–467.

Rebagliati, M. R., Weeks, D. L., Harvey, R. P., and Melton, D. A., 1985, Identification and cloning of localized maternal RNAs from Xenopus eggs, *Cell* **42**:769–777.

Recanzone, G., and Harris, W. A., 1985, Demonstration of neural induction using nuclear markers in *Xenopus, Roux Arch.* **194**:344–345.

Sargent, T. D., Jamrich, M., and Dawid, I. B., 1986, Cell interactions and the control of gene activity during early development of *Xenopus laevis, Dev. Biol.* **114**:238–246.

Saxen, L., and Toivonen, S., 1958, The dependence of the embryonic induction action of HeLa cells on their growth media, *J. Embryol. Exp. Morphol.* **6**:616–633.

Saxen, L., and Toivonen, S., 1962, *Primary Embryonic Induction*, Logos, London.

Scharf, S. R., and Gerhart, J. C., 1980, Determination of the dorsal–ventral axis in eggs of *Xenopus laevis:* Complete rescue of UV-impaired eggs by oblique orientation before first cleavage, *Dev. Biol.* **79**:181–198.

Scharf, S. R., and Gerhart, J. C., 1983, Axis determination in eggs of *Xenopus laevis:* A critical period before first cleavage, identified by the common effects of cold, pressure and ultraviolet irradiation, *Dev. Biol.* **99**:75–87.

Schwartz, W., Tiedemann, H., and Tiedemann, H., 1981, High performance gel permeation of proteins, *Mol. Biol. Rep.* **8**:7–10.

Sirlin, J. L., 1956, Tracing morphogenetic movements by means of labelled cells, *Roux Arch.* **148**:489–493.

Slack, C., and Warner, A., 1973, Intracellular and intercellular potentials in the early amphibian embryo, *J. Physiol. (Lond.)* **232**:313–330.

Slack, J. M. W., 1983, *From Egg to Embryo*, Cambridge University Press, Cambridge.

Slack, J. M. W., 1985, Peanut lectin receptors in the early amphibian embryo: Regional markers for the study of embryonic induction, *Cell* **41**:237–247.

Slack, J. M. W., and Forman, D., 1980, An interaction between dorsal and ventral regions of the marginal zone in early amphibian embryos, *J. Embryol. Exp. Morphol.* **56**:283–299.

Slack, J. M. W., Dale, L., and Smith, J. C., 1984, Analysis of embryonic induction by using cell lineage markers, *Philos. Trans. R. Soc. Lond. (Biol.)* **307**:331–336.

Slack, J. M. W., Cleine, J. H., and Smith, J. C., 1985, Regional specificity of glycoconjugates in *Xenopus* and axolotl embryos, *J. Embryol. Exp. Morphol.* **89**(suppl.):137–153.

Slack, J. M. W., Darlington, B. G., Heath, J. K., and Godsave, S. F., 1987, Mesoderm induction in early *Xenopus* embryos by heparin-binding growth factors, *Nature* (Lond.) **326**:197–200.

Smith, J. C., 1985, Mechanisms of pattern formation in the early amphibian embryo, *Sci. Prog.* **69**:511–532.

Smith, J. C., 1987, A mesoderm-inducing factor is produced by a *Xenopus* cell line, *Development* **99**:3–14.

Smith, J. C., and Malacinski, G. M., 1983, The origin of the mesoderm in an anuran, *Xenopus laevis* and a urodele, *Ambystoma mexicanum, Dev. Biol.* **98**:250–254.

Smith, J. C., and Slack, J. M. W., 1983, Dorsalization and neural induction: properties of the organizer in *Xenopus laevis, J. Embryol. Exp. Morphol.* **78**:299–317.

Smith, J. C., and Watt, F. M., 1985, Biochemical specificity of *Xenopus* notochord, *Differentiation* **29**:109–115.

Smith, J. C., Dale, L., and Slack, J. M. W., 1985, Cell lineage labels and region-specific markers in the analysis of inductive interactions, *J. Embryol. Exp. Morphol.* **89**(suppl.):317–331.

Smith, L. J., 1980, Embryonic axis orientation in the mouse and its correlation with blastocyst relationships to the uterus. I. Relationships between 82 hours and 4.25 days, *J. Embryol. Exp. Morphol.* **55**:257–277.

Smith, L. J., 1985, Embryonic axis orientation in the mouse and its correlation with blastocyst relationships to the uterus. II. Relationships from 4.25 to 9.5 days, *J. Embryol. Exp. Morphol.* **89**:15–35.

Spemann, H., 1938, *Embryonic Development and Induction*, reprinted in 1967 by Hafner, New York.

Spemann, H., and Mangold, H., 1924, Uber Induktion von Embryonenanlagen durch Implantation artfremder Organisatoren, *Roux Arch.* **100**:599–638.

Stiles, C. D., 1983, The molecular biology of platelet-derived growth factor, *Cell* **33**:653–655.

Sudarwati, S., and Nieuwkoop, P. D., 1971, Mesoderm formation in the Anuran *Xenopus laevis* (Daudin), *Roux Arch.* **166**:189–204.

Symes, K. and Smith, J. C., 1987, ...

Vincent, J-P., Oster, G. F., and Gerhart, J. C., 1986, Kinematics of gray crescent formation in *Xenopus* eggs: The displacement of subcortical cytoplasm relative to the egg surface, *Dev. Biol.* **113**:484–500.

Vogt, W., 1929, Gestaltunganalyse am Amphibienkeim mit ortlicher Vitalfarbung. II. Teil. Gastrulation und Mesodermbildung bei Urodelen und Anuren, *Roux Arch.* **120**:384–706.

Waddington, C. H., 1932, Experiments on the development of chick and duck embryos cultivated *in vitro*, *Philos. Trans. Roy. Soc. Lond. Biol.* **221**:179–230.

Waddington, C. H., 1933, Induction by the endoderm in birds, *Roux Arch.* **128**:502–521.

Weeks, D. L. and Melton, D. A., 1987, A maternal mRNA localized to the vegetal hemisphere in Xenopus eggs codes for a growth factor related to TGFβ, *Cell* **51**:861–867.

Wilson, C., Cross, G. S., and Woodland, H. R., 1986, Tissue specific expression of actin genes injected into *Xenopus* embryos, *Cell* **47**:589–599.

Wylie, C. C., Brown, D., Godsave, S. F., Quarmby, J., and Heasman, J., 1985, The cytoskeleton of *Xenopus* oocytes and its role in development, *J. Embryol. Exp. Morphol.* **89**(suppl.):1–15.

Yamada, T., 1958, Embryonic induction, in: *A Symposium on Chemical Basis of Development* (W. McElroy and B. Glass, eds.), pp. 217–238, Johns Hopkins Press, Baltimore.

Yamada, T., and Takata, K., 1961, A technique for testing macromolecular samples in solution for morphogenetic effects on the isolated ectoderm of the amphibian gastrula, *Dev. Biol.* **3**:411–423.

Chapter 4

A New Look at Spemann's Organizer

JOHANNES F. HOLTFRETER

the question: What are the relative contributions that the nucleus and the cytoplasm make, respectively, in the determination of the course of animal development? Weismann had postulated that the determinants of development are localized in the nucleus, particularly in the chromosomes. According to Weismann's theory, diversified differentiation of the organism is the result of the (postulated) fact that as the egg divides into blastomeres and then into a large population of cells, the nuclei become nonequivalent, successively losing some of their determinants at each step of cell division.

We now know that, as is the case for all the morphological and behavioral characteristics of the adult organism, the features of development, too, are determined by directives that originate in the genes. But we have also learned that with progressive cell differentiation the nuclei do not lose their equivalence. This insight was derived in part from Spemann's early experiments (1901–1903). Using a fine hair loop, Spemann constricted the *Triturus* embryo in the 2-cell stage into halves. The result depended on the plane of constriction. If the ligature separated a right from a left egg half, each half developed into a completely normal individual, containing a neural system and an elongated axial system of notochord and somites. The lateral halves had regulated into a whole, and a pair of monozygotic twins had been produced (Fig. 1a–d). But if the egg had been divided into a dorsal and a ventral half, only the dorsal half gave rise to a complete individual, whereas the ventral half formed an epidermis-covered roundish body designated the belly piece, which lacked an axial system (Fig. 1e–g). It was not entirely undifferentiated, however, as some commentators referred to it later. From Spemann's own words and illustrations, we learn that the belly piece exhibited the movements of invagination and differ-

JOHANNES F. HOLTFRETER • Department of Biology, University of Rochester, Rochester, New York 14627.

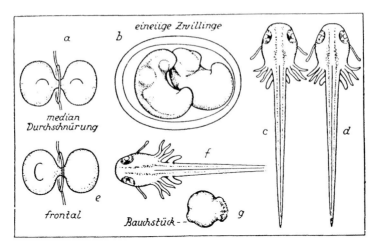

Figure 1. Constriction of a *Triturus* egg at the 2-cell stage into two lateral halves (**a**), resulting in the formation of a pair of twins (**b–d**). (**e–g**) Constriction of the egg into a dorsal and a ventral half (**e**), resulting in a complete embryo (**f**) and a belly piece (**g**). (From Mangold, 1953.)

entiating into epidermis, mesenchyme, blood cells, parts of a pronephric duct, and a lump of endoderm. All these differentiations are now known to arise from the ventral part of the embryo under normal conditions. It has also become known that the chorda–mesoderm axial system normally arises in the dorsal region of the early embryo (gastrula) and could therefore not be expected to be present in the belly piece.

The primordial chorda–mesoderm corresponds to what Spemann called the organizer district. It was obviously this material that engendered the regulation of the lateral halves into a whole individual. But these constriction experiments also demonstrated the limitations of the action radius of the organizer: The tissues normally derived from the ventral half of the egg proved capable of differentiating independently of the chorda–mesoderm.

In a slight modification of the constriction experiments, Spemann (1928) divided a *Triturus* embryo that had not yet divided into two blastomeres. Only one of the halves contained the nucleus. While the nucleated half underwent a series of cleavages, the other half, which had been connected with the nucleated half by a bridge of cytoplasm, remained inert. Sometimes, at the 8- or 16-cell stage, one of the nuclei slipped through the bridge into the anucleate half. This belatedly nucleated half then also started dividing into daughter cells; if it happened to be a lateral half, it developed into a complete individual. It could be concluded that, contrary to Weismann's theory, the nuclei of the 16-cell stage are equipotent.

It is clear that without a nucleus the egg could not develop at all, but of about equal importance is the cytoplasm, which is necessary to activate or inactivate the function of genes. Within this context, I should mention that further evidence for the equivalence of the nuclei has been furnished by the

nuclear transplantations performed by Briggs and King, Gurdon, and others. These researchers first removed the nucleus of an amphibian egg and replaced it with a nucleus taken from cells of the gastrula and even more advanced stages. This surrogate nucleus took over fully the functions of the egg nucleus, and many of the operated specimens developed into larvae. These experimental results attest to the fact that with progressive cellular differentiation there is no diminution of Weismann's nuclear determinants—genes, in modern terminology.

2. The Experiment of Spemann and Hilde Mangold Reconsidered

The following is a detailed and critical account of the paper of Spemann and Mangold (1924), which marked the beginning of a new era of embryological research. Their investigations led to the discovery of the so-called **organizer.**

The experiments of these investigations were done solely by Hilde Mangold (at that time still known by her maiden name, Hilde Proschold), whereas large parts of the text were doubtlessly formulated by Spemann. Hilde was a very intelligent person. Viktor Hamburger (1984), her close friend, tells us that Hilde was well aware of the theoretical importance of her findings. I, who was also laboring on my doctoral thesis, shared a laboratory room with her and thus had the opportunity to witness the birth of the organizer.

Spemann had given Hilde the assignment of performing an experiment that he had performed himself years before (1918) with rather inconclusive results. The experiment consisted of excising a small piece of the upper blastoporal lip of a *Triturus* gastrula and transplanting it into the ventral ectodermal surface of another gastrula. Great manual skill was required to do this.

Like the rest of Spemann's students, Hilde had great trouble raising her operated embryos. They perished at an early stage because of microbial infection. Thus, of the hundreds of experimental cases produced, only six lived to a stage at which they were sufficiently differentiated to be fixed and studied histologically.

Hilde followed Spemann's good advice to work in heteroplastic combinations, using as donors, gastrulae of *Triturus cristatus*, whose cells are unpigmented, and as hosts *Triturus taeniatus*, whose embryonic cells are well pigmented. These species-specific differences in pigmentation made it possible to distinguish later, in the sectioned material, the donor- and the host-derived tissues.

What the authors had referred to as upper blastoporal lip corresponded, as was shown later, to a region that normally gives rise to notochord, head mesoderm, and somites. This dorsal region of the gastrula will henceforth be referred to as the **chorda–mesoderm district (CMD).**

Among the six experimental cases described in this paper in great detail, by far the most instructive one was UM132. It displayed the following features:

1. *Self-differentiation of the graft:* The transplanted blastoporal material performed the morphogenetic movements it would have performed normally. It moved into the depth (invaginated) beneath the ectoderm and then stretched into a long strip, which subsequently subdivided or segregated into a median rod of notochord and bilaterally arranged somites. In other words, the graft differentiated, true to its prospective fate, into a somewhat imperfect but nevertheless clearly recognizable axial system possessing anterior–posterior polarity and bilaterality.

   Strangely enough, the graft also differentiated into a patch of neural tissue, a tissue that is normally derived from the ectoderm and not from the CMD. This could not have happened if the fate of this district had been firmly determined at the gastrula stage. Thus, the experiment had brought to light regulatory potencies that this district does not express normally.

2. *The graft acting as a neural inductor:* Equally startling was the fact that the graft not only produced neural tissue out of its own substance, but that it also stimulated, or **induced,** the overlying host's ectoderm to form neural tissue. It was observed that at the flank of the embryo a secondary neural plate arose that, simultaneously with the primary neural plate, folded inward and became a tube that had the characteristics of a spinal cord. It ran parallel to the graft-derived chorda–somite axial system and possessed, like the latter, an anterior–posterior polarity: Its anterior end was thickened and flanked by two ear vesicles, and it terminated, posteriorly, in an attenuated tail tip. A true brain was not induced in these experiments. These observations suggested that the development of the neural system of the intact embryo also involves such induction. This view was fully supported by evidence discussed later.

3. *Assimilatory induction:* The graft also affected the host's lateroventral mesoderm. Instead of pursuing its normal course of differentiating into lateral plate, it became converted into notochordal and somitic cells, which became harmoniously incorporated into the tissue pattern of the axial system of the graft (which had been somewhat imperfect). The induced notochord cells (unpigmented) joined the graft's (pigmented) notochord, and the induced somitic cells became integral parts of the graft-derived somites. Thus, this chimeric axial system attained a perfect bilaterality. The graft had, so to speak, extended its own tissue organization into the adjacent host's mesoderm. Spemann referred to this phenomenon as **assimilatory induction.**

It should be realized that there are notable differences between neural induction and assimilatory induction. In the former situation, the induced structure (neural plate) is histologically quite different from its inductors (the chorda–mesoderm). Here, induction is heterotypic. Furthermore, the induced neural plate emancipates itself from its inductors and forms an organ system of

its own. Assimilatory induction, however, is homotypic, for in this instance, the induced and the inducing tissue are of the same histological type, and they become closely associated to form a common axial system.

Thus, the term induction has here been used for two quite different phenomena. We have learned, meanwhile, that heterotypic inductions are of widespread occurrence in amphibian development, but there is no clear evidence that the experimentally obtained assimilatory inductions also take place in normal embryogenesis.

If we now take a look at the other cases described in the Spemann-Mangold paper, we encounter a wide range of results. Within the group of six embryos described, the grafts produced notochord in five cases, somitic tissue in two cases, and rather small amounts of neural tissue in three cases. The grafts induced the following tissues: notochord in one clear and one doubtful case, recognizable somitic cells in two cases, mesodermal tissue of an as-yet-undifferentiated state in three cases. Only neural tissue of a rather undefined character was induced in all six cases, but nowhere could typical brain structures be recognized.

It is difficult to explain these erratic results. Perhaps the inconsistencies were the result of different sizes of the grafts or to differences in the blastoporal regions from which the grafts had been derived. It must also be said that some of the embryos had been fixed too early for them to have attained well-defined differentiations.

Fortunately, there was UM132, the paragon, which best exemplified the complexities of the organizer syndrome. Spemann's concept of the organizer was largely based on the information gathered from this single case.

The Spemann–Mangold experiment was repeated by several investigators, but most of them failed to take the precaution of working heteroplastically. Therefore, it was impossible to tell which of the tissues observed in the microscopic sections stemmed from the transplant and which from the host. This considerably diminished their instructional value and makes them unsuitable for further discussion.

However, I wish to draw attention to Fig. 2a–c. Three cases are depicted in which different parts of the dorsal chorda–mesoderm (alias Spemann's organizer district) had been grafted into another early gastrula (Holtfreter, 1933a).

Figure 2a shows the induction of a head containing a perfectly structured brain associated with a pair of eyes. It had been induced by an implant taken from a dorsal region close to the blastoporal lip. Figure 2b depicts an embryo that displays all along its belly side an elongated trunk–tail axial system. Similar to UM132, it is furnished with a spinal cord but lacks a brain. In this case, the graft had been taken from a region farther away from the dorsal lip.

The most spectacular case is illustrated in Fig. 2c. In this case, almost the entire area of the CMD had been transplanted. Interacting with the host tissues, it produced an almost complete secondary embryo, with head, trunk, and tail. Its individuation was accentuated by the fact that its longitudinal axis ran in a direction opposite to that of the host and that, as it developed, it almost separated itself from the host embryo.

Figure 2. Examples of the induction capacities of implanted chorda–mesodermal tissues. (**a**) Induction of a complete head. (**b**) Induction of a secondary embryo with a rudimentary head stretching all along the ventral side of the host. (**c**) Induction of a nearly complete embryo, which almost emancipated itself from the host. (From Holtfreter, 1933a.)

These results demonstrated that the CMD harbors at least two kinds of inductors: head and trunk–tail inductors, the former lying closer to the blastoporal lip than the latter.

3. Misinterpretations That Led to Misconceptions

The developmental biologist endeavors to arrive at an understanding of the processes and mechanisms upon which development is based. To attain this goal, plain observation is not good enough. As the founders of modern embryology, Wilhelm Roux and Hans Driesch, have pointed out and demonstrated personally, a deeper insight into the underlying mechanisms of ontogenesis can be attained only if the embryo is subjected to properly devised experimental interference. The results obtained ought to be properly interpreted, and the inferences drawn from them should, if possible, be related to normal development. This is not always an easy task. The results may be so confusing or equivocal that they cannot be readily rationalized. This is not necessarily the fault of an experimenter who does not know how to ask the proper questions (i.e., those the embryo is capable of answering unequivocally). Rather, it is because of the innate high complexity of the developmental processes themselves that the experimental data are often so complex and difficult to interpret. A single series of experiments may produce a wide range of different results

that, at first sight, seem enigmatic. This was the case in the Spemann–Mangold experiments that we have just discussed.

The discovery of the organizer was like taking a first glimpse of a *terra incognita*, full of mysteries waiting to be explored. It is true that the phenomena there encountered were solid facts that could be described in so many words, but the findings were so novel and unexpected that they could not be explained on the basis of the information available at that time. No wonder that Spemann's early concept of the organizer was vague, tentative, rather metaphorical, and, to some extent, incorrect.

It should be realized that in the interpretation of their experimental data, Spemann and Mangold (1924) were seriously handicapped by the fact that, at that time, a fate map of the *Triturus* gastrula was not yet available. These workers, therefore, did not really know the prospective fate, or significance, of the blastoporal lip that had been transplanted. Like his contemporaries, Spemann believed, erroneously, that the area of the presumptive neural plate reaches close to the dorsal rim of the blastopore (as does the neural plate in the neurula stage). Therefore, he was not greatly surprised when such transplants produced, by themselves, not only notochord and somites, but also some neural tissue. Spemann assumed that the transplants had contained some material of the presumptive neural plate, and that the blastoporal tissues had differentiated according to their normal fate. This would imply that their fate had been determined at the time of transplantation.

This was a misconception. Meanwhile, Walter Vogt (1929) constructed a map of the topographic distribution of the prospective organ districts at the early gastrula stage of *Triturus*. This map informs us that the area normally destined to become neural plate (known as presumptive neural plate) is separated from the dorsal blastoporal groove by a broad belt of prospective chorda–mesoderm and pharyngeal endoderm. As gastrulation proceeds, it is first the pharyngeal endoderm and then the CMD that involute over the blastoporal lip. Thus, the material content of the so-called upper blastoporal lip completely changes in the course of gastrulation.

Apparently, the transplants performed by Mangold did not produce any pharyngeal tissue, indicating that they had been excised from a somewhat advanced gastrula, in which the area above the lip consisted of chorda–mesoderm. But even then it was quite unlikely that the transplants had contained any part of the presumptive neural plate. Therefore, as said before, the observed appearance of graft-derived neural tissue was contrary to the prospective fate of the transplanted area. It follows that the fate of the CMD (Spemann's organizer) is not determined at the gastrula stage.

This conclusion is amply supported by other experiments which showed that transplanted or explanted pieces of the CMD give rise not only to notochord and somites, but also to various structures, such as brain, spinal cord, or epidermis, which are normally derived from the ectoderm (Holtfreter, 1936, 1938). The appearance of these structures signaled that the transplanted CMD had undergone morphogenetic regulation. It is precisely because of its lack of determination that the CMD can exhibit regulation.

Spemann was also on the wrong track when he argued that the dominant role of the organizer is based on its advanced stage of self-determination. He surmised that the fate of the organizer tissues is determined at the gastrula stage, and that the fate of the rest of the embryonic tissues is not. He then proposed that it is by virtue of its early determination that the organizer is endowed with the prerogative to operate as the leader, or instructor, that determines the fate of the other, as yet undetermined or indifferent, tissues of the embryo (Spemann and Mangold, 1924; Spemann, 1927).

This notion has become outdated. We now know that the organizer tissues are neither earlier nor more firmly determined than the other tissues, and that there is no correlation between the state of determination of a tissue and its capacity to act as inductor. There are all kinds of tissues, whether embryonic or adult, alive or dead, which, when inserted into an amphibian gastrula, can operate as powerful inductors (see Section 9).

4. Further Reexamination of Spemann's Concept of the Organizer

Let us not forget that when the organizer phenomena were first encountered, the sciences of biochemistry, cell physiology, and genetics were still in their infancy. Although it was generally recognized that the genes express themselves in terms of cell differentiation and morphogenesis, the molecular biology and chemistry of the genes and their modes of action were entirely unknown. Spemann and the other embryologists of his time seemed to believe that the realms of chemistry and of living beings were two categorically different worlds, each governed by its own laws and principles. No one could have foreseen that, someday, biochemical methods and concepts would be most profitably applied to the exploration of problems of development.

It was only natural that when Spemann was confronted with the strange performances of a transplanted piece of the upper blastoporal lip, he was at a loss as to how to explain them. So he ventured into lofty speculation. There was the fact that the transplant caused the emergence of a so-called "secondary embryo," composed in part of transplanted and in part of host cells. It was this act of individuation (the creation of a new, complex individual) that impressed Spemann most of all. Certainly, cellular interactions of various kinds must have been involved in the processes of induction, but there seemed to be no way of further analyzing them.

In 1921, even before Mangold had finished her experiments, Spemann, realizing the general importance of Hilde's results, commented on them as follows:

> At the beginning of gastrulation, the individuality of the embryo is so to speak represented by the cells of the upper blastoporal lip which operate as a center of organization from which the other, most important parts of the embryo are formed. Such a piece of the *organisationszentrum* may be briefly called an "organisator". It creates within the indifferent material in which it lies, or into which it has been artificially transplanted, a field of organization of a definite direction and extent.

We can no longer subscribe to the Spemann credo without reservations. His idea that individuality is localized in a particular region of the embryo was intuitively conceived and not based on factual evidence. Probably, contemporary geneticists would not agree with this notion. They would insist that the factors that determine individuality—like all the other features of the organism—are localized in the genes. It also has become clear that the so-called organizer is not as powerful as Spemann thought. Large parts of the embryo can develop normally without having been affected by determining effects coming from the organizer region, as shown in the differentiation of the belly piece.

However, the term "field" has remained in the literature as a useful label for certain embryonic districts endowed with the capacity of morphogenetic regulation, whose mechanisms we still do not understand. And yet, embryonic fields are not just the product of our imagination; they are a reality, and they play a crucial role in amphibian development.

Spemann adopted the view that, above and beyond the cellular events, there exists some kind of supracellular, uniquely vital force, or agency, which acts as the supreme controller of development. This notion, however dimly conceived, was reflected in the following pronouncement (Spemann and Mangold, 1924):

> The designation *Organisator* (rather than *Determinator*) is meant to signify not only that the effect emanating from this privileged embryonic part (the organizer) is a determining one, but also that it possesses all those enigmatic attributes which we know only from living organisms.

Spemann did not make clear what he meant by those enigmatic attributes of living organisms. If I understand his rather cryptic remarks correctly, it seems to me that he assigned to the organizer tissues two different functions: first, that of determining the developmental fate, i.e., the direction of differentiation of the induced individual cells, and second, the function of arranging or disposing of the induced cells so as to form a complex organ system. To Spemann, the truly vital and most enigmatic aspect of the induction phenomenon was this second feature, the emergence of a typical pattern of organogenesis—hence, the designation *organizer*. To him, inducing a tissue was synonymous with organizing it. He clung to this opinion for the rest of his life, expressing it again in his last, comprehensive publication (Spemann, 1938).

This concept of the organizer has not stood up to further analysis. There are indications that Spemann's organizer does not deserve its prestigious appellation. The organizer does not actually organize the neural plate that it induces. It has become clear that the mechanisms operating in induction and cell differentiation are totally different from those that are instrumental in the establishment of complex tissue and organ configurations. As far as the neural plate is concerned, there is no doubt that its appearance is caused by inducing agents, or factors, which are transmitted from the underlying CMD to the ectoderm. But the next step, the organization of the neural plate into a complex central neural system, is no longer under the control of the inducing agents. Once it has been induced, the neural plate organizes itself, independently of further inductive stimuli coming from its former inductors. According to this new concept, the designation *organizer* is improper. The CMD has become merely the seat of the primary inductors.

Unable to explain the organizer syndrome in mechanistic terms, Spemann resorted to speaking in metaphors. Here I am referring to the speech he delivered at the solemn occasion when, for a year's duration, he was elected *Rektor* (President) of the University of Freiburg. There, in the academic attire of a ceremonial gown and a medieval beret of black velvet, Herr Professor Spemann addressed a large audience, including us, his devoted disciples. He reported on the exciting research that was going on in his laboratory and then concluded his talk with the following remark: "In animal development, nature proceeds just like an artist who creates a drawing or sculpture—indeed, like any organizer who disposes of a given material, be it living or nonliving" (Spemann, 1923).

Spemann also spoke of the embryo's striving to reach its preconceived final goal (*vorgestecktes Endziel*), i.e., attainment of a typically structured organism. When development is interfered with by the experimenter, the embryo overcomes this disturbance by calling upon its secret emergency powers of regulation, restoring normal development. In short, the embryo seems to be endowed with a "wisdom", akin to that of humans, enabling it to plan and make decisions. Spemann's attitude toward the embryo was that of an inquisitor, who asks the embryo delicate questions about its private inclinations and potentialities and is then amazed at the embryo's response, expressed in terms of morphogenesis.

Spemann's inaugural address was enthusiastically applauded, and I, the young and naive novice of science, fell in with the crowd. To me, this speech was a kind of messianic revelation that touched me deeply. How pleased I was to hear that the artist and the dumb little embryo have much in common, particularly the urge to create! Meantime, while I became a laborer in the vineyard of science, I became more skeptical and wondered whether Spemann's anthropomorphic–psychistic–teleological interpretations of the processes of embryogenesis had any validity and whether they could be of any help in our endeavor to gain a deeper insight into these processes. My critical considerations were—and still are—as follows.

Spemann's pronouncements sound like the contemplations of a *Naturphilosoph* or like the lyrics of a poet who indulges in metaphorical comparisons, but not like the language of a biologist who wishes to explore, analyze, and understand the mysterious workings of the developing organism. Spemann's chief message seemed to be that the enigmatic, vitalistic attributes of living matter spoken of above are shared by both the embryo and the human adult, specifically the artist. They are of a spiritual nature. But is there any truth to this notion? It rather appears that the artistic or managerial activities of man have nothing to do with the processes of ontogenesis. The two cannot be compared. I may point out some of the discrepancies between them.

The creations of the artist are expressions of his inner self, his feelings, imagination, and daydreams, which he projects in the form of paintings, sculptures, or other works of art. These products of the human mind become detached from their creator and may continue living a life of their own. Artists differ from scientists in that they do not wish to dissect, analyze, and rationalize the events of nature, only to end up with abstract schemes or formulas that bear no resemblance to the phenomena as they were originally perceived.

The embryo, on the other hand, is a self-contained dynamic entity that continuously creates itself and undergoes a series of gradual transformations, until it reaches the more stable condition of adulthood. We have learned that embryogenesis involves a great diversity of biological processes, many of which are interdependent, and all of which are harmoniously integrated into a common, unified whole, known as the individual organism.

The physicochemical mechanisms that lie at the basis of these processes are poorly understood. But this ignorance should not serve as an excuse for veering off into fiction and invoking a spiritual agency that is supposed to monitor these decidedly materialistic processes. Actually, there are no factual indications for the existence of such an agency. It is also evident that the embryo does not behave like an artist who creates things outside himself.

Man's thinking takes place in a special organ, the brain. But the early amphibian embryo of which we are speaking does not yet possess a functional brain. Where, then, is its hypothetical psychoid power located? Apparently, Spemann did not assume that this power pervades the whole embryo. He thought that it resides in the upper blastoporal region, a region that he personified and called the organizer.

These considerations have led me to conclude that Spemann's comparison of the activities of the embryo with the mind-controlled activities of an artist or a human organizer was a dubious exercise that confused the issues and did not explain anything.

In an attempt to characterize the organizer phenomena in more concrete terms, Spemann suggested that they are somehow related to certain structural properties of the inducing tissues. He had been impressed by the fact that an implanted piece of the blastoporal lip moved beneath the host's ectoderm and then stretched into a long ribbon that subsequently broke up into a median rod of notochord and bilateral rows of somites. This indicated that the primordial chorda–mesoderm has the inherent capacity of forming an axial system, complete with an anterior–posterior axis and bilateral symmetry. Then, there was the other observation that the neural system, which runs parallel to the chorda–somite system, is similarly structured, exhibiting likewise bilaterality and a longitudinal axis. Was this sheer coincidence? To Spemann it was not. It seemed to indicate that the axial configuration of the chorda–somite system had been somehow projected onto the induced neural system. To him, axial configuration was tantamount to tissue organization. What this projection meant in more concrete terms was not made clear. At any rate, this was a truly vital process in which, apparently, some unspecified forces were involved. But Spemann, a master in the art of suggesting alternatives, did not exclude the possibility that chemicals might play a role in the induction process.

To test the validity of these ideas, Spemann proposed the following experiment: excise the organizer tissue of a gastrula and kill it by crushing, freezing, or heating. This would not only destroy the intrinsic structure but also abolish the (hypothetical) vital spirits of the cell material. Then implant this material beneath the ventral ectoderm of a live gastrula and see whether the dead organizer would still have inductive and organizing effects.

Such experiments were actually performed, first by Spemann (1931a) and then by Marx (1931). But the results of these workers were inconclusive. Then this writer appeared on the scene and tried his luck. In an extensive series of experiments, I tested the inductive capacity of devitalized pieces of the embryo and of all sorts of foreign tissues (Holtfreter, 1932, 1933b, 1934a,b). The results were overwhelmingly positive. Implants that had been devitalized and possessed no internal structure were just as effective as the genuine live inductors (see Section 9). In fact, these investigations signified a turning point in the history of the organizer.

5. Questions of Priority

It is of historical interest that Hilde Mangold was not the first to graft a piece of the dorsal blastoporal region of a gastrula into another embryo. As early as 1907, the American investigator, Warren Lewis, working with *Rana palustris*, excised a piece of this region, stemming from a late gastrula (yolk plug stage), and transplanted it homoplastically beneath the epidermis of an early tail-bud stage. The transplant formed notochord, somites, and neural tissue. Lewis interpreted this as indicating that these three tissues had developed by way of self-differentiation true to the prospective significance of the grafted area. The graft could not have elicited a neural induction because, according to subsequent investigations, the host's epidermis was too old to react to neuralizing inductors.

Years later, Spemann (1918) working with *Triturus taeniatus*, performed an experiment that was almost identical with the one that, under his supervision, was later repeated by Hilde Mangold. However, his transplantations were performed homoplastically and not, as those of Hilde, in heteroplastic combinations between pigmented and unpigmented *Triturus* species. Like Hilde (Spemann and Mangold, 1924), he found that, at the site of transplantation, there appeared a complex of notochord, somites, and neural tissue. However, the crucial question: where did the neural tissue come from?—from the transplant by way of self-differentiation, or from the host, by way of induction—could not be answered, because in this homoplastic combination the tissues derived from the transplant could not be distinguished from those of the host. Nevertheless, Spemann (1918) had an inkling that in the embryo, some influence emanates from the chorda–mesoderm, which perhaps travels from cell to cell and which has a determining effect on the presumptive neural plate.

In this paper, Spemann (1918) briefly discussed the results of Lewis (1907). Spemann agreed with the interpretation that Lewis gave to his results, and he stated explicitly that his own results confirmed those of Lewis. There was no squabble about priority.

Therefore, it came as a great surprise to Spemann when, after the paper of

Spemann and Mangold had been published, De Beer (1927) and Hörstadius (1928) declared that it was Lewis (1907) who had discovered the organizer. Spemann (1931b) rightfully refuted this claim (which Lewis himself had never made). Spemann pointed out that in the experiments of Lewis, inductions could not have been obtained because, as I have just mentioned, the epidermis of the host (tail-bud stage) was too old to react to neural induction. Besides, if inductions had occurred, they would have gone unnoticed because Lewis's experiments were not performed heteroplastically either.

Thus, we arrive at the conclusion that it was neither Lewis nor Spemann, but Hilde Mangold who was the first to provide clear evidence of the inductive capacity of the chorda–mesoderm.

6. Other Investigations of the 1920s

Hans Spemann, director (Ordinarius) of the Institute of Zoology, the University of Freiburg since 1919, was fortunate to have around him a group of disciples and collaborators who were devoted to him, the Master, and whose work contributed greatly to his rising fame. But Hilde Proschold, one of his most promising students, who became the wife of Otto Mangold, did not live to see the day when Spemann, in 1935, was awarded the Nobel prize. Soon after the Spemann–Mangold paper had appeared in print, Hilde perished in a terrible accident. While she was warming the milk for her recently born baby on an alcohol burner, the alcohol spilled over her body and she burned to death.

Otto Mangold held the position of an assistant. Together with the other assistant, Bruno Geinitz, he was in charge of the Grosse and the Kleine Praktikum, that is, of the laboratory courses offered to the students in zoology and medicine. In 1925, as a rather young man, Mangold became director of the Department of Embryology, at the Biological Institute of the prestigious Kaiser Wilhelm Society (now Max-Planck Society) in Berlin–Dahlem. This position had been held before by Spemann during World War I (1914–1919). After some 7 years of fruitful research, Mangold left Dahlem in 1933 to become professor of zoology at the University of Erlangen. He then accepted a call from Freiburg and became Spemann's successor. After Spemann died, in 1944, Mangold (1953) wrote a booklet in memory of Spemann's personality and his scientific achievements.

The old Zoological Institute in Freiburg, dating back to the times of August Weismann, was totally destroyed in World War II, but it has been replaced by a larger and more modern building. At its entrance you are greeted by two busts, the one showing the long-bearded serene face of Weismann, the other representing Spemann, with a short moustache and frowning eyebrows, as though he were still assailing Weismann's theory of the nonequivalence of the nuclei.

But to return to the 1920s. At the time that Hilde Proschold was engaged with the organizer, Otto Mangold (1923) conducted the following interesting experiment: He transplanted a piece of gastrula ectoderm heteroplastically into the dorsolateral region of the gastrula. The transplant was transformed into mesodermal tissues, the type of which depended on the graft's position within the embryo. When it came to lie next to the host's presumptive somites, it formed somites; when positioned close to the pronephros, it formed pronephric tubules; and when occupying the region of the lateral trunk mesoderm, it developed into side plate tissues.

Clearly, the gastrula ectoderm proved to be undetermined as well as pluripotent. Its developmental fate seemed to depend on the kind of tissue with which it became associated at the time of gastrulation. But it must also be realized that these results were experimental artifacts. Under normal conditions, such transformations of ectoderm into mesodermal tissues do not occur.

A somewhat similar experiment was performed by Geinitz (see Spemann and Geinitz, 1927). In this case, a piece of ectoderm derived from the unpigmented gastrula of *T. cristatus* was placed into the midst of the blastoporal lip of a gastrula of *T. taeniatus* (pigmented). Some of the transplant remained on the outside and became neural tissue that was incorporated into the host's spinal cord. Most of the graft invaginated together with the surrounding chorda–mesoderm and was transformed into notochordal cells and a series of somites that, in a complementary fashion, became integral parts of the host's imperfect axial system. As in the case of the Spemann–Mangold experiment (1924), the CMD had extended its field of action into the undetermined, artifically supplied adjacent tissue. But, in this case, it was added ectoderm and not host mesoderm, that was induced and organized so as to become a fully integrated part of the chorda–mesoderm field. In this kind of assimilatory induction, the pluripotency of the ectoderm was once more demonstrated.

The preceding heteroplastic experiments had shown that the inducing agents are not species specific but are shared by different species of *Triturus*. The question arose whether they are also shared by the representatives of different classes and orders of the amphibians. In other words, can organizer tissue taken from a salamander gastrula induce the formation of a neural plate when it is grafted xenoplastically into a frog gastrula?

Geinitz (1925) tackled this question by making reciprocal transplantations between different species of urodeles and anurans. But the mortality rate of his operated embryos was very high. He obtained the relatively best results in experiments in which he had transplanted a piece of blastoporal lip from a gastrula of the anuran *Bombinator pachypus* into a gastrula of the urodele *T. taeniatus*. He reported that in at least one of such cases, URU133, the graft produced, by itself, notochord, somitic mesoderm, and a neural tube, and that it induced the host ectoderm to form another neural tube plus some mesoderm.

These results were similar to those obtained in heteroplastic experiments. There were, however, no indications of a field of action of the grafted tissues, and the graft-derived neural tube remained separated from the induced neural tube.

Although these results of Geinitz were rather skimpy, they did show that the gastrula ectoderm of a urodele can be induced by the chorda–mesoderm of an anuran to form neural tissue. Apparently then, the neuralizing factor is the same in these two, taxonomically quite unrelated species. On a wider scale, xenoplastic transplantations between anurans and urodeles have been performed by Schotté, by Holtfreter, and by Wagner. They were informative in that the representatives of these taxonomically different groups share the same inductive agents; however, the reaction to these agents is strictly species-specific.

Geinitz was, by nature, a field zoologist. He seemed to derive little satisfaction from working all day long in the laboratory, experimenting on amphibian embryos. And so, after having finished the aforementioned investigations, he foresook embryology and devoted the following years to studies in entomology (bees).

A more persevering worker in the field of embryology was Hermann Bautzmann. He was actually a student of medicine, but felt so strongly attracted by what went on in Spemann's institute that he came to join us in our efforts to solve the riddles of the organizer.

In one of his projects, Bautzmann (1926) proceeded to determine the extent of the dorsal region in the early *Triturus* gastrula, which, when transplanted into another gastrula, can induce a neural plate. His results indicated that not only the small area that Spemann loosely defined as upper blastoporal lip, but also fragments taken from the entire CMD possess neuralizing capacities. In broad outline, Bautzmann's results were subsequently verified by other workers.

Bautzmann (1929) also showed that small pieces consisting solely of primordial notochord can induce a small neural tube. Later, Holtfreter (1933c) demonstrated that somitic mesoderm, by itself, can likewise operate as the inductor of a neural tube. It was concluded that, in normal development, the trunk portion of the neural plate that becomes spinal cord arises under the inductive influence of these two quite different tissue primordia, those of the notochord and those of the bilateral rows of somites. These primordia cooperate in a synergistic fashion, forming an inductor system whose function is to bring forth the typically shaped structure of a spinal cord. It has been assumed, though not proven, that the neuralizing effect of these two primordia is based upon the action of one and the same chemical agent. Bautzmann became professor of anatomy at the University of Kiel, where he continued his studies on the problem of induction.

7. Exogastrulation and Its Consequences

In the case of neural, as well as in various other kinds of inductions, the inductor is generally of mesodermal derivation, whereas the reacting tissue is usually represented by ectoderm. The question arose: What would become of the ectoderm if it is prevented from establishing contact with the mesoderm? A clear-cut answer to this question was provided by the following experiment (Holtfreter, 1933d). Early gastrulae of the axolotl were deprived of all their gelatinous and membranous coverings and then raised in a slightly hypertonic physiological salt solution, later to be known as Holtfreter solution. To my great surprise, these embryos failed to gastrulate. The cell material of the prospective mesoderm and endoderm, instead of moving into the blastocoel, beneath the ectoderm, moved in an opposite direction, away from the ectoderm, which was eventually left behind as an empty bag (Fig. 3a–d). This resulted in a complete separation of the entire layer of ectoderm from the evaginated mass of endomesoderm (Fig. 4). How about the differentiations that were achieved by the exogastrulated endomesodermal complex? It turned out that this material differentiated very much according to its prospective fate, although the anatomical arrangement of the tissues became quite abnormal.

With the help of localized vital staining, it was found that the entire mesoderm moved into the core of the endoderm. The endoderm failed to form an intestinal tube but spread over the mesoderm, with its secreting surface facing the outer medium. Nevertheless, in this abnormal position all the sections of the gut (pharynx, stomach, mid- and hindgut) became well differentiated.

The differentiation of the dislocated layer of mesoderm was likewise quite normal. There appeared an elongated notochord, a series of somites, head mesoderm, pronephric tubules, and so forth.

By contrast, the self-isolated layer of ectoderm failed to produce any of the structure (neural plate, sense organs, neural crest derivatives) normally derived from the ectoderm. Yet, it had not stopped developing altogether. Deprived of inductive stimulation, the ectoderm continued digesting its yolk and differentiated uniformly into a mass of epidermal cells that lacked any kind of organization. The cells would become ciliated and mucus-producing, but (for lack of a

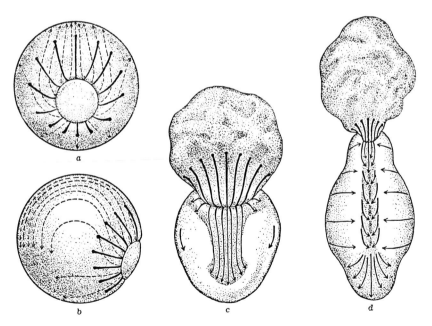

Figure 3. Exogastrulation in axolotl. Diagrams showing the mass movements (**c,d**) compared with those of normal gastrulation (**a,b**). (After Holtfreter, 1933d.)

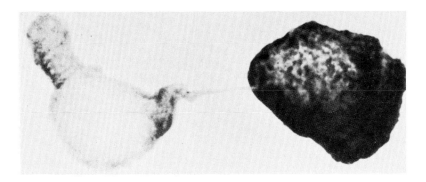

Figure 4. Exogastrulated axolotl embryo, 8 days old. (Right) Self-isolated ectoderm. (Left) Exogastrulated endomesoderm. (From Holtfreter, 1933*a*.)

substratum) they failed to spread as an epithelial sheet. Eventually, this mass of cells dissociated into single, still mobile cells.

Clearly, then, without invagination of the CMD beneath the ectoderm, there can be no induction of a neural plate. The ectoderm would have to follow its own intrinsic trend of differentiation, namely, into epidermis.

This notion was confirmed in another series of experiments, in which the *Triturus* gastrula was subdivided into small fragments that were then cultured as explants (Holtfreter, 1938). At this point, only the differentiation capacity of the isolated ectoderm is of interest to us. It turned out that whether they had been derived from the presumptive epidermis or the presumptive neural plate (the "neurectoderm"), the fragments invariably formed a mass of irregularly arranged epidermal cells, which later disaggregated.

We may then conceive of the gastrula ectoderm as a homogeneous layer of equivalent cells that are genetically predisposed to differentiate exclusively into epidermis. However, the fate of the ectoderm is not determined at this stage. The ectoderm is pluripotent. Experiments have shown that any part of the ectodermal layer can be transformed into actually any kind of tissue, including mesodermal as well as endodermal ones, provided it is subjected to the proper inductive stimuli. It is important to realize that these inductive factors are normally of exogenous derivation. In most cases of induction, it is the mesoderm that produces the inducing agents, and it is the adjacent ectoderm that reacts to them. These extrinsic inductive factors are of a truly determining nature. In order to be effective, they must be able to suppress the action of the genetic directives originally present in the ectodermal cells themselves. It then seems justified to refer to the exogenous inductive factors as cytoplasmic determinants of cell differentiation.

But this leaves us with the questions of how and where do these cytoplasmic determining agents originate. No precise information on this point is available, but it can hardly be doubted that further analysis will show that it is, in the last resort, the genes that hold the key to the answer to this question.

8. Head and Trunk–Tail Inductors

In his last experiments, Spemann (1931b) tested, by means of implantation, the inductive capacities of (1) upper blastoporal lip taken from an early gastrula, in which invagination had just begun, and (2) blastoporal lip from an advanced gastrula, in which the anterior parts of the CMD had already invaginated. The inductive power of these two kinds of implants varied. The former induced brainlike structures, irrespective of whether the implant was positioned in the head or trunk region of the host. The latter would elicit the formation of an axial system when lying in the host's trunk region but could also induce brain structures when lying in the host's head region. These results seemed to indicate that the material that invaginates first acts as a head organizer, whereas the material that invaginates later operates as a trunk–tail organizer. But the aberrant cases (i.e., when the latter induced a brain in the head region) did not fit into this scheme, and altogether these experiments suffered from the shortcoming that they were performed homoplastically. It was not possible to tell, therefore, which of the tissues observed had come from the implant and which represented induced host tissues.

This shortcoming did not exist in the experiments of Holtfreter (1936), which were performed in xenoplastic combinations between *Bombinator* and *Triturus*. The embryonic tissues of these species can be readily distinguished because of differences in the size and stainability of the nuclei. I also used a new method to test the inductive capacity. Instead of transplanting the material to be tested into an intact gastrula, it was wrapped in an isolated piece of ectoderm—the sandwich method. Thus, any possible determining effects derived from the host were avoided.

I first tested the effect of the upper blastoporal lip of an early gastrula on the ectodermal wrapping. The explants were cultured long enough to permit full differentiation of their tissues. The 16 cases of this experiment, which were studied histologically, gave very much the same results. The piece of blastoporal lip invariably differentiated into a great variety of tissues, only some of which corresponded to the prospective fate of the blastoporal region. These tissues were represented by pharyngeal endoderm, head mesoderm, and bits of anterior notochord and somites. But the implant also produced a considerable amount of neural tissue, a tissue normally derived from the ectoderm. On the other hand, the ectodermal envelope was induced to form a large mass of brain tissue that could be associated with a well-shaped eye and other structures of a cephalic nature. These results confirmed Spemann's conclusion that the blastoporal material of an early gastrula is capable of inducing brain and other head structures.

In a second experiment, the ectodermal envelope likewise consisted of ectoderm taken from an early *Triturus* gastrula. But the enclosed material represented a piece of the already invaginated chorda–mesoderm from the trunk region of an advanced neurula of *Bombinator*. The implants of *Bombinator* differentiated into the tissues they would have formed normally (i.e., into a fragment of notochord and a number of somites). Unlike the pieces taken from the CMD of a gastrula, they failed to produce neural tissue or to grow out into a tail. This indicated that the fate of the chorda–somite system of the advanced neurula is determined and hence has lost the regulatory capacity of forming neural tissue.

As for the ectoderm, one could have expected that it would be induced to form merely a piece of spinal cord, but this was not the case. Instead, the ectoderm was induced to form a complete, well-proportioned axial system consisting of a rod of notochord, bilateral rows of somites, and a spinal cord, all of which grew out into a tail provided with a mesenchyme-containing fin. These structures were of a uniquely ectodermal derivation. They were in no way related to the axial configuration of the inducing implant, and there was no trace of assimilatory induction.

There is then sufficient evidence for the notion that the inducing stimuli are regionally specific: *Normally* the cell material that invaginates first and that comes to underlie the anterior part of the neural plate induces cephalic structures, whereas the material that invaginates later and becomes the chorda–somite axial system induces the ectoderm to form a spinal cord. But why, then, in these *experimental* conditions, does the chorda–somite system induce more than that, i.e., a whole axial system? I cannot answer this question. One may speculate, however, that normally the embryo disposes of factors that to some extent interfere with the free expression of the principle of induction. What was induced was not a mosaic of different tissues but a pluripotent trunk–tail field, which then differentiated (independently of any further directives from the inductors) into the complex pattern of a typically organized axial system.

9. In Quest of the Nature of the Inducing Agents

The investigations of the induction phenomenon took a dramatic turn during the early 1930s, when efforts were made to identify the nature of the inducing factors. This quest expanded in different directions, and researchers from all over the globe became engaged in it. That was the heyday of experimental embryology. But, as time passed, the initial euphoria faded. An avalanche of experimental data was collected, but rather than enlightening us, they led to increasing confusion. Thus, this campaign in search of the inducing agents that started out with great expectations petered out during the years of the last world war and ended, during the 1960s, on a note of disappointment and despair.

More or less comprehensive reviews of this fascinating scientific episode have been written by Spemann (1938), Needham (1942, 1968), Holtfreter (1951), Holtfreter and Hamburger (1955), and Saxén and Toivonen (1962). I shall content myself with pointing out only some of the highlights of this story.

Following Spemann's suggestions, several workers, including myself (Holtfreter, 1932, 1934a), subjected the organizer tissues derived from the early gastrula to various treatments that destroyed their vital structure and, as Spemann seemed to expect, their inducing capacity as well. My experimental results did not support his notion. I found that when pieces of the blastoporal lip, containing the head inductors, which had been subjected to drying at 60°C, to freezing, or to temperatures of up to 120°C, were implanted into a live gastrula, they were still capable of inducing the ectoderm to form voluminous portions of a composite brain, sometimes associated with sense organs. This neuralizing capacity gradually disappeared if the blastoporal material had been exposed for a few minutes to heat of above 140°C. This indicated that the neuralizing effect is caused by a chemical that is relatively heat stable.

It also became apparent that treatment of the genuine brain inductors with ethanol, xylene, or ether failed to reduce their neuralizing capacity. More surprising still, I found that parts of the amphibian embryo that normally do not operate as inductors (such as the gastrula ectoderm and endoderm) *became* powerful neural inductors, in consequence of having been subjected to desiccation, heat, or other devitalizing treatments (Holtfreter, 1934a). Furthermore, all parts of a fertilized amphibian egg that had been heat-coagulated proved to be effective neural inductors when transplanted into a live gastrula.

We had held to the basic notion that development is an orderly event, consisting of innumerable part-processes, including the various processes of induction. Order is sustained because these processes normally occur predictably at certain developmental stages and only in certain parts of the embryo. There would be chaos if it were otherwise.

Obviously, although the presence of neuralizing agents in all parts of the egg and embryo is indisputable, it also seems clear that the organism normally disposes of mechanisms that permit the inducing agents to express themselves exclusively in certain tissues (the certified inductors) and not in others. Therefore, one may assume that not only inductive but also inhibitory signals pervade the organism.

The confusion grew further when it was found that inducing substances were present in every tissue or organ derived from various animals that were tested (Holtfreter, 1934b). The specimens employed were the adult forms of amphibians, fish, reptiles, birds, mammals (including man), and even inverte-

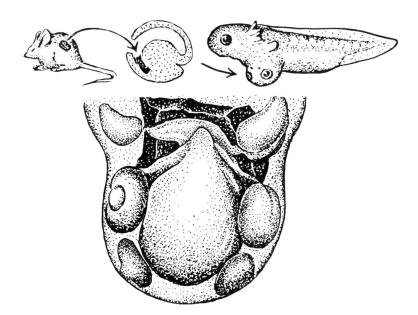

Figure 5. After implantation of a fragment of mouse kidney into a *Triturus* gastrula, a brain flanked by nasal placodes, eyes, and ear vesicles is induced. (From Chuang, 1939.)

brates. The tissues implanted into a *Triturus* gastrula were liver, kidney, muscle, thyroid, brain, ovary, and others. It turned out that most of these foreign tissues (whether fresh or heat-treated) were capable of inducing neural structures of varying size and complexity, whereas others brought forth the appearance of a whole trunk–tail axial system. In other words, some of these foreign tissues could perfectly imitate the action of either the genuine head or trunk-tail inductors.

This was convincingly demonstrated in experiments carried out by Chuang (1938, 1939), a former student of mine and now professor at the Academia Sinica, Shanghai. After implanting a piece of mouse kidney into a gastrula, Chuang obtained a composite brain with eyes, which was bilaterally symmetrical and was associated with secondarily induced pairs of otocysts and olfactory grooves (Fig. 5). In another case in which he enclosed a fragment of salamander liver in an ectodermal jacket, Chuang obtained a well-organized tail axial system (Fig. 6).

These findings, spectacular as they were, left us with many open questions. What, for instance, is the biological significance of the fact that inducing factors are present in such a great diversity of adult tissues, if these tissues are evidently far beyond the stage in which embryonic inductions take place? What role, if any, do these inducing substances play in the physiology of the fully differentiated tissues?

These experimental findings with foreign tissues did not elucidate what goes on in normal embryogenesis, but they refuted Spemann's hypothesis that a certain vital structure of the inducing tissues plays an essential role in the

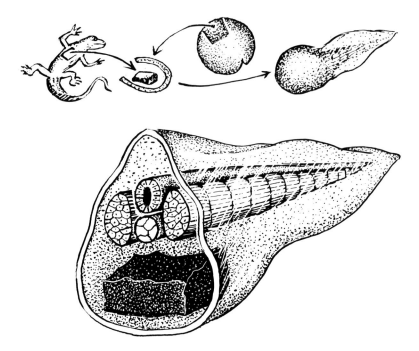

Figure 6. Implantation of a fragment of *Triturus* liver into a jacket of gastrula ectoderm causes the emergence of a typical self-organized tail. (From Chuang, 1938.)

induction process. If a piece of adult mouse kidney can induce a complete brain with sense organs, and a piece of salamander liver can bring forth a complete axial system, this was evidently not because of their specific inner structure but because of some chemical agent(s) that they transmitted to the ectoderm.

Subsequently, great efforts were made to identify the nature of the inducing agents. The trouble was that too many chemically unrelated substances were found that, when applied to the competent ectoderm, had a neuralizing (or mesodermalizing) effect. Effective inducing agents included fatty acids, various polycyclic hydrocarbons (some of which were known as carcinogens), sulfhydryl compounds, nucleoproteins or proteins extracted from homogenized chicken embryos or bone marrow, or cyclic adenosine monophosphate (cAMP). But neuralizations were also obtained by subjecting the isolated ectoderm to shock treatment with Holtfreter solution of high pH or free of calcium, or containing 10% ethanol. More recently, it was found that culturing ectodermal explants in LiCl solution brings about their differentiation into neural and other cell types. It is an open question as to which, if any, of these chemically disparate substances is instrumental in the normal processes of embryonic induction.

10. Conclusions

The inquiries into the chemical nature of inducing agent(s), although inevitable and desirable, have led us astray. The fundamental principles underlying embryonic induction have remained obscure. There is, however, this consolation: Animal development is predictable, repeating itself in every generation. We can be sure that the eggs of frogs and salamanders will always develop according to the very same blueprint that has guided development of their ancestors for eons. How fortunate for the future generations of embryologists. They will still have the opportunity to study and solve the old riddles of development that have defied scientists up till now.

References

Bautzmann, H., 1926, Experimentelle Untersuchungen zur Abgrenzung des Organisationszentrums bei *Triton taeniatus*, *Wilhelm Roux Arch. Entwicklungsmech.* **108**:283–321.

Bautzmann, H., 1929, Über Induktion durch vordere und hintere Chorda der Neurula, *Wilhelm Roux Arch. Entwicklungsmech.* **119**:1–46.

Chuang, H. H., 1938, Spezifische Induktionsleistungen von Leber und Niere im Explantationsversuch, *Biol. Zentralbl.* **58**:472–480.

Chuang, H. H., 1939, Induktionsleistungen von frischen und gekochten Organteilen (Niere, Leber) nach ihrer Verpflanzung in Explantate und verschiedene Wirtsregionen von Tritonkeimen, *Wilhelm Roux Arch. Entwicklungsmech.* **139**:556–638.

DeBeer, G. R., 1927, The mechanics of vertebrate development, *Biol. Rev.* **2**:137–197.

Geinitz, B., 1925, Embryonale Transplantation zwischen Urodelen und Anuren, *Wilhelm Roux Arch. Entwicklungsmech.* **106**:357–408.

Hamburger, V., 1984, Hilde Mangold, Co-Discoverer of the Organizer, *J. Hist. Biol.* **17**:1–11.

Holtfreter, J., 1932, Versuche zur Analyse der Induktionsmittel in der Embryonalentwicklung, *Naturwissenschaften* **30**:791–794.

Holtfreter, J., 1933a, Organisierungsstufen nach regionaler Kombination von Entomesoderm mit Ektoderm. *Biol. Zentralbl.* **53**:404–431.

Holtfreter, J., 1933b, Nachweis der Induktionsfähigkeit abgetöteter Keimteile, *Wilhelm Roux Arch. Entwicklungsmech.* **127**:584–633.

Holtfreter, J., 1933c, Der Einfluss von Wirtsalter und verschiedenen Organbezirken auf die Differenzierung von angelagertem Gastrulaektoderm, *Wilhelm Roux Arch. Entwicklungsmech.* **127**:620–775.

Holtfreter, J., 1933d, Die totale Exogastrulation, eine Selbstablösung des Ektoderms vom Entomesoderm, *Wilhelm Roux Arch. Entwicklungsmech.* **129**:669–793.

Holtfreter, J., 1934a, Der Einfluss thermischer, mechanischer und chemischer Eingriffe auf die Induzierfähigkeit von Tritonkeimteilen, *Wilhelm Roux Arch. Entwicklungsmech.* **132**:225–306.

Holtfreter, J., 1934b, Über die Verbreitung induzierender Substanzen und ihre Leistungen im Triton-Keim, *Wilhelm Roux Arch. Entwicklungsmech.* **132**:307–383.

Holtfreter, J., 1936, Regionale Induktionen in xenoplastisch zusammengesetzten Explantaten, *Wilhelm Roux Arch. Entwicklungsmech.* **134**:466–550.

Holtfreter, J., 1938, Differenzierungspotenzen isolierter Teile der Urodelengastrula, *Wilhelm Roux Arch. Entwicklungsmech.* **138**:522–656.

Holtfreter, J., 1951, Some aspects of embryonic induction, *Growth* **10**(suppl.):117–152.

Holtfreter, J., and Hamburger, V., 1955, Embryogenesis: Progressive differentiation. Amphibians, in: *Analysis of Development.* (B. H. Willier, P. A. Weiss, and V. Hamburger, eds.), pp. 230–296, W. B. Saunders, Philadelphia.

Hörstadius, S., 1928, Über die Determination des Keimes bei Echinodermen, *Acta Zool.* **9:**1–192.

Lewis, W. H., 1907, Transplantation of the lips of the blastopore in *Rana palustris*, *Am. J. Anat.* **7:**137–143.

Mangold, O., 1923, Transplantationsversuche zur Frage der Spezifitat und der Bildung der Keimblätter bei Triton, *Arch. Mikrosk. Anat. Entwicklungsmech.* **100:**198–301.

Mangold, O., 1953, Hans Spemann, J. F. Steinkopf, Stuttgart.

Marx, A., 1931, Über Induktionen durch narkotisierte Organisatoren, *Roux Arch. Entwichlungsmech.* **123:**333–388.

Needham, J., 1942, *Biochemistry and Morphogenesis*, Cambridge University Press, Cambridge.

Needham, J., 1968, Organizer phenomena after four decades: A retrospect and prospect, in *Haldane and Modern Biology* (K. R. Dronamraju, ed.), pp. 277–298, John Hopkins Press, Baltimore.

Saxén, L., and Toivonen, S., 1962, *Primary Embryonic Induction*, Logos Press, London,–Prentice-Hall, Englewood Cliffs, New Jersey.

Spemann, H., 1901, Entwicklungsphysiologische Studien am Tritonei I, *Wilhelm Roux Arch. Entwichlungsmech.* **12:**224–264.

Spemann, H., 1902, Entwicklungsphysiologische Studien am Tritonei II, *Wilhelm Roux Arch. Entwichlungsmech.* **15:**448–534.

Spemann, H., 1903, Entwicklungsphysiologische Studien am Tritonei III, *Wilhelm Roux Arch. Entwichlungsmech.* **16:**551–631.

Spemann, H., 1918, Über die Determination der ersten Organanlagen des Amphibienembryo. I–VI, *Wilhelm Roux Arch. Entwichlungsmech.* **43:**448–555.

Spemann, H., 1921, Die Erzeugung tierischer Chimaeren durch heteroplastische embryonale Transplantation zwischen *Triton cristatus* u. *taeniatus*, *Roux Arch. Entwichlungsmech.* **48:**533–570.

Spemann, H., 1923, Zur Theorie der tierischen Entwicklung, Freiburg Br. Speyer und Kaerner, Universitatsbuchhandlung.

Spemann, H., 1927, Neue Arbeiten über Organisatoren in der tierischen Entwicklungsmech. *Naturwiss.* **15:**946–951.

Spemann, H., 1928, Die Entwicklung seitlicher und dorso-ventraler Keimhälften bei verzögerter Kernversorgung, *Z. Wiss. Zool.* **132:**105–134.

Spemann, H., 1931a, Das Verhalten von Organisatoren nach Zerstörung ihrer Struktur, *Verh. D. Zool. Ges.* **1931:**129–132.

Spemann, H., 1931b, Über den Anteil von Implentat und Wirtskeim an der Orientierung und Beschaffenheit der induzierten embryonalanlage, *Wilhelm Roux Arch. Entwichlungsmech.* **123:**390–516.

Spemann, H., 1938, *Embryonic Development and Induction*, Yale University Press, New Haven, Connecticut.

Spemann, H., and Geinitz, B., 1927, Über Weckung organisatorischer Fähigkeiten durch Verpflanzung in organisatorische Umgebung, *Wilhelm Roux Arch. Entwichlungsmech.* **109:**129–175.

Spemann, H., and Mangold, H., 1924, Über Induktion von Embryonalanlagen durch Implantation artfremder Organisatoren, *Arch. Mikrisk. Anat. Entwicklungsmech.* **100:**599–638; reprinted in 1974, in: *Foundations of Experimental Embryology*, 2nd ed. (B. Willier and J. Oppenheimer, eds.), pp. 146–184, (English transl.), Hafner Press, New York.

Vogt, W., 1929, Gestaltungsanalyse am Amphibienkeim mit örtlicher Vitalfarbung, *Wilhelm Roux Arch. Entwichlungsmech.* **120:**384–706.

Chapter 5

The Molecular Biology of Pattern Formation in the Early Embryonic Development of *Drosophila*

MATTHEW P. SCOTT

> In spite of ourselves the question is forced upon us whether we must not assume the existence, even in the unsegmented egg, of a quite definite and orderly grouping and distribution of the protoplasmic particles and molecules.
>
> C. Rabl (1879)

1. Central Problems of Early Development

The techniques of molecular biology have been brought to bear on the mysteries of early development in the hope that new approaches would lead to a clearer understanding of how overt pattern is formed from its invisible antecedents. The molecules involved in pattern formation have been made visible by molecular probes and have been manipulated by mutation, genetic engineering, and injection. In the study of *Drosophila* development, the key discoveries made using molecular techniques have been heavily dependent on the genetic analyses that provided the molecular biologists with working material. In many ways, genetic studies continue to lead the way, guiding molecular biologists to the genes that matter most.

What are the key issues of *Drosophila* developmental genetics? One major goal is to understand how genes govern the formation of morphology. This problem is being attacked from two directions. The first approach is to identify structural components of cells that appear to be important for developmental processes, to map the genes for such components, and to induce mutations in the genes to see what the developmental effects are. The second approach is to obtain mutations in the developmental process of interest and then use mo-

MATTHEW P. SCOTT • Department of Molecular, Cellular, and Developmental Biology, University of Colorado, Boulder, Colorado 80309-0347.

lecular techniques to identify the gene products and their mechanism of action. It is to be hoped that the two approaches will converge, such that the control of the assembly of structural components, and of cell movements and divisions, will be understood in terms of the actions of the central regulatory loci that have been discovered. Thus, part of the understanding of early development must be sought in an exploration of the basis for cell asymmetries, such as polarized distributions of surface molecules, organized cytoskeletal arrays that alter cell shapes, and the tendency of cells to orient their mitoses in certain directions. Another class of answers will come from studying genes in which mutations alter the course of development by reorganizing, as opposed to destroying, developmental processes.

The original choice of *Drosophila* as an experimental organism was based primarily on its short life cycle and the many morphological features that could serve for detecting mutations. It is not in all respects an ideal organism for developmental studies. It is rather small for easy observation, and it is difficult to obtain large amounts of any of its tissues. These drawbacks are offset, however, by some important advantages. For the developmental biologist, the fly embryo offers surprising insensitivity to abuses such as injections, ligations, transplantations, and ablations, permitting many critical experiments to be done that would probably be fatal to many organisms. Development occurs twice, in a sense—once during embryogenesis and once during metamorphosis. The early embryo offers a relatively (and deceptively) simple structure consisting of only about 6000 cells. For the molecular biologist the advantages are a small genome, polytene chromosomes as guides for gene cloning, the opportunity to reintroduce genes into the germline in a functional form, and a large and rapidly growing collection of already cloned loci. Most importantly, more than 70 years of genetic analysis have led to the identification of many genes that appear to regulate various aspects of development (see review by Mahowald and Hardy, 1985). Two classes of genes that have been especially useful for studying the regulation of development are the **homeotic genes,** recognized by mutations that transform one part of the fly into another (reviewed in Gehring, 1986; Scott, 1987), and the **segmentation genes,** identified by mutations that cause changes in the pattern of body segmentation (reviewed in Gergen *et al.,* 1985 and Scott and O'Farrell, 1986). Both classes of genes are involved in gene networks, and the hierarchical organization of each network is an area of active current research. The genes that have been described to date are listed, with their phenotypes, in Table I; they are organized according to their class and whether they are active maternally or zygotically. The compilation includes about 60 genes, about 0.6–1.2% of the estimated 5–10,000 genes in the *Drosophila* genome. New genes are still being discovered, but systematic screens of the genome suggest that most genes that regulate segmentation and segmental identity have already been found. About one third of the identified genes have been cloned, and the pace of gene cloning continues to accelerate. The work is entering a new phase in which a primary emphasis will be on the molecular interactions among the genes and their products.

Table I. Genes That Control Early Embryogenesis[a–c]

Mutation	Phenotype	Reference
Maternal effect genes controlling anterior–posterior polarity		
bicaudal	Embryo development with two posterior ends	Nüsslein-Volhard (1979)
dicephalic	Embryo development with two anterior ends or, rarely, two posterior ends	Lohs-Schardin (1982)
Maternal-effect genes controlling dorsal–ventral differentiation		
dorsal	Ventral structures missing from embryo; only dorsal structures formed	Nüsslein–Volhard *et al.* (1980); Steward *et al.* (1985)
Toll	Dorsalization of embryo	Anderson *et al.* (1985)
gastrulation defective	Dorsalization of embryo	Anderson and Nüsslein-Volhard (1984)
nudel	Dorsalization of embryo	Anderson and Nüsslein-Volhard (1984)
tube	Dorsalization of embryo	Anderson and Nüsslein-Volhard (1984)
pipe	Dorsalization of embryo	Anderson and Nüsslein-Volhard (1984)
snake	Dorsalization of embryo	Anderson and Nüsslein-Volhard (1984)
easter	Dorsalization of embryo	Anderson and Nüsslein-Volhard (1984)
spätzle	Dorsalization of embryo	Anderson and Nüsslein-Volhard (1984)
pelle	Dorsalization of embryo	Anderson and Nüsslein-Volhard (1984)
Maternal-effect segmentation genes		
staufen	Central abdominal segments deleted; no pole cells	Schüpbach and Wieschaus (1986)
tudor	Most abdominal segments deleted; no pole cells	Boswell and Mahowald (1985)
vasa	Most abdominal segments deleted; no pole cells	Schüpbach and Wieschaus (1986)
valois	Most abdominal segments deleted; no pole cells	Schüpbach and Wieschaus (1986)
torso	Anterior and posteriormost structures of embryo missing; remaining pattern elements spread out; pole cells present	Schüpbach and Wieschaus (1986)

(continued)

Table I. (*Continued*)

Mutation	Phenotype	Reference
trunk	Like *torso*	Schüpbach and Wieschaus (1986)
fs(1)Nasrat211	Like *torso*	Degelmann *et al.* (1986)
oskar	Deletions of abdominal segments; no pole cells	Lehmann (1985)
exuperantia	Anterior pattern deletions like *torso*; posterior development largely normal; pole cells present	Schüpbach and Wieschaus (1986)
Maternal-effect homeotic gene		
fs(1)h	One or more thoracic or abdominal segments missing; some escaping survivors have third thoracic segments transformed into second thoracic segments; enhanced by mutations in *trithorax*	Digan *et al.* (1986)
Homeotic genes active maternally and zygotically		
Polycomb	Transformations of all segments toward eighth abdominal-like pattern	Duncan and Lewis (1982)
Polycomblike 1(4)29	Like *Polycomb*	Duncan (1982)
	Posteriorly directed homeotic transformations; some anteriorly directed transformations as well	Gehring (1970); Denell and Hummels (1985)
Extra sex combs	Like *Polycomb*, but stronger maternal effect	Struhl (1981b)
trithorax	Homeotic transformations opposite to those produced by the *Polycomb* group: abdominal segments are transformed to first abdominal-like segments; first and third thoracic segments transformed into second thoracic segmentlike structures	Ingham and Whittle (1980)
super sex combs	Posteriorly directed transformations of the thoracic and abdominal segments	Ingham (1984)
Additional sex combs	Posteriorly directed transformations; enhanced by mutations in *Posterior sex combs* or *Sex comb on midleg*	Jürgens (1985)
Posterior sex combs	Like *Additional sex combs*; interacts with it and with *Sex comb on midleg*	Jürgens (1985)

Table I. (*Continued*)

Mutation	Phenotype	Reference
Sex comb on midleg	Like *Additional sex combs*; interacts with it and with *Posterior sex combs*	Jürgens (1985)
polyhomeotic	Adult homeotic transformations that suggest gene belongs in the class of negative regulators of the *bithorax* complex and *Antennapedia* complex (however, there is no embryonic lethal phenotype)	Dura *et al.* (1985)
Segmentation gene active maternally and zygotically		
hunchback (also called *Regulator of postbithorax*)	Embryos missing gnathal, thoracic, and most posterior abdominal segments	Lehmann and Nüsslein-Volhard (1986); Bender *et al.* (1986)
Zygotically active genes involved in dorsal–ventral differentiation		
zerknüllt	Embryos have defects in gastrulation; some dorsal structures missing	Wakimoto *et al.* (1984); Doyle *et al.* (1986)
twist	Dorsalization of embryo; interacts with *dorsal*	Simpson (1983)
snail	Similar to *twist*	Simpson (1983)
Zygotically active segmentation genes		
knirps	Central abdominal segments deleted from pattern	Nüsslein-Volhard and Wieschaus (1980)
Krüppel	Thoracic and anterior abdominal segments replaced by a mirror image of posterior abdominal structures	Wieschaus *et al.* (1984b); Priess *et al.* (1985); Knipple *et al.* (1985); Rosenberg *et al.* (1985, 1986)
giant	Defects in labial segment, first and second thoracic segments, and fifth through seventh abdominal segments	Wieschaus *et al.* (1984a)
tailless	Eighth through tenth abdominal segments missing; remaining abdominal segments expanded; pole cells present	Strecker *et al.* (1986)
fushi tarazu	Pair-rule defects: deletions of anterior parts of odd-numbered abdominal segments and posterior parts of even-numbered abdominal segments	Wakimoto and Kaufman (1981); Wakimoto *et al.* (1984); Weiner *et al.* (1984); Kuroiwa *et al.* (1984); Hafen *et al.* (1984a); Carroll and Scott (1985, 1986)

(*continued*)

Table I. (*Continued*)

Mutation	Phenotype	Reference
	Corresponding parts of thoracic segments (e.g., anterior even-numbered) also deleted; phenotypes below also described in terms of abdominal segments but corresponding thoracic segments implied as well	
hairy	Pair-rule defects: deletions of anterior region of even-numbered abdominal segments and posterior region of odd-numbered segments	Nüsslein-Volhard and Wieschaus (1980); Holmgren (1984); Ingham *et al.* (1985); Ish-Horowicz *et al.* (1985)
even-skipped	Pair-rule defects like *hairy* with reduced function alleles; no segments formed with null alleles	Nüsslein-Volhard *et al.* (1984)
paired	Pair-rule defects like those of *hairy*	Nüsslein-Volhard and Wieschaus (1980); Kilchherr *et al.* (1986)
odd-skipped	Pair-rule defects: anterior parts of odd-numbered abdominal segments deleted	Nüsslein-Volhard and Wieschaus (1980)
sloppy-paired	Central parts of odd-numbered segments deleted; extreme anterior parts of even-numbered segments also missing	Gergen *et al.* (1985)
runt	Pair-rule defects: like those of *ftz* but shifted somewhat more anteriorly; missing parts replaced by mirror-image duplications of remaining regions	Gergen and Wieschaus (1985, 1986)
odd-paired	Pair-rule defects: like *runt*, but without duplications	Gergen *et al.* (1985)
engrailed	Segment polarity defects with some characteristics of pair-rule phenotype; defects in posterior part of every segment, but often see pair-rule-like deletion patterns in either of the two alternate frames	Kornberg (1981); Kuner *et al.* (1985); Kornberg *et al.* (1985); DiNardo *et al.* (1985)
patched	Segment polarity defects: central part of each segment replaced by mirror-image of remaining structures; segment borders duplicated	Nüsslein-Volhard and Wieschaus (1980)
gooseberry	Segment polarity defects: posterior part of each segment	Nüsslein-Volhard and Wieschaus (1980)

Table I. (*Continued*)

Mutation	Phenotype	Reference
	replaced by a mirror image of remaining structures	
fused	Like *gooseberry*	Nüsslein-Volhard and Wieschaus (1980)
wingless	Segment polarity defects: posterior region of each segment, including the segment boundary, replaced with a mirror image of remaining anterior structures; therefore no segment boundaries left	Nüsslein-Volhard and Wieschaus (1980)
hedgehog	Similar to *wingless*	Nüsslein-Volhard and Wieschaus (1980)
cubitus interruptus	Similar to *gooseberry*	Nüsslein-Volhard and Wieschaus (1980)
armadillo	Segment polarity defects: posterior region of each segment replaced by a mirror image of remaining structures	Gergen *et al.* (1985)
Zygotically active homeotic genes		
Ultrabithorax	Parasegments 5 and 6 transformed into parasegment 4	Lewis (1978); Bender *et al.* (1983); Sanchez-Herrero *et al.* (1985); Hogness *et al.* (1985)
iab 2 (abd A)	Abdominal segments transformed into anterior first abdominal segment in the anterior and posterior third thoracic segment in the posterior; effect weaker posterior to the fourth abdominal segment	Lewis (1978); Karch *et al.* (1985); Sanchez-Herrero *et al.* (1985)
iab 7 (Abd B)	Fifth through eighth abdominal segments transformed into fourth or fifth abdominal segments	Lewis (1978); Karch *et al.* (1985); Sanchez-Herrero *et al.* (1985)
Antennapedia	Second and third thoracic segments transformed into hybrid first thoracic/gnathal segments	Wakimoto and Kaufman (1981); Garber *et al.* (1983); Scott *et al.* (1983); Levine *et al.* (1983); Kaufman and Abbott (1984); Martinez-Arias (1986); Abbott and Kaufman (1986); Schneuwly *et al.* (1986); Laughon *et al.* (1986); Stroeher *et al.* (1986)
Sex combs reduced	First thoracic segment transformed into second; labial segment transformed into maxillary segment	Wakimoto and Kaufman (1981); Scott *et al.* (1983); Kuroiwa *et al.* (1985)

(*continued*)

Table I. (*Continued*)

Mutation	Phenotype	Reference
Deformed	Effects on the mandibular segment	Kaufman and Abbott (1984); Regulski *et al.* (1985)
proboscipedia	Proboscis transformed into first legs or antennae	Kaufman (1975)

[a]Loci are grouped according to function and time of action. In some cases, not all times of action have been established. Only embryonic presumed loss of function phenotypes are mentioned, and the details of phenotypes are not included here. Consult the original references for more complete descriptions.
[b]In addition to the loci listed here, about 40 loci that may be in the *Polycomb*-like class have been mentioned but not described (Jürgens, 1985).
[c]Some segmentation genes that have maternal effects are being analyzed in germline clones, since the genes are also necessary for zygotic development and therefore embryos from uniformly homozygous females cannot be analyzed. This class of genes may amount to about 40 loci (Perrimon *et al.*, 1985). Therefore, the total number of genes currently suspected to be involved in regulation of segmentation and related developmental processes is about 150.

2. Early *Drosophila* Development

The appealingly simple structure of the early *Drosophila* embryo (Fig. 1) makes it a prime target of experimentation. The embryo is about 0.5 mm long by about 0.3 mm in diameter. At the blastoderm stage the embryo is composed of a monolayer of cells surrounding a central region of yolk (Turner and Mahowald, 1976). The cells are columnar, have elongated nuclei that are active in transcription, and are ultrastructurally uniform. At the posterior end of the embryo is a group of cells destined to become the germ-line cells (Fig. 1d). These pole cells are the earliest cells to form and are the only morphologically distinct blastoderm cells. Despite the apparent uniformity of the cells, the cells at the blastoderm stage are not uniform in their developmental programming. Ablation and transplantation experiments demonstrated some years ago that, by the blastoderm stage, cells are determined to become parts of certain body segments (Lohs-Schardin *et al.*, 1979; Underwood *et al.*, 1980; Hartenstein *et al.*, 1985).

Before the cellular blastoderm stage, *Drosophila* embryos go through a cleavage stage that is different from vertebrate or echinoderm cleavage, in that the nuclei divide without cell division. After fertilization, the nuclei divide relatively synchronously and very rapidly eight times in the central part of the embryo and then begin to migrate toward the surface of the egg (Fig. 1a,b). Five more, progressively slower, divisions follow, the last four occurring while the nuclei are located just below the surface of the egg (Foe and Alberts, 1983) (Fig. 1c–e). Membranes then grow in from the outer surface to divide the monolayer of nuclei into cells, forming the cellular blastoderm at about three hours after fertilization.

The process of gastrulation follows quickly after cellular blastoderm formation (Fig. 1f–h) and initially does not involve any cell division. Most cells, except for the neuroblasts, divide only two or three times during the entire course of embryogenesis. The previously invisible differentiation of the blastoderm cells becomes apparent in the different movements of the cells as they participate in the invaginations that occur to form the internal pools of endoderm and mesoderm cells. After the initial invaginations, cells begin to migrate toward the ventral midline of the embryo, moving around the sides. Rather than accumulating along the ventral midline, the cells move posteriorly and around the caudal end of the embryo. The cells destined to form the most posterior structures thus end up positioned on the dorsal side of the embryo just behind the head. This process of **germ band elongation** takes about two and a half hours (ending at about 6 hr after fertilization), during which time the primordial neuroblasts begin to form and move to the interior of the embryo. The first signs of body segmentation appear while the germ band is fully extended. The division of the developing trunk into three gnathal (head), three thoracic, and about nine abdominal segments can be seen as a series of grooves. However, the grooves do not demarcate segments; rather, they demarcate **parasegments.** A parasegment is composed of the posterior of one segment plus the anterior of the next most posterior segment and is therefore offset from the segmental repeat unit. Parasegments are important both because they are the first visible body divisions and because they are the spatial units of expression of some homeotic genes (Hayes *et al.,* 1984; Struhl, 1984; Martinez-Arias and Lawrence, 1985) (see Section 6). As the germ band shortens (between about 7 and 9 hr after fertilization) to return the cells destined to make posterior structures to the posterior pole of the embryo, cells move through the grooves, like water through a standing wave on a river, so that the deep divisions seen later (and in larvae) do correspond to segmental units.

3. Determination of Blastoderm Cell Fates

Detailed fate maps have been constructed of the cellular blastoderm (Lohs-Schardin *et al.,* 1979; Underwood *et al.,* 1980; Hartenstein *et al.,* 1985). Since nuclei transplanted before the cellular blastoderm stage take on the fate of their new surroundings, whereas transplanted cells maintain the fate bestowed on them by their original position (see Illmensee in Lawrence, 1976; Simcox and Sang, 1983), a central question is: What mechanism underlies the commitment and makes it irreversible? The position of a cell (or, perhaps, nucleus) in the blastoderm appears to determine its fate, although the fates of the progeny of each cell must be further refined as new cells are born. Two types of ideas have been proposed to explain how cells could become determined for a certain pathway, depending on their position in the embryo. One idea is that localized determinants, placed in the embryo during oogenesis or becoming localized during the early period of development, serve as factors that influence the nuclei that move to particular positions (see Chapters 1 and 2). The influence could induce the expression of certain genes, for example, in this way beginning the process of differentiation. The factors could be held in place by cytoskeletal anchors, or the cytoskeleton could play an active role in organizing

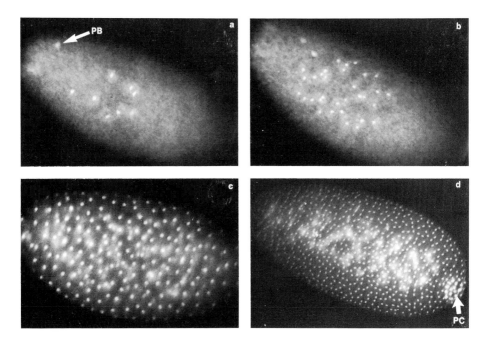

Figure 1. Early *Drosophila* development observed with a DNA stain. Embryos were permeabilized and fixed, then stained with the intercalating dye DAPI, as described by Carroll and Scott (1985). Anterior is to the left in each micrograph, ventral is down, except as noted. (**a**) Syncytial nuclei in an 8-nucleus cleavage-stage embryo. The nuclei are still located near the center of the embryo. PB, polar bodies (the product of meiosis). (**b**) Syncytial nuclei at the 32-nucleus stage. (**c**) Syncytial nuclei after their migration to the cortex of the embryo, after nine divisions. (**d**) Embryo at late anaphase of the eleventh division. Note the pairing of the nuclei. The pole cells (the germline precursor cells; PC) are visible at the posterior end. (**e**) Completed blastoderm stage embryo, after the thirteenth division. Cell membranes are forming. The bright out-of-focus spots in the center of the embryo are nuclei that have remained in the yolk (vitellophages). (**f**) Gastrulation-stage embryo, which has initiated invaginations that will lead to formation of the mesoderm and endoderm. (**g**) Embryo during germ band elongation, stained with DAPI. The pole cells (PC) have moved from the posterior tip toward the anterior along the dorsal part of the embryo. The folds and invaginations of gastrulation can be seen. (**h**) The same embryo as in (**g**), stained with an antibody against an acetylated form of tubulin. Nuclei are unstained, and the cytoplasmic meshwork of tubulin in each cell can be seen. Cells undergoing mitosis (as in the patches marked M) do not stain with the antibody. The pole cells can be seen to disappear into the interior of the embryo. (Panels **g** and **h** courtesy of Nurit Wolf and Margaret Fuller.)

the factors during the earliest stages of development. A second type of idea is that nuclei learn their positions through reading gradients of substances that run through the embryo. One gradient might run in the dorsal–ventral axis, another in the anterior–posterior axis (Nüsslein-Volhard, 1979). Cells would become determined by reading the concentration of the gradient substance and responding with an appropriate pattern of differential gene expression. A vari-

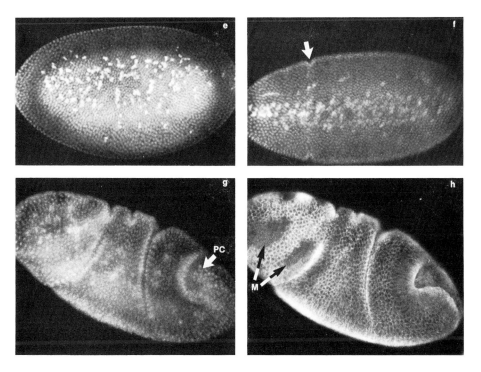

Figure 1. *(continued)*

ant form of this model is to propose that cell communication is important, but that it takes the form of contact between cells rather than a diffusible gradient.

4. Molecular Components of the Cytoskeleton of the Early Embryo

The cytoskeletal components of the early embryo have been examined using antibody probes specific for molecules such as tubulin and actin. A key technological advance was the use of permeabilized whole-mount embryos in immunofluorescence studies. The probes demonstrate that the cortical cytoplasm of the embryo is a highly organized meshwork of filaments (Karr and Alberts, 1986). The dividing, migrating nuclei in the early syncytium are also each surrounded by an organized cloud of cytoplasm that excludes yolk. After the nuclei reach the surface, microtubules extend at least 40 μm into the embryo from the surface. An actin cap lies over each nucleus just below the plasma membrane. During mitosis, the microtubules rearrange to form spindles, and the actin spreads out under the membrane. Transient pseudocleavage furrows form in association with an actin-rich layer. The regulatory basis for all these events is not understood, and little genetic analysis of these early pro-

6

cesses has been done. The meshwork of tubulin in an early gastrula can be seen in Fig. 1G,H.

What could be the functions of the embryonic cytoskeleton? One function is presumably to participate in syncytial nuclear divisions and to move the nuclei to the cortical cytoplasm. Another function may be to convey molecules between nuclei, establishing communication channels that could inform nuclei of their positions along the embryo. The communication could involve direct contact or electrical interactions, rather than, for example, transport of proteins or of morphogen molecules. A third function of the cytoskeleton is likely to be restraint in diffusion of molecules such as mRNAs. As the localized accumulation of segmentation gene messenger RNAs (mRNAs) occurs before the formation of cell membranes, the mRNAs are not free to move about the embryo.

5. Segmentation Genes

The segmental divisions are the most obvious morphological feature of the early embryo and have therefore been a focus of attention in studies of mutants. Segmentation mutants have been categorized into three classes, all of which have deficiencies in the normal pattern of segments. It is important to note that the changes caused by these mutants rarely affect precisely segmental units. Rather, parts of segments, multiple segments, or segment-sized units offset from segments (like parasegments), are deleted from the normal pattern. In some instances, cell death is involved in the pattern deficiencies; in other cases, cells that would have formed some of the missing structures appear to be incorporated into other structures. Therefore, the genetic mechanisms that regulate pattern formation in the *Drosophila* embryo need not be rigidly specific to a segmented organism; they could apply to any situation in which cells need to be organized and programmed according to their position. The degree of general applicability of the principles being learned from *Drosophila* studies remains to be seen.

Much of what we now know about segmentation genes has its origins in the first systematic search for such genes, a genetic screen for new mutations done by Nüsslein-Volhard and Wieschaus and their collaborators (Nüsslein-Volhard and Wieschaus, 1980; Nüsslein-Volhard et al., 1984; Jürgens et al., 1984; Wieschaus et al., 1984a). These investigators looked for recessive lethal mutations that altered the segmentation pattern and did so in such a way as to identify all such genes in the entire *Drosophila* genome. They found about 20 genes that act during early development to form the normal segmentation pattern (reviewed in Gergen et al., 1985) (see Table I). Other searches have identified additional genes required for the segmentation process but active during oogenesis (e.g., Nüsslein-Volhard, 1979; Anderson and Nüsslein-Volhard, 1984; Konrad et al., 1985; Schüpbach and Wieschaus, 1986; Perrimon and Mahowald, 1986). There are at least 30 of these maternally active segmentation genes, but the total number will not be known until the systematic searches of the genome, now in progress, are completed. Some of these maternal effect genes regulate the overall polarities of the embryo. Mutations in such genes lead, for example, to embryos with two abdominal ends or only dorsal tissues. One of the genes that give the dorsalizing effect—*dorsal*—has been cloned. The transcripts from the gene are uniformly distributed in the oocyte (Steward *et al.*, 1985), suggesting that the position-specific effects of *dorsal* mutations are not due to asymmetric transcript distribution. Another large set of maternal effect genes is recognized by mutations that cause gaps in the segmental pattern (see Table I). The phenotypes are reminiscent of the gap mutations in the zygotically active segmentation genes.

The 30 or so maternal-effect genes all can be carried as homozygous mutant alleles in adult female flies, permitting the effects on the progeny to be observed. An additional class of genes might be expected, and indeed has been identified, that is required both during oogenesis and at other times during development (Perrimon *et al.*, 1985). The effects of mutations in such genes on embryogenesis can only be studied in flies that are genetic mosaics, in which the germline is homozygous for the mutation and the rest of the fly is heterozygous for the mutation. Initial studies of the maternal effects of such lethal genes have led to estimates that about 40 loci may be needed both during oogenesis to direct normal segmentation and during other phases of development. In total, about 100 genes may be required during oogenesis in order for the embryos to develop with normal segments. The job that lies ahead is to learn the molecular functions of the maternal gene products and to learn how the products of the maternal-effect genes influence the zygotically active genes that function later.

A total of six of the zygotically active segmentation genes have been isolated as DNA clones. Four of them are from the group of genes known as **pair-rule** genes (Weiner *et al.*, 1984; Kuroiwa *et al.*, 1984; Holmgren, 1984; Kilchherr *et al.*, 1986; Harding *et al.*, 1986). Their phenotypes are such that the homozygous embryos have about half the normal number of segments (cf. Fig. 2a and 2c). One half the segments are deleted from the pattern, although the units of deletion are offset from, or out of frame with, segments. The deletions caused by mutations in the four genes are different, each gene being required for particular parts of the pattern. The four genes are called *fushi tarazu* (*ftz*), *hairy* (*h*), *even-skipped* (*eve*), and *paired* (*prd*). All four genes are expressed at the blastoderm stage in a pattern of seven transverse stripes (Hafen *et al.*, 1984a; Ingham *et al.*, 1985; Kilchherr *et al.*, 1986; Harding *et al.*, 1986) (Fig. 2b). The stripes appear to correspond, at least approximately, to the positions of the primordial cells that give rise to the structures that are missing in mutant embryos.

The stripes are detected with either of two methods: *in situ* hybridization of nucleic acid probes to sectioned embryos or immunofluorescent localization of proteins in permeabilized whole-mount embryos. The *in situ* hybridizations have the advantage that mRNA can be detected earlier than protein, and precursor RNA molecules, or untranslated molecules, can be detected. Double-label experiments can be done by hybridizing different probes to consecutive sections of an embryo. The immunofluorescence technique avoids the use of sec-

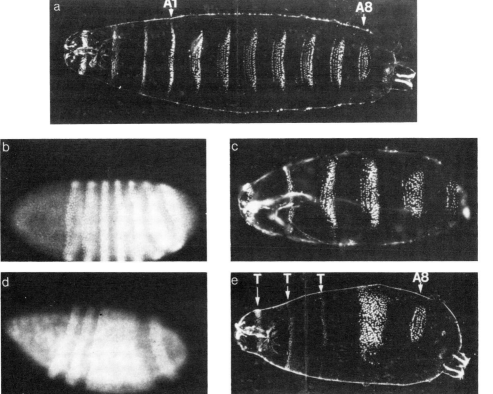

Figure 2. Cuticular pattern and *ftz* stripes in wild-type and mutant embryos. (**a**) Cuticle of a wild-type larva at the completion of embryogenesis. The prominent bands of hairs are known as denticle belts or setal belts. Each group of setae marks the anterior part of a segment. The finer setae at the left mark the three thoracic segments; the coarser setae mark the eight abdominal segments. (**b**) Pattern of *ftz* protein at the blastoderm stage. The protein is located within the nuclei in seven stripes. The striped pattern anticipates the segmental pattern by several hours and in fact disappears before the segments become visible. Each stripe corresponds to the primordium for a segment-size unit of pattern, but the stripes are out of frame with the segment primordia. (**c**) Larval cuticle of an embryo that is homozygous for a mutation in the pair-rule segmentation gene *ftz*. Approximately one half the normal cuticular pattern is missing, and one half the usual number of segmental borders are present. (**d**) The expression of *ftz* in an embryo homozygous for a *knirps* mutation. The anterior two *ftz* stripes are normal, as is the most posterior stripe, but, instead of the usual remaining four stripes, a single broad band of *ftz* protein is observed. The normal pattern of expression of *ftz* in the anterior abdomen is therefore dependent on *knirps* function. (**e**) Larval cuticle of an embryo homozygous for a mutation in the gap segmentation gene *knirps*. The thoracic segments are normal, but all the abdominal segments except the eighth are deleted and replaced with a single broad denticle belt. (Preparations courtesy of Gary Winslow.)

tions, permits observation of the subcellular location of the proteins, and permits the use of double-label detection of two proteins in the same embryo. The antibodies are prepared against proteins made in bacteria, usually by fusing a cDNA clone to the β-galactosidase gene of *Escherichia coli*, expressing the hybrid gene in bacteria, and purifying the fusion protein. The fly protein-specific antibodies are purified from the serum using affinity adsorption, removing the anti-β-galactosidase antibodies by their affinity for pure β-galactosidase.

Both the mRNAs (Hafen *et al.*, 1984a) and the proteins (Carroll and Scott, 1985) of the *ftz* gene have been observed, whereas only the mRNAs of the *h* (Ingham *et al.*, 1985), *eve* (Harding *et al.*, 1986), and *prd* (Kilchherr *et al.*, 1986) genes have been studied. Analysis of the *ftz* and *h* mRNAs has demonstrated that the genes are initially expressed not in stripes but in solids (Weir and Kornberg, 1985). For example, *h* transcripts are first detected throughout the embryo after the twelfth syncytial nuclear division (Ingham *et al.*, 1985); *ftz* transcripts detected at about the same time are found to be already restricted to only part of the embryo: 15–65% of egg length. (These commonly used coordinates are based on the posterior pole of the embryo being defined as 0% egg length and the anterior pole 100%.) During the subsequent approximately 30 min, the solid patterns of both *h* and *ftz* transcripts evolve into their striped patterns. It is not known whether the change from a solid to a striped pattern involves (1) RNA transport, (2) the position-specific stabilization of RNA, or (3) general turnover of RNA accompanied by position-specific transcription. There is a tantalizing feeling that if this pattern refinement process could be understood, we would have moved a long way toward understanding the process of pattern formation.

How important are the striped patterns? Could the genes work equally well if their products were less precisely arranged? This sort of question can be approached by manipulating the genes so as to express them in inappropriate places. The experiments make use of the P-element transformation system (Spradling and Rubin, 1982; Rubin and Spradling, 1982), which is perhaps the most important advance in *Drosophila* genetics in many years. The system uses P-element transposons to carry DNA into the chromosomes. The DNA of interest is inserted into a defective P element, which is then mixed with an element that can provide transposase activity. The mixed DNA is injected into embryos at the syncytial blastoderm stage, permitting the DNA access to the dividing nuclei. With a rather high frequency, transposition into the chromosomes of the pole cells—the presumptive germ-line cells—will occur. Progeny developing from those germline cells will carry the injected DNA, integrated into a chromosome at a random position and usually in a single copy. Up to about 50% of the surviving fertile flies can give rise to progeny carrying the injected DNA integrated into a chromosome. Many genes have been reinserted into the fly genome using P elements, and many have been found to be expressed normally despite their novel chromosomal locations.

The *ftz* gene is small enough to manipulate easily and has been the first gene used in constructions designed to cause abnormal patterns of expression.

A heat-shock promoter was joined to the coding region of *ftz*, and the novel gene was transferred into the fly genome using the P-element transformation system (Struhl, 1985). Heat-shock promoters are believed to be expressed in all cells. Therefore, when embryos carrying the heat shock–*ftz* construction were treated with heat shocks at the blastoderm stage, it was expected that *ftz* protein would be produced in all cells instead of only in the normal striped array. The results were striking: The heat-shocked embryos had pair-rule segmentation phenotypes, but the defects were approximately complementary to the defects seen in an embryo lacking *ftz* function. For example, in embryos lacking *ftz* function, the posterior part of the first thoracic segment (pT1) and the anterior part of the second thoracic segment (aT2) are missing. Both pT1 and aT2 are present in embryos in which *ftz* is expressed everywhere under heat shock control, but pT2 and aT3 (both present in embryos without *ftz* function) are missing. Therefore, it is as harmful to have *ftz* products where they should not be as to lack them where they should be.

Not only the position of segmentation gene expression but its amount is important, suggesting that the balance of different gene products may be involved in setting cell fates. Working with genetic duplications of the pair-rule gene *runt* (*run*), it was shown that extra doses of the gene caused defects in segmentation that, again, were complementary to the defects caused by a lack of *run* (Gergen and Wieschaus, 1986). The defects were out of phase with those caused by *run* mutations. Thus, whereas insufficient *run* function causes defects in aA1, aA3, aA5, and so on, extra doses of *run* cause defects in aA2, aA4, aA6, and so on. Extra doses of genes in *Drosophila* generally cause overproduction of the gene products. Whether the hyperdosage of *run* causes an imbalance among different segmentation gene products, or alternatively production of some *run* product in the wrong places, is not yet known.

If the position-specific expression of pair-rule genes is critical for their function, and if in the absence of their function the pattern of development is dramatically different, the problem of genetic control of pattern formation can be restated as two questions: How is the position-specific expression of genes such as the pair-rule genes attained? What are the molecular functions of the gene products? The first question is being approached in two ways: one molecular and one genetic. The molecular approach is to analyze the cis-regulatory elements of the segmentation genes to determine how transcription is controlled in a spatially and temporally precise manner. The genetic approach is to ask what upstream genes are involved in regulating the pair rule genes to produce striped expression.

The genetic approach has been applied to *ftz*. The objective is to identify any genes that regulate the pattern of *ftz* expression. Among these genes will be, perhaps, the genes that encode proteins that bind to the cis-acting DNA sequences that regulate *ftz*. However, all the genes that regulate the genes that regulate *ftz* will be identified as well. The experiments are straightforward: *ftz* expression is observed in embryos that are mutant for a variety of other genes using *in situ* hybridization or immunofluorescence. But what genes should be tested? There are some excellent candidates, in particular the other segmentation genes.

The segmentation genes other than the pair-rule genes have been classified into two groups: the **segment polarity genes** and the **gap genes** (Nüsslein-Volhard and Wieschaus, 1980). One gene from each class has been cloned: the *Krüppel* gap gene (Priess *et al.*, 1985) and the *engrailed* segment polarity gene (Kuner *et al.*, 1985; Fjose *et al.*, 1985). The gap genes are identified by mutations that cause multiple contiguous segments to be missing from the pattern, like the maternal effect gap genes mentioned earlier. As expected, the *Kr* transcripts are found in broad regions (Knipple *et al.*, 1985), but the regions are not as broad as the part of the embryo affected by the mutations, suggesting that the mutations may affect cells in which the gene is not expressed. Mutations in segment polarity genes cause parts of every segment to be missing. The missing parts are replaced by mirror-image duplications of some of the remaining structures, which accounts for the name of the group of genes. The *engrailed* segment polarity gene (Kornberg, 1981) is expressed in 14 stripes, one per segment, as could be expected from its phenotype (Kornberg *et al.*, 1985; Fjose *et al.*, 1985; DiNardo *et al.*, 1985).

Intuitively, it seems that the different classes of zygotically active genes could function in sequence to subdivide the embryo into finer and finer pattern elements. The gap genes could divide the embryo into broad regions, the pair-rule genes could subdivide the embryo further into segment-size units, and the segment polarity genes could direct formation of subsegmental patterns. This type of hypothesis is difficult to test using traditional genetic techniques. Double mutants give complex, difficult-to-interpret phenotypes that often do not clearly show which mutation is epistatic over the other. Furthermore, many of the events of early embryogenesis happen so quickly and invisibly that phenotypes observed after segments form (or try to form) later do not clearly show what happened. The molecular probes have therefore made a real contribution. In essence, the probes dissect the problem by allowing the investigator to observe one gene's expression at a time. In the long run, each gene's pattern of expression can be examined in embryos that are mutant for each of the other segmentation genes. This process should permit the recognition of which interactions are direct and which are indirect.

The first results of such experiments have shown that the intuitive idea (i.e., a progressive subdivision of the embryo by the gap, then the pair-rule, then the segment polarity genes) may reflect the order of action of the different genes and their hierarchical relationships. For example, the pair-rule gene *ftz* is dependent on all four gap genes for its normal pattern of expression (Carroll and Scott, 1986). An example is shown in Fig. 2d, in which the change in the *ftz* pattern in an embryo mutant for the gap locus *knirps* (*kni*) can be seen. The alterations in the *ftz* pattern occur in the part of the embryo that will later develop abnormally as a result of the *kni* mutation (see Fig. 2e). As expected from the hierarchy model, mutations in the segment polarity genes do not affect the initial pattern of *ftz* expression. Mutations in some of the maternal-effect gap-segmentation genes do affect *ftz* expression in ways that suggest the maternal genes are acting on *ftz* through the zygotically active genes that affect *ftz* (Carroll *et al.*, 1986*b*).

The segment polarity gene *engrailed* (en) requires the functions of the gap genes and the seven pair-rule genes in order for it to be expressed in its normal pattern of 14 stripes (Howard and Ingham, 1986; Harding *et al.*, 1986; S. DiNardo and P. H. O'Farrell, personal communication). Thus, the expression of segment polarity genes is dependent upon gap and pair-rule genes, and the expression of pair-rule genes is dependent on the gap genes. Some of the interactions may regulate initiation of patterned gene expression, as seems to be the case for the effects of the pair-rule gene *eve* upon the segment polarity gene *en*, whereas other interactions affect maintenance of patterns. For example, the expression of *ftz* looks normal in mutant *eve* embryos initially but becomes aberrant later in development, suggesting that *eve* has a role in maintaining proper *ftz* expression (Harding *et al.*, 1986); *eve* function is also required for *en* expression to occur in the central region of the embryo. There are feedback loops as well: *en* function is required to keep *eve* off where it should be off (Harding *et al.*, 1986).

Not all the genes within one class have equal ranking. For example, mutations in the pair-rule gene *hairy* (*h*) alter the pattern of *ftz* expression, but mutations in *ftz* do not alter the pattern of *h* expression (Howard and Ingham, 1986; Carroll and Scott, 1986). Similarly, mutations in some of the other segment polarity genes affect *en* expression (S. DiNardo and P. H. O'Farrell, personal communication). Thus, within and between classes of genes, a hierarchy of interactions can be defined by such experiments.

In addition to the effects of the zygotically active segmentation genes on each other, the maternal effect segmentation genes (see Table I) also control the expression of genes such as *ftz* (Mohler and Wieschaus, 1985; Degelmann *et al.*, 1986; Carroll *et al.*, 1986b). Clarification of the complete genetic hierarchy will require understanding how the maternal-effect genes interact with each of the zygotically active genes. Are the effects of the maternally active genes on *ftz*, for example, mediated by the gap genes that are upstream of (i.e., expressed earlier than) *ftz*? It is not yet clear how all these interactions lead to the formation of the dramatic and precise striped patterns. It does seem clear that many genes act in concert to create the pattern and that each stripe may be dependent on a different array of upstream activities for its formation. The challenge for the molecular biologist is to discover the molecular mechanisms underlying the interactions between the genes. Before discussing the little that is known about molecular functions of segmentation gene products, another group of genes, the homeotic genes, must be described.

6. Homeotic Genes

Homeotic genes are frequently described as regulators of segment identity. Mutations transform one body segment into a duplicate of another. This description is only approximately correct in that often only a part of a segment is transformed, or parts of two adjacent segments. Some homeotic genes are expressed in parasegmental units, and the parts of the fly that are transformed by mutations are in those cases parasegmental. The segmental and parasegmental boundaries respected by the homeotic genes are precisely the boundaries set up by the segmentation genes, suggesting that the homeotic genes respond to the segmentation genes in their spatial patterns of expression. This view has been confirmed in certain instances by finding mutations in segmentation genes that cause homeotic transformations, presumably by causing changes in homeotic gene expression (Weiner *et al.*, 1984; Lehmann and Nüsslein-Volhard, 1986; Duncan, 1986).

Many homeotic genes are found in two clusters called the ***bithorax* complex** (BX-C) (Lewis, 1978, 1982) and the ***Antennapedia* complex** (ANT-C) (Kaufman *et al.*, 1980). Each cluster is located within about 400 kilobases (kb) of DNA, and each has been cloned using **chromosome walks** (Bender *et al., 1983*; Garber *et al.*, 1983; Scott *et al.*, 1983; Karch *et al.*, 1985). Chromosome walks involve the isolation of overlapping pieces of genomic DNA until a long stretch of the chromosome is represented in the collection of cloned segments. The function of the clustering of the homeotic genes in the two complexes, if any, is unknown, aside from its enormous value to experimentalists interested in cloning multiple homeotic genes. Although the definition of a gene becomes somewhat difficult for loci that are as complex as some of the homeotic loci, three complementation groups are defined by lethal alleles in the BX-C (Sanchez-Herrero *et al.*, 1985) and at least four homeotic genes are in the ANT-C (Kaufman and Abbott, 1984). The BX-C also contains several functions defined by nonlethal mutations, at least some of which may be cis-acting regulatory elements that act on the genes defined by lethal alleles (Lewis, 1978). The different genes in the two complexes each affect different parts of the fly, during embryogenesis and during meta-morphosis. Curiously, the order of the genes along the chromosome corre-sponds, for the most part, to the order of the body parts they affect (Lewis, 1978). Thus, genes in the BX-C that affect the posterior abdomen are adjacent to genes that affect the anterior abdomen, which in turn are adjacent to genes that affect the thorax.

As might be expected, when the distribution of BX-C and ANT-C gene products are examined by *in situ* hybridization and immunofluorescence, the positions of the products correspond to the positions of the segments affected by the absence of gene function (Akam, 1983; Levine *et al.*, 1983; White and Wilcox, 1984, 1985*a,b*; Beachy *et al.*, 1985; Kuroiwa *et al.*, 1985; Martinez-Arias, 1986; Carroll *et al.*, 1986). However, cells in any particular position along the embryo often contain the products of more than one homeotic gene. It appears that cells learn their identities from the arrays of homeotic genes active within them (Lewis, 1978; Struhl, 1982). This sort of model predicts that ex-pression of a homeotic gene in which it is normally silent would lead to changes in cell fates and that it is the combination of active homeotic genes, and not the activity of any one gene alone, that determines cell fate.

Certain homeotic mutations are consistent with the expectation that ex-pression of a homeotic gene in a novel place would cause transformations. For example, dominant *Antennapedia* mutations transform antennae into legs. The loss of *Antennapedia* function does not lead to such a transformation, nor is *Antennapedia* function normally required in the head (Denell *et al.*, 1981; Struhl, 1981*a*; Hazelrigg and Kaufman, 1983). The gene normally is required and is expressed in the thorax and abdomen. The gene normally directs thoracic development in the thorax; the dominant alleles cause the gene to exert a similar effect in the head.

The pattern of expression of each of the two most intensively studied homeotic genes, *Antennapedia* (Levine *et al.*, 1983; Carroll *et al.*, 1986) and *Ultrabithorax* (Akam, 1983; White and Wilcox, 1984, 1985a; Beachy *et al.*, 1985; Akam and Martinez-Arias, 1985)—the former from the ANT-C and the latter from the BX-C—is temporally and spatially complex (Fig. 3). The highest levels of mRNA (or protein) are reached in the ventral nervous system (Fig. 3B), but position-specific expression is also observed in the epidermal ectoderm, in the mesoderm, and in the peripheral nervous system (Fig. 3C). Each gene in the *bithorax* complex and in the *Antennapedia* complex has a discrete part of the embryo in which its product accumulates to a high level. In addition, each gene is expressed at a lower level in more posterior regions (Regulski *et al.*, 1985; Harding *et al.*, 1985).

The high-level expression of the homeotic gene in the ventral nervous system was unexpected, although effects of homeotic mutations on the nervous system had been discovered. It appears likely that the homeotic genes play an important role in differentiation of the embryonic nervous system. At least some of the segmentation genes are also expressed in the developing nervous system (Carroll and Scott, 1985) (Fig. 3A), suggesting that the earliest stages of development of neural patterns may depend on the segmentation genes.

Homeotic gene expression is limited precisely to certain segments or parasegments, and in addition the proportion and arrangement of cells expressing either gene within each segment varies, both in the epidermis and the nervous system. An example of the complexity of the pattern is shown in Fig. 3C, in which the expression of *Antp* in the peripheral nervous system can be seen. Only a particular subset of the peripheral nervous system cells produce *Antp* protein. The complexity of the pattern means that each cell of a segment could be programmed differently because of the combination of genes active within it. The genes may be involved not only in controlling differentiation of segments from one another but in differentiating cells within each segment from each other as well.

The homeotic genes must be responding to quite elaborate control signals in order to attain such intricate patterns of expression. Genetic analysis of homeotic gene function was successful in predicting in which segments the genes would function, but molecular probes were required to obtain such a precise picture of the complexities of the expression patterns. It had not been predicted, for example, that the *Antennapedia* gene would be expressed in the peripheral nervous system, or which particular subsets of cells within the abdominal segments would express *Antennapedia*.

Part of the complexity of homeotic gene expression is attributable to control of some homeotic genes by others. This is most clearly seen by examining the distribution of homeotic gene products in embryos mutant for other homeotic genes. Thus, *Antp* is normally expressed at high levels in parts of the thorax and at much lower levels in the abdominal segments. If the embryo lacks the BX-C, *Antp* RNA (Hafen *et al.*, 1984b) and protein (Carroll *et al.*, 1986) are found at high levels in most of the abdominal segments. Therefore, BX-C genes prevent high-level *Antp* expression in the posterior of the embryo. Similarly, the BX-C gene *Ubx* has been found to be negatively regulated by two other BX-C genes, *abd A* and *Abd B*, both of which are active in the abdomen (Struhl and White, 1985). The *abd A* gene also negatively regulates *Antp*. The emerging picture is that genes active in more posterior regions reduce the levels of expression in the posterior of genes that are primarily active in more anterior regions. This relatively simple sort of relationship had not been deduced from genetic studies, although there were hints of gene interactions, and demonstrates how useful the molecular probes are in demonstrating the effects of some of the genes in the hierarchy upon others.

One group of genes, listed in Table I as homeotic genes active both maternally and zygotically, have been proposed to be regulators of the BX-C and ANT-C genes. *Polycomb* (*Pc*) and *extra sex combs* (*esc*), for example, appear to behave as negative regulators of homeotic genes in both complexes (Duncan and Lewis, 1982; Struhl, 1981b). The molecular probes demonstrate that a subtle correction of these models is necessary: The *initial* pattern of a BX-C gene, *Ubx*, is normal in an embryo lacking *esc* function (and derived from a mother also lacking the function), but *Ubx* fails to remain off where it should be off (Struhl and Akam, 1985). *Pc* mutations also lead to general derepression of other *Antennapedia* complex and *bithorax* complex genes (Wedeen *et al.*, 1986). Thus, genes such as *esc* may be required for maintenance of determined cell states rather than for initiation of them. (This idea may or may not apply to the other genes in the group.) That the situation is even more subtle is shown by the result that *esc* function is required at a discrete early time during embryogenesis and not later in development (Struhl and Brower, 1982). This suggests that the maintenance function is exercised at a specific time and is not required continuously.

7. Complex Patterns of Expression Require Complex Genes

Chromosome walking, the process of cloning large sections of a chromosome by isolating overlapping cloned pieces of DNA, has been a powerful tool for characterizing developmentally interesting *Drosophila* genes. Both homeotic gene complexes were isolated as series of recombinant phage clones, and many of the segmentation genes have been obtained in the same way. The goals are to understand how position-specific gene expression is attained and to determine the structures and molecular functions of the RNA and protein products of the genes.

The first segmentation gene to be isolated, *ftz*, at first appeared to have a simple structure (Laughon and Scott, 1984). The *ftz* gene is, for unknown reasons, located in the ANT-C cluster. The gene has a small transcription unit of 1.95 kb, a single intron, and a single major protein-coding region. However, this apparent simplicity is deceptive, as might be expected from the number of

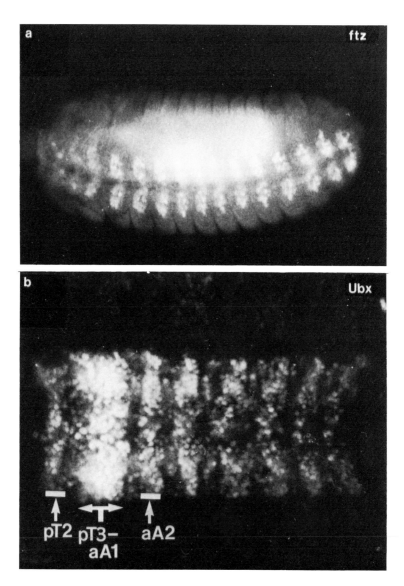

Figure 3. Expression of homeotic and segmentation genes in the developing nervous system. Permeabilized whole-mount embryos were stained using indirect immunofluorescence. Anterior is to the left in each panel. (a) Embryo just after germ-band shortening. Ventral view. Antibodies specific for the protein product of the *fushi tarazu* (*ftz*) segmentation gene stain a subset of the nuclei in each of the 15 paired units in the developing nervous system. The general fluorescence in the center of the embryo is due to yolk autofluorescence. The grooves demarcating the body segments can be seen. The expression of *ftz* in the nervous system is transient and disappears soon after the stage shown. The nervous system expression in every segment should be contrasted with the expression in alternate segment-size units seen earlier at the blastoderm stage (see Fig. 2). (b) Accumulation of the *Ultrabithorax* homeotic protein in the segments of the ventral nervous system. As in the case of *ftz* and *Antp* protein, the *Ubx* protein is located within nuclei. The micrograph is a ventral view of the nervous system of an embryo that has already undergone germ band shortening.

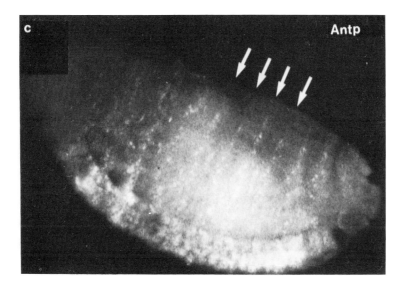

Figure 3. *Ubx* protein is found at the highest level in the posterior third thoracic segment (pT3) and anterior first abdominal segment (aA1). Lower levels are observed in posterior T2 and in the anterior parts of the other abdominal segments. (**C**) Like *Ubx* protein, *Antennapedia* (*Antp*) protein is found at high levels in the ventral nervous system, as can be seen at the bottom (ventral) side of the embryo. *Antp* protein is also present in a subset of the cells of the peripheral nervous system, here seen as fine stripes of cells extending around the sides of the embryo. The arrows point to four of the rows of peripheral nervous system cells. The function of *Antp* in the peripheral nervous system is unknown. (Micrographs courtesy of Sean Carroll.)

genes that impinge on *ftz* expression; *ftz* is not only expressed in the striped pattern at the blastoderm stage but is also expressed transiently in a different pattern in the developing ventral nervous system (see Fig. 3A). The multiple cis-acting DNA sequences responsible for the control of *ftz* expression were first shown by experiments in which the cloned *ftz* gene was introduced back into the fly genome. An important first question was whether *ftz* could function away from its normal location in the ANT-C.

A 10-kb fragment of DNA was found to provide *ftz* pair-rule gene function, as tested by complementation with *ftz* mutant alleles (Hiromi *et al.*, 1985). Therefore, *ftz* need not be in its normal position to function. The upstream (5′) region of the *ftz* gene was fused to the *E. coli* β-galactosidase gene and tested in flies. A striped pattern of β-galactosidase was observed that corresponds exactly to the normal pattern of *ftz* expression, except that the β-galactosidase is more stable than *ftz* protein and lasts longer in the embryos. These experiments demonstrate that 5.5 kb of upstream DNA is needed for blastoderm stage and neural expression, thus ridding the gene of any image of simplicity. Subsequent deletion experiments narrowed down the DNA relevant to the striped pattern to an upstream element of 2.7 kb plus a 680-base-pair (bp) region just 5' of the transcription start site. Another section of the upstream region is required for *ftz* expression in the nervous system. Presumably, the 5' *ftz* DNA contains the target sites for regulatory DNA binding proteins; the next generation of experiments will lead to the identification of the precise regulatory sequences and the proteins that bind to them. The requirement for an unusually large 5' flanking region thus appears to reflect the complex regulation of *ftz* by a large number of other genes.

Another segmentation gene that has been cloned is the *engrailed* (en) gene (Kuner et al., 1985); *en* appears to be a complex gene with a relatively small (~5 kb) transcription unit and a large regulatory region. Breakpoints of chromosome rearrangements that affect *en* function are spread over a region of 70 kb, and breakpoints that do lethal damage to the gene are distributed over ~40 kb, in the center of which is the transcription unit. Current models of *en* gene structure suggest that most of the 70 kb is composed of cis-regulatory sequences, perhaps intermingled with nonessential spacer sequences. A study of the evolutionary conservation of *en* sequences demonstrated that many sequences across the 70-kb region are conserved and so are, curiously, the spacer distances between the conserved regions (Kassis et al., 1985); 33 regions totaling 20 kb are detectably conserved. The 33 regions are in the same order in both *Drosophila melanogaster* and *D. virilis*, which are estimated to have diverged more than 60 million years ago—long enough that unimportant DNA sequences would no longer cross-hybridize. Many regulatory elements of *en* may therefore have been highly conserved. Their functions remain to be discovered.

A member of the third class of segmentation genes, the gap genes, has also been cloned: the *Krüppel* (Kr) gene (Preiss *et al.*, 1985). Embryos lacking *Kr* gene function develop without the thoracic and anterior abdominal segments. In their place, a mirror-image duplication of some of the remaining abdominal segments is observed (Wieschaus et al., 1984b). The phenotype is somewhat reminiscent of the double-abdomen embryo resulting from the maternal effect mutation *bicaudal* (Nüsslein-Volhard, 1979). The initial confirmation of the isolation of the recombinant *Kr* clones used a unique approach: DNA from the λ phage clone thought to contain the *Kr* gene was injected into mutant *Kr/Kr* embryos. A remarkable local partial rescuing effect due to the injected DNA showed that at least a functional part of the gene was contained in the injected cloned DNA (Preiss et al., 1985). The opposite type of experiment has also been done with the *Kr* gene, inhibiting its function by injection of antisense RNA (Rosenberg et al., 1985). Therefore, some part of the *Kr* mRNA appears to be sufficiently accessible to the injected antisense RNA to hybridize and thereby lose function.

The most startling gene structures have been those of the homeotic genes *Antennapedia* (Antp) and *Ultrabithorax* (Ubx). Both have enormous transcription units: 103 kb for *Antp* (Garber et al., 1983; Scott et al., 1983) and 73 kb for *Ubx* (Bender et al., 1983; Hogness et al., 1985). *Ubx* is flanked on its 5' end by a smaller transcription unit, the *bithoraxoid* (bxd) 25-kb transcription unit (Fig. 4, top) (Hogness et al., 1985). The *bxd* unit appears to be a complex cis-acting

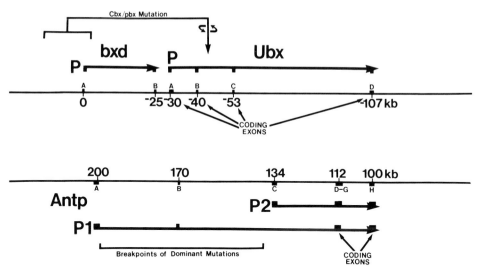

Figure 4. Structures of the homeotic genes *Antennapedia* and *Ultrabithorax*. The *Ubx* gene (**top**) is flanked on the 5' side (**left**) by the *bithoraxoid* (*bxd*) transcription unit. The *bxd* unit appears to contain cis-acting elements that affect the expression of *Ubx*. The *Cbx/pbx* mutation is a double mutation: The deletion of 17 kb from the *bxd* region results in the *pbx* phenotype, whereas the insertion of the same 17 kb into the *Ubx* intron, as shown, results in the *Cbx* phenotype. The insertion is opposite in orientation relative to its normal position. The coding exons (A–D) of *Ubx* are distributed across the entire 75-kb *Ubx* region. The B and C exons are microexons of only 51 bp each. P, promoter. (Data from Hogness *et al.*, 1985.) The *Antp* transcription unit (**bottom**) is about the size of *bxd* and *Ubx* together. However, in contrast to *Ubx*, the coding exons of *Antp* (A–H) are located near one end of the gene. The A, B, C, and D *Antp* exons form noncoding leader sequences in the mRNAs. The *Antp* protein is encoded in exons E, F, G, and H. The H exon contains the homeobox. *Antp* has two promoters, P1 and P2, and mRNAs derived from the P1 promoter lack the C exon. Dominant mutations that transform antennae into legs are often associated with chromosome rearrangements, the breakpoints of which are located within the indicated 60-kb region. The breakpoints therefore leave the P2 transcription unit and the coding exons apparently intact. (Data from Schneuwly *et al.*, 1986; Laughon *et al.*, 1986.)

regulatory region for *Ubx*. *Antp* actually has two transcription units driven by two promoters (Fig. 4, bottom) (Schneuwly *et al.*, 1986; Laughon *et al.*, 1986; Stroeher *et al.*, 1986). Both *Antp* transcription units regulate the expression of the same set of protein-coding exons. The smaller transcription unit is completely contained within the larger and is 36 kb long. As in the case of *en*, *Antp* and *Ubx* both appear to control relatively simple products in complex ways. Each gene's RNA products are only known to have a single large open reading frame, in each case slightly larger than 1 kb. There is some variable RNA splicing of the coding exons of both genes that leads to multiple, slightly different, encoded proteins (Hogness *et al.*, 1985; J. R. Bermingham and M. P. Scott, unpublished data) (see Chapter 11 for a discussion of this mechanism). Since the mRNAs for both genes are in the range of only 3–5 kb, most of the genes' DNA sequences are either involved in cis regulation, are active in RNA precursor molecules, or are unimportant for gene function. Little is yet known about cis-acting control elements for either gene, although many mutations in the *bithoraxoid* region appear to exert their influence through altering *Ubx* expression (Bender *et al.*, 1985; Beachy *et al.*, 1985). Other regulatory elements may be located within the *Ubx* gene itself.

Chromosome rearrangements that interrupt either the *Antennapedia* or *Ultrabithorax* transcription units cause recessive lethal mutations. This leads to transformations of some of the embryonic segments or parasegments into others. Without *Ubx* function, for example, parasegments 5 and 6 (posterior second thoracic segment through anterior first abdominal segment) are transformed into two parasegment 4s (each posterior first thoracic segment plus anterior second thoracic segment). There are also minor effects on the more posterior abdominal segments.

Some chromosome breaks, however, do not merely destroy gene function. The classic dominant *Antennapedia* mutations that transform antennae into legs provide an example. Most of these mutations are associated with inversions or translocations that interrupt the *Antennapedia* gene. One effect is to cause a recessive lethal mutation; a second is to cause the dominant transformation. Interrupting the gene causes new gene activity. How could this work? The gene structure helps to solve the paradox. The coding exons (see Fig. 4) are all located at the 3′ end of the gene, in only 13 kb of the 103-kb gene. The chromosome breaks that cause dominant mutations leave the coding exons intact. In addition, the smaller (P2) transcription unit is also left apparently intact. The basis for the mutations may therefore be that new control elements are brought into juxtaposition with the small transcription unit in a way that leads to expression of the normal *Antennapedia* protein(s) in the head. This in turn leads to a thoracic transformation of the antennae.

There are also mutations of *Ubx* that appear to cause expression of the gene where it should be off. One of these, *Contrabithorax* (*Cbx*), is associated with an insertion of 17 kb of DNA into one of the *Ubx* introns (Bender *et al.*, 1983) (see Fig. 4). The resulting phenotype is a transformation of part of the wing into halterelike structures, in other words, a transformation of second thoracic segment into third. The prediction is that *Ubx* is being expressed more anteriorly than it should be; normally it is expressed only at a low level in the second thoracic segment. This prediction is borne out by experiments using molecular probes. Both *in situ* hybridization and immunofluorescence techniques demonstrate *Ubx* products at high levels in second thoracic tissues (White and Akam, 1985; Cabrera *et al.*, 1985). How could the 17-kb insertion cause this effect? The 17-kb segment is not from an unknown part of the genome. It is from the *bxd* region just upstream of the *Ubx* gene (see Fig. 4, top). The 17 kb in the *Cbx* allele is inverted and inserted into the *Ubx* intron. When the regulatory elements located in the *bxd* region are better understood, the basis for the effect of the 17-kb insert will undoubtedly become clearer, but it seems possible, for example, that the insert carries elements that, when rearranged, direct expression in the second thoracic segment. An alternative is that the insertion interferes with the function of a negative regulatory element that normally keeps *Ubx* off in the second thoracic segment.

In addition to the effect of the 17-kb insertion, the original mutation also had another effect, which is caused by the deletion of the 17 kb from its normal position (Bender *et al.*, 1983). The separability of the two phenotypes was seen by separating the deletion from the insertion using recombination (Lewis, 1955). The deletion alone causes a transformation of posterior haltere into posterior wing: a *postbithorax* (*pbx*) phenotype that is opposite in nature to the *Cbx* phenotype. The deletion presumably removes regulatory DNA sequences that normally activate *Ubx* in the developing haltere. The *pbx* function does not seem to be important to embryonic development.

These few examples selected from the many homeotic mutations are intended to convey the types of effects that are observed, and how molecular biology is beginning to explain the basis for some of the mutants. The mutations are not just curiosities, as they indicate control elements that are normally important for gene function, as well as elements that normally may be parts of other genes. In most cases, the homeotic mutants tell us more about cis regulation than about the functions of the gene products. Molecular analysis of the transcripts and the proteins has recently begun to suggest molecular mechanisms that may be involved in allowing the homeotic and segmentation genes to exert their extraordinarily powerful influence over developmental processes.

8. Protein Products of the Homeotic and Segmentation Genes

Discovering the molecular function of a protein known only from the sequence of a complementary DNA (cDNA) clone can be a difficult problem. Even purified proteins can sometimes remain refractory to any sort of functional analysis. Clues to function can be obtained from knowing the subcellular location of the protein, and from homology of the protein's sequence with known proteins. To date, the sequences of three segmentation gene protein products and two homeotic protein products (one of them partial) have been reported. Many others will soon be known.

One exceptionally useful unifying theme among the products of homeotic and segmentation loci has been a 180-bp sequence of DNA called the **homeobox** (McGinnis *et al.*, 1984; Scott and Weiner, 1984). Three copies of the homeobox are located in the BX-C; another at least six copies are in the ANT-C (Regulski *et al.*, 1985; unpublished data of A. Laughon and M. P. Scott). The sequence is highly conserved, even from *Drosophila* to humans (Levine *et al.*, 1984), and encodes an even more highly conserved 60-amino acid sequence, the **homeodomain.** A homeobox therefore is an indicator of a protein coding region. There is one homeobox for each of the three lethal complementation groups of the BX-C, suggesting that each gene so defined encodes a protein that contains a homeodomain. Of the genes that have already been mentioned, *Antennapedia*, *Ultrabithorax*, *fushi tarazu*, and *engrailed* have one homeobox apiece. The homeotic loci *Sex combs reduced* and *Deformed* of the *Antennapedia* complex also have homeoboxes, as do the homeotic loci *abdA* and *AbdB* of the *bithorax* complex. The *engrailed* and *fushi tarazu* loci are segmentation genes rather than homeotic genes, suggesting that there may be some biochemical functions common to products of both classes of loci. There is reason to believe that one such function may be DNA binding.

The first suggestion that the homeodomain may be a DNA-binding domain came from the observation that it is composed of about 30% basic amino acids. However, the sort of nonspecific interaction implied by this observation was strengthened by the finding that a more subtle homology with bacterial DNA-binding proteins could be seen (Laughon and Scott, 1984). A 20-amino acid region of the homeodomain has the correct types of residues in the correct positions to form a α-helix–β-turn/α-helix structure like that found in λ repressor, λ cro, lac repressor, and other bacterial DNA-binding proteins (reviewed in Pabo and Sauer, 1984). In the bacterial proteins, one of the helices sits in the major groove of the DNA and makes sequence-specific contacts, while the other helix stabilizes the structures and interactions of the first. On the basis of sequence homology, a similar structure is proposed to exist in two protein products of the mating-type locus of yeast (Shepherd *et al.*, 1984; Laughon and Scott, 1984). The known locations of *Antp*, *Ubx*, *en*, and *ftz* proteins in the nucleus are consistent with a role for these proteins as DNA-binding factors.

The possible structural homology led quickly to experiments to test the role of the homeodomain in DNA binding. The first experiments employed proteins produced in transformed bacteria. The proteins were hybrid, consisting partly of β-galactosidase and partly of the *engrailed* protein product (Desplan *et al.*, 1985). It was found that the partially purified protein was capable of binding specifically to certain DNA fragments from a mixture of many fragments *in vitro*. This capability was found to be due to the homeodomain itself; the complete protein was not required. Several binding sites for the homeodomain were identified near the *engrailed* gene and others near the *ftz* gene. There is no evidence yet, however, that any of these sites are biologically meaningful or even that the homeodomain-containing proteins bind to DNA *in vivo*. The results obtained with en protein have now been replicated with other homeodomain-containing proteins; current work is directed at proteins not linked to β-galactosidase or any other bacterial carrier and produced in insect cells. The sequences recognized by the proteins *in vitro* will soon be known; their importance for function *in vivo* can then be assessed with P-element transformation experiments.

The homeodomain (~6000 M_r in size) provides a useful clue about the function of one small part of each of the proteins that contain it, but the proteins whose sequences are known so far are in the range of 40–60 M_r. Nothing is known about the roles of the other parts of the proteins. The homeodomain is generally near the C termini of the proteins. One notable characteristic of all of the proteins is that they contain regions of homopolymer i.e., stretches very rich in amino acids such as glutamine, serine, alanine, glycine, or asparagine. Since the putative DNA-binding part of the proteins (and specifically the α-helix postulated to be involved in sequence recognition) is nearly identical in all the 25 or so homeodomains of which the sequences are known, one role of the remainder of the proteins may be to in some way confer specificity of DNA-binding so that the proteins can be functionally distinct (Laughon *et al.*, 1985). Alternatively, the proteins may all bind to the same sequences but act differently once they are there.

A cDNA clone from one of the gap segmentation genes, *Krüppel* (*Kr*), has been sequenced (Rosenberg *et al.*, 1986). Like some of the other segmentation genes, *Kr* does not have a homeobox. The sequence of the *Kr* protein suggests a different sort of homology. The *Kr* protein contains a region that is predicted to form the so-called zinc fingers, which were first proposed as a structure for the 5S ribosomal RNA transcription factor TFIIIA. TFIIIA has been proposed to bind to both DNA and RNA through loops of about 28 amino acids (the fingers), which are stabilized by interacting with one zinc atom each (Miller *et al.*, 1985). The base of each loop is formed by two histidine residues and two cysteine residues, and the presence of these amino acids every 30 or so residues provides an indicator that a finger protein may be encoded. The TFIIIA protein has nine fingers, which are believed to contact the DNA at about 5-bp intervals. The *Kr* protein has four (or possibly five) fingers and may therefore have a smaller recognition site. Thus, the *Kr* protein may also be a DNA (or RNA) binding protein, although of a different sort than the homeodomain-containing proteins. (Other *Drosophila* proteins have also been found to have sequences that suggest the zinc finger motif.) Position-specific expression of *ftz* is regulated by *Kr:* In the absence of *Kr* function, the posterior two *ftz* stripes (and perhaps the most anterior one) are normal, but the other four are replaced by an abnormally wide and an abnormally narrow stripe (Carroll and Scott, 1986). Since *Kr* encodes a protein that may bind to DNA, the possibility exists that *Kr* protein binds directly to *ftz* DNA (or RNA). Two other equally viable possibilities are that *Kr* protein interacts with *ftz* RNA or that the effect of *Kr* on *ftz* is indirect and mediated by a series of other proteins.

9. Conclusions

It appeared in the past that the homeotic genes and the segmentation genes were two classes of genes, each comprising a hierarchy of interacting regulators. The relationships between the two categories of genes are now rapidly emerging, and the relationships between genes within each class are becoming much clearer through the use of molecular probes. However, the basic mysteries that have existed for many years still stand. We know now that visible pattern is preceded by exquisite patterns of gene expression, but it is still not known how the patterns of gene expression are attained. It will be important to learn how the cis-acting regulatory elements of genes are able to direct such fine patterning. In addition, the next generation of experiments will be directed at understanding how the biochemical functions of the proteins encoded by the regulatory genes participate in the interactions that have been found.

Although only a few protein sequences are known, and much remains to be learned about the interactions between the homeotic and segmentation genes, the molecular analysis of the protein products has already reached the point of testing how some of the interactions between segmentation genes may work at the molecular level. The prospects for rapid progress are excellent, although undoubtedly some of the apparent simplicity of the models described above will disappear rapidly as more of the genes are analyzed. With a better idea of how development is controlled by genes in *Drosophila*, we may be able to ask the right questions about the developmental genetics of other organisms as well.

ACKNOWLEDGMENTS.   I thank the members of my laboratory group, especially Dr. Sean Carroll and Dr. Allen Laughon, for many interesting discussions. Research in my laboratory is supported by the National Institutes of Health, the American Cancer Society, a March of Dimes Basil O'Connor award, and a Searle Scholar's Award. I also thank Sean Carroll, Gary Winslow, Nurit Wolf, and Margaret Fuller for their contributions of previously unpublished figures and Cathy Inouye for help with the manuscript.

References

Abbott, M. K., and Kaufman, T. C., 1986. The relationship between the functional complexity and the molecular organization of the *antennapedia* locus of *Drosophila melanogaster*, *Genetics* **114**:919–942.

Akam, M. E., 1983, The location of *Ultrabithorax* transcripts in Drosophila tissue sections, *EMBO J.* **2**:2075–2084.

Akam, M. E., and Martinez-Arias, A., 1985, The distribution of *Ultrabithorax* transcripts in Drosophila embryos, *EMBO J.* **4**:1689–1700.

Anderson, K. V., and Nüsslein-Volhard, C., 1984, Information for the dorsal–ventral pattern of the *Drosophila* embryo is stored as maternal mRNA, *Nature (Lond.)* **311**:223–227.

Anderson, K. V., Bokla, L., and Nüsslein-Volhard, C., 1985, Establishment of dorsal-ventral polarity in the Drosophila embryo: The induction of polarity by the *Toll* gene product, *Cell* **42**:791–798.

Beachy, P. A., Helfand, S. L., and Hogness, D. S., 1985, Segmental distribution of bithorax complex proteins during *Drosophila* development, *Nature (Lond.)* **313**:545–551.

Bender, W., Akam, M. A., Beachy, P. A., Karch, F., Peifer, M., Lewis, E. B., and Hogness, D. S., 1983, Molecular genetics of the bithorax complex in *Drosophila melanogaster*, *Science* **221**:23–29.

Bender, W., Weiffenbach, B., Karch, F., Peifer, M., 1985, Domains of cis-interaction in the bithorax complex, *Cold Spring Harbor Symp. Quant. Biol.* **50**:173–180.

Bender, M., Turner, F. R., and Kaufman, T. C., 1986, A developmental genetic analysis of the gene *Regulator of postbithorax* in *Drosophila melanogaster*, *Dev. Biol.* **119**:418–432.

Boswell, R. E., and Mahowald, A. P., 1985, *tudor*, a gene required for assembly of the germ plasm in *Drosophila melanogaster*, *Cell* **43**:97–104.

Cabrera, C. V., Botas, J., and Garcia-Bellido, A., 1985, Distribution of *Ultrabithorax* proteins in mutants of *Drosophila* bithorax complex and its transregulatory genes, *Nature (Lond.)* **318**:569–571.

Carroll, S. B., and Scott, M. P., 1985, Localization of the *fushi tarazu* protein during Drosophila embryogenesis, *Cell* **43**:47–57.

Carroll, S. B., and Scott, M. P., 1986, Zygotically-active genes that affect the spatial expression of the *fushi tarazu* segmentation gene during early Drosophila embryogenesis, *Cell* **45**:113–126.

Carroll, S. B., Laymon, R. A., McCutcheon, M. A., Riley, P. D., and Scott, M. P., 1986a, Localization and regulation of *Antennapedia* protein expression in Drosophila embryos, *Cell* **47**:113–122.

Carroll, S. B., Winslow, G. A., Schüpbach, T., and Scott, M. P., 1986*b*, Maternal control of segmentation genes in Drosophila, *Nature (Lond.)* **323**:278–280.

Denell, R. E., Hummels, K. R., Wakimoto, B. T., and Kaufman, T. C., 1981, Developmental studies of lethality associated with the Antennapedia gene complex in *Drosophila melanogaster, Dev. Biol.* **81**:43–50.

Degelmann, A., Hardy, P. A., Perrimon, N., and Mahowald, A. P., 1986, Developmental analysis of the torso-like phenotype in *Drosophila* produced by a maternal effect locus, *Dev. Biol.* **115**:479–489.

Desplan, C., Theis, J., and O'Farrell, P. H., 1985, The *Drosophila* developmental gene, engrailed, encodes a sequence-specific DNA binding activity, *Nature (Lond.)* **318**:630–635.

Digan, M. E., Haynes, S. R., Mozer, B. A., and Dawid, I., 1986, Genetic and molecular analysis of *fs(1)h*, a maternal effect homeotic gene in Drosophila, *Dev. Biol.* **114**:161–169.

DiNardo, S., Kuner, J. M., Theis, J., and O'Farrell, P. H., 1985, Development of embryonic pattern in *D. melanogaster* as revealed by accumulation of the nuclear *engrailed* protein, *Cell* **43**:59–69.

Duncan, I., 1986, Control of bithorax complex functions by the segmentation gene *fushi tarazu* of *Drosophila melanogaster. Cell* **47**:297–309.

Duncan, I. M., 1982, Polycomblike: A gene that appears to be required for the normal expression of the *bithorax* and *Antennapedia* gene complexes of *Drosophila melanogaster, Genetics* **102**:49–70.

Duncan, I. M., and Lewis, E. B., 1982, Genetic control of body segment differentiation in *Drosophila,* in: *Developmental Order: Its Origin and Regulation* (S. Subtelny, ed.), pp. 533–554, Alan R. Liss, New York.

Dura, J-M., Brock, H. W., and Santamaria, P., 1985, *Polyhomeotic:* A gene of *Drosophila melanogaster* required for correct expression of segmental identity, *Mol. Gen. Genet.* **198**:213–220.

Fjose, A., McGinnis, W. J., and Gehring, W. J., 1985, Isolation of a homeobox-containing gene from the *engrailed* region of *Drosophila* and the spatial distribution of its transcript, *Nature (Lond.)* **313**:284–289.

Foe, V. E., and Alberts, B., 1983, Studies of nuclear and cytoplasmic behavior during the five mitotic cycles that precede gastrulation in *Drosophila* embryogenesis, *J. Cell Sci.* **61**:31–70.

Garber, R. L., Kuroiwa, A., and Gehring, W. J., 1983, Genomic and cDNA clones of the homeotic locus *Antennapedia* in Drosophila, *EMBO J.* **2**:2027–2034.

Gehring, W. J., 1986, Homeotic genes and the homeobox, *Annu. Rev. Genetics* **20**:147–173.

Gergen, J. P., and Wieschaus, E. F., 1985, The localized requirements for a gene affecting segmentation in *Drosophila:* Analysis of larvae mosaic for *runt, Dev. Biol.* **109**:321–335.

Gergen, J. P., and Wieschaus, E., 1986, Dosage requirements for *runt* in the segmentation of Drosophila embryos, *Cell* **45**:289–299.

Gergen, J. P., Coulter, D., and Wieschaus, E. F., 1986, Segmental pattern and blastoderm cell identities, in: *Gametogenesis and the Early Embryo* (J. G. Gall, ed.), pp. 195–220, Liss, New York.

Hafen, E., Kuroiwa, A., and Gehring, W. J., 1984*a*, Spatial distribution of transcripts from the segmentation gene *fushi tarazu* of Drosophila, *Cell* **37**:825–831.

Hafen, E., Levine, M., and Gehring, W. J., 1984*b*, Regulation of Antennapedia transcript distribution by the bithorax complex in Drosophila, *Nature (Lond.)* **307**:287–289.

Harding, K., Wedeen, C., McGinnis, W., and Levine, M., 1985, Spatially regulated expression of homeotic genes in Drosophila, *Science* **229**:1236–1242.

Harding, K., Rushlow, C., Hoyle, H. J., Hoey, T., and Levine, M., 1986, Cross-regulatory interactions among pair-rule genes in Drosophila, *Science* **233**:953–959.

Hartenstein, V., Technau, G. M., and Campos-Ortega, J. A., 1985, Fate-mapping in wild-type *Drosophila melanogaster.* III. A fate map of the blastoderm, *Roux Arch. Dev. Biol.* **194**:213–216.

Hayes, P. H., Sato, T., and Denell, R. E., 1984, Homeosis in Drosophila: The *Ultrabithorax* larval syndrome, *Proc. Natl. Acad. Sci. USA* **81**:545–549.

Hazelrigg, T. I., and Kaufman, T. C., 1983, Revertants of dominant mutations associated with the Antennapedia gene complex of *Drosophila melanogaster:* Cytology and genetics, *Genetics* **105**:581–600.

Hiromi, Y., Kuroiwa, A., and Gehring, W. J., 1985, Control elements of the *Drosophila* segmentation gene *fushi tarazu*, *Cell* **43**:603–613.

Hogness, D. S., Lipshitz, H. D., Beachy, P. A., Peattie, D. A., Saint, R. A., Goldschmidt-Clermont, M., Harte, P. J., Gavis, E. R., and Helfand, S. L., 1985, Regulation and products of the *Ubx* domain of the *bithorax* complex, *Cold Spring Harbor Symp. Quant. Biol.* **50**:181–194.

Holmgren, R., 1984, Cloning sequences from the *hairy* gene of *Drosophila*, *EMBO J.* **3**:569–573.

Howard, K., and Ingham, P., 1986, Regulatory interactions between the segmentation genes *fushi tarazu*, *hairy*, and *engrailed* in the *Drosophila* blastoderm, *Cell* **44**:949–957.

Ingham, P. W., 1984, A gene that regulates the *bithorax* complex differentially in larval and adult cells of Drosophila, *Cell* **37**:815–823.

Ingham, P., and Whittle, R., 1980, *Trithorax*: A new homeotic mutation of *Drosophila melanogaster* causing transformants of abdominal and thoracic imaginal segments, *Mol. Gen. Genet.* **179**:607–614.

Ingham, P. W., Howard, K. R., and Ish-Horowicz, D., 1985, Transcription pattern of the *Drosophila* segmentation gene *hairy*, *Nature (Lond.)* **318**:439–445.

Ish-Horowicz, D., Howard, K. R., Pinchin, S. M., and Ingham, P. W., 1985, Molecular and genetic analysis of the *hairy* locus in *Drosophila*, *Cold Spring Harbor Symp. Quant. Biol.* **50**:135–144.

Jürgens, G., 1985, A group of genes controlling the spatial expression of the *bithorax* complex in *Drosophila*, *Nature (Lond.)* **316**:153–155.

Jürgens, G., Wieschaus, E., Nüsslein-Volhard, C., and Kluding, H., 1984, Mutations affecting the pattern of the larval cuticle in *Drosophila melanogaster*. II. Zygotic loci on the third chromosome, *Wilhelm Roux Arch. Entwicklungsmech. Org.* **193**:283–295.

Karch, F., Weiffenbach, B., Peifer, M., Bender, W., Duncan, I., Celniker, S., Crosby, M., and Lewis, E. B., 1985, The abdominal region of the *bithorax* complex, *Cell* **43**:81–96.

Karr, T. L., and Alberts, B. M., 1986, Organization of the cytoskeleton in early *Drosophila* embryos, *J. Cell Biol.* **102**:1494–1509.

Kassis, J. A., Wong, M. L., and O'Farrell, P. H., 1985, Electron microscopic heteroduplex mapping identifies regions of the engrailed locus that are conserved between *Drosophila melanogaster* and *Drosophila virilis*, *Mol. Cell. Biol.* **5**:3600–3609.

Kaufman, T. C., 1978, Cytogenetic analysis of chromosome 3 in *Drosophila melanogaster*: Isolation and characterization of four new alleles of the proboscipedia (pb) locus, *Genetics* **90**:579–596.

Kaufman, T. C., and Abbott, M. K., 1984, Homoeotic genes and the specification of segmental identity in the embryo and adult thorax of *Drosophila melanogaster*, in: *Molecular Aspects of Early Development* (G. M. Melacinski and W. H. Klein, eds.), pp. 189–217, Plenum, New York.

Kaufman, T. C., Lewis, R., and Wakimoto, B., 1980, Cytogenetic analysis of chromosome 3 in *Drosophila melanogaster*: The homoeotic gene complex in polytene chromosome interval 84A,B, *Genetics* **94**:115–133.

Kilchherr, F., Baumgartner, S., Bopp, D., Frei, E., and Noll, M., 1986, Isolation of the *paired* gene of *Drosophila* and its spatial expression during early embryogenesis, *Nature (Lond.)* **321**:493–499.

Knipple, D. C., Seifert, E., Rosenberg, U. B., Priess, A., and Jäckle, H., 1985, Spatial and temporal patterns of *Krüppel* gene expression in early *Drosophila* embryos, *Nature (Lond.)* **317**:40–44.

Konrad, K. D., Engstrom, L., Perrimon, N., and Mahowald, A. P., 1985, Genetic analysis of oogenesis and the role of maternal gene expression in early development, in: *Developmental Biology: A Comprehensive Synthesis*, Vol. 1: *Oogenesis* (L. W. Browder, ed.), pp. 577–617, Plenum, New York.

Kornberg, T., 1981, *engrailed*: A gene controlling compartment and segment formation in *Drosophila*, *Proc. Natl. Acad. Sci. USA* **78**:1095–1099.

Kornberg, T., Siden, I., O'Farrell, P., and Simon, F., 1985, The *engrailed* locus of *Drosophila*: In situ localization of transcripts reveals compartment-specific expression, *Cell* **40**:45–53.

Kuner, J. M., Nakanishi, M., Ali, Z., Drees, B., Gustaven, E., Theis, J., Kauvar, L., Kornberg, T., and O'Farrell, P. H., 1985, Molecular cloning of *engrailed*: A gene involved in the development of pattern in *Drosophila melanogaster*, *Cell* **42**:309–316.

Kuroiwa, A., Hafen, E., and Gehring, W. J., 1984, Cloning and transcriptional analysis of the segmentation gene *fushi tarazu* of *Drosophila*, *Cell* **37**:825–831.

Kuroiwa, A., Kloter, U., Baumgartner, P., and Gehring, W. J., 1985, Cloning of the homeotic *Sex combs reduced* gene in *Drosophila* and *in situ* localization of its transcripts, *EMBO J.* **4**:3757–3764.

Laughon, A., and Scott, M. P., 1984, Sequence of a *Drosophila* segmentation gene: Protein structure homology with DNA-binding proteins, *Nature (Lond.)* **310**:25–31.

Laughon, A., Carroll, S. B., Storfer, F. A., Riley, P. D., and Scott, M. P., 1985, Common properties of proteins encoded by the *Antennapedia* complex genes of *Drosophila melanogaster*, *Cold Spring Harbor Symp. Quant. Biol.* **50**:253–262.

Laughon, A., Boulet, A. M., Bermingham, J. R., Laymond, R. A., and Scott, M. P., 1986, The structure of transcripts from the homeotic *Antennapedia* gene of *Drosophila*: Two promoters control the major protein-coding region, *Mol. Cell Biol.* **6**:4676–4689.

Lawrence, P. A. (ed.), 1976, *Insect Development*, Halsted Press, London.

Lehmann, R., 1985, Regionspezifische Segmentierungamutanten bei *Drosophila melanogaster* Meigen, Doctoral dissertation, Tübingen.

Lehmann, R., and Nüsslein-Volhard, C., 1986, *hunchback*, a gene required for segmentation of an anterior and posterior region of the *Drosophila* embryo, *Dev. Biol.* **119**:402–417.

Levine, M., Hafen, E., Garber, R. L., and Gehring, W. J., 1983, Spatial distribution of *Antennapedia* transcripts during *Drosophila* development, *EMBO J.* **2**:2037–2046.

Levine, M., Rubin, G. M., and Tjian, R., 1984, Human DNA sequences homologous to a protein-coding region conserved between homeotic genes of *Drosophila*, *Cell* **38**:667–673.

Lewis, E. B., 1954, The theory and application of a new method of detecting chromosomal rearrangements in *Drosophila melanogaster*, *Am. Nat.* **88**:225–239.

Lewis, E. B., 1955, Some aspects of position pseudoallelism, *Am. Nat.* **80**:73–89.

Lewis, E. B., 1978, A gene complex controlling segmentation in *Drosophila*, *Nature (Lond.)* **276**:565–570.

Lewis, E. B., 1982, Control of body segment differentiation in *Drosophila* by the *bithorax* gene complex, in: *Embryonic Development: Genes and Cells* (M. Burger, ed.), pp. 269–288, Alan R. Liss, New York.

Lohs-Schardin, M., Cremer, C., and Nüsslein-Volhard, C., 1979, A fate map for the larval epidermis of *Drosophila melanogaster*: Localized cuticle defects following irradiation of the blastoderm with an ultraviolet laser microbeam, *Dev. Biol.* **73**:239–255.

Mahowald, A. P., and Hardy, P. A., 1985, Genetics of *Drosophila* embryogenesis, *Annu. Rev. Genet.* **19**:149–177.

Martinez-Arias, A., 1986, The *Antennapedia* gene is required and expressed in parasegments 4 and 5 of the *Drosophila* embryo, *EMBO J.* **5**:135–141.

Martinez-Arias, A., and Lawrence, P. A., 1985, Parasegments and compartments in the *Drosophila* embryo, *Nature (Lond.)* **313**:639–642.

McGinnis, W., Garber, R. L., Wirz, J., Kuroiwa, A., and Gehring, W. J., 1984, A homologous protein-coding sequence in *Drosophila* homoeotic genes and its conservation in other metazoans, *Cell* **37**:403–408.

Miller, J., McLachlan, A. D., and Klug, A., 1985, Repetitive zinc-binding domains in the protein transcription factor IIIA from *Xenopus* oocytes, *EMBO J.* **4**:1609–1614.

Mohler, J., and Wieschaus, E., 1985, *Bicaudal* mutations of *Drosophila melanogaster*: Alteration of blastoderm cell fate, *Cold Spring Harbor Symp. Quant. Biol.* **50**:105–112.

Nüsslein-Volhard, C., 1979, Maternal effect mutations that alter the spatial coordinates of the embryo of *Drosophila melanogaster*, in: *Determinants of Spatial Organization* (S. Subtelny and I. R. Konigsberg, eds.), pp. 185–211, Academic, New York.

Nüsslein-Volhard, C., and Wieschaus, E., 1980, Mutations affecting segment number and polarity in *Drosophila*, *Nature (Lond.)* **287**:795–801.

Nüsslein-Volhard, C., Lohs-Schardin, M., Sander, K., and Cremer, C., 1980, A dorso-ventral shift of embryonic primordia in a new maternal-effect mutant of *Drosophila*, *Nature (Lond.)* **283**:474–476.

Nüsslein-Volhard, C., Wieschaus, E., and Kluding, H., 1984, Mutations affecting the pattern of the larval cuticle in *Drosophila melanogaster*. I. Zygotic loci on the second chromosome, *Wilhelm Roux Arch. Dev. Biol.* **193**:267–282.

Pabo, C. O., and Sauer, R. T., 1984, Protein-DNA recognition, *Annu. Rev. Biochem.* **53:**293–321.

Perrimon, N., Engstrom, L., and Mahowald, A. P., 1985, A pupal lethal mutation with a paternally influenced maternal effect on embryonic development in *Drosophila melanogaster*, *Dev. Biol.* **110:**480–491.

Perrimon, N., and Mahowald, A. M., 1968, X-linked female-sterile loci in *Drosophila melanogaster*, *Genetics* **113:**695–712.

Preiss, A., Rosenberg, V. B., Keinlin, A., Seifert, E., and Jäckle, H., 1985, Molecular genetics of *Krüppel*, a gene required for segmentation of the *Drosophila* embryo, *Nature (Lond.)* **313:**27–32.

Rabl, C., 1879, *Ueber die Entwicklung der Morphologisches* Jahrbuch, Leipzig.

Regulski, M., Harding, K., Kostriken, R., Karch, F., Levine, M., and McGinnis, W., 1985, Homeobox genes of the *Antennapedia* and *bithorax* complexes of *Drosophila, Cell* **43:**71–80.

Rosenberg, U. B., Knipple, D., Preiss, A., Seifert, E., and Jäckle, H., 1985, Production of phenocopies by *Krüppel* antisense RNA injection into *Drosophila* embryos, *Nature (Lond.)* **313:**703–706.

Rosenberg, U. B., Schröder, C., Priess, A., Kienlin, A., and Kote, S., 1986, Structural homology of the product of the *Drosophila* Krüppel gene with *Xenopus* transcription factor IIIA, *Nature (Lond.)* **319:**336–339.

Rubin, G. M., and Spradling, A. C., 1982, Genetic transformation of *Drosophila* with transposable element vectors, *Science* **218:**348–353.

Sanchez-Herrero, E., Vernos, I., Marco, R., and Morata, G., 1985, Genetic organization of *Drosophila bithorax* complex, *Nature (Lond.)* **313:**108–113.

Schneuwly, S., Kuroiwa, A., Baumgartner, P., and Gehring, W. J., 1986, Structural organization and sequence of the homeotic gene *Antennapedia* of *Drosophila melanogaster, EMBO J.* **5:**733–739.

Schüpbach, T., and Wieschaus, E., 1986, Maternal-effect mutations altering the anterior–posterior pattern of the *Drosophila* embryo, *Wilhelm Roux Arch. Dev. Biol.* **195:**302–317.

Scott, M. P., 1987, Complex loci of *Drosophila, Annu. Rev. Biochem.* **56:**195–227.

Scott, M. P., and O'Farrell, P. H., 1986, Spatial programming of gene expression in early *Drosophila* embryogenesis, *Annu. Rev. Cell Biol.* **2:**49–80.

Scott, M. P., and Weiner, A. J., 1984, Structural relationships among genes that control development: Sequence homology between the *Antennapedia, Ultrabithorax*, and *fushi tarazu* loci of *Drosophila, Proc. Natl. Acad. Sci. USA* **81:**4115–4119.

Scott, M. P., Weiner, A. J., Polisky, B. A., Hazelrigg, T. I., Pirrotta, V., Scalenghe, F., and Kaufman, T. C., 1983, The molecular organization of the *Antennapedia* locus of *Drosophila, Cell* **35:**763–76.

Shepherd, J. C. W., McGinnis, W., Carrasco, A. E., DeRobertis, E. M., and Gehring, W. J., 1984, Fly and from homeodomains show homologies with yeast mating type regulatory proteins, *Nature (Lond.)* **310:**70–71.

Simcox, A. A., and Sang, J. M., 1983, When does determination occur in *Drosophila* embryos?, *Dev. Biol.* **97:**212–21.

Simpson, P., 1983, Maternal–zygotic gene interactions during formation of the dorsoventral pattern in *Drosophila* embryos, *Genetics* **105:**615–632.

Spradling, A. C., and Rubin, G. M., 1982, Transposition of cloned P elements into *Drosophila* germ line chromosomes, *Science* **218:**341–347.

Steward, R., Ambrosio, L., and Schedl, P., 1985, Expression of the dorsal gene, *Cold Spring Harbor Symp. Quant. Biol.* **50:**223–228.

Strecker, T. R., Kongsuwan, K., Lengyel, J. A., and Merriam, J. R., 1985, The zygotic mutant *Tailless* affects the anterior and posterior ectodermal regions of the *Drosophila* embryo, *Dev. Biol.* **113:**64–76.

Stroeher, V. L., Jorgensen, E. M., and Garber, R. L., 1986, Multiple transcripts from the *Antennapedia* gene of *Drosophila, Mol. Cell Biol.* **6:**4667–4675.

Struhl, G., 1981a, A homoeotic mutation transforming leg to antenna in *Drosophila, Nature (Lond.)* **292:**635–638.

Struhl, G., 1981*b*, A gene product required for correct initiation of segmental determination in *Drosophila*, *Nature (Lond.)* **293**:36–41.

Struhl, G., 1982, Genes controlling segmental specification in the *Drosophila* thorax, *Proc. Natl. Acad. Sci. USA* **79**:7380–7384.

Struhl, G., 1984, Splitting the *bithorax* complex of *Drosophila*, *Nature (Lond.)* **308**:454–457.

Struhl, G., 1985, Near-reciprocal phenotypes caused by inactivation or indiscriminate expression of the *Drosophila* segmentation gene *ftz*, *Nature (Lond.)* **318**:677–680.

Struhl, G., and Akam, M., 1985, Altered distributions of *Ultrabithorax* transcripts in *extra sex combs* mutant embryos of *Drosophila*, *EMBO J.* **4**:3259–3264.

Struhl, G., and Brower, D., 1982, Early role of the *esc*⁺ gene product in the determination of segments in *Drosophila*, *Cell* **31**:285–292.

Struhl, G., and White, R. A. H., 1985, Regulation of the *Ultrabithorax* gene of *Drosophila* by other bithorax complex genes, *Cell* **43**:507–519.

Turner, F. R., and Mahowald, A. P., 1976, Scanning electron microscopy of *Drosophila* embryogenesis. I. The structure of the egg envelopes and the formation of the cellular blastoderm, *Dev. Biol.* **50**:95–108.

Underwood, E. M., Turner, F. R., and Mahowald, A. P., 1980, Analysis of cell movements and fate mapping during early embryogenesis in *Drosophila melanogaster*, *Dev. Biol.* **74**:286–301.

Wakimoto, B. T., and Kaufman, T. C., 1981, Analysis of larval segmentation in lethal genotypes associated with the *Antennapedia* gene complex in *Drosophila melanogaster*, *Dev. Biol.* **81**:51–64.

Wakimoto, B. T., Turner, F. R., and Kaufman, T. C., 1984, Defects in embryogenesis in mutants associated with the *Antennapedia* gene complex of *Drosophila melanogaster*, *Dev. Biol.* **102**:147–172.

Wedeen, C., Harding, K., and Levine, M., 1986, Spatial regulation of *Antennapedia* and *bithorax* gene expression by the *Polycomb* locus in *Drosophila*, *Cell* **44**:739–748.

Weiner, A. J., Scott, M. P., and Kaufman, T. C., 1984, A molecular analysis of *fushi tarazu*, a gene in *Drosophila melanogaster* that encodes a product affecting embryonic segment number and cell fate, *Cell* **37**:843–851.

Weir, M. P., and Kornberg, T., 1985, Patterns of *engrailed* and *fushi tarazu* transcripts reveal novel intermediate stages in *Drosophila* segmentation, *Nature (Lond.)* **318**:433–439.

White, R. A. H., and Akam, M. E., 1985, *Contrabithorax* mutations cause inappropriate expression of *Ultrabithorax* products in *Drosophila*, *Nature (Lond.)* **318**:567–569.

White, R. A. H., and Wilcox, M., 1984, Protein products of the *bithorax* complex in *Drosophila*, *Cell* **39**:163–171.

White, R. A. H., and Wilcox, M., 1985*a*, Distribution of *Ultrabithorax* proteins in *Drosophila*, *EMBO J.* **4**:2035–2043.

White, R. A. H., and Wilcox, M., 1985*b*, Regulation of the distribution of *Ultrabithorax* proteins in *Drosophila*, *Nature (Lond.)* **318**:563–567.

Wieschaus, E., Nüsslein-Volhard, C., and Jürgens, G., 1984*a*, Mutations affecting the pattern of the larval cuticle in *Drosophila melanogaster*. III. Zygotic loci on the X chromosome and fourth chromosome, *Wilhelm Roux Arch. Dev. Biol.* **193**:296–307.

Wieschaus, E., Nüsslein-Volhard, C., and Kluding, H., 1984*b*, Krüppel, a gene whose activity is required early in the zygotic genome for normal embryonic segmentation, *Dev. Biol.* **104**:172–186.

II

Developmental Regulation of Gene Expression in Animals

Chapter 6

Regulation of Actin Gene Expression during Sea Urchin Development

WILLIAM R. CRAIN, JR.

1. Introduction

Differentiation of the early embryo results from the orchestration of a complex series of molecular and cellular events. To dissect and understand this process, it can be useful to focus on a subset of these events, which may yield a detailed and understandable set of observations and relationships. The examination of actin gene expression in the sea urchin embryo is an excellent example of a focused molecular analysis that is yielding information that may have general significance for understanding embryonic development. Interpretation of this kind of analysis in the sea urchin can be particularly informative because of the availability of a large literature characterizing the events of early development, including a description of the embryonic cell lineages (Hörstadius, 1939). In this animal, the opportunity thus exists to view the molecular details of the expression of these genes within the context of the morphological changes that occur within the embryo. In the broadest sense, the goals of this work are to understand how the regulation of expression of these genes as well as the functions of the proteins they encode influence, or are influenced by, determination and induction in the developing embryo.

Several features of actins and the genes that encode them make study of their expression in embryos particularly likely to yield information of general interest. First, actins are exceptionally conserved proteins that are present and abundant in all eukaryotic organisms (Pollard and Weihing, 1974; Goldman et al., 1976; Vandekerckhove and Weber, 1978). In addition, they are encoded by small gene families in all metazoans examined, and the individual family members often display tissue-specific expression, indicating that individual genes are independently regulated (McKeown and Firtel, 1981; Minty et al., 1982; Fyrberg et al., 1983; Gunning et al., 1983; Sanchez et al., 1983; Garcia et al., 1984; Mayer et al., 1984; Shott et al., 1984). Furthermore, actins function in several cellular processes universally important to eukaryotic organisms, in-

WILLIAM R. CRAIN, JR. • Cell Biology Group, Worcester Foundation for Experimental Biology, Shrewsbury, Massachusetts 01545.

cluding contractility, motility, cell division, and the organization of the cytoskeleton. Because these processes are fundamental components of embryonic development, the actins must play important roles in the morphological and physiological differentiation of embryos.

This chapter describes the progress that has been made in the examination of actin gene expression in sea urchin early development. Most of the discussion presented here concerns experimental analyses conducted with the species *Strongylocentrotus purpuratus*. When relevant, information from other species is included. An attempt is made to convey not only the fascinating set of observations that have been made in this experimental system, but also how these observations might give us insight into some aspects of embryonic development. In addition to specifically discussing actin gene expression in sea urchin embryos, this chapter illustrates how the tools of modern molecular biology can be applied to classic questions of embryonic development.

2. Actin Synthesis during Embryogenesis

Actins are known to be important in contraction, motility, and the structure of the cytoskeleton. Within the developing sea urchin embryo, actins are likely to participate in some aspect of each of these functions. Whether particular actin isoforms encoded by different actin genes are exclusively involved in particular functions is unknown. The synthesis of actin in the sea urchin embryo has, however, been examined, and several observations important to understanding the expression of this gene family are discussed. Analysis of total protein from eggs and embryonic stages on two-dimensional polyacrylamide gels has demonstrated that a single actin isoform is predominant in all stages (Durica and Crain, 1982; Bedard and Brandhorst, 1983). This isoform comigrates with mammalian β-actin (Spudich and Spudich, 1979; Mabuchi and Spudich, 1980; Durica and Crain, 1982) and does not change in relative amount throughout embryonic development (Bedard and Brandhorst, 1983). It is thus possible that all the actin protein required for early development is stored within the unfertilized egg, at least in a quantitative sense. The apparent static nature of total embryonic actin, however, belies a much more complex pattern of embryonic actin synthesis.

Analysis of proteins labeled *in vivo* with [^{35}]methionine demonstrates striking quantitative and qualitative changes in the synthesis of actin during embryogenesis. Synthesis of actin in the unfertilized egg is detectable but low, and the only detectable isoform comigrates with mammalian β-actin, referred to as **form I** by Durica and Crain (1982). The synthesis of actin begins to increase sharply between the 64-cell stage and hatching, increasing approximately 100-fold by pluteus (Durica and Crain, 1982; Bedard and Brandhorst, 1983). By 13 hr of development (late cleavage/early blastula), two additional isoforms begin to appear; one (**form II**) is more basic than β-actin and one (**form III**) more acidic. These variants continue to be synthesized throughout the remainder of embryonic development, although they remain as quantitatively minor variants, never comprising more than 10–20% of the total actin synthesized (Durica and Crain, 1982). Analysis of the products of *in vitro* translation of total RNA from eggs and blastula-stage embryos further indicates that the changes in actin synthesis are largely the result of changes in actin mRNA levels (Fig. 1). Both the change in relative amount of actin synthesized and the shift from a single isoform to three isoforms are evident in the *in vitro* translation products (Durica and Crain, 1982).

It is reasonable to speculate that the *de novo* synthesis of actin is important to the development of the sea urchin embryo because the relative amount of actin synthesized increases impressively. It cannot be concluded with certainty, however, that the new synthesis is important *quantitatively*, because the total amount of actin protein does not change detectably as embryogenesis proceeds. How, then, is the new actin important? Several possibilities should be considered. First, the newly synthesized actin may be qualitatively different from the predominant stored form. New actins might thus perform cell-specific functions that are unique and required for the differentiation of specific structures within the embryo. The question of whether extremely similar, but different, actin variants perform different functions is interesting and unresolved. A second possibility is that the newly synthesized actin is replacing stored actin that is unstable during embryogenesis. This possibility does not seem likely, because the only actin isoform that is detectable by staining at any stage is the β-like form, whereas two additional isoelectric variants are being synthesized as development proceeds. A third possibility is that new actins synthesized in the embryo are not functionally important at the time when they first appear. Instead, they may reflect the beginning of differentiation processes that culminate later and that will require the new actins at that time, either quantitatively or qualitatively, or both.

3. Actin mRNA Levels during Embryonic Development

Actin synthesis is regulated temporally during sea urchin embryogenesis (see Section 2). Furthermore, *in vitro* translation of RNA from embryos demonstrates that the amount of actin messenger RNA (mRNA) in the embryo also increases during development. A more complete description of the increase in total actin mRNA content during sea urchin embryogenesis has been obtained using Northern (or RNA) blotting to assay directly for the size and relative abundance of actin mRNAs (Fig. 2). This type of analysis has been performed with cloned actin-coding DNA sequences as hybridization probes to define the general pattern of accumulation of actin messages in sea urchin embryos. Because the protein-coding portions of these genes are highly conserved, it is possible to examine all actin-coding mRNA at once with a single hybridization probe.

Several key features of actin gene expression in the sea urchin S. *purpuratus* emerge from this type of analysis. First, all actin mRNAs fall within two broad size classes—2.2 and 1.8 kilobases (kb)—and the relative abundance

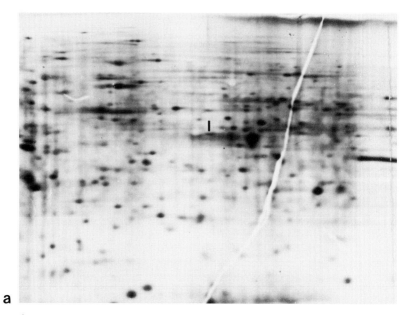

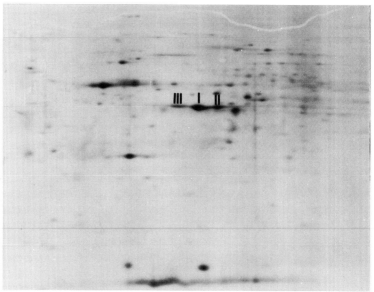

Figure 1. *In vitro* translation products of egg and blastula RNA. (**a**) Total egg RNA was translated *in vitro* in a message-dependent reticulocyte lysate. The labeled proteins were then electrophoresed on a two-dimensional polyacrylamide gel and autoradiographed. The faint spot labeled I (isoform I) comigrates with stained β-actin from the reticulocyte lysate; 208,000 cpm, 10-day exposure. (**b**) Total polysomal RNA from 18-hr embryos was translated *in vitro* and analyzed as described above. The predominant actin is isoform I. The spot labeled isoform II migrates to the basic side of reticulocyte γ actin, which is visible in the stained gel. The spot-labeled isoform III migrates to the acidic side of a mouse α actin; 185,000 cpm, 4-day exposure. (From Durica and Crain, 1982.)

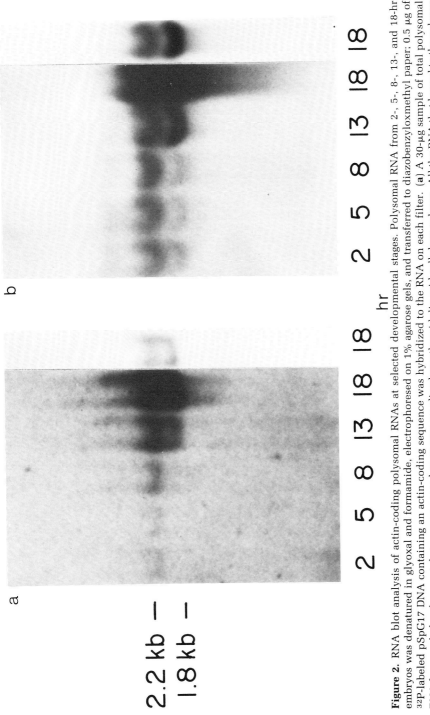

Figure 2. RNA blot analysis of actin-coding polysomal RNAs at selected developmental stages. Polysomal RNA from 2-, 5-, 8-, 13-, and 18-hr embryos was denatured in glyoxal and formamide, electrophoresed on 1% agarose gels, and transferred to diazobenzyloxmethyl paper; 0.5 μg of ^{32}P-labeled pSpG17 DNA containing an actin-coding sequence was hybridized to the RNA on each filter. (**a**) A 30-μg sample of total polysomal RNA from each developmental stage was passed once over an oligodeoxythymidylic acid cellulose column. All the RNA that bound to the column [poly(A)]$^+$ enriched] was then loaded into one lane on the gel. The last lane in this panel is a shorter exposure (1/10) of the next to last lane. (**b**) A 30-μg sample of total polysomal RNA from each developmental stage was loaded into one lane of the gel. The last lane is a shorter exposure (1/5) of the next to last lane. (From Crain *et al.*, 1981.)

of both classes increases sharply beginning 8–13 hr after fertilization. The accumulation of total actin mRNA roughly parallels the increase in total actin synthesis (Crain *et al.*, 1981; Merlino *et al.*, 1981; Shott *et al.*, 1984). Also, actin mRNA accumulates on polysomes with kinetics similar to its accumulation in the cytoplasm (Crain *et al.*, 1981, 1982). These observations imply that there is no major translational regulation of actin gene expression in the embryo after new mRNA begins to accumulate. Evidence suggesting the possibility of some translational regulation before detectable accumulation of new actin message is discussed in Durica and Crain (1982). A second important feature is that both classes of actin mRNA are present in the stored RNA population of the unfertilized egg (Crain *et al.*, 1981). Furthermore, the amount of these mRNA classes in the embryo does not change detectably at fertilization and is constant through the first 8 hr of development. This observation, together with data indicating that transcription of two actin genes that produce most of these messages is activated between 7–11 hr after fertilization (see Section 6), indicate that any actin synthesis in the early cleavage stages is supported by stored maternal actin mRNA.

If the general characteristics of embryonic actin gene expression are associated with differentiation pathways important to the development of the embryo, they might also be found in embryos from related species. Indeed, two major features of actin gene expression are found in five different echinoderm species, including four sea urchins (*S. purpuratus, S. droebachiensis, Arbacia punctulata,* and *Lytechinus variegatus*) and a sand dollar (*Echinarachnius parma*) (Bushman and Crain, 1983). In each of these species, the relative abundance of actin mRNA is low in early embryos and accumulates throughout embryonic development beginning at late-cleavage/early-blastula stages. In addition, each species expresses a major actin mRNA size class of 2.0–2.2 kb. This conservation of some general features of embryonic actin gene expression in several echinoderms suggests that the *de novo* synthesis of actin mRNA is required for the differentiation of the embryo. It further suggests that members of the gene family might be associated with programs of embryonic gene expression that emerged early in echinoderm evolution and are similar in different modern species.

In addition to the characterization of actin gene expression in embryos of several different species, the expression of these genes also has been examined in three sets of interspecific sea urchin embryo hybrids. These experiments demonstrate that some members of this gene family are transcribed and produce mature mRNA in hybrid embryos of *S. purpuratus* × *L. variegatus* and *S. purpuratus* × *S. droebachiensis,* and that in the reciprocal crosses of *S. purpuratus* × *L. variegatus,* both the maternally and paternally contributed alleles of at least two different actin genes are expressed as mRNA (Crain and Bushman, 1983). An extensive analysis of actin mRNA accumulation throughout early development in hybrid embryos of *S. purpuratus* and *Lytechinus pictus* has revealed temporally correct expression of four different actin genes (Bullock, Nisson, and Crain, in preparation). The relative amounts of mRNA from these *S. purpuratus* actin genes were measured in hybrid embryos constructed in both directions. In each case the message accumulated with the kinetics normally found in S. purpuratus embryos. These experiments demonstrate that expression of these S. purpuratus genes is correctly regulated when they are introduced into L. pictus eggs, and that the introduction of the L. pictus paternal genome into S. purpuratus eggs does not disrupt their regulated expression. If expression of these four actin genes is regulated by effector molecules present in the egg at fertilization, then these molecules must be sufficiently conserved for those from L. pictus to regulate the genes from the distantly related species, S. purpuratus. Approximate divergence of these species estimated from the fossil record is 150–200 million years ago (Durham, 1966). If, on the other hand, the molecules that regulate the S. purpuratus actin genes are synthesized from the S. purpuratus genome in hybrid embryos, then their expression also must have been correctly regulated by L. pictus factors. These analyses therefore indicate that at least some mechanisms for regulating embryonic gene expression are highly conserved between these sea urchins.

4. The Sea Urchin Actin Genes Comprise a Small Gene Family

An extensive examination of the actin gene family in the sea urchin S. purpuratus has been carried out to establish the framework necessary for the characterization of actin gene expression in this animal. Recombinant DNA clones carrying various members of this gene family were originally identified and characterized in several laboratories (Durica et al., 1980; Merlino et al., 1980; Overbeek et al., 1981; Scheller et al., 1981; Schuler and Keller, 1981). It was evident from this early work that 5–20 actin genes are present in the S. purpuratus genome. Thorough characterization of many cloned representatives from this gene family, in association with genome Southern blot analysis, has now shown that there are at least six active actin genes, one pseudogene, and one actinlike sequence that may or may not be an active gene.

As the examination of these genes has proceeded at different laboratories, the many cloned representatives (both genomic and cDNA) have been designated with a variety of different names. A coherent and unifying system for naming these genes has now been presented by Lee et al. (1984). Table I summarizes all the genes according to this system; it also indicates each of the other names that appear in the literature. Rather than attempt to describe the experimental analyses that resulted in the current description of this gene family, this discussion reviews briefly the characteristics of these genes which are essential to a thorough examination of their expression.

The experimental basis for defining and distinguishing different actin genes in the sea urchin genome is the diversity in the 3′ untranslated mRNA sequences associated with these genes. As is the case for actin genes from a variety of other species, the actin-coding portion of these genes is highly conserved, whereas the 3′ untranslated message sequences are not. Six active actin genes with distinct restriction site maps have been identified in recombinant

Table I. *Strongylocentrotus purpuratus* Actin Gene Subtypes[a]

Actin gene type[b]	Subtype designations[c]	Individual gene designations	Gene designations in previous studies
Cytoskeletal	CyI	CyI	Gene C,[e] gene 2,[f] pSpG17,[g] λSA16,[i] λSA22[i]
	CyII	CyIIa	Gene G/F,[e] gene 1[f]
		CyIIb	λSA11,[i] gene I/H,[e] pSA38,[h]
		(CyIIc)[d]	—
	CyIII	CyIIIa	pSpec4[j]
		CyIIIb	Gene K[e]
		CyIIIc	—
Muscle	M	M	Gene J,[e] pSpG28[g]

[a]Adapted from Lee *et al.* (1984).

[b]The actin genes are classified as cytoskeletal or muscle types on the basis of their representation in the RNA of various adult and embryo sea urchin tissues (Garcia *et al.*, 1984; Shott-Akhurst *et al.*, 1984). The cytoskeletal actin genes are represented in the RNAs of eggs, or pregastrula embryos, or in adult coelomocytes, none of which include muscle. Muscle actin gene transcripts are not detectable in any of these cell types, but are found in adult sea urchin lantern muscle, gut, and tubefeet.

[c]Subtype designation (Cy, cytoskeletal; M, muscle) refers to the transcribed, but untranslated, 3'-terminal sequence region of the gene. Genes of a given subtype are those that display reaction with the respective 3'-terminal subcloned probe at a hybridization criterion of 55°C, 0.75 M Na$^+$.

[d]The existence of this gene is only inferred from genome blot reactions with the CyIIb 3'-terminal probe sequence. However, it is not yet demonstrated that this sequence is associated with sequences homologous to actin gene-coding regions.

[e]From Scheller *et al.* (1981).

[f]Schuler *et al.* (1983) isolated gene 1 and gene 2 from the same recombinant DNA library as used by Scheller *et al.* (1981). This library was described by Anderson *et al.* (1981).

[g]From Durica *et al.* (1980).

[h]From Merlino *et al.* (1980). This clone is a complementary DNA (cDNA) clone. Its identification as a cloned transcript of the CyIIb gene is based on 3'-terminal sequence data. The clone was sequenced by Schuler *et al.* (1983). Comparison of the data of Lee *et al.* (1984) for the CyIIb gene shows the two sequences to be almost identical.

[i]From Overbeek *et al.* (1981). These identifications are tentative, as they are based only on restriction map homologies. λSA11 probably contains both the CyIIa and the CyIIb genes. It is probably identical to λSpG2-39 of Scheller *et al.* (1981). The isolates of Overbeek *et al.* (1981) are also derived from the genome library of Anderson *et al.* (1981).

[j]From Bruskin *et al.* (1981).

DNA clones and have been shown to fall into four categories of subtypes designated **CyI, CyII, CyIII,** and **M** by Lee *et al.* (1984) on the basis of their 3' untranslated sequences. The Cy designation indicates a cytoskeletal actin gene, and M indicates a muscle actin gene. Hybridization probes derived from the 3' regions of each gene subtype do not cross-react with the other subtype genes under nonstringent hybridization conditions (55°C, 0.75 M Na$^+$). The CyI and M subtypes each contain a single gene, whereas the CyII and CyIII subtypes are each composed of two cloned intact genes whose 3' sequences cross-react at low stringency (CyIIa and CyIIb and CyIIIa and CyIIIb, respectively). At higher hybridization stringency (68°C, 0.75 M Na$^+$), the 3' hybridization probes of CyIIa and CyIIb show no detectable cross-reaction, making it possible to distinguish experimentally these genes and their transcripts. The 3' untranslated regions of the two CyIII genes appear to be more closely related, because the probe from CyIIIb will cross-react with CyIIIa under the higher stringency conditions. A subclone of a 131-nucleotide fragment from the 3′ untranslated region of CyIIIa has been shown, however, to react specifically with this gene (Lee *et al.*, 1986). Two additional actin-coding sequences classified as CyIIc and CyIIIc have been identified (Lee *et al.*, 1984). Blotting and sequence analysis of a cloned representative of CyIIIc suggest that it is a pseudogene that is missing sequences at its 5′ end (Akhurst *et al.*, 1987). The existence of CyIIc is inferred from genome blot analysis, but, because no recombinant clone of this gene has been identified, it has not been determined whether it is an active gene. These eight sequences seem to account for all actin-coding sequences present in the sea urchin genome as determined by the Southern blot technique.

Several of the six characterized actin genes are linked in the genome. One linkage group consists of the cytoskeletal genes CyI–CyIIa–CyIIb arranged in the same orientation (transcription from left to right) and spaced several kilobases apart (Lee *et al.*, 1984). The second linkage group contains the cytoskeletal genes CyIIIa and CyIIIb, which are also separated by several kilobases and arranged in the same transcriptional orientation (Akhurst *et al.*, 1987). The sixth gene, M (the single muscle actin gene), does not appear to be closely linked to any other actin gene. Whether these linkage arrangements are important in the regulation of expression of these genes cannot be conclusively stated. However, examination of expression of these genes (see Section 5) demonstrates that only two of the linked genes (CyI and CyIIb) exhibit coordinate expression (Shott *et al.*, 1984; Cox *et al.*, 1986).

The nucleotide sequences of four members of this gene family have been determined: CyI (Cooper and Crain, 1982; Schuler *et al.*, 1983), CyIIa (Schuler *et al.*, 1983), M (Crain *et al.*, 1987) and CyIIIa (Akhurst *et al.*, 1987). These sequences have revealed several interesting properties of these genes and their encoded proteins. First, the protein-coding sequences of the two linked cytoskeletal genes CyI and CyIIa are closely related, differing at only 4.2% of their nucleotides, while the unlinked cytoskeletal gene CyIIIa is more distantly related, differing from CyI at 6.3% of the nucleotides. The sequence of the muscle gene, on the other hand, is much more divergent relative to each of the cytoskeletal genes (12.1%, 11.7%, and 13.7% nucleotide sequence difference compared with CyI, CyIIa and CyIIIa, respectively). Furthermore, the coding regions of each of the cytoskeletal genes are split by introns between codons 121 and 122 and within codon 204. The muscle gene contains two introns at precisely these same positions and two additional introns between codons 41 and 42 and within codon 267. The close linkage and the related structure and sequence of CyI and CyIIa suggest that they (and probably CyIIb) are derived from a common ancestral actin gene that was duplicated fairly recently to produce the three linked genes. The similar structure but somewhat more distantly related sequence of CyIIIa indicates that CyI and CyIIIa must have diverged from a common ancestral gene before the duplication that yielded the CyI–CyIIa–CyIIb linkage group. The different intron structure and more divergent sequence of the muscle gene, when compared with the cytoskeletal genes, indicate that it diverged from the ancestral cytoskeletal gene substantially before the cytoskeletal gene duplications. Another feature of the evolution of this gene family, indicated by comparison of these sequences, is that a fairly recent gene conversion event seems to have matched the sequence of the muscle actin gene with that of either CyI or CyIIa in the region from codons 61–120 (Crain et al., 1987). Gene conversion has therefore played a role in maintaining homogeneity within this gene family. This particular gene conversion lies entirely within the protein coding portion of these genes. If such a conversion were to extend into regions of a gene that are involved in regulation of its expression, then its expression could be converted to an entirely different pattern. Such changes would have important implications for the evolution of developmental pathways.

The actins that are encoded by the cytoskeletal genes CyI, CyIIa and CyIIIa are like vertebrate cytoskeletal actins; i.e., when the amino acids at the most diagnostic positions are compared, CyI, CyIIa and CyIIIa contain, respectively, 73%, 75%, and 59% cytoskeletal-like residues; 14%, 15%, and 23% muscle-like residues; and 14%, 10%, and 18% that fall into neither category. Somewhat surprisingly, the muscle gene also encodes a more cytoskeletal-like than muscle-like actin. Of the diagnostic positions, 62% encode cytoskeletal-like amino acids, 24% muscle-like residues, and 14% fall into neither category (Crain et al., 1987). All other invertebrate muscle actins reported so far also are predominantly cytoskeletal-like (Fyrberg et al., 1983; Sanchez et al., 1983; Vandekerckhove and Weber, 1984). This again raises the question of whether actins expressed in specific cells or tissues can perform only the specific functions required in these tissues, or whether they exhibit specificity of expression because of an association with units of genes whose coordinate expression is required for particular developmental pathways. Certainly, the precise actin sequence conserved among the vertebrate muscle actins is not required for muscle function within invertebrates.

5. Distinct and Characteristic Patterns of Specific Actin Gene Expression in Embryos and Adult Tissues

The 3′ untranslated mRNA sequences of the different actin genes are sufficiently diverse to serve as gene-specific hybridization probes (see Section 4). Using these probes to detect mRNA from individual actin genes in the RNA from a variety of embryonic stages and adult tissues, several studies have described the expression of six members of this gene family (Garcia et al., 1984; Shott et al., 1984; Lee et al., 1986). The expression of these genes is summarized in Table II. The gene that shows the most striking specificity of expression is the muscle gene (M). The probe for this gene detects an abundant 2.2-kb RNA in three adult muscle-containing tissues and pluteus-stage embryos; there is no detectable expression of this gene in coelomocytes (a non-muscle–containing adult cell type) or early embryos. Both ovary and testis also express low amounts of this mRNA and probably contain small amounts of muscle tissue.

Table II. Qualitative Estimates of Expression of Actin Genes in Sea Urchin Embryo and Adult Tissues[a]

Gene	Transcript length (kb)	First appearance in embryogenesis	Expression in adult tissues relative to pluteus embryos[b]						
			Testis	Ovary	Coelomocytes	Intestine	Tubefoot	Lantern muscle	Pluteus
CyI	2.2	Maternal mRNA	+++++	++++++	+++++	+++++	+++++	+++++	+++++
CyIIa	2.2	Gastrula (40 hr)	−	−	+/−	+	+/−	+/−	+
CyIIb	2.1	Early blastula (14 hr)	+/−	+/−	+	+	+	+	+
CyIIIa	1.8	Maternal mRNA	−	+	−	−	−	−	+++++
CyIIIb	2.1	Early blastula (14 hr)	−	+	−	−	−	−	+
M	2.2	Early pluteus (62 hr)	+/−	+/−	−	+	+++	+++++	+

[a]From Shott et al. (1984).
[b]+, relative abundance of each gene transcript in different tissues; −, undetectable (believed to mean less than two transcripts per average cell at the pluteus stage. +/−, detectable transcript but only on long exposure.

Several studies now indicate that the coelomic pseudopod-forming cells of the sea urchin pluteus are authentic muscle cells (Gustafson, 1975; Ishimoda-Takagi *et al.*, 1984). Their appearance during embryo differentiation coincides with the appearance of the muscle actin mRNA (i.e., late prism/early pluteus), further supporting the notion that this gene encodes a muscle-specific actin. The other five genes appear to encode cytoskeletal actins, based on the observation that they are expressed in embryonic stages and adult tissues that lack muscle. The three genes in the linkage group CyI–CyIIa–CyIIb display the least restricted expression of these six actin genes. CyI is the most widely expressed member of the gene family, exhibiting abundant 2.2-kb mRNA in all embryonic stages and adult tissues examined. Expression of mRNA from CyIIb (2.1 kb) corresponds qualitatively to that of CyI, but its abundance is generally lower than that of CyI. CyIIa, the middle gene in this linkage group, is expressed (2.2-kb RNA) at lower levels than CyI and CyIIb in embryos and several adult tissues; its messenger is undetectable in testis and ovary. The two genes in the second linkage group CyIIIa–CyIIIb produce mRNAs of 1.8 and 2.1 kb, respectively, and are expressed only in embryos. Each of these mRNAs increases in abundance during embryogenesis, with the sharpest rise occurring between the late cleavage and mesemchyme blastula stages. At every stage the amount of CyIIIa mRNA is substantially greater than that of the CyIIIb mRNA (Bullock, Nisson and Crain, submitted). That these actins are expressed exclusively in embryos raises the possibility that their function is specific for structures within the embryo. Alternatively, they may have the potential to function in a more general way but simply are associated with a developmental program of gene expression that is specific for particular differentiation pathways of the embryo.

It is evident from the analyses discussed here that expression of the members of the actin gene family of the sea urchin is independently regulated. This is true despite the fact that these genes encode proteins whose sequences differ very little. The physical basis for the differential expression of these genes is unknown and is the question underlying much of the research currently under way on these genes. The available evidence indicates that close linkage within the genome is not sufficient to cause coordinate expression within this gene family. This follows from the observation that CyIIa displays a distinct pattern of expression when compared to CyI and CyIIb to which it is linked and that the closely linked CyIIIa and CyIIIb genes show quite different levels of expression in the developing embryo. It seems that the most likely mechanism for regulating the expression of these genes independently is the interaction of embryonic effector molecules with specific sequences associated with each gene.

6. Embryonic Expression of at Least Three Actin Genes Results from Transcriptional Activation

The observation that the mRNAs from actin gene family members accumulate with distinct patterns in developing embryos could be explained by differential transcription of these genes or by differential stability of the RNAs. The

Table III. Transcription of Three *S. purpuratus* Actin Genes in Isolated Nuclei[a]

Gene	Relative transcription at different stages[b]					
	7 hr	9 hr[c]	11 hr	Blastula	Gastrula	Pluteus
CyI	−	+/−	+	+	+	+ + + +
CyIIIa	−	+	+ + + + +	+ +	+ +	+ +
M	−	−	−	−	−	+

[a]From Hickey *et al.* (1987).
[b]The relative amount of transcription at each stage is shown, with the lowest detectable hybridization signal set at (+). These relative values were determined by densitometer quantitation of hybridization bands. Comparisons are made only within a single stage; no valid comparisons can be made between stages.
[c]The hybridization signals of the CyI- and CyIIIa-specific bands at 9 hr after fertilization were very weak and were not quantitated. In this case (+) means detectable and (+/−) means barely detectable visually on the original films.

distinction between these possibilities is important. These two alternatives would require either different types or distributions of effector molecules within the embryo, or both, and would thus suggest different molecular mechanisms for embryonic determination or induction of these genes. Evidence derived from transcription in nuclei isolated from embryos (runoff transcription) at different stages demonstrates a close parallel between the appearance of mRNA and activation of transcription of three of these genes, CyI, CyIIIa, and M (summarized in Table III, from Hickey *et al.*, 1987). This analysis, which detects the presence of newly synthesized nuclear RNA that hybridizes to gene-specific probes, demonstrates that transcription of both CyI and CyIIIa is activated 7–11 hr after fertilization, and that at the early stages the CyIIIa gene is severalfold more active per embryo than CyI. The relative transcription of these two genes changes as development proceeds so that by the pluteus stage the CyI gene is at least twice as active per embryo as the CyIIIa gene. Measurement of levels of mRNA from these two genes has shown that new messenger begins to accumulate in the embryo 8–13 hr after fertilization (Crain *et al.*, 1981; Shott *et al.*, 1984). The CyIIIa mRNA accumulates more rapidly than CyI mRNA in early embryos but then levels off at the blastula stage, while CyI mRNA continues to accumulate through pluteus (Shott *et al.*, 1984; Lee *et al.*, 1986). This tight temporal coupling of the accumulation of these cytoskeletal actin mRNAs and the transcription of their genes in isolated nuclei suggest that their expression is regulated largely, if not entirely, at the level of transcription. Transcription of the muscle actin gene is first detected in pluteus-stage embryos. This corresponds closely with the appearance of stable mRNA from this gene, suggesting that regulation of its expression also is determined largely or entirely at the level of transcription.

The evidence discussed above demonstrates that transcriptional regulation must play a significant role in controlling expression of at least three actin genes, which are located in different regions of the genome and whose patterns of expression are quite distinct. A further question that remains is whether transcription of the genes that are closely linked to each other is also independently regulated. If transcriptional activation results from rearrangement of the structure of large chromatin domains, closely linked genes such as CyI–CyIIa–CyIIb and CyIIIa–CyIIIb may be transcribed at the same time. If this were the case, the differential appearance in embryos of mRNAs from the linked actin genes (CyIIa and CyI/CyIIb, CyIIIa and CyIIIb) would result from an additional level of regulation, perhaps mRNA stability.

7. Cell Lineage-Specific Expression of Actin Genes

Expression of members of the actin gene family in the embryo and adult tissues of the sea urchin is diverse and indicates that these genes are regulated independently. The potential therefore exists for each to be embedded in a distinct regulatory program of gene expression that is essential for a specific program of differentiation within the embryo. To address this possibility, Cox *et al.* (1986) applied the powerful technique of *in situ* hybridization of gene-specific probes to RNA transcripts in the sea urchin embryo. This procedure permits detection of mRNA from specific genes in regions and, in the best case, in specific cells of the embryo, making it possible to define precisely the spatial expression of genes during embryogenesis. In the case of the sea urchin embryo, these analyses can be interpreted within the context of considerable knowledge of the cell lineages that give rise to the structures of the larva (discussed in Angerer and Davidson, 1984), which were depicted in some detail by Hörstadius (1939). The data reported by Cox *et al.* (1986) describe an impressively complex set of cell- and region-specific expression patterns for the different actin genes. These expression patterns further appear to correspond to previously described morphological lineages, and in some cases, lineage-specific actin gene expression precedes (and thus foretells) morphological differentiation.

An example of *in situ* hybridization of the CyI-specific probe to embryos from the mesenchyme blastula through the pluteus stages is shown in Fig. 3. Expression of CyI and CyIIb were found to be indistinguishable both temporally (Shott *et al.*, 1984; Lee *et al.*, 1986) and spatially (Cox *et al.*, 1986), and thus the description of CyI expression is a description of CyIIb expression as well. CyI mRNA appears to be uniformly distributed in early embryos ($\leq$18 hr), but by 23 hr it begins to accumulate preferentially in the cells of the vegetal pole of the embryo. Cox and co-workers speculate that this transition is due to decay (and cessation of synthesis) of CyI mRNA in the aboral ectoderm lineage as its differentiation occurs. As development procedes, expression of this gene becomes more complex. At gastrula, mRNA appears in the secondary mesenchyme cells (Fig. 3e), in the archenteron (Fig. 3e,f), and in the presumptive oral region of the ectoderm (Fig. 3f). By the pluteus stage, expression is most pronounced in the gut but is also evident over the oral ectoderm and some of the mesenchyme cells.

Messenger RNA from the CyIIa gene, which lies between CyI and CyIIb in the genome, is not detected in the embryo until the early blastula stage (20 hr) and is less abundant in total embryo RNA than any other actin message. In blastula-stage embryos, mRNA from the gene is present only in the vegetal pole. At early mesenchyme blastula (23 hr), some expression is detected in primary mesenchyme and presumptive secondary mesenchyme cells. Several hours later, CyIIa mRNA seems to have disappeared entirely from the ingressed primary mesenchyme cells. From mid-blastula to late gastrula, this mRNA is detected only in secondary mesenchyme cells. By the pluteus stage, CyIIa mRNA is confined to the gut, indicating a shift in expression from mesenchyme to gut in the later embryo.

CyIIIa and CyIIIb are both expressed only in embryos, and their mRNAs accumulate during development beginning at late cleavage stages. Cox *et al.* (1986) concluded that expression of both of these genes is restricted to the aboral ectoderm lineage throughout embryogenesis. This localization is seen by 18 hr as newly synthesized message begins to accumulate and continues to be present in gastrula and pluteus stage embryos.

The muscle gene (M) is expressed only late in embryogenesis, at stages when embryonic muscle structures are formed. *In situ* hybridization analysis localizes this expression to two small clusters of cells at the late gastrula and pluteus stages (Cox *et al.*, 1986). In the late gastrula, these cells are in the wall of the archenteron, whereas at pluteus the labeled cells are associated with the paired coelomic rudiments. Analyses by Gustafson and Wolpert (1967) and Ishimoda-Takagi *et al.* (1984) demonstrated that the embryonic muscle structures are formed from pseudopodia of cells in the coelomic sacs. The location of mRNA from the muscle actin gene is therefore consistent with expression in muscle cells.

Expression of three of these genes (CyIIIa, CyIIIb, and M) is confined to single cell lineages and thus constitutes clear examples of lineage-specific gene expression (Cox *et al.*, 1986). The other three genes (CyI, CyIIa, and CyIIb), which are linked to each other in the genome, display a more complex pattern of expression. The appearance of mRNAs from each of these genes appears to shift between at least two of the morphologically defined lineages as development proceeds. In some lineages, these mRNAs appear during development, whereas in others they disappear.

The complex spatial expression of the different actin genes within the embryo again returns us to the question of the functional role of these genes and proteins in development. Does each actin perform a unique function, or are genes encoding functionally similar proteins expressed differently because of their association with separate regulatory units of gene expression? Cox *et al.* (1986) point out some distinctive morphological properties of the lineages expressing the different actin genes that suggest the possibility of functional distinctions among the actins. First, the CyI and CyIIb mRNAs seem to be found mainly in dividing cells and absent from nondividing cells. By contrast, expression of the CyIII genes seems to reflect the differentiated state of the cells within the aboral ectoderm. The remaining cytoskeletal gene, CyIIa, appears to be present only in cells that use contractile pseudopodia. Finally, expression of the muscle gene appears to be associated directly with muscular structures of

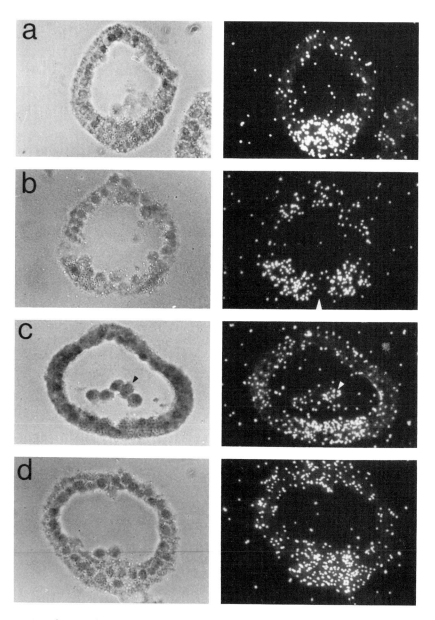

Figure 3. Distribution of CyI mRNA in the sea urchin embryo. The CyI probe (1.3×10^8 dpm/μg) was hybridized at a concentration estimated to be 85% that required for saturation. Auto-radiographic exposure was for 68 days (**a–e**) or for 42 days (**f,g**). (**a–c**) Mesenchyme blastula, 23 hr; (**d**) late mesenchyme blastula, 29 hr; (**e**) gastrula, 35 hr; (**f**) late gastrula/early prism, 48 hr; (**g**) pluteus, 82 hr. Sections in (**a,b,d,e**) pass approximately through the animal–vegetal axis, and the vegetal pole is at the bottom; the section in (**c**) is probably slightly oblique and in (**f**) is approximately perpendicular to this axis. The arrows in different panels indicate: (**b**) the small gap in

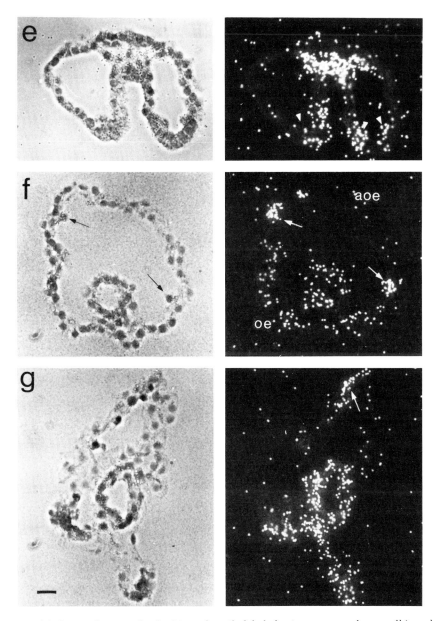

Figure 3. labeling at the vegetal pole; (c) one heavily labeled primary mesenchyme cell in a cluster of five of these cells; (e) three mesenchyme cells on the left that are unlabeled and the two on the right that are moderately labeled; (f) two highly labeled mesenchyme cells, probably secondary; and (g) labeled mesenchyme cells, which characteristically appear at the vertex of the embryo closely applied to the aboral ectoderm. aoe, aboral ectoderm; oe, oral ectoderm. All sections are shown at the same magnification; scale bar in (g): 10 μm. (From Cox et al., 1986.)

the embryo. The generality of these relationships are not unequivocally demonstrated by the data, but they are intriguing.

8. Summary

The progress that has been made in the last several years toward an understanding of the expression of the actin genes of the sea urchin is impressive. It serves as an excellent example of how the application of modern molecular biological techniques to a classic experimental system (the sea urchin embryo) can begin to give us insight into the processes of embryological development. There is reason to hope that general principles will emerge from studies such as these, but many questions are unanswered. With specific regard to the actin genes and proteins, there are some obvious questions. Are the actins encoded by the different genes functionally distinct, and what roles do they play in differentiation and development? How is the expression of each of these genes regulated; i.e., what molecules participate, how do they work, where are they located in the embryo, and when do they appear? The more general question is: How are these (and other) genes and proteins affected by, or how do they contribute to, determination and induction in early development? We hope that answers to the specific questions posed will provide important steps toward answers to the general question.

References

Akhurst, R. J., Calzone, F. J., Lee, J. J., Britten, R. J., and Davidson, E. H., 1987, Structure and organization of the CyIII actin gene subfamily of the sea urchin, *Strongylocentrotus purpuratus*, *J. Mol. Biol.* **194**:193–203.

Anderson, D. M., Scheller, R. H., Posakony, J. W., McAllister, L. B., Trabert, S. G., Beall, C., Britten, R. I., and Davidson, R. H., 1981, Repetitive sequences of the sea urchin genome: distribution of members of specific repetitive families, *J. Mol. Biol.* **145**:5–28.

Angerer, R. C., and Davidson, E. H., 1984, Molecular indices of cell lineage specification in sea urchin embryos, *Science* **226**:1153–1160.

Bedard, P. A., and Brandhorst, B. P., 1983, Patterns of protein synthesis and metabolism during sea urchin embryogenesis, *Dev. Biol.* **96**:74–83.

Bruskin, A. M., Tyner, A. L., Wells, D. E., Showman, R. M., and Klein, W. H., 1981, Accumulation in embryogenesis of five mRNAs enriched in the ectoderm of the sea urchin pluteus, *Dev. Biol.* **87**:308–318.

Bushman, F. D., and Crain, W. R., 1983, Conserved pattern of embryonic actin gene expression in several sea urchins and a sand dollar, *Dev. Biol.* **98**:429–436.

Cooper, A. D., and Crain, W. R., 1982, Complete nucleotide sequence of a sea urchin actin gene, *Nucl. Acids Res.* **10**:4081–4092.

Cox, K. H., Angerer, L. M., Lee, J. J., Davidson, E. H., and Angerer, R. C., 1986, Cell lineage-specific programs of expression of multiple actin genes during sea urchin embryogenesis, *J. Mol. Biol.* **188**:159–172.

Crain, W. R., and Bushman, F. D., 1983, Transcripts of paternal and maternal actin gene alleles are present in interspecific sea urchin embryo hybrids, *Dev. Biol.* **100**:190–196.

Crain, W. R., Durica, D. S., and VanDoren, K., 1981, Actin gene expression in developing sea urchin embryos, *Mol. Cell. Biol.* **1**:711–720.

Crain, W. R., Durica, D. S., Cooper, A. D., VanDoren, K., and Bushman, F. D., 1982, Structure and developmental expression of actin genes in the sea urchin, in: *Muscle Development: Molecular and Cellular Control* (M. Pearson and H. F. Epstein, eds.), pp. 97–105, Cold Spring Harbor, New York.

Crain, W. R., Boshar, M. F., Cooper, A. D., Durica, D. S., Nagy, A., and Steffen, D., 1987, The sequence of a sea urchin muscle actin gene suggests a gene conversion with a cytoskeletal actin gene, *J. Mol. Evol.* **25**:37–45.

Durham, J. W., 1966, Echinoids, classification, in: *Treatise on Invertebrate Paleontology* Part II, *Echinodermata 3*, Vol. I, (R. C. Moore, ed.), p. 280, Geological Society of America and University of Kansas Press, Lawrence.

Durica, D. S., and Crain, W. R., 1982, Analysis of actin synthesis in early sea urchin development, *Dev. Biol.* **92**:428–439.

Durica, D. S., Schloss, J. A., and Crain, W. R., 1980, Organization of actin gene sequences in the sea urchin: Molecular cloning of an intron-containing DNA sequence coding for a cytoplasmic actin, *Proc. Natl. Acad. Sci. USA* **77**:5683–5687.

Fyrberg, E. A., Mahaffey, J. W., Bond, B. J., and Davidson, N., 1983, Transcripts of the six *Drosophila* actin genes accumulate in a stage and tissue-specific manner, *Cell* **33**:115–123.

Garcia, R., Paz-Aliaga, B., Ernst, S. G., and Crain, W. R., 1984, Three sea urchin actin genes show different patterns of expression: Muscle specific, embryo specific and constitutive, *Mol. Cell. Biol.* **4**:840–845.

Goldman, R., Pollard, T., and Rosenbaum, J. (eds.), 1976, *Cell Motility*, Vol. B, Cold Spring Harbor Laboratory, Cold Spring Harbor, New York.

Gunning, P., Ponte, P., Blau, H., and Kedes, L., 1983, α-Skeletal and α-cardiac actin genes are coexpressed in adult human skeletal muscle and heart, *Mol. Cell Biol.* **3**:1985–1995.

Gustafson, T., 1975, Cellular behavior and cytochemistry in early stages of development, in: *The Sea Urchin Embryo* (G. Czihak, ed.), pp. 233–266, Springer-Verlag, New York.

Gustafson, T., and Wolpert, L., 1967, Cellular movement and contact in sea urchin morphogenesis, *Biol. Rev.* **42**:447–498.

Hickey, R. J., Boshar, M. F., and Crain, W. R., 1987, Transcription of three actin genes and a repeated sequence in isolated nuclei of sea urchin embryos, *Dev. Biol.*, **124**:215–227.

Hörstadius, S., 1939, The mechanics of sea urchin development, studied by operative methods, *Biol. Rev. Camb. Philos. Soc.* **14**:132–179.

Ishimoda-Takagi, T., Chino, I., and Sato, H., 1984, Evidence for the involvement of muscle tropomyosin in the contractile elements of the coelom-esophagus complex in sea urchin embryos, *Dev. Biol.* **105**:365–376.

Lee, J. J., Shott, R. J., Rose, S. J., Thomas, T. L., Britten, R. J., and Davidson, E. H., 1984, Sea urchin actin gene subtypes: Gene number, linkage and evolution, *J. Mol. Biol.* **172**:149–176.

Lee, J. J., Calzone, F. J., Britten, R. J., Angerer, R. C., and Davidson, E. H., 1986, Activation of sea urchin actin genes during embryogenesis: measurement of transcript accumulation from five different genes in *Strongylocentrotus purpuratus*, *J. Mol. Biol.* **188**:173–183.

Mabuchi, I., and Spudich, J. A., 1980, Purification and properties of soluble actin from sea urchin eggs, *J. Biochem. (Tokyo)* **87**:785–802.

Mayer, Y., Czosnek, H., Zeelon, P. E., Yaffe, D., and Nudel, U., 1984, Expression of the genes coding for the skeletal muscle and cardiac actins in the heart, *Nucl. Acids Res.* **12**:1087–1100.

McKeown, M., and Firtel, R. A., 1981, Differential expression and 5′ end mapping of actin genes in *Dictyostelium*, *Cell* **24**:799–807.

Merlino, G. T., Water, R. D., Chamberlain, J. P., Jackson, D. A., El-Gewely, M. R., and Kleinsmith, L. J., 1980, Cloning of sea urchin actin gene sequences for use in studying the regulation of actin gene transcription, *Proc. Natl. Acad. Sci. USA* **77**:765–769.

Merlino, G. T., Water, R. D., Moore, G. P., and Kleinsmith, L. J., 1981, Change in expression of the actin gene family during early sea urchin development, *Dev. Biol.* **85**:505–508.

Minty, A. J., Alonso, S., Caravatti, M., and Buckingham, M. E., 1982, A fetal skeletal muscle actin mRNA in the mouse and its identity with cardiac actin mRNA, *Cell* **30**:185–192.

Overbeek, P. A., Merlino, G. T., Peters, N. K., Cohn, V. H., Moore, G. P., and Kleinsmith, L. J., 1981, Characterization of five members of the actin gene family in the sea urchin, *Biochim. Biophys. Acta* **656**:195–205.

Pollard, T. D., and Weihing, R. R., 1974, Actin and myosin and cell movement, *Crit. Rev. Biochem.* **2:**1–65.

Sanchez, F., Tobin, S. L., Rdest, U., Zulauf, E., and McCarthy, B. J., 1983, Two *Drosophila* actin genes in detail: Gene structure, protein structure and transcription during development, *J. Mol. Biol.* **163:**533–551.

Scheller, R. H., McAllister, L. B., Crain, W. R., Durica, D. S., Posakony, J. W., Thomas, T. L., Britten, R. J., and Davidson, E. H., 1981, Organization and expression of multiple actin genes in the sea urchin, *Mol. Cell. Biol.* **1:**609–628.

Schuler, M. A., and Keller, E. B., 1981, The chromosomal arrangement of two linked actin genes in the sea urchin S. purpuratus, *Nucl. Acids. Res.* **9:**591–604.

Schuler, M. A., McOsker, P., and Keller, E. B., 1983, DNA sequence of two linked actin genes of sea urchin, *Mol. Cell. Biol.* **3:**448–456.

Shott, R. J., Lee, J. J., Britten, R. J., and Davidson, E. H., 1984, Differential expression of the actin gene family of *Strongylocentrotus purpuratus*, *Dev. Biol.* **101:**295–306.

Shott-Akhurst, R. J., Calzone, F. J., Britten, R. J., and Davidson, E. H., and Davidson, E. H., 1984, Isolation and characterization of a cell lineage-specific cytoskeletal actin gene family of *Strongylocentrotus purpuratus*, in: *Molecular Biology of Development* (E. H. Davidson and R. A. Firtel, eds.), pp. 119–128, Liss, New York.

Spudich, A., and Spudich, J. A., 1979, Actin in triton-treated cortical preparations of unfertilized and fertilized sea urchin eggs, *J. Cell Biol.* **82:**212–226.

Vandekerckhove, J., and Weber, K., 1978, Mammalian cytoplasmic actins are the products of at least two genes and differ in primary structure in at least 25 identified positions from skeletal muscle actins, *Proc. Natl. Acad. Sci. USA* **75:**1106–1110.

Vandekerckhove, J., and Weber, K., 1984, Chordate muscle actins differ distinctly from invertebrate muscle actins, *J. Mol. Biol.* **179:**391–413.

Chapter 7

Regulation of the Mid-Blastula Transition in Amphibians

LAURENCE D. ETKIN

1. Introduction

Development from a fertilized egg to an adult organism requires expression of the genetic program, which is coordinated both spatially and temporally. The genome must provide the information for the orderly control of the cell cycle and of morphogenic events such as cleavage and gastrulation, as well as of molecular processes such as protein synthesis, DNA replication, and transcription. Cleavage may result in the asymmetric distribution of cytoplasmic components into daughter cells establishing specific cell lineages (see Chapter 1). Cellular interactions that occur during early development add another dimension to the functional regionalization of the embryo (see Chapter 3). During gastrulation, the germ layers become juxtaposed in patterns necessary for such inductive cellular interactions.

Many organisms have evolved different strategies for the synthesis and utilization of components necessary for morphological and biochemical differentiation. In the mouse, the translational products of RNAs synthesized during oogenesis are needed during early embryogenesis. However, the results of studies with α-amanitin (an inhibitor of RNA synthesis) demonstrate that normal development also depends on new transcription from the zygotic genome, which is initiated by the two-cell stage (Mintz, 1964; Golbus et al., 1973). The sea urchin egg also contains a quantity of maternally derived products and initiates zygotic gene expression during early cleavage stages. However, in contrast to the mouse, maternal products in the sea urchin will support normal development through the blastula stage, even in the absence of new transcription (Davidson, 1986). The amphibian embryo synthesizes many molecular components (both protein and RNA) during oogenesis for use in the developing embryo. However, unlike the case in mouse and sea urchin development, the frog embryo relies exclusively on these products of the maternal program dur-

LAURENCE D. ETKIN • Department of Molecular Genetics, The University of Texas Cancer Systems Center, M.D. Anderson Hospital and Tumor Institute, Houston, Texas 77030.

ing early embryogenesis and does not initiate transcription until the mid-blastula stage of development. The mid-blastula stage is therefore a crucial period in amphibian embryogenesis. Besides the onset of zygotic transcription, a series of coordinated events occur, including changes in the cell cycle and the initiation of cell movement. This transitional period during amphibian development is referred to as the **transition blastuleene** (Signoret and Lefresne, 1971), or the **mid-blastula transition** (**MBT**) (Gerhart, 1980; Newport and Kirschner, 1982*a,b*). The purpose of this chapter is to discuss the changes that occur at the MBT in amphibians, in particular the South African clawed frog *Xenopus laevis*, as well as how they may be regulated. Section 2 discusses the events that occur during development of the amphibian up to the point of the MBT.

2. Development from Oogenesis through the Mid-Blastula Stage

The amphibian oocyte is a virtual storehouse of products that are utilized during early embryogenesis. This maternal legacy consists of polymerases, nucleoplasmin, nuclear lamins, histones, actin, and many other proteins and mRNAs of known and unknown function (Davidson, 1986). Among the stored proteins and transcripts are undoubtedly components that serve a regulatory function and control processes, such as the establishment of body form or the overt manifestations of the differentiated phenotype. Some of these regulatory components are localized in specific regions of the oocyte and become segregated into different blastomeres during cleavage. There are several examples of such putative regulatory components, such as the germ plasm (Smith *et al.*, 1983), the o$^+$ factor in the Mexican axolotl (Briggs, 1979), a factor that may affect the tissue-specific expression of actin genes (Gurdon *et al.*, 1985), several proteins that become localized in the nuclei of specific cell types during development (Dreyer *et al.*, 1983), and several localized mRNAs (Weeks and Melton, 1987).

The bulk of synthesis of poly(A)$^+$ RNA occurs in previtellogenic oocytes (stage 2) (Dumont, 1972), before the maximal lampbrush chromosome stage, although synthesis of poly(A)$^+$ RNA continues at a low rate throughout oogenesis (Dolecki and Smith, 1979). In *Xenopus*, approximately 30% of the poly(A)$^+$ RNA represents functional messenger RNA (mRNA), whereas the other 70% is in the form of apparently untranslatable RNA sequences containing interspersed repetitive elements (Anderson *et al.*, 1980). The synthesis of 5S ribosomal RNA and 4S transfer RNA (tRNA) occurs during the early stages of oogenesis (stages 1 and 2), whereas the bulk of the 18S and 28S ribosomal RNA synthesis occurs slightly later (stage 3). The net result of transcription during oogenesis is the accumulation in full-grown oocytes of approximately 10^{12} ribosomes and enough stored RNA to support development of the embryo through the late blastula to gastrula stages.

When the oocyte has completed its growth phase in the ovary, it remains in the diplotene stage of the first meiotic prophase. Upon stimulation with progesterone, the oocyte undergoes a maturation process that results in the dissolution of the germinal vesicle (GV), continuation of meiosis to metaphase of the second meiotic division, and the secession of transcriptional activity.

During oocyte maturation, the overall level of protein synthesis from maternal mRNAs is stimulated two- to fourfold (Smith and Richter, 1985). Ballantine *et al.* (1979) observed that no new types of proteins are produced after maturation of oocytes *in vitro.* However, they found quantitative differences in the synthesis of some of the proteins detected by two-dimensional gel electrophoresis. Analysis of the four core histone proteins showed that their rate of synthesis increases dramatically at maturation. However, H1 does not show the same dramatic increase in synthesis at maturation as do the other four histones (Adamson and Woodland, 1977; Flynn and Woodland, 1980).

The mature oocyte passes through the oviduct, picking up several layers of jelly coating, and is then capable of being fertilized. Fertilization results in a series of physiological changes, including completion of the second meiotic division, DNA replication, the start of the mitotic cell cycle, changes in ooplasmic organization, and various metabolic events.

Transcriptional activity during early development has been studied by the analysis of the incorporation of radioactive RNA precursors into newly synthesized RNA using autoradiography of histological sections (Bachvarova and Davidson, 1966) and polyacrylamide gels (Newport and Kirschner, 1982). These studies have demonstrated that little, if any, radioactive precursor is incorporated into newly synthesized RNA from fertilization through the blastula stage of development, suggesting that there is no transcription during this period. However, since there are so few nuclei at early cleavage stages, low rates of transcriptional activity may be impossible to detect. Kimelman *et al.* (1987), using high specific activity radioactive precursors and long exposure times, detected low levels of transcription at the 128 cell stage, well before the MBT. The significance of this transcriptional activity is unclear.

Translational regulation of protein synthesis contributes significantly to the pattern of gene expression from fertilization through the blastula stage. The qualitative pattern of protein synthesis, as determined by two-dimensional gel electrophoresis of radiolabeled proteins and fluorography, does not change during this period (Ballantine *et al.*, 1979). After fertilization, there is a twofold increase in the polysome content in *Xenopus* (Woodland, 1974). Of 18 arbitrarily selected RNA sequences that are nonpolysomal in the mature oocyte, 13 have become enriched on polysomes by the 16-cell stage (Dworkin *et al.*, 1985). One of the remaining sequences was found equally distributed in the polysomal and in the nonpolysomal fractions at this time, whereas four others became polysomal later than the 16-cell stage. Ruderman *et al.* (1979), using an *in vitro* cell-free translation system, showed that the amount of translatable histone mRNA in *Xenopus* oocytes, eggs, and embryos is approximately constant. There is, however, a dramatic change in the polyadenylation of histone transcripts among the oocyte, the egg, and the embryo, which probably involves deadenylation of preexisting maternal transcripts at the time of maturation. During early cleavage, the four core histones (H2a, H2b, H3, and H4) are synthesized on maternal transcripts. H1 histone is translated later during cleavage, although the mRNA is present throughout the early cleavage stages. These results indicate the importance of translational regulation during early amphibian development.

During the period between fertilization and the blastula stage, the cell cycle is short (20–30 min), consisting primarily of mitosis and the DNA synthetic phases with no G1 or G2 phases. Hara *et al.* (1980) observed surface contraction waves in the egg cortex preceding each cleavage division during this period. A contraction precedes every division through the twelfth cleavage cycle. Eggs prevented from cleaving with antimitotic drugs also undergo a series of surface contraction waves that correspond to the time of cleavage in control nontreated eggs. Surface contractions also occur in activated eggs, enucleated whole eggs, and in egg fragments lacking any functional genome, suggesting that the *Xenopus* egg contains a cytoplasmic clock or oscillator involved in the timing of the cell cycle. The oscillator appears to be independent of the presence of the nucleus and the sperm centriole and is triggered by activation or fertilization.

3. General Events Associated with the Mid-Blastula Transition

The twelfth cleavage cycle (4000–8000 cells) marks the mid-blastula transition (MBT). At this time, reliance on the maternal program is decreased, and the zygotic genomic program begins to unfold. Several characteristic events occur at the MBT, which include (1) an increase in transcriptional activity; (2) a decrease in the rate of DNA synthesis; (3) a slowdown of the cell cycle; which now incorporates the G1 and G2 phases; and (4), the beginning of cell movement (summarized in Fig. 1).

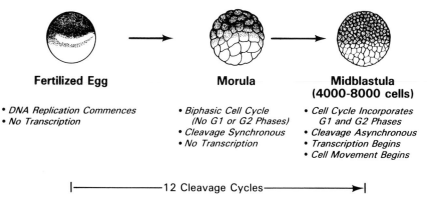

Figure 1. Events at the mid-blastula transition in *Xenopus laevis*. Schemmatic representation of the cellular events that occur at the various stages during development from the point of fertilization to the mid-blastula transition. While the number of cleavage cycles is the number occurring in *Xenopus laevis*, all other events listed under the specific stages are similar in all amphibians analyzed.

Table I. Genes Transcribed at or Near the Mid-Blastula Transition

Transcript	Quantity of new transcription[a]	Polymerase	Stage analyzed[b]	Reference
rRNA	2.0 ng	I	10	Brown and Littna (1966)
5SRNA	pg quantities	III	9	Wormington and Brown (1983)
tRNA	40–50 ng	III	10	Brown and Littna (1966)
snRNA(u1)	1.6 ng	II	8–9	Forbes et al. (1984)
ribosomal proteins	N.D.	II	10	Baum and Wormington (1985)
poly(A)$^+$ RNA	30–50 ng	II	9–10	Shiokawa et al. (1981)
DG clones		II	10	Sargent and Dawid (1983)
4	5 pg			
10	15 pg			
17	9 pg			
42	48 pg			
56	1 pg			
70	110 pg			
81	12 pg			
61	22 pg			
GS 17	fg–pg quant	II	8–9	Kreig and Melton (1985)

[a]Based on the accumulation of newly synthesized transcripts from stage 8 until the stage measured.
[b]From Nieuwkoop and Faber (1956).

3.1. Transcription at the MBT

The amphibian egg contains a complement of maternal RNA sufficient to permit development through the late blastula stage of development. Transcriptional activity is not detected during early cleavage stages and is first detectable when the embryo reaches the 4000–8000-cell stage (Bachvarova and Davidson, 1966; Brown and Littna, 1964, 1966; Newport and Kirschner, 1982).

Genes transcribed by all three classes of RNA polymerases (I, II, and III) are activated at or near the MBT. Table I summarizes the varieties of RNAs detected as newly synthesized transcripts at or near the MBT. These include transcripts of genes coding for tRNA, 5SRNA, several small nuclear RNAs, mitochondrial RNAs, histone RNAs, several ribosomal protein RNAs (see Chapter 8), and 18S and 28S ribosomal RNA as well as others discussed below. It is quite obvious that all genes in the genome are not active, since many cell-specific products are not detectable, although one could argue that this effect is at the mRNA stability level. This is unlikely, however. The following discussion will focus attention on the quantitative and qualitative aspects of transcription at the MBT and assess evidence for a common regulatory mechanism involved in the transcription of the various classes of RNAs.

At the blastula stage, the *Xenopus* embryo contains approximately 4 μg total RNA, of which 40–80 ng is poly(A)$^+$ RNA representing on the order of 20,000 different RNA species (Perlman and Rosbash, 1978). Approximately

40% of the mass of poly(A)$^+$ RNA is newly synthesized between the mid-blastula and gastrula stages of development and consists of the same or similar sequences present in the unfertilized egg (Shiokawa *et al.*, 1981). Most of the newly synthesized poly(A)$^+$ transcripts therefore probably represent turnover of the preexisting maternal stores, since the overall content of poly(A)$^+$ RNA does not increase (Sagata *et al.*, 1980). Dworkin and Dawid (1980) analyzed more than 200 cloned sequences from recombinant DNA libraries of *Xenopus* embryos by colony hybridization and found that the abundant poly A$^+$ RNA population is relatively constant from fertilization through the gastrula stage.

Unique RNA transcripts synthesized at or near the MBT have been analyzed by isolating cDNA clones for individual RNAs (Sargent and Dawid, 1983; Kreig and Melton, 1985). The unique RNA species may be important in the establishment of body form and function, germ layer differentiation, and/or control of gastrulation, since they are first detected just before these events. Sargent and Dawid (1983) isolated a group of 84 cDNA clones from a subtracted cDNA library prepared by depleting poly A$^+$ RNA sequences that are in common between the egg and the gastrula. They called these **DG clones,** because their transcripts are differentially expressed in the gastrula. The DG RNAs represent transcripts first transcribed at the MBT or shortly thereafter. The RNA abundance levels of eight of these sequences reach a peak after gastrulation and then decline in abundance by the neurula stage. All but one of the transcripts are undetectable in 3-day-old tadpoles. In quantitative terms, the new DG transcripts range from 1 to 110 pg/embryo at the gastrula stage (Dawid *et al.*, 1985) (Table I). DG42 codes for a polypeptide of unknown function whose RNA accumulates immediately after the MBT and then declines by the neurula stage of development. Other genes that exhibit a longer span of expression have been identified as members of the keratin gene family. Sargent *et al.* (1986) showed that some of the DG genes are expressed in a regionally specific manner. For example, DG42 is first expressed in animal pole cells at stage 8–9, but by stage 10 (gastrula) its transcripts are uniformly distributed in the embryo, and by stage 14–15 (neurula) the transcripts are localized in endodermal cells. Other clones such as DG56 are expressed equally in all regions of the embryo. Kreig and Melton (1985) also isolated a gene (GS 17) coding for a protein of unknown function that is expressed first at the MBT, whose transcripts are undetectable after the neurula stage of development. It is important to note that none of these gene products transcribed first at the MBT persist or have cognate RNAs expressed in adult frogs.

An example of differential gene expression at the MBT is found in the genes coding for small nuclear RNAs (snRNA). Forbes *et al.* (1984) detected several species of U1 small nuclear RNAs (snRNA) synthesized in *Xenopus* tissue culture cells as well as several species expressed in oocytes. They found that snRNA genes are transcribed at a high rate at the MBT. The snRNA transcripts expressed at the MBT differ from the U1 RNA subset synthesized in stage 6 oocytes, indicating that there is differential expression of the snRNA gene family at the MBT.

Many other genes are expressed for the first time after the MBT, in some instances, their products persist into later developmental stages. Dworkin *et al.* (1984) analyzed the expression of a group of genes expressed first at the late gastrula to early neurula stage. The transcripts of these genes begin to accumulate at moderately abundant levels at the neurula stage, with some exhibiting a tissue-specific pattern of expression in the tadpole (Dworkin-Rastl *et al.*, 1986). The mRNA species with the highest titers at the late gastrula–neurula stage probably code for proteins specific to the developing organ systems of the embryo.

It is apparent that transcription at the MBT is selective in that only a limited set of genes is transcribed, many of which represent products present in the maternal stores and some of which represent products that are expressed for the first time. It is therefore an important component of any model accounting for the restoration of transcription at the MBT to include a mechanism for this selectivity. It is possible that genes that are expressed at the MBT are important in the programming or the activation of genes that will be expressed later in the lifetime of the organism, resulting in a cascade effect, or what had been termed **progressive differentiation** (Holtfreter and Hamburger, 1955).

There are several ways to perturb experimentally the normal timing of transcription at the MBT. It is hoped that such experimental manipulations will provide insights into the mechanism of gene regulation at this stage. One way involves microinjection of exogenous DNA into eggs (Newport and Kirschner, 1982). Exogenous yeast tRNA genes showed activation of transcription at the MBT after injection into fertilized *Xenopus* eggs. Transcription of the exogenous DNA occurred whether the DNA was injected into cleaving embryos or into coenocytic embryos produced by blocking cleavage with gentle centrifugation. However, when the yeast tRNA-containing plasmid was injected along with increasing amounts of the plasmid pBR322, Newport and Kirschner (1982) observed that 24 ng (the mass of genomic DNA present in normal embryos at the MBT) of the pBR322 resulted in transcription of the tRNA genes before the mid-blastula transition. They also detected premature transcription of endogenous snRNA genes after co-injection of pBR322 in the cleavage-arrested embryos. In addition, they were able to induce transcription before the MBT in polyspermic eggs that initially contained 6–10 extra sperm nuclei. These results are consistent with the hypothesis that the egg contains an inhibitor of general transcription that can be titrated by adding additional DNA to the developing system, resulting in the induction of transcription before the mid-blastula stage. All these experiments were performed in coenocytic embryos, since large amounts of exogenous DNA microinjected into cleaving embryos resulted in developmental arrest. However, careful examination of the data in Newport and Kirschner (1982) shows that there is transcription before the MBT in coenocytic embryos, making interpretation of these results difficult.

Shiokawa *et al.* (1985) showed that 18S and 28S rRNA transcription was inhibited in *Xenopus* blastula and neurula cells that were exposed to ammonium salts. Ammonium chloride treatment, however, did not affect the activation of the transcription of hnRNA, 5sRNA, or snRNAs, nor did it inhibit protein synthesis or DNA replication. The inhibition of 18S and 28S RNA synthesis was at the step of transcription of the 40S rRNA precursor. Shiokawa and co-workers also measured the change in ammonia concentration within the developing embryo and found that the level is 3 mM (the same concentration that inhibits rRNA synthesis in neurula cells) in the egg through early cleavage stages and then decreases abruptly to 1.2 mM at the mid-blastula stage, when rRNA synthesis is first detectable (Shiokawa *et al.*, 1986). These investigators do not claim that ammonia is the primary step in the regulation of rRNA gene expression; it may be only a secondary agent. It is possible that exposure to ammonia may affect conditions such as intracellular pH and that transcription of the ribosomal genes may be more sensitive to these conditions than is transcription of polymerase II or polymerase III type genes. However, these results suggest that the activation of transcription at the mid-blastula transition may not be a general phenomenon in the sense that all genes are regulated by the same mechanism; instead, a variety of strategies may be used to regulate different sets of genes.

It is likely that unique promoters or DNA consensus sequences acting in *cis* may play a fundamental role in determining which genes are activated at the MBT. Studies have demonstrated that a variety of exogenous DNAs injected into fertilized eggs are not transcribed until the mid-blastula transition (reviewed by Etkin *et al.*, 1986). These include polymerase II-type genes with viral promoters (Etkin and Balcells, 1985), *Xenopus* promoters (Bendig and Williams, 1984; Kreig and Melton, 1985), sea urchin promoters (Bendig, 1981; Etkin *et al.*, 1984), rabbit promoters (Rusconi and Schaffner, 1981), polymerase III-type genes with yeast promoters (Newport and Kirschner, 1982), and polymerase I-type genes (Busby and Reeder, 1983). The role of *cis*-acting DNA sequences in the transcription of genes at the MBT was tested by microinjection of cloned genes into *Xenopus* embryos. Recent evidence suggests that such sequences are associated with a gene that is expressed at the MBT (GS17) (Kreig *et al.*, 1986). The data indicate that a relatively short 74 base pair DNA sequence located 700 base pairs upstream from the GS17 promoter and having enhancer-like properties, may activate this gene at the MBT (Kreig and Melton, 1987).

It is also not clear whether transcription occurs simultaneously in all cells of the embryo at the MBT. There is evidence, however, that rRNA synthesis becomes detectable in endodermal cells of *Xenopus* later than in the remainder of the embryo (Woodland and Gurdon, 1968). It is known that several of the DG genes are transcribed in a regionally specific manner. The regional patterns of expression of the DG genes differ in that some are expressed in animal hemisphere cells, others in vegetal or endodermal cells, and still others are expressed ubiquitously in all regions of the embryo. The regional and selective expression of genes at the MBT argues against a common mechanism regulating the expression of all genes transcribed at the MBT.

It is possible that transcription cannot take place before the MBT due to the rapid rate of DNA synthesis occurring during the early cleavage divisions and that transcription may begin on DNA templates as a result of the elongation of the cell cycle after the MBT. The lack of detectable transcription may be the result of the inability of transcription complexes to read through replicating regions of DNA. After the MBT, the cell cycle incorporates the G1 and G2 interphase periods, permitting transcription to take place in the absence of rapid cycles of replication. There is evidence for this type of mechanism in the developing *Drosophila* embryo, where it has been observed that transcriptional activation is not directly controlled by the nuclear–cytoplasmic ratio but may be an effect of the elongation of the interphase period of the cell cycle at the blastoderm stage of development (Edgar *et al.*, 1986). Recently, Kimelman *et al.* (1987) used a series of inhibitors to analyze the role of protein synthesis, DNA replication, and microtubule assembly on the initiation of transcription at the MBT. They observed that inhibition of DNA synthesis with amphidicolin or protein synthesis with cycloheximide (which arrests cleavage and also results in inhibition of DNA synthesis), resulted in activation of transcription at the MBT even though these embryos did not have the equivalent amount of DNA as untreated control embryos at the MBT. This result suggests that transcription can be initiated in embryos prior to reaching the proper DNA/cytoplasmic ratio normally associated with the MBT. This is directly contradictory to the results of Newport and Kirschner (1982). They also found, however, that eggs treated with the drug nocodazole (which blocks cleavage but does not affect DNA replication or the surface contraction waves that normally occur before each cleavage) do not activate transcription at the MBT, but do so at a later time. These results tend to complicate the interpretation that the nuclear–cytoplasmic ratio is the crucial factor in regulation of the events of the MBT and support the contention that the major factor in the initiation of transcription may be the restoration of the complete cell cycle that occurs at the MBT. It is likely that the regulation of the expression of genes at the MBT is complex and involves not only the presence of external *trans*-acting factors, but also involves the presence of unique regulatory DNA sequences, which are associated with various classes of genes.

3.2. Protein Synthesis at the MBT

The general qualitative pattern of protein synthesis as demonstrated by two-dimensional gel electrophoresis changes little between the fully grown oocyte and the blastula stage of development. Many new species of proteins are first detectable at the gastrula stage (Ballantine *et al.*, 1979; Mohun *et al.*, 1981; Brock and Reeves, 1978; Bravo and Knowland, 1979; Smith, 1986). Since there is some delay in the detection of translational products of new transcripts that first appear at the MBT, it is probable that some of the proteins detected at the gastrula stage represent products of genes first transcribed at the MBT.

There are specific examples of the recruitment of maternal mRNAs onto polysomes at or near the mid-blastula transition. These include some of the small nuclear U-ribonucleoproteins (snRNPs) (Fritz *et al.*, 1984), nuclear lamina proteins LIII and LI (Stick and Hausen, 1985), and fibronectin (Lee *et al.*,

1984). The increase in fibronectin synthesis occurs in eggs activated artifically and maintained until the time of MBT, suggesting that the recruitment of this mRNA onto polysomes is probably not dependent on fertilization, cell division, or nucleocytoplasmic ratio (Lee et al., 1984). There are also examples of transcripts recruited onto polysomes at early developmental stages other than the MBT. These include histone H1, which is translated just before the MBT (Woodland et al., 1979), and nuclear lamin LII, which is recruited at the gastrula stage of development (Stick and Hausen, 1985). Baum and Wormington (1985) demonstrated that newly synthesized transcripts from genes coding for ribosomal proteins are first detected at the gastrula stage but are not recruited onto polysomes until stage 30 (see Chapter 8). It is apparent that other parameters are involved in the recruitment of mRNAs onto polysomes that appear to be independent of the timing of the MBT.

3.3. Cell Cycle Regulation and the MBT

After fertilization, the Xenopus egg begins to divide at a rapid rate, with a typical cell cycle of approximately 30 min, depending on the temperature. Cell divisions through the twelfth cycle are synchronous or metachronous and biphasic, lacking both the G1 and G2 phases of the cell cycle. At the time of the MBT (twelfth cleavage), cleavage becomes asynchronous, and the cell cycle elongates. The cycles now include the G1 and G2 phases. Numerous attempts have been made to analyze the mechanism responsible for the change in cell cycle at the MBT.

A number of laboratories have found that a component present in unfertilized eggs is important in the induction of cells to enter mitosis (Meyerhof and Masui, 1979; Reynhout and Smith, 1984; Wu and Gerhart, 1980; Miake-Lye et al., 1983; Newport and Kirshner, 1984; Gerhart et al., 1984). This mitotic factor has been termed **maturation promoting factor** (**MPF**), since it causes meiotic maturation after microinjection into fully grown oocytes (Masui and Markert, 1971). It is postulated that oscillation of MPF activity is responsible for regulating the embryonic cell cycle during early cleavage stages, based on the observation that MPF activity is present during the M-phase and undetectable during S phase (Gerhart et al., 1984; Newport and Kirshner, 1984). Newport et al. (1985) demonstrated that extracts prepared from unfertilized eggs can induce exogenously added nuclei to enter mitosis (i.e., nuclear envelope breakdown and chromosome condensation) in vitro. These workers found that 3000–6000 nuclei/egg equivalent of extract could respond to the extract but, as they increased the number of nuclei/egg equivalents of extract, the time required to induce nuclear envelope breakdown and chromosome condensation increased. Thus, it was apparent that the mitotic activities in these extracts were rate limiting with regard to the induction of nuclear envelope breakdown and chromosome condensation. To determine whether the rate-limiting mitotic factors bind to DNA or other nuclear components, nuclear ghosts were prepared by treating nuclei with DNAses and extracting with detergent. These nuclear ghosts lacked

95% of the DNA and consisted of nuclear envelopes containing primarily the nuclear lamin proteins. The mitotic factor could be titrated by addition of DNA-free nuclear envelope ghosts before the addition of whole responding nuclei, which suggests that the factor binds to the nuclear envelope. The quantity of this inducer is estimated to be sufficient to last until the blastula stage of development, suggesting that there is a maternal store of an activity (perhaps MPF itself) involved in the induction of mitosis, or at least in nuclear envelope breakdown and chromosome condensation. Newport *et al.* (1985) hypothesize that the titration of this mitotic factor by nuclear envelope components at the 4000–8000-cell stage (MBT) might be responsible for the slowing down of the rate of induction into the M phase of the cell cycle. The titration of the mitotic factor would result in the elongation of the cell cycle and the incorporation of the G1 and G2 phases and possibly the induction of other events associated with the MBT. It should also be pointed out that microinjected exogenous DNA is sequestered into nucleus-like structures complete with nuclear lamins, and other nuclear components (Forbes *et al.*, 1983). Thus, in Newport and Kirschner's (1982*a,b*) experiments, in which they hypothesized the titration of a negative regulator of transcription by injected exogenous DNA, they may have been titrating a mitotic factor, resulting in elongation of the cell cycle. The initiation of transcription may be the result of the change in the cell cycle.

In another series of experiments in the Japanese newt, *Cynops pyr-rhogaster*, fertilized eggs were divided into half- or quarter-embryos with glass rods before first cleavage (Kobayakawa and Kubota, 1981). The development of the partial embryos containing the nuclei was monitored for the synchrony of cell division for the first 15 divisions. It was observed that desynchronization of cleavage started two divisions earlier in quarter-embryos and one division earlier in half-embryos compared with control whole embryos. The variation in the timing of desynchronization in partial embryos is consistent with the presence of a cytoplasmic component that is titratable by nuclei and is involved with the maintenance of cell synchrony before the MBT.

3.4. Cell–Cell Interactions and the MBT

Inductive interactions between cells are involved in the determination and regional subdivision of the amphibian embryo (see Chapter 3). Cell interactions undoubtably are involved in the activation of genes in the processes of cellular determination and differentiation. Recently, Sargent *et al.* (1986) and Jamrich *et al.* (1986) attempted to define the requirements for cell–cell interaction in the activation of several genes (specifically the DG genes) that are expressed during early development of *Xenopus*. They analyzed the expression of DG 42 (which is endoderm specific), DG 81 (which is an ectoderm-specific cytokeratin), and α-actin (whose expression is restricted to mesodermally-derived cells). DG42 and DG 81 are expressed at or near the MBT (Sargent and Dawid, 1983), whereas the actin gene is expressed at the late gastrula–early neurula stage (Gurdon *et al.*, 1984). Cells were in Ca^{2+}- and Mg^{2+}-free medium and either cultured dispersed, thus preventing contact-mediated cell interactions, or dissociated, thus permitting loose cell–cell contacts. They also cultured embryos in Ca^{2+}- and Mg^{2+}-free media, within the vitelline envelope. The cells in these embryos lacked the tight apical adhesions, causing a collapse of the blastocoel cavity and resulting in abnormal contacts between the animal and vegetal cells. The cells were dispersed during early cleavage stages, after which they were permitted to reaggregate for a period of time equivalent to stage 12 (late gastrula). The DG genes 42 and 81 were expressed in cells that were completely dispersed; thus these genes exhibited cell autonomous expression. Conversely, the actin genes required cell interaction for transcriptional activation. Interestingly, however, DG81 expression was inhibited in embryos in which abnormal animal–vegetal cell contacts occurred. Therefore, for at least two of the genes expressed at the MBT, cell interactions during cleavage stages are unnecessary for their activation. However, abnormal interactions may affect the expression of at least one of these genes. When blastomeres of early embryos are disassociated they remain nonmotile until the MBT, at which time active cytoplasmic blebbing and pseudopodia formation begins. Also, in dissociated cells, the normal surface contraction waves occur at the proper intervals. These morphological events appear to be cell autonomous and not dependent on cell–cell interactions. Thus it is clear that cell–cell interactions are not crucial to elicit the increase in transcription and the morphogenetic events that accompany the MBT.

4. Mechanisms for Regulation of the Mid-Blastula Transition

A major issue is whether the constellation of events surrounding the MBT is a result of a cleavage-counting mechanism (which counts rounds of DNA replication), is dependent on transcription (which is activated at the MBT), or is independent of any of these processes. Newport and Kirshner (1982) determined the role of a cleavage-counting mechanism by inhibiting cleavage in fertilized eggs with cytochalasin or by centrifugation and assaying for the time of initiation of transcription, changes in the rate of DNA synthesis, and the occurrence of cortical contractions. In eggs in which cleavage was blocked by either means, surface contractions occur concomitant with cleavage cycles in control untreated eggs. In cytochalasin-treated eggs, DNA is synthesized at an exponential rate for the first 6 hr and then slows abruptly. This rate of DNA synthesis compares favorably to the rate of DNA synthesis in cleaving embryos. Both cytochalasin-treated and centrifuged eggs exhibit a change in DNA synthesis rate at the time of MBT. The time and rate of RNA synthesis in cleavage-arrested embryos is comparable to control embryos. These results suggest that changes in RNA and DNA synthesis associated with the MBT are independent of cytokinesis.

When cleaving embryos are injected with α-amanitin at concentrations that inhibit 80% of RNA polymerase III transcription and completely block RNA polymerase II transcription, neither is activated at the time of MBT. Inhibition of RNA transcription apparently does not affect either the pattern of cleavage through cleavage twelve or the slowdown and desynchronization of the cell cycle that follow the MBT. Also, dissociated blastomeres from α-amanitin-treated embryos become motile at MBT (Newport and Kirschner, 1982). These results suggest that the onset of transcription at the MBT is probably not a cause of the other events such as changes in the cell cycle and the initiation of cell movement that occur at the MBT.

Newport and Kirschner (1982) also tested the role of timing or counting mechanisms in the control of the MBT by performing the hair ligature experiment first used by Spemann (1938). In this experiment, the migration of one of the cleavage nuclei was delayed in entering one half of a partially constricted egg, while the nucleated half commenced cleavage. Constriction resulted in two developing half-embryos, one of which had received a nucleus two or three division cycles later than the other. If a timing mechanism was involved, both halves should undergo the MBT at the same time. The result was that the MBT was delayed two to three divisions in the half-embryo receiving the nucleus later, suggesting that the control of the MBT is not dependent on the total number of rounds or DNA synthesis or on a strict timing mechanism. Thus, it appears that the MBT is autonomous in the sense that the associated events are not entirely dependent on a timing mechanism activated at fertilization, RNA transcription, a cleavage-counting mechanism, or counting rounds of DNA replication.

5. Summary

The cleavage cycles during early amphibian development are synchronous, rapid, and biphasic. There is no transcription and no growth of the embryo, and the nuclear cycle is independent of the cytoplasmic cleavage cycle. All components necessary for development through the blastula stage are provided by maternal stores. At the twelfth cleavage division, a major transition occurs that involves initiation of transcription, an elongation of the cell cycle, an increase in cell movement, and asynchrony of cell division.

It is probable that the major controlling factor in the regulation of all the aforementioned events is the cell cycle. During early cleavage stages, the cell cycle is both rapid (30–35 min) and synchronous. There is evidence that the cycling time may be controlled by the presence of several mitotic factors, such as MPF, CSF (cytostatic factor), and a titratable component that binds to nuclear membranes. The rapid rate of DNA synthesis may inhibit the formation of transcription complexes, resulting in the absence of detectable transcription before the MBT. Cellular movement may also be inhibited in the rapidly dividing cell.

As the cell cycle elongates (possibly due to the functional loss or sequestration of one or more of the mitotic control factors), the G1 and G2 phases are incorporated into the cell cycle. Under conditions of slower rates of DNA replication and the presence of the G1 and G2 phases, the transcriptional machinery becomes functional. It is apparent that at the MBT, not all classes of transcripts are activated simultaneously in every cell, nor is their expression regulated by a common mechanism. Incorporation of the G1 and G2 phases in the cell cycle may also permit the synthesis and assembly of microtubules and cytoskeletal components necessary for the initiation of the cell movements characteristic of this stage of development. The role of the cell cycle in controlling events at the MBT is supported by evidence from studies in which perturbation of the cell cycle that results in its elongation or arrest produces subsequent initiation of events that occur normally at the MBT.

The MBT therefore appears to be a window in the developmental time frame, during which a number of molecular and morphogenetic events occur independently of one another, but all are necessary for subsequent morphogenesis and cellular differentiation. It is a transition from the strict reliance on the maternal program to a dependence on the new transcription from the embryonic genetic program. It is probable that the major regulatory mechanism involved in the occurrence of this constellation of cellular events is the change in the cell cycle. Further characterization of the mitotic factors (e.g., MPF, CSF) that regulate the transit through the cell cycle will aid in our understanding of how the cell cycle is regulated, as well as the mechanism of regulation of the other events at the MBT. Also, when we learn more about the possible *cis*-acting elements involved in activating genes at the MBT and about the regulatory factors that may interact with them, our understanding of the role of the cell cycle change in transcriptional activation at the MBT may be clearer. The frog MBT provides an interesting and advantageous system for looking at the interrelationships between the control of the cell cycle and transcription, cell movement, and various developmental processes that occur subsequent to this period.

References

Adamson, E., and Woodland, H. R., 1977, Changes in the rate of histone synthesis during oocyte maturation and very early development of *Xenopus laevis*, *Dev. Biol.* **43**:159–174.

Anderson, D. M., Richter, J. D., Chamberlain, M. E., Price, D. H., Britten, R. J., Smith, L. D., and Davidson, E. H., 1982, Sequence organization of the poly(A) RNA synthesized and accumulated in lamp-brush chromosome stage *Xenopus laevis* oocytes, *J. Mol. Biol.* **155**:281–309.

Bachvarova, R., and Davidson, E. H., 1966, Nuclear activation at the onset of amphibian gastrulation, *J. Exp. Zool.* **163**:285–296.

Ballantine, J. E. M., Woodland, H. R., and Sturgess, E. A., 1979, Changes in protein synthesis during the development of *Xenopus laevis*, *J. Embryol. Exp. Morphol.* **51**:137–153.

Baum, E. Z., and Wormington, W. M., 1985, Coordinate expression of ribosomal protein genes during *Xenopus* development, *Dev. Biol.* **111**:488–498.

Bendig, M. M., 1981, Persistence and expression of histone genes injected into *Xenopus Laevis* eggs in early development, *Nature (Lond.)* **292**:65–67.

Bendig, M. M., and Williams, J. G., 1984, Differential expression of the *Xenopus laevis* tadpole and adult β-globin genes when injected into fertilized *Xenopus laevis* eggs, *Mol. Cell. Biol.* **4**:567–570.

Bravo, R., and Knowland, J., 1979, Classes of proteins synthesized in oocytes, eggs, embryos and differentiated tissues of *Xenopus laevis*, *Differentiation* **13**:101–108.

Briggs, R., 1979, Genetics of cell type determination, *Int. Rev. Cytol.* **9**:107–127.

Brock, H. W., and Reeves, R., 1978, An investigation of *de novo* protein synthesis in the South African clawed toad, *Xenopus laevis, Dev. Biol.* **66**:128–141.

Brown, D. D., and Littna, E., 1964, RNA synthesis during the development of *Xenopus laevis*, the South African Clawed Toad, *J. Mol. Biol.* **8**:669–687.

Brown, D. D., and Littna, E., 1966, Synthesis and accumulation of low molecular weight RNA during embryogenesis of *Xenopus laevis, J. Mol. Biol.* **20**:95–112.

Busby, S. J., and Reeder, R. H., 1983, Spacer sequences regulate transcription of ribosomal gene plasmids injected into *Xenopus laevis* embryos, *Cell* **34**: 989–996.

Davidson, E. H., 1986, *Gene Activity in Early Development*, 3rd ed., Academic, Orlando.

Dawid, I. B., Haynes, S. R., Jamrich, M., Jonas, E., Miyatani, S., Sargent, T., and Winkles, J., 1985, Gene expression in *Xenopus* embryogenesis. *J. Embryol. Exp. Morph.* **89**(suppl.):113–124.

Dolecki, G. J., and Smith, L. D., 1979, Poly(A)$^+$ RNA metabolism during oogenesis in *Xenopus laevis, Dev. Biol.* **69**:217–236.

Dreyer, C., Wang Hui, Y., Wedlich, D., and Hausen, P., 1983, Oocyte nuclear proteins in the development of *Xenopus* in: *Current Problems in Germ Cell Differentiation* (A. McLaren and C. C. Wylie, eds.), pp. 329–352, Cambridge University Press, New York.

Dumont, J. N., 1972, Oogenesis in *Xenopus laevis*. I. Stages of oocyte development in laboratory maintained animals, *J. Morphol.* **136**:153–179.

Dworkin, M. B., and Dawid, I. B., 1980, Use of a cloned library for the study of abundant adenylated RNA during *Xenopus laevis* development, *Dev. Biol.* **76**:449–464.

Dworkin, M. B., Shrutkowski, A., Baumgarten, M., and Dworkin-Rastl, E., 1984, The accumulation of prominent tadpole mRNAs occurs at the beginning of neurulation in *Xenopus laevis* embryos, *Dev. Biol.* **106**:289–295.

Dworkin, M. B., Shrutkowski, A., Dworkin-Rastl, E., 1985, Mobilization of specific maternal RNA species into polysomes after fertilization in *Xenopus laevis, Proc. Natl. Acad. Sci. USA* **82**:7636–7640.

Dworkin-Rastl, E., Kelley, D. B., and Dworkin, M. B., 1986, Localization of specific mRNA sequences in *Xenopus laevis* embryos by *in situ* hybridization, *J. Embryol. Exp. Morphol.* **91**:153–168.

Edgar, B. A., Kiehle, C. P., and Schubiger, G., 1986, Cell cycle control by the nucleo-cytoplasmic ratio in early *Drosophila* development, *Cell* **44**:365–372.

Etkin, L. D., and Balcells, S., 1985, Transformed *Xenopus* embryos as a transient expression system to analyze gene expression at the mid-blastula transition, *Dev. Biol.* **108**:173–178.

Etkin, L. D., Pearman, B., Roberts, M., and Bektesh, S., 1984, Replication, integration, and expression of exogenous DNA injected into fertilized eggs of *Xenopus laevis, Differentiation* **26**:194–202.

Etkin, L. D., and Pearman, B., and Balcells, S., 1986, Regulation of heterologous genes injected into oocytes and eggs of *Xenopus laevis*, in: *Molecular Genetics of Mammalian Cells* (Malacinski, G. ed.), pp. 247–268, Macmillan, New York.

Flynn, J. M., and Woodland, H. R., 1980, The synthesis of histone H1 during early amphibian development, *Dev. Biol.* **75**:222–230.

Fritz, A., Parisot, R., Newmeyer, D., and DeRobertis, E. M., 1984, Small nuclear U-Ribonucloproteins in *Xenopus laevis* development, *J. Mol. Biol.* **178**:273–285.

Forbes, D. J., Kirschner, M. W., and Newport, J. W., 1983, Spontaneous promotion of nucleus-like structures around bacteriophage DNA microinjected into *Xenopus* eggs, *Cell* **34**:13–23.

Forbes, D. J., Kirschner, M. W., Caput, D., Dahlberg, J. E., and Lund, E., 1984, Differential expression of multiple U1 small nuclear RNAs in oocytes and embryos of *Xenopus laevis, Cell* **38**:681–689.

Gerhart, J., 1980, Mechanisms regulating pattern formation in the amphibian egg and early embryo, in: *Biological Regulation and Development*, Vol. 2 (R. F. Goldberger, ed.), pp. 133–315, Academic, New York.

Gerhart, J. C., Wu, M., and Kirschner, M., 1984, Cell cycle dynamics of an M-phase specific cytoplasmic factor in *Xenopus laevis* oocytes and eggs. *J. Cell Biol.* **98**:1247.

Golbus, M. S., Calarco, P. G., and Epstein, C. J., 1973, The effects of inhibitors of RNA synthesis (α-amanitin and actinomycin D) on preimplantation mouse embryogenesis, *J. Exp. Zool.* **157:**207–216.

Gurdon, J. B., Brennan, S., Fairman, S., and Mohun, T., 1984, Transcription of muscle-specific actin genes in early *Xenopus* development: Nuclear transplantation and cell dissociation, *Cell* **38:**691–700.

Gurdon, J. B., Mohun, T. J., Fairman, S., and Brennan, S., 1985, All components required for the eventual activation of muscle-specific actin genes are localized in the subequatorial region of an uncleaved amphibian egg, *Proc. Natl. Acad. Sci. USA* **82:**139–143.

Hara, K., Tydeman, P., and Kirschner, M., 1980, A cytoplasmic clock with the same period as the division cycle in *Xenopus* eggs, *Proc. Natl. Acad. Sci. USA* **77:**462–466.

Holtfreter, J., and Hamburger, V., 1955, Embryogenesis: Progressive Differentiation, in: *Analysis of Development* (B. Willier, P. Weiss, and V. Hamburger, eds.), pp. 230–296, W. B. Saunders, New York.

Jamrich, M., Sargent, T. D., and Dawid, I. B., 1985, Altered morphogenesis and effects of gene activity in *Xenopus laevis* embryos, *Cold Spring Harbor Symp. Quant. Biol.* **50:**31–36.

Kimelman, D., Kirschner, M., and Scherson, T., 1987, The events of the midblastula transition in *Xenopus* are regulated by changes in the cell cycle, *Cell* **48:**399–407.

Kobayakawa, Y., and Kubota, H. Y., 1981, Temporal pattern of cleavage and the onset of gastrulation in amphibian embryos developed from eggs with the reduced cytoplasm, *J. Embryol. Exp. Morph.* **62:**83–94.

Kreig, P. A., and Melton, D. A., 1985, Developmental regulation of a gastrula specific gene injected into fertilized eggs, *EMBO J.* **4:**3463–3471.

Kreig, P., and Melton, D., 1987, An enhancer responsible for activating transcription at the midblastula transition in *Xenopus* development, *Proc. Natl. Acad. Sci. USA* **84:**2331–2335.

Kreig, P. A., Rebagliati, M. R., Weeks, D. L., and Melton, D. A., 1986, Gene activation during *Xenopus* embryogenesis, in: *Gametogenesis and the Early Embryo* (J. Gall, ed.), pp. 357–370, Alan R. Liss, New York.

Lee, G., Hynes, R., and Kirschner, M., 1984, Temporal and spatial regulation of fibronectin in early *Xenopus* development, *Cell* **36:**729–740.

Masui, Y., and Markert, C., 1971, Cytoplasmic control of nuclear behavior during meiotic maturation of frog oocytes, *J. Exp. Zool.* **177:**129–140.

Meyerhof, P. G., and Masui, Y., 1979, Chromosome condensation activity in *Rana pipiens* eggs matured *in vivo* and in blastomeres arrested by cytostatic factor (CSF), *Exp. Cell Res.* **123:**345.

Miake-Lye, R., Newport, J., and Kirschner, M., 1983, Maturation promoting factor induces nuclear envelope breakdown in cycloheximide arrested embryos of *Xenopus laevis, J. Cell Biol.* **97:**81–91.

Mintz, B., 1964, Synthetic processes and early development in the mammalian egg, *J. Exp. Zool.* **157:**85–100.

Mohun, T., Brownson, S., and Wylie, C. C., 1981, Protein synthesis in interspecies hybrid embryos of the amphibian *Xenopus, Exp. Cell Res.* **132:**281–288.

Newport, J., and Kirschner, M., 1982a, A major developmental transition in early *Xenopus* embryos. I. Characterization and timing of cellular changes at the midblastula stage, *Cell* **30:**675–686.

Newport, J., and Kirschner, M., 1982b, A major developmental transition in early *Xenopus* embryos. II. Control of the onset of transcription, *Cell* **30:**687–696.

Newport, J., and Kirschner, M., 1984, Regulation of the cell cycle during early *Xenopus* development, *Cell* **37:**731.

Newport, J., Spann, T., Kanki, J., and Forbes, D., 1985, The role of mitotic factors in regulating the timing of the midblastula transition in *Xenopus, Cold Spring Harbor Symp. Quant. Biol.* **50:**651–656.

Nieuwkoop, P. D., and Faber, J., 1956, Normal tables of *Xenopus laevis* Daudin, North-Holland, Amsterdam.

Perlman, S., and Rosbash, M., 1978, Analysis of *Xenopus laevis* ovary and somatic cell polyadenylated RNA by molecular hybridization, *Dev. Biol.* **63:**197–212.

Reynhout, J. K., and Smith, L. D., 1974, Studies on the appearance of a maturation inducing factor in the cytoplasm of amphibian oocytes exposed to progesterone, *Dev. Biol.* **38**:394–400.

Ruderman, J. V., Woodland, H. R., and Sturgess, E. A., 1979, Modulation of histone messenger RNA during the early development of *Xenopus laevis, Dev. Biol.* **71**:71–82.

Rusconi, S., and Shaffner, W., 1981, Transformation of frog embryos with a rabbit beta globin gene, *Proc. Natl. Acad. Sci. USA* **78**:5051–5055.

Sagata, N., Shiokawa, K., and Yamana, K., 1980, A study of the steady-state population of poly(A)$^+$ RNA during early development of *Xenopus laevis, Dev. Biol.* **77**:431–448.

Sargent, T. D., and Dawid, I. B., 1983, Differential gene expression in the gastrula of *Xenopus laevis, Science* **222**:135–139.

Sargent, T. D., Jamrich, M., and Dawid, I. B., 1986, Cell interactions and control of gene activity during early development of *Xenopus laevis, Dev. Biol.* **114**:238–246.

Shiokawa, K., Tashiro, K., Misumi, Y., and Yamana, K., 1981, Non-coordinated synthesis of RNAs in pre-gastrular embryos of *Xenopus laevis, Dev. Growth Diff.* **23**:589–597.

Shiokawa, K., Kawazoe, Y., and Yamana, K., 1985, Demonstration that inhibitor of rRNA synthesis in "charcoal-extracts" of *Xenopus* embryos is artifactually produced ammonium perchlorate, *Dev. Biol.* **112**:258–260.

Shiokawa, K., Kawazoe, Y., Nomura, H., Miura, T., Nakakura, N., Horiuchi, T., and Yamana, K., 1986, Ammonium ion as a possible regulator of the commencement of rRNA synthesis in *Xenopus laevis* embryogenesis, *Dev. Biol.* **115**:380–391.

Signoret, J., and Lefresne, J., 1971, Contribution à l'étude de la segmentation de l'oeuf d'axolotl. I. Definition de la transition blastuleene, *Ann. Embryol. Morphogen.* **4**:113–123.

Smith, L. D., Michael, P., and Williams, M. A., 1983, Does a predetermined germ line exist in amphibians?, in: *Current Problems in Germ Cell Differentiation* (A. McLaren and C. Wylie, eds.), pp. 19–40, Cambridge University Press, London.

Smith, L. D., and Richter, J. D., 1985, Synthesis, accumulation, and utilization of maternal macromolecules during oogenesis and oocyte maturation, in: *Biology of Fertilization* (A. Monroy and C. Metz, eds.), pp. 141–160, Academic Press, New York.

Smith, R. C., 1986, Protein synthesis and messenger RNA levels along the animal–vegetal axis during early *Xenopus* development, *J. Embryol. Exp. Morphol.* **95**:15–35.

Spemann, H., 1938, *Embryonic Development and Induction*, Yale University Press, New Haven, Connecticut.

Stick, R., and Hausen, P., 1985, Changes in the nuclear lamina composition during early development of *Xenopus laevis, Cell* **41**:191–200.

Weeks, D. L. and Melton, D. A., 1987, A maternal mRNA localized to the vegetal hemisphere in *Xenopus* eggs codes for a growth factor related to TGFβ, *Cell* **51**:861–867.

Woodland, H. R., 1974, Changes in the polysome content of developing *Xenopus laevis* embryos, *Dev. Biol.* **40**:90.

Woodland, H. R., and Gurdon, J. B., 1968, The relative rates of synthesis of DNA, sRNA, and rRNA in the endodermal region and other parts of *Xenopus laevis* embryos, *J. Embryol. Exp. Morphol.* **19**:363.

Woodland, H. R., Flynn, J. M., and Wylie, A. J., 1979, Utilization of stored mRNA in *Xenopus* embryos and its replacement by newly synthesized transcripts: Histone H1 synthesis using interspecies hybrids, *Cell* **18**:165–171.

Wormington, M. W., and Brown, D. D., 1983, Onset of 5 sRNA gene regulation during *Xenopus* embryogenesis, *Dev. Biol.* **99**:248–257.

Wu, M., and J. G. Gerhart, 1980, Partial purification and characterization of maturation promoting factor from eggs of *Xenopus laevis, Dev. Biol.* **79**:465.

Chapter 8

Expression of Ribosomal Protein Genes during *Xenopus* Development

W. MICHAEL WORMINGTON

1. Introduction

Biosynthesis of the eukaryotic ribosome encompasses the expression of genes encoding approximately 60 integral proteins and two distinct classes of rRNA genes in the nucleus and the nucleolus. The regulation of ribosome production is an important aspect of gene expression during *Xenopus laevis* development. During oogenesis in *Xenopus* and other amphibia, ribosomes are accumulated at least 1000 times more rapidly than in the most synthetically active somatic cells (Korn and Gurdon, 1981). This massive stockpile of 10^{12} ribosomes within a single egg is sufficient to support protein synthesis through development of the swimming tadpole, which consists of approximately 10^6 cells (Brown and Gurdon, 1964). The enormous synthesis of ribosomes during oogenesis and the requirement to impose somatic regulation of ribosome production during embryogenesis renders *Xenopus* an exquisite model system for the analysis of ribosome biogenesis during vertebrate development.

The levels of developmental control of the rRNA genes are well defined. The genes encoding 18S and 28s rRNA are selectively amplified during early oogenesis (Brown and Dawid, 1968; Gall, 1968). The transcriptional basis for nucleolar dominance of these genes in *Xenopus* hybrids has been determined (Reeder and Roan, 1984); 5S rRNA synthesis is achieved by the differential expression of two distinct multigene families. The developmental regulation of *Xenopus* 5S rRNA genes has been extensively presented by Brown and Schlissel (1985) and by Krämer (1985) in Volume 1 of this series. An excellent monograph that integrates structural features of the nucleolus with ribosome biogenesis has been presented recently by Hadjilov (1985).

This chapter provides an overview of the current knowledge concerning the expression of the genes encoding ribosomal proteins. The mechanisms that coordinate the expression of these genes, whose products are represented in equimolar amounts in the mature 40S and 60S cytoplasmic ribosomal subunits,

W. MICHAEL WORMINGTON • Department of Biochemistry, Rosenstiel Basic Medical Sciences Research Center, Brandeis University, Waltham, Massachusetts 02254.

are specifically addressed. Several unresolved aspects of regulation superimposed on this basic problem of coordinate gene expression are considered. The most notable is the uncoupling of ribosomal protein and rRNA synthesis from assembly of ribosomal subunits at distinct stages during *Xenopus* oogenesis and embryogenesis. The analysis of ribosomal protein genes provides insight into the regulated expression of a related set of genes of known function during development.

2. Isolation and Structure of *Xenopus* Ribosomal Protein Genes

2.1. Isolation of cDNAs for Ribosomal Protein mRNAs

Initial studies of ribosomal protein synthesis during oogenesis by Hallberg and Smith (1975) indicated that production of these proteins constitutes as much as 20–30% of total protein synthesis in small vitellogenic oocytes (stages II–III) (Dumont, 1972). The small size of eukaryotic ribosomal protein mRNAs enabled enrichment for these transcripts by size fractionation of poly(A)$^+$ RNA isolated from stage II–III oocytes. Screening of complementary DNA (cDNA) libraries constructed from the 10S–12S fraction of mRNA has resulted in the isolation of cDNAs corresponding to 10 different mRNAs encoding cytoplasmic 60S (L1, L13, L14, L15, L23, and L32) and 40S (S1, S8, S19, and S22) ribosomal proteins (Bozzoni *et al.*, 1981; Baum and Wormington, 1985). Hybrid select translation of ovarian poly(A)$^+$ RNA using each of the ten cDNAs reveals that each clone corresponds to an individual ribosomal protein messenger RNA (mRNA). Northern blot analysis confirms that each cDNA hybridizes to a single predominant mRNA species sufficiently large to encode its cognate ribosomal protein.

2.2. Structural Analysis of Ribosomal Protein Genomic Clones

Partial-length cDNAs corresponding to ribosomal protein mRNAs have been used as probes to determine the genomic organization of these genes. The copy numbers for ribosomal protein genes in the *X. laevis* genome approximate those found in such diverse eukaryotes as yeast (Fried *et al.*, 1981) and mammals (Monk *et al.*, 1981). The genes encoding individual *Xenopus* ribosomal proteins are not highly reiterated and are present in 1 to 10 copies (Bozzoni *et al.*, 1981; Baum and Wormington, 1985). Thus, the elevated level of ribosomal protein synthesis during oogenesis is not directed by the expression of highly repeated genes.

The structures of genomic clones encoding two 60S ribosomal proteins, L1 and L14, have been described (Bozzoni *et al.*, 1982). Hybrid select translation experiments have revealed that sequences within the L1 and L14 genomic clones cross-hybridize to several mRNAs that encode translation products comigrating with authentic ribosomal proteins. Further characterization of these homologous sequences or assignment of these additional ribosomal proteins to individual subunits have not been reported. Initial analysis of genomic clones corresponding to three of the four L14 genes indicates that the genes are not tightly linked. Each L14 gene has seven introns containing highly repeated sequences.

The two genes encoding ribosomal protein L1 have been characterized in greatest detail (Bozzoni *et al.*, 1982). Both the L1a and L1b genes are functional, as demonstrated by the isolation of specific cDNAs corresponding to each of these genes from an ovarian poly(A)$^+$ cDNA library (Loreni *et al.*, 1985). The two L1 coding regions are highly conserved. A small duplication within the L1b mRNA results in synthesis of an L1b protein that is five amino acid residues longer than the L1a peptide at its carboxy terminus. These two species of L1 protein contribute unequally to the L1 pool in oocytes. The L1a protein is more abundant than L1b in 60S subunits. The basis for this unequal representation of the two L1 proteins has not been determined. The unequal contribution of two genes encoding a single ribosomal protein has been observed for the rp51A and rp51B genes in yeast (Abovich and Rosbash, 1984).

The 3' untranslated regions are highly conserved between the L1a and L1b mRNAs. The authors propose that such conservation outside of the L1 coding region may indicate the possible involvement of these sequences in the post-transcriptional or translational regulation of L1 mRNA that is observed during embryogenesis. An additional example of conserved sequences is observed within four of the nine introns of the L1a gene. A 60-base pair (bp) region is present with 80% sequence homology in introns 2, 4, 7, and 8. These homologous sequences exclude both the exon–intron junction consensus sequences and the branch sites adjacent to the 3' splice acceptor sequences. These regions represent, therefore, a novel example of high-sequence homology within the individual introns of a single gene. Loreni *et al.* (1985) have proposed that these conserved intron sequences may be involved in mediating the splicing efficiency of L1 transcripts. These examples of highly conserved sequences within the introns and 3' untranslated regions of L1 transcripts provide initial candidate sequences for functional analyses of their contribution to L1 gene expression.

3. Ribosomal Protein Gene Expression during Oogenesis

3.1. Temporal Accumulation of Ribosomal Protein Transcripts

The translational regulation of maternal mRNAs during oogenesis has been the focus of extensive studies by Smith and colleagues (Smith and Richter, 1984). These analyses have demonstrated that a progressive increase in the polysomal recruitment of maternal mRNAs is responsible for the dramatic elevation in total protein synthesis that occurs as oogenesis proceeds. Thus, while maximal steady state levels for nuclear encoded poly(A)$^+$ RNAs are attained in stage II oocytes and maintained throughout oogenesis (Rosbash and

Ford, 1974; Golden *et al.*, 1980), fully grown stage VI oocytes are fifty times more active in protein synthesis than stage II oocytes (Taylor and Smith, 1985). Individual ribosomal protein mRNAs reach maximal steady-state levels in stage II oocytes; the mRNA levels are reduced in fully grown oocytes (Pierandrei-Amaldi *et al.*, 1982; Baum and Wormington, 1985). The steady-state levels for individual ribosomal protein mRNAs, however, vary by as much as fivefold, suggesting that coordinate accumulation of ribosomal proteins does not depend upon the presence of equimolar concentrations of the corresponding mRNAs (Baum and Wormington, 1985).

3.2. Translational Regulation of Ribosomal Protein Synthesis during Oogenesis

The synthesis of ribosomal proteins and translational utilization of their transcripts during oogenesis have been the subjects of several studies. These analyses indicate that as a class, ribosomal protein mRNAs exhibit several features of translational regulation that distinguish them from the overall oocyte mRNA population.

Initial studies by Hallberg and Smith (1975), as well as more recent investigations by Dixon and Ford (1982), have shown that ribosomal protein synthesis is negligible in previtellogenic stage I–II oocytes. 5S rRNA, however, accumulates to its maximal levels in these immature oocytes, indicating that its synthesis is uncoupled from both production of ribosomal proteins and ribosome assembly (Krämer, 1985). Ribosomal proteins comprise as much as 20% of total protein synthesis in vitellogenic stage III oocytes, concomitant with the onset of 18S and 28S rRNA accumulation. The contribution of ribosomal proteins to total protein synthesis is diminished in fully grown stage VI oocytes. Thus, ribosomal protein synthesis fails to increase proportionately with the elevated rate of overall translation as oogenesis proceeds. A comparison between the translational recruitment of ribosomal protein mRNAs and the total oocyte mRNA population addresses the basis for this distinct regulation.

Taylor and Smith (1985) determined that stage III oocytes have less than 1 ng of mRNA associated with polysomes, representing approximately 1–2% of the total poly(A)$^+$ RNA. By contrast, at least 50% of each of the ribosomal protein mRNAs examined is polysomal in stage III oocytes (Baum and Wormington, 1985). This finding suggests that ribosomal protein mRNAs appear to maintain a higher degree of polysomal association relative to the overall stage III oocyte mRNA population. The functional basis for this distinction remains to be determined. In fully grown stage VI oocytes, 4 ng of mRNA is found associated with polysomes, whereas the total proportion of ribosomes engaged in protein synthesis remains equal to that observed in stage III oocytes (Taylor and Smith, 1985). This translational recruitment of additional maternal mRNAs is responsible for the five- to eightfold increase in total protein synthesis between stages III and VI. By contrast, ribosomal protein synthesis between these stages remains constant. This may partly be due to the reduced levels of ribosomal protein mRNAs in fully grown oocytes, but is principally because of a lack of recruitment of additional ribosomal protein mRNAs onto polysomes in stage VI oocytes (Baum and Wormington, 1985). Thus, the same proportion of ribosomal protein mRNA is associated with polysomes in stage III and VI oocytes, but translational recruitment of additional maternal mRNAs in effect dilutes translation of ribosomal protein mRNAs, diminishing their contribution to overall protein synthesis.

A unique aspect of oogenesis concerns the organization of as much as 70% of the poly(A)$^+$ RNA into interspersed repeated transcripts that are translationally inactive (Anderson *et al.*, 1982; Richter *et al.*, 1984). A consequence of this structural organization for the majority of oocyte mRNA would be the exclusion of these transcripts from the polysomal fraction of mRNAs. The presence of interspersed repeated sequences within ribosomal protein transcripts has been addressed by both structural and functional criteria. Pierandrei-Amaldi and Beccari (1980) first observed that polysomal and nonpolysomal poly(A)$^+$ RNA preparations from stage III oocytes are nearly equivalent in template activity to direct protein synthesis *in vitro*. Both mRNA preparations saturate wheat germ extract at 40 μg/ml RNA, although overall stimulation of translation was 25% lower for non-polysomal mRNA relative to the polysomal fraction. Baum and Wormington (1985) analyzed ribosomal protein transcripts in stage III and VI oocytes. These studies have indicated that no gross structural or functional distinctions exist between ribosomal protein mRNAs that are actively translated *in vivo* (as demonstrated by their association with polysomes) and the nonpolysomal fraction. Ribosomal protein mRNAs isolated from both fractions serve as functional templates for ribosomal protein synthesis *in vitro*. Northern blot analysis of RNA isolated from polysomal and nonpolysomal fractions reveals the presence of a single predominant species of mRNA hybridizing to each ribosomal protein cDNA probe. These results indicate that ribosomal protein mRNAs are represented exclusively within the translationally competent fraction of oocyte poly(A)$^+$ RNA.

3.3. Coordination of Ribosomal Protein Synthesis with Ribosome Assembly

Analyses of ribosomal protein mRNAs have focused on measurements of transcript accumulation and translational recruitment during oogenesis. The levels at which synthesis of individual ribosomal proteins and rRNAs are integrated with ribosome assembly remain to be elucidated. Oogenesis is unusual in that synthesis of ribosomal constituents is in part uncoupled from actual subunit assembly. The synthesis of 5S rRNA best illustrates this phenomenon. 5S rRNA accumulates to maximal levels in immature previtellogenic oocytes and is sequestered in a nonribosome-associated 7S storage particle before the synthesis of 18S and 28S rRNA and ribosomal proteins (Picard and Wegnez, 1979; Pelham and Brown, 1980; Krämer, 1985) (see Section 3.2).

The fate of newly synthesized ribosomal proteins that are not assimilated into ribosomal subunits has not been determined definitively in oocytes. The contribution of ribosomal proteins to overall protein synthesis is maximal in stage III oocytes, which have accumulated less than 25% of the levels of 18S and 28S rRNA attained in fully grown oocytes (Taylor and Smith, 1985). Dixon and Ford (1982) observed that stage II–III oocytes, in contrast to fully grown oocytes, sequester the majority of newly synthesized ribosomal proteins into nonribosomal subunit complexes. Neither the precise composition of these subribosomal particles nor their stability and cellular localization throughout oogenesis has been investigated. The role of these particles as a source of free ribosomal proteins in oocytes or in developing embryos and the potential exchange of ribosomal proteins from these particles into mature subunits remain to be determined.

Autogenous translational regulation of ribosomal protein mRNAs by a subset of these proteins in *Escherichia coli* has been demonstrated by a combination of genetic and biochemical analyses (Nomura et al., 1984). Pierandrei-Amaldi et al. (1985a) were unable to demonstrate repression of endogenous ribosomal protein synthesis in stage VI oocytes microinjected with exogenous proteins purified from ovarian ribosomes. The lack of translational repression may be an indication that ribosomal protein mRNAs associated with polysomes are refractory to inhibition and that the nonpolysomal fraction may not be accessible for binding to the microinjected proteins. However, neither the RNA binding activity of the purified ribosomal proteins nor their ability to be incorporated into subunits was ascertained. Thus, conclusive evidence concerning autogenous regulation of translation or other potential levels of ribosomal protein gene expression in oocytes awaits further investigation.

4. Ribosomal Protein Gene Expression during Embryogenesis

4.1. Utilization of the Maternal Ribosome Pool

The maternal ribosome pool present in the fertilized egg is used for protein synthesis through development of the stage 40 swimming tadpole consisting of one million cells. Woodland (1974) measured the recruitment of these maternal ribosomes onto polysomes throughout early development and has observed that, whereas fewer than 3% of the ribosomes in the unfertilized egg are in polysomes, more than 70% of the ribosomes present in a stage 40 tadpole are actively engaged in protein synthesis. Brown and Littna (1964) found that the total ribosome content changes little between the unfertilized egg and stage 40. These findings, in conjunction with the observation by Brown and Gurdon (1964) that homozygous anucleolate embryos develop to stage 40 in the absence of 18S and 28S rRNA synthesis, indicate that there is no requirement for new ribosomes before the swimming tadpole stage of development.

4.2. Regulation of Ribosomal Protein mRNA Accumulation

The mechanisms regulating the onset of *de novo* ribosome biosynthesis during embryogenesis have been addressed through studies of both ribosomal protein and rRNA genes. The cessation of ribosome synthesis upon maturation of fully grown oocytes is due, at least in part, to the transcriptional repression of rRNA genes during germinal vesicle breakdown (Smith and Richter, 1984; Wormington and Brown, 1983). Maternal ribosomal protein mRNAs are dead-enylated at maturation (Hyman *et al.*, 1984) and are then degraded during early cleavage stages (Pierandrei-Amaldi *et al.*, 1982; Baum and Wormington, 1985). Deadenylation has been observed for several other maternal mRNAs that are subsequently released from polysomes and rapidly degraded after fertilization (Sagata *et al.*, 1980; Colot and Rosbash, 1983). Thus, degradation of the majority of maternal ribosomal protein transcripts in the absence of transcription before the midblastula stage represents the principal contribution to the lack of ribosomal protein synthesis throughout early cleavage stages.

New ribosomal protein transcripts are first detected during gastrulation. The levels of these transcripts increase dramatically, reaching maximal ac-cumulation by late neurula stages (Weiss *et al.*, 1981; Pierandrei-Amaldi *et al.*, 1982; Baum and Wormington, 1985). As in stage VI oocytes, the steady-state levels of individual ribosomal protein mRNAs are not tightly coordinated in stage 40 tadpole poly(A)$^+$ RNA (Baum and Wormington, 1985). Similar varia-tion in the relative amounts of individual ribosomal protein mRNAs in differ-ent mouse cell lines has been reported by Meyuhas and Perry (1980). Although a correlation between the levels of ribosomal protein mRNAs and synthesis of the cognate proteins has not been determined, these results further suggest that equimolar accumulation of ribosomal proteins does not depend upon equi-molar levels of these transcripts.

4.3. Synthesis of Early Ribosomal Proteins in the Absence of Ribosome Assembly

The absence of ribosome production before the midblastula stage can be attributed to the degradation of the majority of maternal ribosomal protein mRNAs and the absence of *de novo* rRNA synthesis. However, metabolic label-ing of embryonic proteins by microinjection of [^{35}S]methionine demonstrates that three ribosomal proteins, S3, L17, and L31, are synthesized continuously throughout early cleavage (Pierandrei-Amaldi *et al.*, 1982; Baum and Worm-ington, 1985). Because of the absence of detectable transcription before the mid-blastula stage of development (see Chapter 7 for a discussion of the mid-blastula transition), synthesis of these three early ribosomal proteins is most likely directed by maternal mRNAs that apparently escape the deadenylation and degradation observed for most maternal ribosomal protein mRNAs. The synthesis of an additional ribosomal protein, L5, is first detected during the mid-blastula transition. The onset of L5 accumulation, therefore, occurs con-currently with the transcriptional activation of 5S rRNA genes (Wormington and Brown, 1983). L5 is associated with 5S RNA in the 60S subunit (Picard and Wegnez, 1979).

The expression of a small subset of ribosomal proteins before synthesis of 18S and 28S mRNAs during gastrulation and *de novo* ribosome assembly in stage 30 tailbud embryos has precedence in other eukaryotes. A set of early ribosomal proteins is synthesized during early *Drosophila* development in the absence of ribosome assembly (Santon and Pelligrini, 1980). Three 60S proteins in HeLa cells—L10, L19, and L28—are synthesized in the absence of rRNA synthesis and can exchange with fully assembled subunits (Lastick and Mc-Conkey, 1976). Thus, further analysis of the expression of the four "early" ribosomal protein genes of *Xenopus* is warranted to address several aspects of ribosomal protein production during embryogenesis. First, what discriminates the mRNAs encoding S3, L17, L31, and possibly L5, to escape the degradation encompassing the majority of maternal ribosomal protein mRNAs? Second, what is the functional significance for the synthesis and stable accumulation of four ribosomal proteins in the absence of 18S and 28S rRNA synthesis and *de novo* ribosome assembly? Finally, what provides the specificity that enables these early mRNAs to be excluded from the translational regulation that encompasses the majority of ribosomal protein mRNAs? Isolation of the appropriate cDNAs corresponding to these early mRNAs will provide the initial basis by which these analyses can be undertaken.

4.4. Translational Regulation of Ribosomal Protein Synthesis during Embryogenesis

The transcriptional activation of ribosomal protein and rRNA genes after the mid-blastula transition is uncoupled from the actual requirement for new ribosomes in the swimming tadpole. This principle can be further extended to include uncoupling ribosomal protein mRNA synthesis from production of new ribosomal proteins. Metabolic labeling of embryonic proteins by microinjection of [^{35}S]methionine indicates that although ribosomal protein mRNAs begin to accumulate during gastrulation, a complete complement of new ribosomal proteins is not synthesized before development of the stage 30 tailbud embryo (Pierandrei-Amaldi *et al.*, 1982; Baum and Wormington, 1985). This regulation is achieved by exclusion of ribosomal protein mRNAs from polysomes until the tailbud stage. Ribosomal protein transcripts synthesized before this stage, however, are polyadenylated, fully processed mRNAs that are functional templates to direct synthesis of a complete set of ribosomal proteins *in vitro* (Baum and Wormington, 1985).

Translational control of maternal mRNAs during early *Xenopus* development is well documented. The utilization of histone (Woodland *et al.*, 1979), α- and β-actin (Sturgess *et al.*, 1980), and fibronectin mRNAs (Lee *et al.*, 1984) exemplifies this control. By contrast, the uncoupling of ribosomal protein gene transcription from the translation of ribosomal protein mRNAs provides an example of the stage-specific translational recruitment of embryonic transcripts. Proposed mechanisms for the translational control of ribosomal protein mRNAs remain speculative at this time. This regulation could be mediated by the presence of negative regulatory factors specific for these transcripts. The masking of maternal mRNAs in immature oocytes by specific RNA binding proteins illustrates negative control of translation in *Xenopus* (Richter and Smith, 1983, 1984). Similar proteins present throughout early development could act in an analogous manner to repress translation of ribosomal protein mRNAs.

The studies of Pierandrei–Amaldi *et al.* (1982), and Baum and Wormington (1985) have not addressed mechanisms for translational activation of ribosomal protein mRNAs in tailbud embryos. One potential mechanism involves translational activation of old ribosomal protein mRNAs, which have accumulated in a repressed state since gastrulation. Such recruitment may or may not be accompanied by polysomal recruitment of newly synthesized ribosomal protein transcripts in stage 30 embryos. Alternatively, old ribosomal protein mRNAs that are synthesized before stage 30 may never be recruited to polysomes; *de novo* synthesis of ribosomal proteins may be achieved exclusively by translational recruitment of new ribosomal protein transcripts synthesized in the tailbud embryo. The former mechanism involves a translational repression followed by recruitment of the same mRNAs, whereas the latter mechanism requires irreversible repression of ribosomal protein mRNAs synthesized before stage 30 and the absence of such repression following development of the tailbud stage. Elucidation of this control of ribosomal protein synthesis should provide additional insight into the molecular basis for the translational discrimination between old and new mRNAs observed previously by Dworkin and Hershey (1981). These studies, albeit limited to a small number of abundant RNA species, indicate that polysomes from later stages of development contain predominantly newly synthesized transcripts as opposed to transcripts recruited from earlier stages.

4.5. Regulation of Ribosomal Protein Synthesis in Anucleolate Embryos

Homozygous anucleolate (0-nu) embryos, lacking the genes encoding 18S and 28S rRNA, proceed through development of the swimming tadpole by utilization of the maternal stockpile of ribosomes (Brown and Gurdon, 1964). These mutant embryos have been used to determine the levels of ribosomal protein gene expression that are coupled to the accumulation of rRNAs. Initial observations by Hallberg and Brown (1969) showed that stage 40 0-nu tadpoles fail to synthesize new ribosomes and accumulate ribosomal proteins to less than 4% of the levels observed in wild-type (2-nu) tadpoles.

Definitive studies of ribosomal protein synthesis in 0-nu embryos have been presented by Pierandrei-Amaldi *et al.* (1985b). The profiles for synthesis and accumulation of ribosomal protein mRNAs are indistinguishable for 2-nu and 0-nu embryos up to stage 30. Ribosomal protein transcripts in 0-nu embryos decline considerably after this stage relative to 2-nu embryos. The levels of nonribosomal protein transcripts are unaltered in 0-nu stage 30 embryos. The basis for the reduction in ribosomal protein transcripts appears to be posttranscriptional. Nuclear run off transcription reactions have indicated that

rates of synthesis of ribosomal protein transcripts are identical in nuclei isolated from 0-nu and 2-nu stage 33 embryos. Thus, the stability of ribosomal
protein mRNAs appears to be coupled to synthesis of rRNAs and ribosome
assembly.

The translational recruitment of ribosomal protein mRNAs is not altered in
0-nu embryos. As seen in 2-nu embryos, ribosomal protein mRNAs are recruited onto polysomes in stage 30 tailbud embryos. Thus, autogenous translational regulation analogous to the control of ribosomal protein synthesis in the
absence of rRNA synthesis in E. coli (Nomura et al., 1984) does not apparently
operate in 0-nu Xenopus embryos.

The translation of ribosomal protein mRNAs, however, does not lead to the
stable accumulation of new ribosomal proteins. Metabolic labeling of
ribosomal proteins in stage 31 embryos reveals that ribosomal proteins are
synthesized normally in 0-nu embryos but are unstable with a half-life of less
than 1 hr. The synthesis of nonribosomal proteins, such as histones, is unaltered in 0-nu embryos. The combination of normal ribosomal protein gene
activation with subsequent instability of these mRNAs and of free ribosomal
proteins in anucleolate Xenopus embryos is similar to the same features of
ribosomal protein regulation in bobbed mutants of Drosophila, which likewise
are lacking genes encoding 18S and 28S rRNA (Kay and Jacobs–Lorena, 1985).

The studies conducted by Pierandrei-Amaldi et al. (1985b), in addition to
the analysis of 5S rRNA accumulation in 0-nu embryos by Miller (1974), lead to
the conclusion that the transcriptional and translational activation of ribosomal
protein genes and the transcriptional activation of both classes of rRNA genes
appear to be independently regulated events before the assembly of new
ribosomes in the swimming tadpole. These aspects of uncoupled gene expression during early Xenopus development are distinct from the balanced
expression and accumulation of ribosomal constituents, which are tightly coupled with actual ribosome assembly in E. coli.

5. Expression of Exogenous Ribosomal Protein Genes and mRNAs in Oocytes

The mechanisms that balance the accumulation of ribosomal proteins and
rRNAs with subunit assembly can be ascertained most directly through the
combination of genetic analyses in which synthesis of individual ribosomal
constituents is eliminated and by introduction of exogenous genes into intact
cells to assay for their accurate expression. The use of these approaches in
yeast, for example, has provided evidence for multiple levels of control for
different ribosomal protein genes (Abovich and Rosbash, 1984; Abovich et al.,
1985; Warner et al., 1985). In Xenopus, genetic analysis of ribosome biosynthesis has been restricted to the absence of rRNA synthesis in homozygous
0-nu embryos. Thus, the resolution of regulatory mechanisms for Xenopus
ribosomal protein genes depends largely upon functional analyses of cloned

ribosomal protein genes and specifically altered variants of these genes or gene products introduced into oocytes and fertilized eggs.

The use of *Xenopus* oocytes as an expression system for exogenous genes has been extensively documented (Gurdon and Melton, 1981). The initial analysis of cloned L1 and L14 ribosomal protein gene expression in microinjected oocytes by Bozzoni *et al.* (1984) illustrates different levels of control for these two genes. The cloned L14 gene is accurately transcribed and the primary transcript efficiently processed to generate elevated levels of L14 mRNA in microinjected oocytes. Stimulation of L14 synthesis is proportional to the amount of L14 mRNA produced and leads to the stable accumulation of additional L14 protein. By contrast, the cloned L1 gene is subject to a selective block in RNA processing. The second and third introns specifically fail to be removed from the L1 primary transcript, leading to the stable accumulation of a partially spliced intermediate in the oocyte nucleus. Thus, synthesis of elevated levels of L1 protein in response to microinjection of the L1 genomic clone is not observed. The L1 gene used in these studies is actively expressed in oocytes, and there is no evidence that endogenous L1 transcripts are subject to this splicing defect during oogenesis. The incomplete RNA splicing may reflect the high level of precursor RNA produced in response to the increase L1 gene dose. This may indicate that synthesis of L1 is balanced by a specific reduction in splicing efficiency. Alternatively, the elevated concentration of an L1 precursor RNA, which serves as an intrinsically inefficient splicing substrate may exceed the limited splicing capacity of the oocyte. The role of this selective splicing deficiency in the regulation of L1 mRNA production, however, is clearly amenable to further analysis, and its contribution to the control of L1 synthesis during development can be ascertained.

The regulation of L1 synthesis in oocytes has also been analyzed by microinjection of exogenous L1 mRNA synthesized *in vitro* using SP6 RNA polymerase (Baum, 1986). These results reveal that translation of the synthetic L1 mRNA in stage VI oocytes is not subject to dosage regulation. Stimulation of L1 synthesis is proportional to the amount of exogenous L1 mRNA present. This is analogous to the increased synthesis of L14 observed upon microinjection of the L14 genomic clone (Bozzoni *et al.*, 1984). However, in contrast to L14, the increased synthesis of L1 is followed by degradation of excess L1 protein. This suggests that L1 protein synthesis is potentially regulated by RNA splicing of the primary transcript and by post-translational proteolysis, possibly linked to subunit assembly.

6. Summary

The *Xenopus* ribosomal protein genes provide an excellent system to elucidate the complex regulation encompassing 60 functionally related proteins present in equimolar amounts in ribosomal subunits. Oogenesis and embryogenesis provide unique opportunities to investigate ribosome biosynthesis in situations wherein gene activation of individual components is uncoupled from assembly of the ribosomal subunits. This chapter has focused on the basic parameters that control ribosomal protein gene expression during development. Translational control is clearly a major level for coordinating the regulation of these genes during development, as is posttranslational stability of the ribosomal proteins and RNA splicing of the L1 gene. In addition to these levels of control under active investigation, a number of intriguing problems remain to be addressed in any detail. For example, the mechanisms that balance ribosomal protein production with subunit assembly in oocytes remain to be determined. Resolution of these events must also define the processes by which ribosomal proteins, upon synthesis in the cytoplasm, are first translocated to the nucleus and subsequently to the nucleolus for subunit assembly. Functional approaches in which these genes are assayed for accurate developmental control in microinjected oocytes and fertilized eggs will undoubtedly provide information on the synthesis of this eukaryotic organelle and the signals responsible for altering these processes at different developmental stages.

ACKNOWLEDGMENTS. I thank Dr. Ellen Baum, Dr. Joel Richter, Dr. Michael Rosbash, and Dr. Dennis Smith for many insightful discussions concerning translational regulation. I am especially grateful to the members of my laboratory, Linda Hyman, Helen Romanczuk, and Susan Varnum, for greatly improving the chapter with helpful criticisms. The research conducted in the author's laboratory was supported by grants from the National Institutes of Health and the March of Dimes Birth Defects Foundation.

References

Abovich, N., and Rosbash, M., 1984, Two genes for ribosomal protein 51 of *Saccharomyces cerevisiae* complement and contribute to the ribosomes, *Mol. Cell. Biol.* **4**:1871–1879.

Abovich, N., Gritz, L., Tung, L., and Rosbash, M., 1985, Effect of RP51 gene dosage alterations on ribosome synthesis in *Saccharomyces cerevisiae*, *Mol. Cell. Biol.* **5**:3429–3435.

Anderson, D. M., Richter, J. D., Chamberlin, M. E., Price, D. H., Britten, R. J., Smith, L. D., and Davidson, E. H., 1982, Sequence organization of the poly(A) RNA synthesized and accumulated in lampbrush chromosome stage *Xenopus laevis* oocytes, *J. Mol. Biol.* **155**:281–309.

Baum, E. Z., 1986, Developmental regulation of *Xenopus* ribosomal protein genes: Control at the translational level, Ph.D. thesis, Brandeis University, Waltham, Massachusetts.

Baum, E. Z., and Wormington, W. M., 1985, Coordinate expression of ribosomal protein genes during *Xenopus* development, *Dev. Biol.* **111**:488–498.

Bozzoni, I., Beccari, E., Luo, Z. X., and Amaldi, F., 1981, *Xenopus laevis* ribosomal protein genes: Isolation of recombinant cDNA clones and study of the genomic organization, *Nucl. Acids Res.* **9**:1069–1086.

Bozzoni, I., Tognoni, A., Pierandrei-Amaldi, P., Beccari, E., Buongiorno-Nardelli, M., and Amaldi, F., 1982, Isolation and structural analysis of ribosomal protein genes in *Xenopus laevis*, *J. Mol. Biol.* **161**:353–371.

Bozzoni, I., Fragapane, P., Annesi, F., Pierandrei-Amaldi, P., Amaldi, F., and Beccari, E., 1984, Expression of two *X. laevis* ribosomal protein genes in injected frog oocytes: A specific splicing block interferes with L1 RNA maturation, *J. Mol. Biol.* **180**:987–1005.

Brown, D. D., and Dawid, I. B., 1968, Specific gene amplification in oocytes, *Science* **160**:272–280.

Brown, D. D., and Gurdon, J. B., 1964, Absence of rRNA synthesis in the anucleolate mutant of *X. laevis, Proc. Natl. Acad. Sci. USA* **51**:139–146.

Brown, D. D., and Littna, E., 1964, RNA synthesis during the development of *Xenopus laevis* the South African clawed toad, *J. Mol. Biol.* **8**:669–687.

Brown, D. D., and Schlissel, M. S., 1985, The molecular basis of differential gene expression of two 5S RNA genes, *Cold Spring Harbor Symp. Quant. Biol.* **50**:549–554.

Colot, H. V., and Rosbash, M., 1982, Behavior of individual maternal pA+ RNAs during embryogenesis of *Xenopus laevis, Dev. Biol.* **94**:79–86.

Dixon, L. K., and Ford, P. J., 1982, Regulation of protein synthesis and accumulation during oogenesis in *Xenopus laevis, Dev. Biol.* **93**:478–497.

Dumont, J. N., 1972, Oogenesis in *Xenopus laevis* (Daudin). I. Stages of oocyte development in laboratory maintained animals, *J. Morphol.* **136**:153–180.

Dworkin, M. B., and Hershey, J. W. B., 1981, Cellular titers and subcellular distributions of abundant polyadenylate-containing ribonucleic acid species during early development in the frog *Xenopus laevis, Mol. Cell. Biol.* **1**:983–993.

Fried, H. M., Pearson, N. J., Kim, K. H., and Warner, R. J., 1981, The genes for fifteen ribosomal proteins of *Saccharomyces cerevisiae, J. Biol. Chem.* **256**:10176–10183.

Gall, J. G., 1968, Differential synthesis of the genes for rRNA during amphibian oogenesis, *Proc. Natl. Acad. Sci. USA* **60**:553–560.

Golden, L., Schafer, U., and Rosbash, M., 1980, Accumulation of individual pA+ RNAs during oogenesis of *Xenopus laevis, Cell* **22**:835–844.

Gurdon, J. B., and Melton, D. A., 1981, Gene transfer in amphibian eggs and oocytes, *Annu. Rev. Genet.* **15**:189–218.

Hadjiolov, A. A., 1985, *The Nucleolus and Ribosome Biogenesis, Cell Biology Monographs,* Vol. 12, Springer-Verlag, New York.

Hallberg, R. L., and Brown, D. D., 1969, Coordinated synthesis of some ribosomal proteins and ribosomal DNA in embryos of *Xenopus laevis, J. Mol. Biol.* **46**:393–411.

Hallberg, R. L., and Smith, D. C., 1975, Ribosomal protein synthesis in *Xenopus laevis* oocytes, *Dev. Biol.* **42**:40–52.

Hyman, L. E., Colot, H. V., and Rosbash, M., 1984, Accumulation and behavior of mRNA during oogenesis and early embryogenesis of *Xenopus laevis,* in: *Molecular Aspects of Early Development* (G. M. Malacinski and W. Klein, eds.), pp. 142–151, Plenum, New York.

Kalthoff, H., and Richter, D., 1982, Subcellular transport and ribosomal incorporation of microinjected protein S6 in oocytes from *Xenopus laevis, Biochemistry* **18**:4144–4147.

Kay, M. A., and Jacobs-Lorena, M., 1985, Selective translational regulation of ribosomal protein gene expression during early development of *Drosophila melanogaster, Mol. Cell. Biol.* **5**:3583–3592.

Korn, L. J., and Gurdon, J. B., 1981, The reactivation of developmentally inert 5S RNA genes in somatic nuclei injected into *Xenopus* oocytes, *Nature (Lond.)* **289**:461–465.

Krämer, A., 1985, 5S ribosomal gene transcription during *Xenopus* oogenesis, in: *Developmental Biology: A Comprehensive Synthesis,* Vol. 1, Oogenesis (L. W. Browder, ed.), pp. 431–448, Plenum, New York.

Lastick, S. M., and McConkey, E. H., 1976, Exchange and stability of HeLa ribosomal proteins *in vivo, J. Biol. Chem.* **251**:2867–2875.

Lee, G., Hynes, R., and Kirschner, M., 1984, Temporal and spatial regulation of fibronectin in early *Xenopus* development, *Cell* **36**:729–740.

Loreni, F., Ruberti, I., Bozzoni, I., Pierandrei-Amaldi, P., and Amaldi, F., 1985, Nucleotide sequence of the L1 ribosomal protein gene of *Xenopus laevis*: remarkable sequence homology among introns, *EMBO J.* **4**:3483–3488.

Meyuhas, O., and Perry, R. P., 1980, Construction and identification of cDNA clones for mouse ribosomal proteins: Application for the study of r-protein gene expression, *Gene* **10**:113–129.

Miller, L., 1974, Metabolism of 5S RNA in the absence of ribosome production, *Cell* **3**:275–281.

Monk, R. J., Meyuhas, O., and Perry, R. P., 1981, Mammals have multiple genes for individual ribosomal proteins, *Cell* **24**:301–306.

Nomura, M., Gourse, R., and Baughman, G., 1984, Regulation of the synthesis of ribosomes and ribosomal components, *Annu. Rev. Biochem.* **53:**75–118.

Pelham, H. R. B., and Brown, D. D., 1980, A specific transcription factor that can bind either the 5S RNA gene or 5S RNA, *Proc. Natl. Acad. Sci. USA* **77:**4170–4174.

Picard, B., and Wegnez, M., 1979, Isolation of 7S particle from *Xenopus laevis* oocytes: A 5S RNA-protein complex, *Proc. Natl. Acad. Sci. USA* **76:**241–245.

Pierandrei-Amaldi, P., and Beccari, E., 1980, Messenger RNA for ribosomal proteins in *Xenopus laevis* oocytes, *Eur. J. Biochem.* **106:**603–611.

Pierandrei-Amaldi, P., Campioni, N., Beccari, E., Bozzoni, I., and Amaldi, F., 1982, Expression of ribosomal protein genes in *Xenopus laevis*, *Cell* **30:**163–171.

Pierandrei-Amaldi, P., Campioni, N., Gallinari, P., Beccari, E., Bozzoni, I., and Amaldi, F., 1985a, Ribosomal protein synthesis is not autogenously regulated at the translational level in *Xenopus laevis*, *Dev. Biol.* **107:**281–289.

Pierandrei-Amaldi, P., Beccari, E., Bozzoni, I., and Amaldi, F., 1985b, Ribosomal protein production in normal and anucleolate *Xenopus* embryos: Regulation at the post-transcriptional and translational levels, *Cell* **42:**317–323.

Reeder, R. H., and Roan, J. G., 1984, The mechanism of nucleolar dominance in *Xenopus* hybrids, *Cell* **38:**39–44.

Richter, J. D., and Smith, L. D., 1983, Developmentally regulated RNA binding proteins during oogenesis in *Xenopus laevis*, *J. Biol. Chem.* **258:**4864–4869.

Richter, J. D., and Smith, L. D., 1984, Reversible inhibition of translation by *Xenopus* oocyte-specific proteins, *Nature (Lond.)* **309:**378–380.

Richter, J. D., Smith, L. D., Anderson, D. M., and Davidson, E. H., 1984, Interspersed poly(A) RNAs of amphibian oocytes are not translatable. *J. Mol. Biol.* **173:**227–241.

Rosbash, M., and Ford, P. J., 1974, Polyadenylic acid-containing RNA in *Xenopus laevis* oocytes, *Dev. Biol.* **85:**87–101.

Sagata, N., Shiokawa, K., and Yamana, K., 1980, A study on the steady-state population of poly(A)$^{+}$ RNA during early development of *Xenopus laevis*, *Dev. Biol.* **77:**431–438.

Santon, J. B., and Pellegrini, M., 1981, Rates of ribosomal protein and total protein synthesis during *Drosophila* early embryogenesis, *Dev. Biol.* **85:**252–257.

Smith, L. D., and Richter, J. D., 1984, Synthesis, accumulation, and utilization of maternal macromolecules during oogenesis and oocyte maturation, in: *Biology of Fertilization*, Vol. 1 (C. B. Metz, and A. Monroy, eds.), pp. 141–188, Academic, New York.

Sturgess, E. A., Ballantine, J. E. M., Woodland, H. R., Mohun, T. R., Lane, C. D., and Dimitriades, G. J., 1980, Actin synthesis during the early development of *Xenopus laevis*, *J. Embryol. Exp. Morphol.* **58:**303–320.

Taylor, M. A., and Smith, L. D., 1985, Quantitative changes in protein synthesis during oogenesis in *Xenopus laevis*, *Dev. Biol.* **110:**230–237.

Warner, J. R., 1977, In the absence of ribosomal RNA synthesis, the ribosomal proteins of HeLa cells are synthesized normally and degraded rapidly, *J. Mol. Biol.* **155:**315–333.

Warner, J. R., Mitra, G., Schwindinger, W. F., Studeny, M., and Fried, H. M., 1985, *Saccharomyces cerevisiae* coordinates accumulation of yeast ribosomal proteins by modulating mRNA splicing, translational initiation, and turnover, *Mol. Cell. Biol.* **5:**1512–1521.

Weiss, Y. C., Vaslet, C. A., and Rosbash, M., 1981, Ribosomal protein mRNAs increase dramatically during *Xenopus* development, *Dev. Biol.* **87:**330–339.

Woodland, H. R., 1974, Changes in the polysome content of developing *Xenopus laevis* embryos, *Dev. Biol.* **40:**90–101.

Woodland, H. R., Flynn, J. M., and Wylie, A. J., 1979, Utilization of stored mRNA in *Xenopus* embryos and its replacement by newly synthesized transcripts: Histone H1 synthesis using interspecies hybrids, *Cell* **18:**165–171.

Wormington, W. M., and Brown, D. D., 1983, Onset of 5S RNA gene regulation during *Xenopus* embryogenesis, *Dev. Biol.* **99:**248–257.

Chapter 9

Regulation of Expression of *Xenopus* Vitellogenin Genes

JAMSHED R. TATA

The assembly of the developing egg is a fascinating example of division of labor among oocytes and other tissues, with the coordination of activities in different tissues being achieved by hormones. Among nonoocyte tissues are (1) the **follicle cells,** which, under the influence of follicle-stimulating hormone (FSH), synthesize and release the sex steroids estrogen and progesterone; (2) the **oviduct,** which produces egg white proteins, coat proteins, or egg-sustaining fluids and whose activity is regulated by both steroid hormones; and (3) the **liver,** which, under the sole control of estrogen, provides the developing egg with yolk proteins, cholesterol, phospholipids, and lipoproteins. The participation of the oviduct and liver is particularly pronounced in nonmammalian species with free-living embryos, as in birds and amphibia.

During oogenesis in invertebrates and oviparous vertebrates, the gradual deposition of yolk proteins, a process known as **vitellogenesis,** has been commonly used as an index of egg development. With the rapid introduction of methods of recombinant DNA, monoclonal antibodies, and *in situ* localization of macromolecules, the study of vitellogenesis has yielded important information on the developmental and hormonal regulation of gene expression. This chapter deals with diverse aspects of the nature of genes encoding yolk proteins (**vitellogenin genes**), the characteristics of their expression, and how the latter is hormonally regulated. It is largely based on studies on *Xenopus* vitellogenin genes, particularly how their activation *de novo* in hepatocytes from male animals has helped clarify some important mechanisms underlying steroid hormonal regulation of gene expression. The reader is also referred to Volume 1 of this series (Browder, 1985) as well as other general reviews on egg development (Metz and Monroy, 1985; Tata *et al.*, 1986) and to other articles that deal specifically with *Xenopus* vitellogenin gene expression (Tata and Smith, 1979; Wahli *et al.*, 1981; Ryffel and Wahli, 1983; Tata, 1985; Wallace, 1985; Tata *et al.*, 1986a).

JAMSHED R. TATA • Laboratory of Developmental Biochemistry, National Institute for Medical Research, Mill Hill, London NW7 1AA, England.

1. Vitellogenesis

It is essential that the mature egg have all its essential components as-sembled in a precisely timed sequence, if it is to achieve successful fertilization and early embryonic development. In general, the oocyte itself produces all the components of protein and DNA synthesis, including a large repertoire of ma-ternal messenger RNAs (mRNAs), for use immediately after fertilization. All vertebrate and most invertebrate eggs contain liver or fat-body-derived yolk made up of proteins, lipid and carbohydrate, but its content can vary enor-mously from one species to another. As a rule, it represents the major bulk of the egg mass in oviparous organisms, where it serves to nourish both the devel-oping egg and early embryo (Wallace et al., 1983).

1.1. Hormonal Regulation of Vitellogenesis

The first signal for the initiation and regulation of egg development origi-nates in the central nervous system (CNS) in response to environmental stimuli (Tata, 1984). The latter include changes in length of daylight, temperature, and pheromones. These are conveyed by neurotransmitter substances to the hy-pothalamus or other brain cells and, via neurohormones, eventually to endo-crine glands such as the ovary and prothoracic gland, which synthesize and secrete hormones that act directly on cells that make egg proteins. Figure 1 summarizes the cascade of neurohormonal regulatory elements in vertebrate vitellogenesis.

Two important features of induction of vitellogenesis by estrogen in verte-brate liver are worth considering. First, a single administration of estrogen to male animals causes the synthesis and secretion into the blood of large quan-tities of yolk proteins and lipids. This ability to activate the production of egg components in male liver is unique and explains why vitellogenesis is in-creasingly popular as a model for studying hormonal regulation of gene ex-pression. Second, the memory effect, or the primary and secondary induction of vitellogenesis, is particularly amenable to experimental analysis. The initial vitellogenic response to a single dose of estrogen decays upon withdrawal of the hormone. A second dose produces a more rapid and massive formation of yolk proteins and lipids, a phenomenon common to many inducible responses. This facilitates definition of the roles of such factors as hormone receptors, changes in gene configuration, translational efficiency, and secretory mecha-nisms underlying hormonally regulated developmental processes.

1.2. Vitellogenin

Although the egg yolk proteins, **phosvitin** and **lipovitellin,** have been in-tensively studied for the past 50 years, particularly in birds, it is only during the last 10 or 12 years that we know that these are not individually synthesized

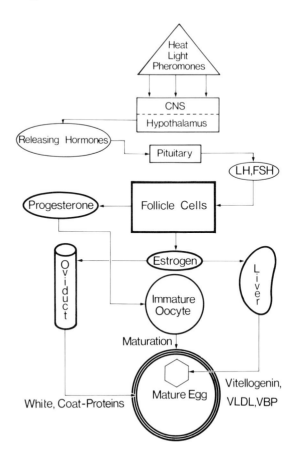

Figure 1. Schematic outline of the integration of various hormone-regulated functions participating in the process of oocyte growth and maturation. CNS, central nervous system; LH, luteinizing hormone; FSH, follicle-stimulating hormone; VBP, vitamin-binding proteins; VLDL, very-low density lipoprotein.

but are derived from a common precursor called **vitellogenin.** The term vitellogenin was first used to refer to all plasma precursors of egg yolk proteins in insects (Pan *et al.*, 1969). Vitellogenins are almost exclusively synthesized in the liver or fat body of oviparous animals, secreted into the blood, and transported to the ovary, where they are cleaved and processed into the yolk components.

   The characteristics of plasma vitellogenins are highly similar in diverse species, as illustrated in Table I for the chicken, *Xenopus*, and locust. In all species, each of the two subunits is $210–250 \times 10^3$ M_r in weight and has a comparable composition in carbohydrate, lipid, and phosphorus (Clemens, 1974; Tata and Smith, 1979). This is true even in invertebrates, whose eggs lack the phosphorus-rich glycoprotein phosvitin and contain only the lipid-rich lipovitellins. The sequencing of complementary DNA (cDNA) to vitellogenin mRNA will eventually provide accurate information on the organization of phosvitin and lipovitellin within the vitellogenin molecule in view of the 50–60% of serine residues in the phosvitin moiety compared with only 5% in lipovitellin. Whatever the internal organization of the vitellogenin molecule, large parts of the molecule have been conserved during evolution, as demon-

Table I. Similarity of Amphibian, Avian, and Insect Vitellogenins[a,b]

Species	Subunit (kDa)	Mol. %		Composition (%)		
		Ser	Met	P	Lipid	CHO
Xenopus	215	12	2	1.6	12	2
Chicken	240	14	2	2.4	12	2
Insect[c]	260	10	2	1.5	15	3

[a]From Tata and Smith (1979).
[b]Abbreviations: Ser, Serine; Met, methionine; P, phosphorus; CHO, carbohydrate.
[c]Average of values for locust and silkmoth.

strated by immunologic cross-reaction and DNA–DNA and RNA–DNA cross-hybridization.

1.3. Advantages of Studying Hormonal Regulation of Vitellogenesis

Among several advantageous features of induced vitellogenesis, perhaps the most important is the possibility of inducing the *de novo* synthesis of vitellogenin with a single administration of estrogen to male animals. This offers a unique opportunity to study the activation of permanently silent genes. Since the process is reversed after withdrawal of the hormone, one can easily study both the induction and deinduction processes. In *Xenopus* hepatocytes, the physiological primary and secondary inductions have been fully reproduced in primary cell cultures, which is another major advantage in analyzing the early events leading to hormonal regulation of gene expression. At the same time, neither the primary nor secondary induction at early stages is complicated by overlapping changes in cell proliferation or DNA synthesis. Finally, the multicomponent nature of vitellogenin and its highly specific uptake and cleavage into egg yolk proteins in the oocyte make it a particularly attractive model for studying endocytosis and precursor-product processing.

2. The Vitellogenin Multigene Family

Many developmentally and hormonally regulated genes belong to small multigene families, such as those coding for globin, ovalbumin, $\alpha_{2\mu}$-globulin, and α-amylase. Vitellogenin genes also exhibit multiplicity, chicken and *Xenopus* vitellogenin genes being encoded by three and four members, respectively (Ryffel and Wahli, 1983). Thanks to the work of Wahli's group (Wahli *et al.*, 1979, 1980, 1981, 1982; Walker *et al.*, 1983, 1984; Germond *et al.*, 1984), more is known about the structure and organization of vitellogenin genes of *Xenopus* than of any other species, although an increasing amount of information is becoming available for chicken and some invertebrate vitellogenin genes.

2.1. The *Xenopus* Vitellogenin Gene Family

The four actively expressed vitellogenin genes of *Xenopus laevis* fall into two groups and are termed A1, B1, A2, and B2 (Wahli *et al.*, 1979) (Fig. 2). There is 80% homology of coding sequence between the two groups and 95% between two member genes of each group. Each is made up of 34 similar exons, but the size of the entire gene varies from 16 to 21 kilobase pairs (kbp) of DNA because of large variations in intron sizes. The A1 and B1 genes are linked by 15 kbp of DNA, a fact of some interest in the context of differential rates of activation of the two genes (see Section 3.3), but it is not known whether the other two genes are similarly linked (Wahli *et al.*, 1982). Sequencing of the 5′ upstream flanking regions in Wahli's laboratory (Germond *et al.*, 1984; Walker *et al.*, 1984) has demonstrated similarities and differences within the gene family that provide clues to the possible regulatory regions with which the estrogen receptor complex may interact (see Section 4.2).

Comparison of vitellogenin genes of *Xenopus laevis* with those of the more ancient *X. tropicalis* has shed some interesting light on the evolution of these genes. The genome size of *X. tropicalis* is one-half that of *X. laevis*, with only two active vitellogenin genes (one each of A and B type). It can be argued that after a very early duplication of the primordial vitellogenin gene, there took place a more recent duplication of the whole genome resulting in four genes in modern *Xenopus* species (Jaggi *et al.*, 1982). A parallel comparison of the vitellogenin genes, their mRNAs, and protein products in the more closely related *X. laevis* and *X. borealis* and the distantly related *X. tropicalis* has also been useful (Baker *et al.*, 1985). Whereas the overall coding sequences were highly conserved in these three species, restriction endonuclease digestion analysis of the DNA and peptide mapping of the plasma vitellogenins demonstrates significant differences in the organization of the genes. It was concluded that there must have occurred significant rearrangement within the vitellogenin gene family during evolution and that there is room for flexibility within the protein itself.

2.2. Comparative Aspects

The coding and 5′ flanking sequences of *Xenopus* and chicken vitellogenin genes have been compared; it is not altogether surprising that there is a high degree of overall sequence conservation (James *et al.*, 1982; Germond *et al.*, 1984; Walker *et al.*, 1983, 1984). What is more remarkable is the similarity in gene size, structure, and cross-hybridization one observes in vitellogenin genes among insects, nematodes, birds, amphibia, and fish (Ryffel and Wahli, 1983; Wyatt *et al.*, 1984; Spieth *et al.*, 1985). Certain domains of the protein, such as sites interacting with oocyte receptors, cleavage points, and tertiary structure, may vary in position within the molecule but have to be retained with a high degree of fidelity. Small stretches of DNA sequences in the 5′ upstream flanking regions have also been conserved, both as to number and relative position,

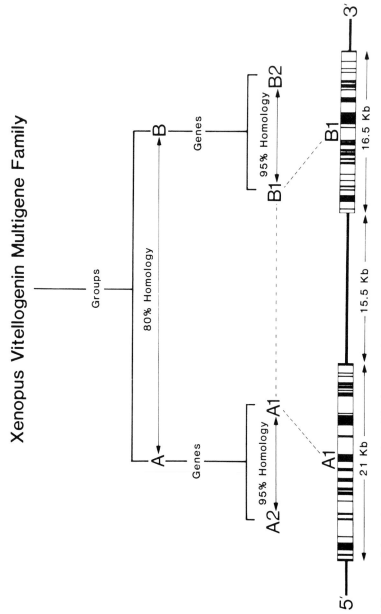

Figure 2. Schematic representation of the four actively expressed vitellogenin genes in *Xenopus laevis* and their classification as two members each of the A and B groups. The percentages denote the coding sequence homology between individual genes and the two groups. (From Wahli *et al.*, 1979.)

in the more actively expressed *Xenopus* A1 and B1 genes and the chicken Vtg II gene (Walker *et al.*, 1984). This indicates the important role of upstream regions in the regulation of gene expression by estrogen (see Section 4).

3. Hormonal Regulation of Expression of *Xenopus* Vitellogenin Genes

3.1. Induction in Whole Animals

Under normal physiological conditions, once the estrogen titer in the blood of female *Xenopus* has been elevated in response to neural stimuli, the liver secretes large amounts of vitellogenin, which is rapidly taken up by the oocytes (see Fig. 1). It is not easy to follow the early stages of hormonal induction of vitellogenin synthesis in the female, however. The complete absence of any vitellogenin-like material or its mRNA in the adult male vertebrate liver or blood and the induction of massive amounts of yolk protein in the male by a single administration of estrogen obviates this difficulty. As in the rooster and other oviparous male vertebrates, a prolonged administration of large doses of estradiol to male *Xenopus* causes the accumulation of such vast quantities of yolk phosphoprotein that it becomes the major protein in the blood (Follett and Redshaw, 1968; Wallace and Jared, 1968).

The first study of the memory effect in roosters was vitiated by the lack of a sensitive assay for vitellogenin in the blood, but a time-course analysis indicated that the secondary response had a shorter lag and reached a higher magnitude than after the first hormonal administration (Beuving and Gruber, 1971; Jailkhani and Talwar, 1972). With the availability of specific antibodies against *Xenopus* vitellogenin, it became possible to demonstrate the memory effect in male *Xenopus* in a more quantitative fashion (Farmer *et al.*, 1976) (Fig. 3). After the primary response dies away, the second administration of estradiol causes a rapid and massive buildup of circulating vitellogenin to greatly exceed the level of circulating albumin. Examination of Fig. 3 also highlights another interesting feature of induced vitellogenesis, namely the virtual disappearance of albumin from the blood. This phenomenon of deinduction of albumin synthesis that accompanies estrogen induction of vitellogenesis in *Xenopus* liver has recently received much attention (Wolffe *et al.*, 1985; Kazameier *et al.*, 1985; Riegel *et al.*, 1986).

When cDNA to vitellogenin mRNA became available, the characteristics of hormonal induction of vitellogenin synthesis could be studied more throughly at the gene level. From a careful hybridization analysis, Baker and Shapiro (1977) established that no vitellogenin mRNA could be detected in the male *Xenopus* liver and that the differences in primary and secondary responses seen in circulating vitellogenin were preceded by parallel changes in total vitellogenin mRNA in male *Xenopus* liver. It was also shown that these mRNA levels exceeded those of albumin and that the kinetics and extent of accumulation of vitellogenin mRNA during secondary induction in male and female

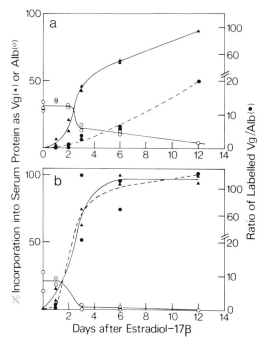

Figure 3. Immunological quantitation of [^{35}S]methionine-labeled vitellogenin and albumin in the serum of male *Xenopus* as a function of time during the onset of primary and secondary responses to estradiol. (●) albumin; (▲) vitellogenin; (○) ratio of incorporation into vitellogenin to that into albumin. (a) Primary response. (b) Secondary response. (From Farmer *et al.*, 1978. Reprinted by permission from *Nature*, Vol. 273, No. 5661, pp. 401–403. Copyright © 1978 Macmillan Journals Limited.)

liver were similar. However, these studies in whole animals have serious drawbacks if one is to address the central issue of the mechanisms underlying the role of hormone–receptor interactions with the induced gene. In particular, it is impossible to control reproducibly the amount of hormone available to the liver and to establish the early events leading to gene activation. For these reasons, the author's laboratory has devoted much attention to studying the regulation of vitellogenin gene expression in primary cell cultures.

3.2. Regulation of Expression of Vitellogenin Genes in Primary Cultures of Male *Xenopus* Hepatocytes

3.2.1. Advantages and Disadvantages of Primary Cell Culture

Among the many advantages of cell culture over whole animals in studying hormonal regulation of gene expression are (1) the ability to control accurately hormone concentration; (2) better analysis of the early events associated with hormone–receptor interaction with regulatory elements of the gene; (3) rapid reversibility of induction by removal of the hormone, enabling the study of the deinduction process; and (4) the analysis of single cell types in heterogeneous tissues pooled from many animals, thus reducing variability and interference by noncompetent cells.

Among the disadvantages of primary cell culture are (1) the inconvenience of preparing fresh cell cultures for each experiment; (2) the variability of responses; and (3) the low level or lack of response during a refractory period in freshly prepared cultures. The first two drawbacks were largely obviated by working at low cell density, plating, and allowing for the rapid metabolism of estrogens by hepatocytes (Tenniswood *et al.*, 1983). The last disadvantage (i.e., the refractory period) is serious and is applicable to all primary cell culture systems in which induction is studied (Wolffe and Tata, 1984). This problem, also termed **culture shock,** has been studied in detail in the author's laboratory. This research has enabled reproduction of the physiological regulation of *Xenopus* vitellogenin gene expression in tissue culture.

3.2.2. Culture Shock and the Physiological Hormonal Response *in Vitro*

Many workers have observed that freshly prepared primary cell cultures respond poorly to various stimuli, including hormones, nutrients, and drugs, (Wolffe and Tata, 1984). The response, as measured by the inducibility of specific cellular products, usually improves after the first 2–3 days in culture, but the cause of this refractoriness remained unknown until quite recently. In the course of work on activation of vitellogenin genes in *Xenopus* hepatocytes by estrogen, it was noticed that the stress of isolation of cells from various tissues results in a large accumulation of stress or heat shock proteins, particularly hsp 70 (Wolffe *et al.*, 1984a). The synthesis of stress proteins declines with time in culture; the ability of estrogen to activate vitellogenin genes in cultured male *Xenopus* hepatocytes is a reciprocal function of the amount of hsps in the cells.

In other studies (Wolffe *et al.*, 1984b), the experimental induction of hsp's by thermal shock was exploited as a tool in manipulating various parameters of activation and regulation of vitellogenin gene transcription and of estrogen receptor levels. But it is the recognition of the culture shock phenomenon that made it possible to reproduce in primary cell cultures the *de novo* activation of vitellogenin genes and to maintain the accumulation of vitellogenin mRNA at high rates for several days in the continued presence of estrogen. The Northern blot analysis (Fig. 4) illustrates this point. In these experiments, in which freshly prepared cells were allowed to overcome culture shock for 3 days, total RNA was extracted from one batch of primary male *Xenopus* hepatocyte cultures at different times after the first addition of 10^{-7} M estradiol (primary induction). Another batch of cells, to which estradiol was added at the same time, was kept without any hormone for the next 2 days when vitellogenin mRNA synthesis had ceased. At this time, the same amount of estradiol was added, and aliquots of cells were taken for RNA extraction at different times after hormone addition (secondary induction). The RNA samples were resolved by gel electrophoresis, the RNA transferred by blotting, and vitellogenin mRNA visualized by hybridization to ^{32}P-labeled cloned *Xenopus* vitellogenin cDNA. In addition to demonstrating the more pronounced secondary response or memory phenomenon, these experiments show that the physiological charac-

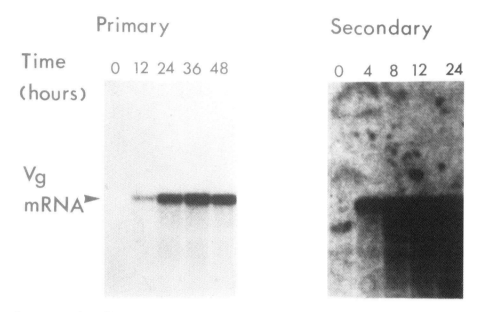

Figure 4. Northern blot analysis of RNA from primary cultures of male *Xenopus* hepatocytes during primary and secondary response to estrogen *in vitro*. RNA was extracted at different times (hours) indicated after the first addition of the hormone, which was then maintained continuously, resolved by electrophoresis, and the vitellogenin mRNA identified by hybridization to all four ^{32}P-labeled *Xenopus* vitellogenin cDNAs, followed by autoradiography.

teristics of primary and secondary inductions are also produced in culture if the cells are permitted sufficient time to recover from the stress of placing them in culture.

3.2.3. Simultaneous Regulation of Transcription and Stability of Vitellogenin mRNA

It has been known for some time from whole animal studies that, following a single administration of estrogen or upon withdrawal of the hormone, both vitellogenin in blood and vitellogenin mRNA in hepatocytes rapidly disappear as the hormone is lost (Farmer *et al.*, 1976; Baker and Shapiro, 1977). Although it was considered most likely that the rapid buildup and decay of vitellogenin mRNA during primary and secondary induction in male hepatocytes were due to a combined effect of the hormone on transcription and stability of the mRNA, the contribution of each process could not be easily established in whole animals but was accurately determined in primary hepatocyte cultures.

The absolute transcription rate and steady-state levels of A and B groups of vitellogenin genes activated *de novo* by estrogen in cultures of *Xenopus* hepatocytes are shown in Fig. 5. In these experiments, the absolute rate of transcription of vitellogenin genes in primary cultures of naive male hepatocytes was measured by labeling RNA with [³H]uridine and hybridization on filter disks to a vast excess of unlabeled A and B group vitellogenin cDNA, as a function of time after the addition of estradiol to the cultures. The steady-state levels, which indicated continuous accumulation of vitellogenin mRNA, were determined by hybridization of total unlabeled RNA, extracted from cells at different times after the addition of estradiol, with ³²P-labeled cloned A and B group vitellogenin cDNA. (For further experimental details, see Wolffe and Tata, 1983.) When the stability of vitellogenin mRNA was measured in the continuous presence or withdrawal of estrogen from cultures, it was obvious that the presence of the hormone stabilized vitellogenin mRNA. Thus, whereas the half-life (t½) of vitellogenin mRNA in the continuous presence of estrogen was >48 hr, the removal of the hormone caused this value to fall to >16 hr (Fig. 6a). Similar results had been described by Brock and Shapiro (1983). What is of particular significance is that the stabilization by estrogen is specific for the induced mRNA. Indeed, estrogen has the opposite effect on the constitutively expressed albumin mRNA (Fig. 6b). In the presence of estrogen, albumin mRNA decays more rapidly than upon its withdrawal (Wolffe et al., 1985). This latter effect may in part explain the deinduction of albumin that accompanies induction of vitellogenin synthesis (Farmer et al., 1976; Tata and Smith, 1979; Kazamaier et al., 1985; Riegel et al., 1986). The mechanism underlying the regulation of mRNA stability remains unknown.

The simultaneous enhancement of both transcription and stabilization of mRNA has also been observed for the induction of ovalbumin mRNA in chicken oviduct by estrogen, that of casein mRNA in rat mammary gland by prolactin, and in other induction systems (Shapiro and Brock, 1985). In general, stabilization of mRNA makes a more substantial contribution to the rapid accumulation and decay of mRNA upon hormone administration and withdrawal

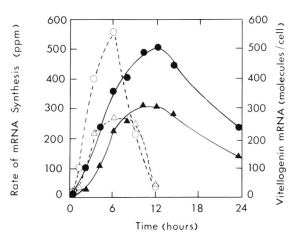

Figure 5. Relative rate of vitellogenin mRNA synthesis compared with the accumulation and disappearance of vitellogenin mRNA upon stimulation by a single dose of 1 μM 17β-estradiol added to naive male *Xenopus* hepatocytes. The relative rate of transcription (- - -) of vitellogenin mRNA corresponding to group A (△) or group B (○) vitellogenin genes was determined by hybridization of ³H-labeled RNA to the cloned A and B group cDNAs. The accumulation curve (—) of vitellogenin mRNA corresponding to the A (▲) or B (●) group was expressed as parts per million (ppm) of total RNA. (From Wolffe and Tata, 1983.)

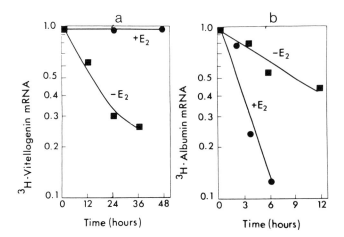

Figure 6. (a) Rate of disappearance of estrogen-induced vitellogenin mRNA in the presence and absence of estrogen. Male hepatocytes were incubated for 12 hr at 26°C with a single addition of 10^{-6} M estradiol (E_2) in culture medium containing [³H]uridine. The cultures were then transferred to fresh medium containing 10^{-6} M estradiol, replenished every 4 hr (●). Another batch of hepatocytes (■) was transferred to fresh estradiol-free culture medium. RNA was extracted at the times indicated, and the amount of radioactive vitellogenin mRNA present per unit mass was determined by hybridization to cloned vitellogenin cDNA. (b) Destabilization of the 74-kDa albumin mRNA in Xenopus hepatocytes exposed to estradiol. Male Xenopus hepatocyte cultures were incubated for 11 hr without estrogen in 2 ml culture medium containing [³H]uridine. The cultures were then transferred to fresh medium either containing 1 μM estradiol, replenished every 4 hr (●), or not (■). RNA was extracted at the times indicated, and the amount of radioactive 74,000-M_r albumin mRNA per unit mass was determined by hybridization to cloned 74,000-M_r albumin cDNA. (Adapted from Wolffe et al., 1985.)

than does the enhancement or cessation of transcription. In any case, there are no examples of massive changes in mRNA levels, such as the induction and deinduction of vitellogenin mRNA, that are exclusively attributable to effects on either transcription or stability of mRNA. Rather, it is becoming increasingly clear that some as yet undiscovered mechanism controlling specific mRNA stability is simultaneously mobilized upon the enhancement or activation of transcription during the presence or absence of the induction signal.

3.3. Differential Hormonal and Developmental Activation of Individual *Xenopus* Vitellogenin Genes

Studies based on *in vitro* translation of mRNA from livers of chronically estrogen-treated female *Xenopus* indicated that the amount of mRNA encoded by the individual vitellogenin genes is the same (Felber *et al.*, 1980). However, it was not known how each of the four genes would be activated by the hormone at the onset of *de novo* synthesis of vitellogenin in male hepatocytes. To test the possibility of a differential activation of these highly homologous genes, a high strigency disk hybridization assay was devised to analyze the

RNA extracted at early periods after the addition of estradiol to primary cultures of male hepatocytes (Ng *et al.*, 1984).

The results of these experiments quite clearly showed that the transcription and steady state levels of individual vitellogenin mRNAs are not regulated coordinately or to the same extent. The rate and extent of accumulation of vitellogenin mRNA vary in the order of B1 > A1 >A2 ≅ B2 after addition of hormone to both adult male and female *Xenopus* hepatocytes (Fig. 7). This pattern is flexible in that the differential activation is enhanced or attenuated by varying the period of exposure to the hormone or its concentration. That the differential rate of accumulation of mRNA at the onset of induction is a reflection of unequal rates of transcription was verified by directly measuring the absolute rate of transcription of the individual vitellogenin genes. The same pattern of B1 > A1 > A2 ≅ B2 was again observed. This is compatible with earlier findings in primary cultures (Wolffe and Tata, 1983) and in nuclear runoff transcription and DNase I sensitivity measurements in whole liver (Williams and Tata, 1983) that the B group genes are activated to a greater extent than A group genes.

How early in development is this pattern of differential activation within a gene family seen in adult cells established? It was known that vitellogenin

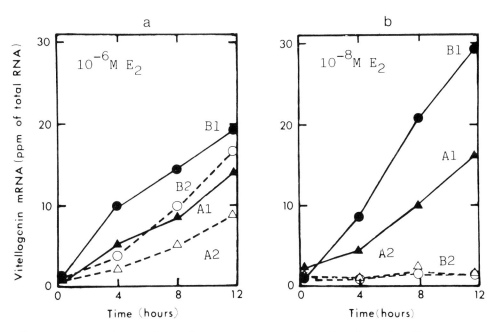

Figure 7. Kinetics of accumulation of transcripts corresponding to the individual vitellogenin genes in male *Xenopus* hepatocyte cultures after primary induction with 17β-estradiol. Vitellogenin mRNA corresponding to genes A1 (▲), A2 (△), B1 (●), and B2 (○) was quantitated by filter disk hybridization to ^{32}P-nick translated HindIII excised cDNA. (a) Estradiol 10^{-6} M was added once at time zero. (b) Estradiol 10^{-8} M was replenished in the culture medium every hour over 12 hr. (From Ng *et al.*, 1984.)

could be detected immunologically in the blood at late metamorphic stages of *Xenopus* tadpoles or froglets immersed in water containing estrogen (Knowland, 1978; Huber *et al.*, 1979). Hybridization analysis showed that larval hepatocytes acquire competence to synthesize vitellogenin mRNA in response to the hormone by Nieuwkoop–Faber stage 61 (i.e., estrogen receptor is present in hepatocytes at least by late metamorphosis; Ng *et al.*, 1984). Measurement of individual vitellogenin gene transcripts in metamorphosing tadpole liver displayed the same relative pattern of expression as in adult hepatocytes, i.e., gene B1 > A1 > A2 ≅ B2, at the earliest stages of activation of these dormant genes. Thus, the unequal pattern of expression is maintained throughout life, although the absolute rate of transcription of each gene increases rapidly between late metamorphic and froglet stages.

What can be the underlying mechanism governing the differential activation of this closely related family of genes? The most likely explanation may be different promoter strengths or variable intensities of interaction between estrogen receptor or other transcription factors and gene sequences bearing regulatory elements. It is therefore most relevant to consider the organization and structure of the 5′ flanking regions of the four vitellogenin genes (Walker *et al.*, 1984). All four *Xenopus* vitellogenin genes have one or two blocks of the palindromic sequences GGTCANNNTGACC between −310 and −375 bp in the 5′ upstream region (Fig. 8). However, the linked genes A1 and B1, which are

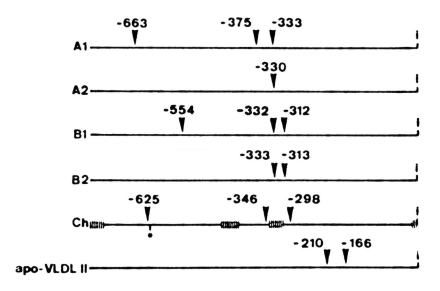

Figure 8. Distribution of GGTCANNNTGACC element in DNA sequence 5′ upstream from transcription initiation site of *Xenopus* vitellogenin genes A1, A2, B1, and B2, as well as in chicken vitellogenin II and apoVLDL genes. The position (bp from transcription initiation site) of the palindromic sequence is indicated by arrows. (Adapted from Walker *et al.*, 1984.)

more strongly expressed than the pair A2 and B2 (see Fig. 7), have an additional element further upstream at -663 and -554 bp, respectively. The chicken vitellogenin gene Vtg II, which is more strongly expressed than the other two vitellogenin genes, also has three such elements at similar locations 5′ upstream from the transcription initiation site. It thus seems that the differential expression of the *Xenopus* vitellogenin genes may arise from the cooperative interaction between more than one regulatory element in the upstream flanking sequences and positive transcription regulatory factors, including estrogen receptors.

4. The Role of Estrogen Receptor in Vitellogenin Gene Transcription

It is widely accepted that the regulation of transcription of specific genes in steroid target cells is a function of the level of hormone receptor (Eriksson and Gustafsson, 1983). However, there is no direct evidence for this to be generalized as a principle. The fact that estrogen activates *de novo* the permanently silent vitellogenin genes in male *Xenopus* hepatocytes and that the process can be reversibly regulated in culture, as occurs physiologically in the female, offers a unique opportunity to obtain direct evidence.

4.1. Estrogen Receptor Levels: Upregulation and Vitellogenin Gene Transcription

Adult male *Xenopus* liver has very low levels of estrogen receptor, comprising less than 200 molecules tightly bound to the nucleus per cell or about 500 molecules per cell in total (Westley and Knowland, 1978; Hayward *et al.*, 1980; Perlman *et al.*, 1984). Treatment of male *Xenopus* with estrogen or addition of the hormone to hepatocyte cultures causes an upregulation or a 5–10-fold increase in specific nuclear receptor, reaching levels found in female *Xenopus* liver. This elevated level of receptor persists for several weeks, perhaps partly explaining the more rapid vitellogenic response to the hormone during secondary induction.

Equally important are the findings correlating nuclear estradiol receptor levels with the absolute rate of transcription of vitellogenin genes as a function of time after estradiol addition to male hepatocyte cultures (Perlman *et al.*, 1984). Figure 9 clearly shows for the first time a stoichiometric relationship between nuclear receptor and activation of dormant vitellogenin genes. If the protein synthesis inhibitor cycloheximide is added at different times to the cell cultures, it is also seen that, after a slight initial activation of transcription due to preexisting receptor in untreated male hepatocytes, the sustained activation of the genes requires continuing protein synthesis. The recent successes in cloning estrogen receptor cDNA (Green *et al.*, 1986; Greene *et al.*, 1986) will not only provide important insights into the nature of the receptor but will also

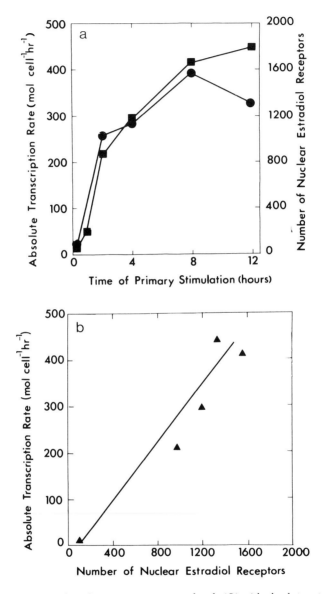

Figure 9. (a) Correlation of nuclear estrogen receptor levels (●) with absolute rates of vitellogenin gene transcription (■) in male *Xenopus* hepatocytes as a function of time (hr) after the addition of estradiol. (b) Relationship between nuclear estrogen receptor and absolute rate of transcription of vitellogenin genes in male *Xenopus* hepatocyte cultures after various periods of primary exposure to the hormone as in (a). (From Perlman et al., 1984.)

make it possible to study how the hormone regulates the expression of the gene encoding its own receptor.

4.2. Steroid Receptor and Regulatory Sequences on Target Genes

Recent studies with cloned genomic DNA or target genes have permitted identification of 5′ upstream regions flanking genes that may be likely sites of interaction with steroid hormone receptor (Eriksson and Gustafsson, 1983; Chambon *et al.*, 1984; Renkawitz *et al.*, 1984; Payvar *et al.*, 1984). Footprinting procedures have revealed blocks of sequences between −100 and −700 bp upstream from transcription sites of genes such as ovalbumin, lysozyme, uteroglobin, and mouse mammary tumor virus. A consensus sequence between −458 and −725 bp in the 5′ flanking region of the chicken vitellogenin gene is the apparent site of interaction with estrogen receptor (Jost *et al.*, 1984), whereas an enhancer sequence at a 5′ upstream DNase hypersensitive site has also been reported in this gene (Burch, 1984). As regards the *Xenopus* vitellogenin gene family, we have already noted (see Fig. 8) that although all four genes of this family have 13-nucleotide-long palindromic sequence blocks at −310 to −375 bp 5′ upstream, only the more actively expressed A1 and B1 genes have an additional element further upstream. It will be most relevant to our understanding of steroid receptor function to determine if this extra site enhances the strength of interaction of receptor with all the possible sites of the regulatory DNA sequences taken together. It is, however, important to note that DNA sequences *per se* may not be the only determinants of steroid hormonal regulation of gene expression. We know very little about such factors as the influence of the higher order organization of genes in chromatin structure or the effect of distribution of different members of gene families on different chromosomes on gene activity.

5. Switching on Silent Vitellogenin Genes *in Vitro*

5.1. Regulation of Transcription in Isolated Nuclei

Recently, there have been successful attempts at mimicking the induction of gene transcription with soluble extracts of nuclei or cytoplasm added to nuclei or nuclear extracts. These include the activation of transcription of genes encoding silk fibroin, globin, heat shock proteins, and adenovirus (Tsuda and Suzuki, 1983; Emerson *et al.*, 1985; Dynan and Tjian, 1985) by extracts of cells in which these genes are actively transcribed. However, in all these systems the test gene is transcribed constitutively at a low rate (i.e., it is in an open configuration), and the soluble extracts modulate the rate of transcription. The completely silent vitellogenin genes in adult male *Xenopus* liver offer a most suitable model to study the process of switching on of silent genes *in vitro*.

For the above reason, and in order to test the roles played by estrogen receptor and other positive transcriptional factors, the author's laboratory has recently undertaken a detailed analysis of the effects of tissue extracts on the *de novo* activation of vitellogenin genes in isolated male *Xenopus* hepatocyte nuclei (Tata and Baker, 1985). By paying particular attention to the amount of a soluble extract (S-100, which is the $100,000 \times$ g supernatant of the postnuclear fraction) added to the nuclei and a period (45–90 min) of pre-incubation of nuclei and S-100, it is possible to demonstrate a specific switching on of silent vitellogenin genes. Preincubation of hormonally untreated male liver nuclei with homologous S-100 leads to the transcription of albumin, but not vitellogenin, genes (Table II). However, preincubation and incubation of the same nuclei with S-100 fractions from estrogen-treated male *Xenopus* induces the transcription of vitellogenin mRNA. Since vitellogenin genes are fully dormant in these nuclei, their transcription represents a *de novo* activation, which was also corroborated by incubation of nuclei in the presence of heparin, which is an inhibitor of initiation of transcription.

The above process of switching on vitellogenin genes *in vitro* is to some extent tissue specific. The S-100 from male liver cells fails to activate vitellogenin mRNA synthesis in nuclei from erythrocytes and oviduct, the latter being also a major estrogen-regulated tissue involved in egg protein synthesis. In other experiments, the liver nuclear transcripts were analyzed for an uncharacterized *Xenopus* oviduct-specific estrogen-inducible mRNA, termed 6G (James *et al.*, 1985). Thus, the hormonally competent liver S-100 fails to activate the dormant vitellogenin genes in oviduct nuclei, although gene 6G continues to be transcribed (Table III). Conversely, an S-100 from adult *Xenopus* oviduct, in which 6G is expressed, fails to induce its transcription in male liver nuclei. Since both *Xenopus* liver and oviduct have estrogen receptors, it is unlikely that the receptor is the only active component in switching on the dormant vitellogenin genes, but some other tissue-specific transcriptional factor(s) may be involved as well.

Table II. S-100 Extract from Estrogen-Treated (E_2) Male *Xenopus* Liver Confers Hormone-Specific *de novo* Transcription of Vitellogenin Genes in Control (C) Male Liver Nuclei[a,b]

| S-100 | Heparin | [32P]-RNA hybridized (ppm) | | Vg/Alb |
		Vitellogenin	Albumin	
C	–	0.5	93	0.005
C	+	0.7	86	0.01
E_2	–	22.0	61	0.37
E_2	+	4.0	70	0.06

[a]From Tata and Baker (1985).
[b]Nuclei from untreated male *Xenopus* liver were preincubated and incubated with S-100 extracts, with and without heparin, and transcription of vitellogenin mRNA then measured.

Table III. Tissue Specificity of Activation of Vitellogenin and 6G Genes as Seen by Co-Incubation of Nuclei and S-100 Fractions from *Xenopus* Oviduct and Male Liver[a,b]

| Nuclei | S-100 | Rate of transcription (ppm) | |
		Vitellogenin	6G[c]
Male liver	Male liver	1	0
Male liver	Oviduct[d]	2	0
Oviduct	Male liver + E_2	0	75

[a]From James *et al.* (1985).
[b]Nuclei and S-100 were preincubated for 90 min before the nucleotides were added and the transcription reaction carried out for a further 45 min.
[c]6G refers to an mRNA encoding an unknown protein that is expressed in *Xenopus* oviduct but not liver and is inducible with estrogen.
[d]The oviduct S-100 strongly inhibited (by about 75%) transcription in liver nuclei; the values are therefore corrected for this inhibition.

5.2. Involvement of DNA-Binding Proteins and Estrogen Receptor

Many transcriptional factors have DNA-binding properties (Yamamoto, 1984); steroid receptors bind to specific sequences flanking the genes they regulate (Eriksson and Gustafsson, 1983). The above studies on activation of silent vitellogenin genes in isolated nuclei were therefore extended to examine the roles of DNA-binding proteins and estrogen receptor in the process.

By using a procedure based on DNA affinity-competition chromatography, whereby species-specific DNA-binding proteins are separated from nonspecific proteins, it is possible to show that *Xenopus* DNA-specific proteins from estrogen-treated liver S-100, but not from control tissue, strongly activate vitellogenin mRNA synthesis (Tata *et al.*, 1988). The procedure also permits substantial enrichment of the active factor(s) in the crude S-100 extracts. Proteins with a high affinity for low-copy-number *Xenopus* DNA, but not for repetitive DNA, cause this gene activation in isolated nuclei. By using defined cloned genomic DNA fragments to fractionate these partially enriched extracts further, it should be possible both to identify DNA sequences involved in this activation and to obtain purified transcriptional factors.

As regards estrogen receptor, since it is easily released from the nucleus upon cell disruption (Gorski *et al.*, 1984; Greene, 1984), the S-100 fraction from estrogen-treated livers should contain a significant amount of receptor. In order to test the possibility that estrogen receptor is involved in the activation of dormant vitellogenin genes in isolated male liver nuclei, the effect of monoclonal antibodies to estrogen receptor was investigated in the same studies. Incubating the competent S-100, or an enriched DNA-binding protein fraction derived from it, completely abolishes its ability to switch on vitellogenin genes without affecting the normal transcription of albumin genes (Table IV). A 40% inhibition of secondary stimulation of chicken vitellogenin genes in isolated

Table IV. Effect of Estrogen Receptor Antibody on the Transcription of Albumin and Vitellogenin Genes by Male *Xenopus* Liver Nuclei after Preincubation with S-100 and *Xenopus* DNA-Binding Proteins from Estrogen-Treated *Xenopus* Liver

Estrogen preparation added[a]	Treated with ER antibody	$[^{32}P]$-RNA (ppm)[b]	
		Albumin	Vg
E_2–S-100	–	235	47
E_2–S-100	+	221	0
E_2–DNA eluate	–	276	53
E_2–DNA eluate	+	189	0

[a]S-100 or total *Xenopus* DNA-binding proteins from estrogen-treated male *Xenopus* were incubated or not with ER antibody before preincubation with control male liver nuclei. When no antibody was used, samples were incubated in parallel with 40 μl phosphate-buffered saline.
[b]The synthesis of ^{32}P-labeled albumin and vitellogenin mRNA was monitored as described by Tata *et al.* (1988).

liver nuclei by nuclear and cytoplasmic extracts treated with antibodies to estrogen receptor has also been reported (Jost *et al.*, 1986). Total inhibition of vitellogenin gene activation by antibody to estrogen receptor does not imply that estrogen receptor is the sole component in the S-100 extracts responsible for the activation. Rather, the specific activation of vitellogenin genes is more likely to require the combined participation of estrogen receptor and tissue-specific transcription factors. Further experiments also showed that exposure of nuclei from estrogen-treated liver to receptor antibodies does not affect the rate of ongoing transcription of vitellogenin genes already activated *in vivo*. Thus, the major role of the receptor seems to be in initiating *de novo* transcription *in vitro* and that the antibody fails to interact with receptor already engaged in a functionally active complex in the nucleus.

6. General Conclusions

Few would consider the liver either an accessory sexual tissue or a factory for egg proteins and lipids. Yet, besides yolk proteins, this tissue contributes all the lipoproteins, vitamin-binding proteins, cholesterol, and phospholipids of the egg. All these are synthesized under estrogen control. The fact that this hormone induces the *de novo* synthesis of the yolk protein precursor vitellogenin in hepatocytes of male oviparous vertebrates has allowed a better analysis of the initial stages of gene activation than for estrogen-regulated genes in female-specific tissues, such as the oviduct. Furthermore, there is the additional advantage of studying the process of vitellogenesis in *Xenopus*, since in primary hepatocyte culture specific gene activation can be studied in the absence of any cellular proliferation. Thus, it has been possible to demonstrate the unequal expression of the four vitellogenin genes in this species, which in turn may be useful for understanding the role of interaction between estrogen receptor and regulatory gene sequences in the action of steroid hormone.

Another special feature of induction of vitellogenesis in male oviparous vertebrates is the low level of estrogen receptor in hepatocytes and the substantial upregulation of receptor level produced by the hormone. This has made it possible to demonstrate a stoichiometric relationship between receptor number and absolute transcription rate of the induced gene *in vivo*. The inactive configuration of vitellogenin genes in male hepatocytes and the low receptor number has also proved highly advantageous in devising an assay for transcription regulatory factors *in vitro*. The specific activation of dormant genes in male hepatocyte nuclei provides a simple test system for characterization and isolation of factors conferring expression on genes. Ultimately, it will be of utmost importance to explain the high degree of tissue specificity of regulation of gene expression by steroid hormones. Figure 10 poses this problem schematically, as

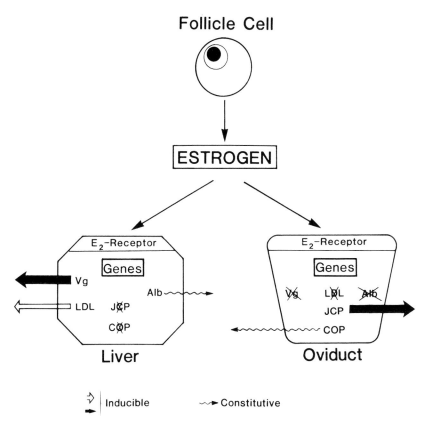

Figure 10. Scheme depicting the tissue specificity of regulation of gene expression by estrogen in *Xenopus* liver and oviduct. Heavy arrows relate to genes induced by estrogen, whereas light arrows indicate genes that are constitutively expressed in each tissue. Genes without an arrow are silent. Vg, vitellogenin; JCP, jelly coat protein; Alb, albumin; LDL, low density lipoprotein; COP, constitutively expressed oviduct protein.

illustrated by the selective activation of different genes in *Xenopus* liver and oviduct by estrogen. Whatever the outcome of these future studies, it is clear that work on the regulation of *Xenopus* vitellogenin genes has already made a significant contribution to our understanding of steroid hormone action, in particular, and of regulation of eukaryotic gene expression, in general.

ACKNOWLEDGMENTS.   The author is grateful to Mrs. Ena Heather and to Mrs. Marie Morris for expert help in the preparation of this article.

References

Baker, H. J., and Shapiro, D. J., 1977, Kinetics of estrogen induction of *Xenopus laevis* vitellogenin messenger RNA as measured by hybridization to complementary DNA, *J. Biol. Chem.* **252:**8428–8434.

Baker, B. S., Steven, J., and Tata, J. R., 1985, Vitellogenin genes and their products in closely and distantly related species of *Xenopus, Comp. Biochem. Physiol.* **82B:**497–505.

Beuving, G., and Gruber, M., 1971, Induction of phosvitin synthesis in roosters by estradiol injection, *Biochim. Biophys. Acta* **232:**529–536.

Brock, M. L., and Shapiro, D. J., 1983, Estrogen stabilizes vitellogenin mRNA against cytoplasmic degradation, *Cell* **34:**207–214.

Browder, L. W. (ed.), 1985, *Developmental Biology: A Comprehensive Synthesis,* Vol. 1, Oogenesis, Plenum, New York.

Burch, J. B. E., 1984, Identification and sequence analysis of the 5′ end of the major chicken vitellogenin gene, *Nucl. Acids Res.* **12:**1117–1135.

Chambon, P., Dierich, A., Gaub, M.-P., Jakowlev, S., Jongstra, J., Krust, A., LePennec, J.-P., Oudet, P., and Reudelhuber, T., 1984, Promoter elements of genes coding for proteins and modulation of transcription by estrogens and progesterone, *Recent Prog. Horm. Res.* **40:**1–39.

Clemens, M. J., 1974, The regulation of egg yolk protein synthesis by steroid hormones, *Prog. Biophys. Mol. Biol.* **28:**69–108.

Dynan, W. S., and Tjian, R., 1985, Control of eukaryotic messenger RNA synthesis by sequence-specific DNA-binding proteins, *Nature (Lond.)* **316:**774–778.

Emerson, B. M., Lewis, C. D., and Felsenfeld, G., 1985, Interaction of specific nuclear factors with the nuclease-hypersensitive region of the chicken adult β-globin gene: Nature of the binding domain, *Cell* **41:**21–30.

Eriksson, H., and Gustafsson, J.-A. (eds.), 1983, *Steroid Hormone Receptors: Structure and Function,* Elsevier, Amsterdam.

Farmer, S. R., Henshaw, E. C., Berridge, M. V., and Tata, J. R., 1978, Translation of *Xenopus* vitellogenin mRNA during primary and secondary induction, *Nature (Lond.)* **273:**401–403.

Felber, B. K., Maurhofer, S., Jaggi, R. B., Wyler, T., Wahli, W., Ryffel, G. U., and Weber, T., 1980, Isolation and translation in vitro of four related vitellogenin mRNAs of estrogen-stimulated *Xenopus laevis, Eur. J. Biochem.* **105:**17–24.

Follett, B. K., and Redshaw, M. R., 1968, The effects of oestrogen and gonadotrophins on lipid and protein metabolism in *Xenopus laevis* Daudin, *J. Endocrinol.* **40:**439–456.

Germond, J.-E., Walker, P., Heggeler, T. B., Brown-Luedi, M., de Bony, E., and Wahli, W., 1984, Evolution of vitellogenin genes: Comparative analysis of the nucleotide sequences downstream of the transcription initiation site of four *Xenopus laevis* and one chicken gene. *Nucl. Acids Res.* **12:**8595–8609.

Gorski, J., Welshons, W., and Sakai, D., 1984, Remodeling the estrogen receptor model, *Mol. Cell. Endocrinol.* **36:**11–15.

Green, S., Walter, P., Kumar, V., Krust, A., Bornert, J.-M., Argos, P., and Chambon, P., 1986, Human oestrogen receptor cDNA: Sequence, expression and homology to v-erb-A, *Nature (Lond.)* **320:**134–139.

Greene, G. L., Sobel, N. B., King, W. J., and Jensen, E. V., 1984, Immunochemical studies of estrogen receptors, *J. Steroid Biochem.* **20**:51–56.

Greene, G. L., Gilna, P., Waterfield, M., Baker, A., Hort, Y., and Shine, J., 1986, Sequence and expression of human estrogen receptor complementary DNA, *Science* **231**:1150–1154.

Hayward, M. A., Mitchell, T. A., and Shapiro, D. J., 1980, Induction of estrogen receptor and reversal of the nuclear/cytoplasmic receptor ratio during vitellogenin synthesis and withdrawal in *X. laevis, J. Biol. Chem.* **255**:11308–11312.

Huber, S., Ryffel, G. U., and Weber, R., 1979, Thyroid hormone induces competence for oestrogen-dependent vitellogenin syntheses in developing *Xenopus laevis* liver, *Nature (Lond.)* **278**:65–67.

Jaggi, R. B., Wyler, T., and Ryffel, G. U., 1982, Comparative analysis of *Xenopus tropicalis* and *Xenopus laevis* vitellogenin gene sequences, *Nucl. Acids Res.* **10**:1515–1533.

Jailkhani, B. L., and Talwar, G. P., 1972, Induction of phosvitin by oestradiol in rooster liver needs DNA synthesis, *Nature (Lond.)* **239**:240–241.

James, T. C., Bond, U. M., Maack, C. A., Applebaum, S. W., and Tata, J. R., 1982, Evolutionary conservation of vitellogenin genes, *DNA* **1**:345–353.

James, T. C., Maack, C. A., Bond, U. M., Champion, J., and Tata, J. R., 1985, *Xenopus* egg jelly coat proteins. 2. Characterisation of messenger RNAs of the oviduct and cloning of complementary DNAs to poly(A)- containing RNA, *Comp. Biochem. Physiol.* **80B**:89–97.

Jost, J.-P., Seldran, M., and Geiser, M., 1984, Preferential binding of estrogen-receptor complex to a region containing the estrogen-dependent hypomethylation site preceding the chicken vitellogenin II gene, *Proc. Natl. Acad. Sci USA* **81**:429–433.

Jost, J.-P., Moncharmont, B., Jiricny, J., Saluz, H., and Hertner, T., 1986, *In vitro* secondary activation (memory effect) of avian vitellogenin II gene in isolated liver nuclei, *Proc. Natl. Acad. Sci. USA* **83**:43–47.

Kazamaier, M., Bruning, E., and Ryffel, G. U., 1985, Post-transcriptional regulation of albumin gene expression in *Xenopus* liver, *EMBO J.* **4**:1261–1266.

Knowland, J., 1978, Induction of vitellogenin synthesis in *Xenopus laevis* tadpoles, *Differentiation* **12**:47–51.

Metz, C. B., and Monroy, A. (eds.), 1985, *Biology of Fertilization*, Vol. 1, Academic, Orlando, Florida.

Ng, W. C., Wolffe, A. P., and Tata, J. R., 1984, Unequal activation by estrogen of individual *Xenopus* vitellogenin genes during development, *Dev. Biol.* **102**:238–247.

Pan, M. L., Bell, W. J., and Telfer, W. H., 1969, Vitellogenic blood protein synthesis by insect fat body, *Science* **165**:393–394.

Payvar, F., DeFranco, D., Firestone, G. L., Edgar, B., Wrange, O., Okret, S., Gustafsson, J.-A., and Yamamoto, K. R., 1983, Sequence specific binding of glucocorticoid receptor to MTV DNA at sites within and upstream of the transcribed region, *Cell* **35**:381–392.

Perlman, A. J., Wolffe, A. P., Champion, J., and Tata, J. R., 1984, Regulation by estrogen receptor of vitellogenin gene transcription in *Xenopus* hepatocyte cultures, *Mol. Cell. Endocrinol.* **38**:151–161.

Renkawitz, R., Schutz, G., Van der Ahe, D., and Beato, M., 1984, Sequences in the promoter region of the chicken lysozyme gene required for steroid regulation and receptor binding, *Cell* **37**:503–510.

Riegel, A. T., Martin, M. B., and Schoenberg, D. R., 1986, Transcriptional and post-transcriptional inhibition of albumin gene expression by estrogen in *Xenopus* liver, *Mol. Cell. Endocrinol.* **44**:201–209.

Ryffel, G. U., and Wahli, W., 1983, Regulation and structure of the vitellogenin genes, in: *Eukaryotic Genes* (N. Maclean, S. P. Gregory, and R. A. Flavell, eds.), pp. 329–341, Butterworths, London.

Shapiro, D. J., and Brock, M. L., 1985, Messenger RNA stabilization and gene transcription in the estrogen induction of vitellogenin mRNA, in: *Biochemical Actions of Hormones*, Vol. XII (G. Litwack, ed.), pp. 139–172, Academic, Orlando.

Spieth, J., Denison, K., Kirtland, S., Cane, J., and Blumenthal, T., 1985, The *C. elegans* vitellogenin genes: Short sequence repeats in the promoter regions and homology to the vertebrate genes, *Nucl. Acids Res.* **13**:5283–5295.

Tata, J. R., 1985, Environmental cues in egg maturation, in: *The Endocrine System and the Environment* (B. K. Follett, S. Ishii, and A. Chandola, eds.), pp. 85–91, Japan Scientific Society Press, Tokyo/Springer-Verlag, Berlin.

Tata, J. R., and Baker, B. S.-, 1985, Specific switching on of silent egg protein genes *in vitro* by an S-100 fraction in isolated nuclei from male *Xenopus, EMBO J.* **4**:3253–3258.

Tata, J. R., and Smith, D. F., 1979, Vitellogenesis: A versatile model for hormonal regulation of gene expression, *Recent Prog. Horm. Res.* **35**:35–47.

Tata, J. R., Ng, W. C., Perlman, A. J., and Wolffe, A. P., 1986a, Activation and regulation of the vitellogenin gene family, in: *Gene Regulation by Steroid Hormones* (A. Roy, ed.), pp. 205–232. Springer-Verlag, Berlin.

Tata, J. R., Sargent, M., Baker, B. S., and Bennett, M., 1988, Participation of DNA-binding proteins and oestrogen receptor in activation of silent vitellogenin genes *in vitro, EMBO J.* (submitted).

Tenniswood, M. P. R., Searle, P. F., Wolffe, A. P., and Tata, J. R., 1983, Rapid estrogen metabolism and vitellogenin gene expression in *Xenopus* hepatocyte cultures, *Mol. Cell. Endocrinol.* **30**:329–345.

Tsuda, M., and Suzuki, Y., 1983, Transcription modulation *in vitro* of the fibroin gene exerted by a 200-base-pair region from the "TATA" box, *Proc. Natl. Acad. Sci. USA* **80**:7442–7446.

Wahli, W., Dawid, I. B., Wyler, T., Jaggi, R. B., Weber, R., and Ryffel, G. U., 1979, Vitellogenin in *Xenopus laevis* is encoded in a small family of genes, *Cell* **16**:535–549.

Wahli, W., Dawid, I. B., Wyler, T., Weber, R., and Ryffel, G. U., 1980, Comparative analysis of the structural organization of two closely related vitellogenin genes in *X. laevis, Cell* **20**:107–117.

Wahli, W., Dawid, I. B., Ryffel, G. U., and Weber, R., 1981, Vitellogenesis and the vitellogenin gene family, *Science* **212**:298–304.

Wahli, W., Germond, J.-E., Heggeler, B. T., and May, F. E. B., 1982, Vitellogenin genes A1 and B1 are linked in the *Xenopus laevis* genome, *Proc. Natl. Acad. Sci. USA* **79**:6832–6836.

Walker, P., Brown-Luedi, M., Germond, J.-E., Wahli, W., Meijlink, F. C. P. W., van het Schip, A. D., Roelink, H., Gruber, M., and AB, G, 1983, Sequence homologies within the 5′ end region of the estrogen controlled vitellogenin gene in *Xenopus* and chicken, *EMBO J.* **2**:2271–2279.

Walker, P., Germond, J.-E., Brown-Luedi, M., Givel, F., and Wahli, W., 1984, Sequence homologies in the region preceding the transcription initiation site of the liver estrogen-responsive vitellogenin and apo-VLDL II genes, *Nucl. Acids Res.* **12**:8611–8626.

Wallace, R. A., 1985, Vitellogenesis and oocyte growth in nonmammalian vertebrates, in: *Developmental Biology: A Comprehensive Synthesis*, Vol. I, *Oogenesis* (L. W. Browder, ed.), pp. 127–177, Plenum, New York.

Wallace, R. A., and Jared, D. W., 1968, Studies on amphibian yolk, VII. Serum phosphoprotein synthesis by vitellogenic females and estrogen-treated males of *Xenopus laevis, Can. J. Biochem.* **46**:953–959.

Wallace, R. A., Opresko, L., Wiley, H. S., and Selman, K., 1983, The oocyte as an endocytic cell, *Ciba Found. Symp* **98**:228–248.

Westley, B., and Knowland, J., 1978, An estrogen receptor from *Xenopus laevis* liver possibly connected with vitellogenin synthesis, *Cell* **15**:367–374.

Williams, J. L., and Tata, J. R., 1983, Simultaneous analysis of conformation and transcription of A and B groups of vitellogenin genes in male and female *Xenopus* during primary and secondary activation by estrogen, *Nucl. Acids Res.* **11**:1151–1166.

Wolffe, A. P., and Tata, J. R., 1983, Coordinate and non-coordinate estrogen-induced expression of A and B groups of vitellogenin genes in male and female *Xenopus* hepatocytes in culture, *Eur. J. Biochem.* **130**:365–372.

Wolffe, A. P., and Tata, J. R., 1984, Primary culture, cellular stress and differentiated function, *FEBS Lett.* **176**:8–15.

Wolffe, A. P., Glover, J. F., and Tata, J. R., 1984a, Culture shock: Synthesis of heat-shock-like proteins in fresh primary cell cultures, *Exp. Cell Res.* **154**:581–590.

Wolffe, A. P., Perlman, A. J., and Tata, J. R., 1984b, Transient paralysis by heat shock of hormonal regulation of gene expression, *EMBO J.* **3**:2763–2770.

Wolffe, A. P., Glover, J. F., Martin, S. C., Tenniswood, M. P. R., Williams, J. L., and Tata, J. R., 1985, Deinduction of transcription of *Xenopus* 74-kDa albumin genes and destabilization of mRNA by estrogen in vivo and in hepatocyte cultures, *Eur. J. Biochem.* **146:**489–496.

Wyatt, G. R., Dhadialla, T. S., and Roberts, P. E., 1984, Vitellogenin synthesis in locust fat body: Juvenile hormone-stimulated gene expression, in: *Biosynthesis, Metabolism and Mode of Action of Invertebrate Hormones* (J. Hoffman and M. Porchet, eds.), pp. 457–484, Springer-Verlag, Berlin.

Chapter 10

Developmental Regulation of the *rosy* Locus in *Drosophila melanogaster*

F. LEE DUTTON, JR. and ARTHUR CHOVNICK

1. Introduction

The *rosy* locus of *Drosophila melanogaster* encodes the enzyme xanthine dehydrogenase (XDH, xanthine : NAD oxidoreductase). The most thoroughly studied gene–enzyme system in higher eukaryotes (Sang, 1984), and long a paradigm for eukaryotic gene organization, the *rosy* locus has recently become a fertile system for analysis of eukaryotic gene regulation. The major reasons for this are:

1. The locus encodes an enzyme (XDH) that has been isolated and characterized and is easily assayed in extracts of small numbers of whole organisms or isolated tissues (see Sections 3 and 6).
2. The locus has been subject to intensive fine-structure genetic analyses that have defined its limits and organization (see Sections 4 and 7).
3. Effective phenotypic selection and screening protocols are available (see Section 2), taking advantage of conditional lethality at the locus, and resulting in the isolation of a wealth of mutants, including regulatory variants (see Section 7).
4. It is clear that expression of *rosy* is regulated at both the stage and tissue levels during development (see Section 6); mutants that affect particular stages and tissues have been identified (see Section 7).
5. The locus has been cloned and sequenced. Thus, molecular probes are available for analyses at the DNA and RNA levels (see Sections 4 and 5).
6. The technique of P element-mediated transformation was initiated using the *rosy* system, and genetic and molecular analyses using transposons are now quite advanced (see Sections 4 and 7). Manipulation of *rosy* transposons also permits analysis of a range of regulatory phenomena not exclusive to XDH, including euchromatic position effects, heterochromatic position effects, and dosage compensation (see Section 7).

F. LEE DUTTON, JR., and ARTHUR CHOVNICK • Molecular and Cell Biology Department, The University of Connecticut, Storrs, Connecticut 06268

Each of these phenomena is now being analyzed at the level of specific tissues throughout development.

For these and other reasons, the *rosy* locus constitutes a useful paradigm for the developmental regulation of eukaryotic gene systems in general. This chapter outlines the molecular biology of *rosy*, with emphasis on our evidence for, and current understanding of, regulatory phenomena associated with the locus.

2. Phenotypes of *rosy* Mutants

2.1. Eye Color and Pigments

The *rosy* locus in *D. melanogaster* [ry:3-52.0] was originally identified by its recessive mutant phenotype of reddish-brown eye color. Because of their lack of XDH (Forrest *et al.*, 1956; Glassman and Mitchell, 1959), *rosy* mutants are deficient in the drosopterin eye pigments, lack isoxanthopterin, and accumulate 2-amino-4-hydroxypteridine (Hadorn and Mitchell, 1951; Hadorn and Schwinck, 1956; Chovnick *et al.*, 1962). Mutants also lack uric acid and accumulate hypoxanthine, xanthine, guanosine, and inosine (Mitchell *et al.*, 1959; Baker, 1973).

2.2. Purine Sensitivity

The *rosy* mutants are conditional lethals that are unable to survive when reared on standard media supplemented with purine (Glassman, 1965; Chovnick *et al.*, 1970). This is the basis of an invaluable selection system that makes possible fine-structure recombination studies and screening of large-scale mutageneses. It has long been reported that purine (7H-imidazo[4,5-d]pyrimidine) is not itself a substrate of XDH. Recently, however, it has been found that XDH does act on a number of purine analogs *in vitro* (F. L. Dutton and A. Chovnick, unpublished data). The lethal effect of purine may be attributable to an accumulation of toxic metabolic products or to derangement of regulation of the purine metabolic pathways, or both. (Levels of purine needed to kill mutants typically also delay development of wild types. Individuals that survive to the larval stages may live for extended periods without completing development.)

2.3. Phenocopy on Allopurinol

Leaky *rosy* mutants retain only very low levels of XDH activity, yet have nonmutant eyes (Gelbart *et al.*, 1976). However, treatment even of wild-type larvae with the XDH inhibitor 4-hydroxypyrazolo[3,4-d]pyrimidine (allopurinol, or HPP) induces a phenocopy of *rosy* eye color in adults (Keller and Glassman, 1965; Boni *et al.*, 1967). Mutants exhibiting low levels of XDH activity are hypersensitive to this inhibitor, producing full *rosy* eye color on much lower levels of HPP than that required for wild-types (Clark *et al.*, 1979). As with purine sensitivity, this effect can be exploited in screening large scale recombination and mutagenesis experiments. Such studies have identified underproducer variants of the locus (e.g., Clark *et al.*, 1979; Lee *et al.*, 1987).

3. The *rosy* Gene Product

3.1. XDH Peptide Encoded by the *rosy* Locus

The genetic information encoding the primary structure of xanthine dehydrogenase was mapped to the *rosy* locus by three lines of evidence. First, it was found that *rosy* mutants lack detectable XDH activity (Forrest *et al.*, 1956; Glassman and Mitchell, 1959). Second, the level of enzyme activity parallels the genetic dosage of the locus in flies carrying one, two, or three doses of the gene (Grell, 1962; Glassman *et al.*, 1962; Chovnick, 1966). Third, unambiguous structural variants of the enzyme map to the locus (Yen and Glassman, 1965; McCarron *et al.*, 1974).

3.2. Reactions of XDH

The reactions known to be catalyzed by XDH are shown in Fig. 1. Each requires the cofactor NAD^+ (Morita, 1958; Glassman and Mitchell, 1959; Forrest *et al.*, 1961; Collins *et al.*, 1971). XDH enzyme activity can be observed in a variety of contexts, each providing one part of the total picture of enzyme expression. We devote some space to a summary of these methods in order to illuminate the advantages and limitations of each.

Conversion of the purine hypoxanthine sequentially to xanthine and uric acid can be linked to the reduction of tetrazolium (Smith *et al.*, 1963), permitting detection of the enzyme immobilized by a variety of methods. For example, electrophoresis of crude extracts in nondenaturing gels (Yen and Glassman, 1965; McCarron *et al.*, 1974) provides a means of recognizing electrophoretic variants and also permits crude estimates of relative enzyme activity in different strains. The same reaction can be employed to detect enzymatic activity in embedded or sectioned tissues.

Enzyme can also be electrophoresed in agarose gels containing anti-XDH serum (rocket electrophoresis) (Laurell, 1966; McCarron *et al.*, 1979), the gels stained for enzyme activity as above, and the heights of the precipitin rockets formed by different samples compared. Rocket heights may be compared more accurately than relative staining intensities and reflect the actual number of XDH molecules present (including inactive forms), rather than possible differences in specific activity of enzyme from different sources.

Where XDH activity is low or absent, crossed-line rocket electrophoresis permits estimation of inactive cross-reacting material (crm) (Lee *et al.*, 1987). In this method, a slot is created perpendicular to the direction of electrophoretic migration and is filled with wild-type enzyme extract. Sample wells may then be loaded with extracts prepared from null mutants. If the mutants produce inactive crm, this will add to the local concentration of antigen. This in turn results in a break or bulge in the precipitin line formed from the wild-type enzyme, revealing additional crm from the mutant despite the absence of any associated enzymatic activity. This is illustrated in Fig. 2.

The pteridine reaction permits assay by following the change in fluorescence associated with conversion of pteridine to isoxanthopterin mediated *in vitro* by crude enzyme preparations (Glassman, 1962). This method provides a

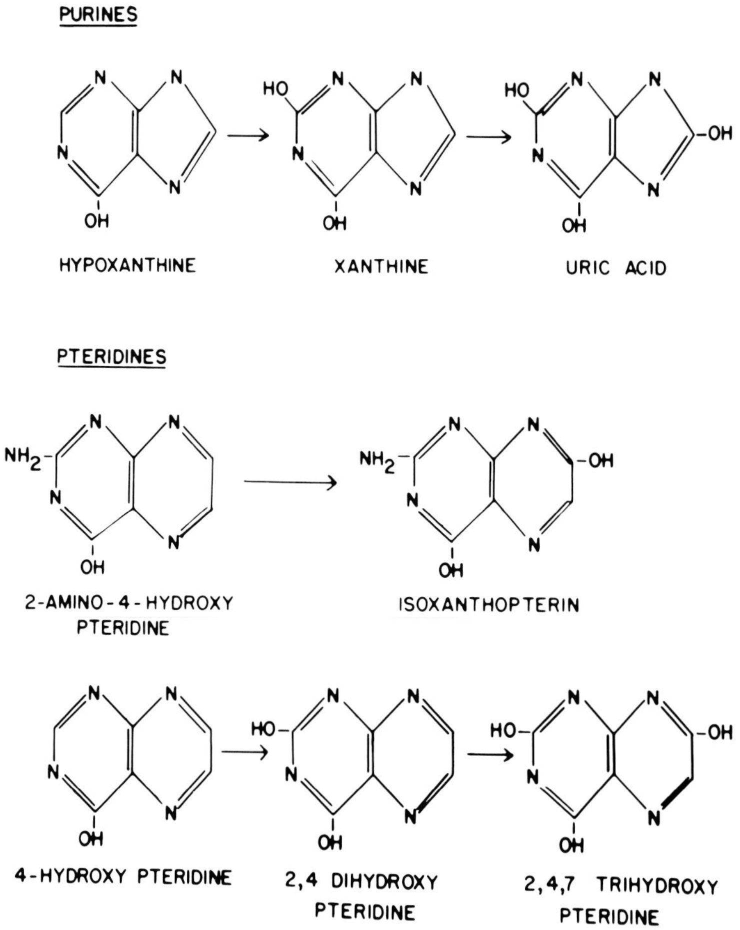

Figure 1. Reactions of xanthine dehydrogenase.

Figure 2. Crossed-line rocket electrophoresis. The precipitin line was formed from a slot containing extract of the wild-type ry^{+5}. Samples were placed in wells below this slot. Extracts were prepared from the following strains: (**A**) ry^{+10}, an underproducer of XDH; (**B**) ry^{5105}, a null mutant; (**C**) ry^{+5}, standard strain; (**D**) ry^{+10}, one-half the amount applied in sample **A.**; (**E**) ry^{5106}, a null mutant; (**F**) ry^{5107}, a low-activity variant. The unbroken precipitin line above sample well **B** indicates that

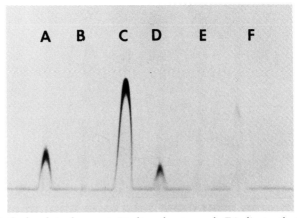

ry^{5105} produces no crm. In contrast, the break in the precipitin line about sample **E** indicates that ry^{5106} produces crm with no enzyme activity. The faint rocket in sample **F** indicates that ry^{5107} produces nearly normal amounts of crm, yet has little enzyme activity, characteristic of a structural alteration in the peptide.

relatively quick and sensitive quantitation of total activity but cannot distinguish between qualitative and quantitative differences in enzyme molecules.

The kinetic constants of XDH from a number of *Drosophila* strains have been characterized (Glassman and Mitchell, 1959; Edwards *et al.*, 1977). It is probable that the purine and pteridine catalytic sites are distinct and interact allosterically (Yen and Glassman, 1967).

3.3. Physical Properties of XDH

The molecular weight of native XDH is estimated to be 250,000 M_r (Glassman *et al.*, 1966). The observation of hybrid electrophoretic forms led Yen and Glassman (1965) to suggest that the enzyme is a multimer of at least two subunits. Gelbart *et al.* (1974) confirmed this finding and proved that both subunits are encoded by the *rosy* locus. The size of the XDH peptide was determined by denaturing acrylamide gel electrophoresis of purified enzyme to be approximately 150,000 M_r (Andres, 1976; Edwards *et al.*, 1977). This estimate supports the conclusion that XDH is multimeric, most probably a dimer with subunits of similar size. Gelbart *et al.* (1974) considered two possibilities: (1) XDH functions as a heterodimer (i.e., is composed of two different subunits of similar size); or (2) XDH is a homodimer (with two identical subunits). The issue was resolved as described below.

Electrophoretic variants identified by Gelbart *et al.* (1974) map into two major clusters with respect to the null mutant map (see Section 4). Slow alleles representing each of these clusters were identified. Likewise, chromosomes with fast alleles at each cluster were found, and pairwise combinations of these

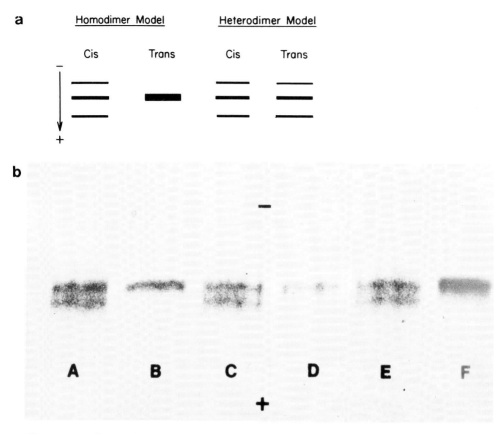

Figure 3. (a) The predictions of the homodimer and heterodimer models. Diagrams of the electrophoretic patterns expected on either model are included. (b) XDH electropherogram of alternating *cis* (A,C,E) and *trans* (B,D,F) heterozygotes. The *trans* heterozygote produces a single electrophoretic band, while the *cis* heterozygote produces three distinct bands. These results contradict the heterodimer model.

two types were generated by intragenic recombination. These constructions facilitated testing of both *cis* and *trans* arrangements of slow and fast alleles in each cluster. The heterodimer and homodimer models yield different predictions in this type of test (Fig. 3a). If two distinct peptides are produced from the locus (heterodimer model), both *cis* and *trans* combinations of slow and fast alleles in different clusters will result in equal numbers of slow and fast monomers, and a 1 : 2 : 1 ratio of three different mobilities will be observed. If the two sites are part of a single peptide (homodimer model), slow–slow/fast–fast (*cis*) heterozygotes will also produce the 1 : 2 : 1 pattern. However, in the *trans* configuration (slow–fast/fast–slow), only one type of peptide will be produced, resulting in a single mobility class. Figure 3b shows the actual XDH mobilities observed in *cis* and *trans* combinations of fast and slow alleles in the two clusters. The observation of a single band of intermediate mobility in the *trans*

configuration confirms that the coding element represents a single long poly-peptide, two identical copies of which dimerize to form active XDH.

4. *rosy* Gene Organization

4.1. Similarities in Prokaryotic and Eukaryotic Genome Structure Shown by Analysis of *rosy*

Gene regulation and gene organization are inseparable topics; analysis of the *rosy* locus has yielded a number of major contributions to our understanding of eukaryotic gene organization and expression. The first of these is the demonstration that eukaryotic and prokaryotic genetic materials, although packaged differently, have similar fine structures; i.e., both are susceptible to intragenic recombination and genetic fine-structure analysis, and both can thus be demonstrated to consist of a linear order of genetic sites (Chovnick et al., 1962; Chovnick et al., 1964). Before the initial demonstration of intragenic recombination at the *rosy* locus (which depended on a genetic crossover selection system), it had been widely speculated that such analyses were precluded by some fundamental physical characteristic unique to higher eukaryotes. The advent of the purine selection system (see Section 2) greatly facilitated further large-scale fine-structure analyses. These analyses in turn made possible the precise definition of control and structural regions of the gene.

4.2. Differences in Eukaryotic and Prokaryotic Genetic Organization Illuminated by *rosy*

Although studies of the *rosy* locus established an underlying physical similarity between bacterial and eukaryotic DNA, the *rosy* system also served to highlight a key difference between eukaryotic and prokaryotic gene organization. Prokaryotic genes with related functions are frequently clustered in the bacterial genome and transcribed as a polycistronic message under the common regulation of a single control region. Since XDH is involved in the purine metabolic pathway, it could reasonably be postulated that the *rosy* locus is part of a complex including several genes having related functions. To address this question, the *rosy* region was saturated with lesions in order to locate all neighboring loci that are either vital or produce visibly mutant phenotypes (Hilliker et al., 1980). A total of 153 recessive lethal mutations were identified and subdivided by *inter se* complementation and deficiency mapping into 20 complementation groups, representing at least 14 separate loci. Figure 4 summarizes these results. Furthermore, all neighboring loci were shown to be functionally as well as spatially distinct from the *rosy* gene. This organization is in contrast to that of polycistronic prokaryotic gene clusters.

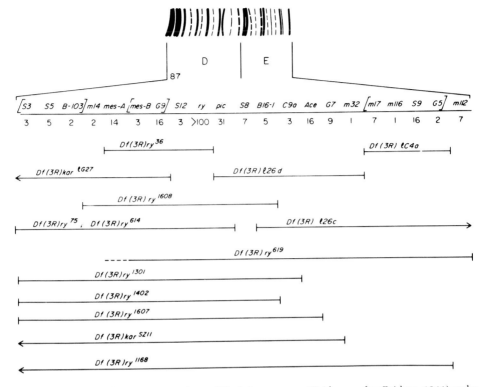

Figure 4. Polytene chromosome region 87DE of chromosome 3R (drawn after Bridges, 1941) and a complementation map of loci identified by Hilliker *et al.* (1980). Each complementation group is represented by the indicated number of alleles, and the extent of deficiencies used in the analysis is presented.

4.3. Genetic Fine Structure of *rosy*

Having established that *rosy* is an independent genetic element, the fine structure of this element was investigated by intragenic recombination. Fine-structure mapping of DNA sites responsible for structural variations in an enzyme can be used to define the genetic limits of a structural gene. (Regulatory variants would be expected to map outside this region.) Several categories of XDH variants can be recognized unambiguously as structural alterations in the peptide. These include (1) electrophoretic variants, (2) mutants that exhibit interallelic complementation, (3) variants that alter thermostability of the molecule, and (4) mutants or variants with lowered levels of XDH activity but normal amounts of protein (e.g., purine-sensitive leaky mutants).

Figure 5 presents the results of fine-structure mapping of several categories of structural variants. Taken together, these yield an estimate for the size of the structural locus of 0.005 map units (Gelbart *et al.*, 1976; Clark *et al.*, 1986a). Although only a handful of control element variants have been found (see Section 7), all of these map to the left, beyond the left-most structural variant.

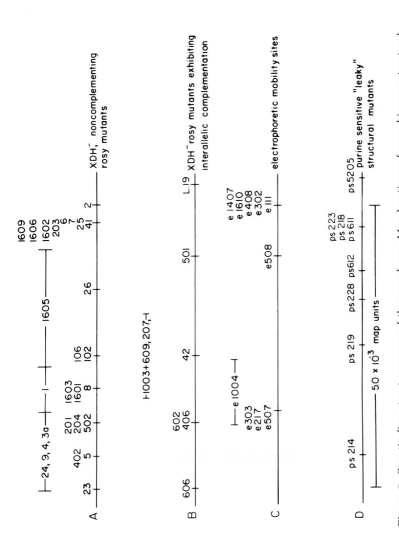

Figure 5. Genetic fine structure maps of the rosy locus. Map locations of unambiguous structural element variants (**B**, **C**, and **D**) are positioned relative to the map of XDH non-complementing mutants (**A**).

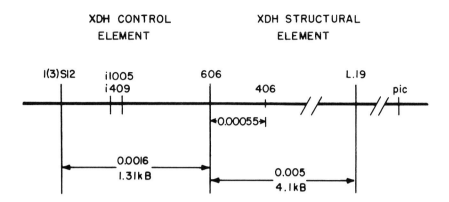

Figure 6. The *rosy* locus. Size estimates of the structural and control elements.

Too few control element variants are available to permit fine-structure mapping of this region comparable to that of the structural element, but an upper limit on the size of the control region is set by the close proximity of the neighboring *1(3)S12* locus (see Section 7). Figure 6 represents the genetic map of control and structural regions of the *rosy* locus. It is clear that the control element must be much smaller than the structural gene, on the order of 0.001 map units.

4.4. Identification, Cloning, and Restriction Mapping of *rosy* Locus DNA

Using the molecular and cytogenetic methodology of chromosomal "walking," Bender *et al.* (1983) isolated DNA segments that together constitute 315 kilobases (kb) of DNA of the 87DE region of the polytene third chromosome of *Drosophila*. This region includes the *rosy* locus. The endpoints of a number of *rosy* region deficiencies were located on the 315-kb molecular map by *in situ* hybridization and whole-genome Southern blot analysis (Spierer *et al.*, 1983). In this way, the *rosy* DNA was localized to a segment of approximately 30 kb. Still more precise localization of the *rosy* DNA was accomplished by whole-genome Southern analysis carried out on a large number of strains carrying various spontaneous and induced *rosy* mutations (Coté *et al.*, 1986) and by large-scale intragenic mapping experiments, which localized insertions and deletion mutants within the genetically defined *rosy* structural element (Clark *et al.*, 1986b). Early progress in this analysis made it possible for Rubin and Spradling (1982) to attempt P element-mediated transformation of *Drosophila*, using an 8.1-kb *Sal* I fragment of DNA, which included the *rosy* locus. Figure 7 compares the physical (restriction enzyme) map of DNA at the *rosy* locus with the genetic map of structural and control elements. The sequence of approximately 5 kb of genomic DNA that includes the *rosy* region is now available (Lee *et al.*, 1987; Kieth *et al.*, 1987) and is presented in Appendix A to this chapter.

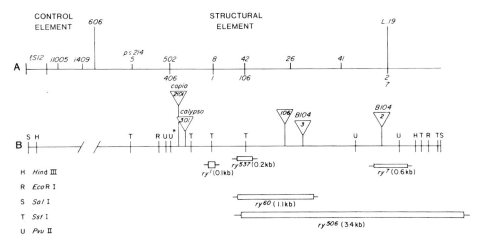

Figure 7. Genetic map of the rosy locus (**A**) aligned with the DNA restriction map of the 8.1-kb Sal I fragment of rosy region DNA (**B**). Insertions are indicated by triangles drawn above the restriction map. Rosy mutant alleles associated with DNA deletions are illustrated below the restriction map.

5. Transcription of *rosy*

5.1. Identification and Size of *rosy* Transcripts

Owing to the availability of cloned *rosy* DNA, transcripts of the locus have been identified in several studies (Covington *et al.*, 1984; Rushlow *et al.*, 1984; Clark *et al.*, 1984; Coté *et al.*, 1986). Figure 8A depicts transcripts detected by northern blot analysis of polyadenylated RNA extracted from young adults of several genotypes. The differences in signal intensity among the three wild-type extracts reflect real differences in transcript abundance in the three (see Section 7). Slight mobility differences may also be detected, with ry^{+4} and ry^{+10} transcripts migrating faster than that of ry^{+5}.

Figure 8b compares the same four genotypes as extracts of late third instar larvae. The transcripts in ry^{+5} again appear to migrate slightly behind those of ry^{+4} and ry^{+10}. Figure 8B also reveals a second signal sometimes seen in larval RNA of ry^{+4}. This signal has not yet been identified. These and other observations suggest that there may be differences in transcription or processing between different strains and/or between different developmental stages.

The size of *rosy* transcripts has been estimated for one wild-type strain by Coté *et al.* (1986). The mature transcript is estimated to be 4.5 kb in length.

5.2. Transcript Abundance

If the *rosy* system has one major disadvantage, it must be the rarity of the transcript(s). Both transcript abundance and enzymatic activity peak in late

A

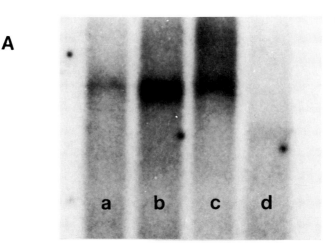

a   b   c   d

B

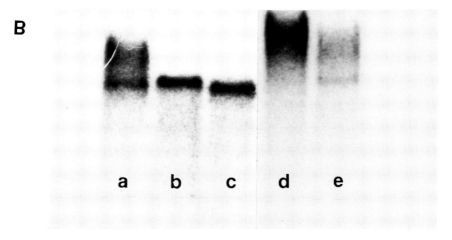

a   b   c   d   e

Figure 8. (**A**) Northern analysis of polyadenylated transcripts from adult flies. Blot was probed with a nick-translate of the 8.1-kb *Sal* I DNA fragment. Samples contained 20 μg of RNA prepared from the following genotypes: (a) ry^{+10}, underproducer variant; (b) ry^{+4}, overproducer variant; (c) ry^{+5}, standard strain; (d) ry^{506}, deletion mutant. (**B**) Northern analysis of polyadenylated transcripts from late third instar larvae. Probe in this case was a 2.7-kb DNA fragment from the *rosy* structural region. Samples contained 20 μg of RNA from the following genotypes. (a) ry^{+4}; (b) ry^{+5}; (c) ry^{+10}; (d) $sd[ry^{2216-547}]$; ry^{506}; (e) $sd[ry^{+2216}]$; ry^{506}.

larval and early adult stages (see Section 6). Even at these stages, however, the messenger has been difficult to analyze due to its low abundance. Isolation of intact transcript has probably also been complicated by its large size and consequent susceptibility to breakage during extraction and purification schemes. On the basis of blot analyses, it is estimated that *rosy* transcripts represent $\geq 10^{-6}$ of polyadenylated RNA (and thus about 10^{-8} of total nucleic acid). Assuming 3 μg total nucleic acid per fly, this would equal ~12,000 molecules per individual. (If transcript abundance parallels crm, this in turn is equivalent to ~10 RNA molecules per cell in Malpighian tubules.)

The rarity of the message has caused RNA level work to lag behind DNA sequencing and complementary DNA (cDNA) analyses. However, several approaches should facilitate transcript analyses in the future. First, by using M13 clones to provide single-stranded DNA for S1 protection analyses, far greater amounts of RNA can be manipulated than can be analyzed by northern blotting. Thus, as long as sufficient material can be accumulated, S1 analyses should become practical. (This approach is demonstrated below in Section 5.4.) Second, hybrid selection protocols of various sorts have been described (Maniatis et al., 1982), and some modification of existing procedures should permit the enrichment of messenger necessary for primer extension analyses. Finally, Henikoff (1983) described a method for convenient cDNA cloning of rare transcripts for which cloned genomic DNA is available. This method should facilitate indirect analyses of transcripts from specific tissues and stages.

5.3. cDNA Analyses

A number of partial *rosy* cDNAs have been analyzed, from which the structure of a large part of the *rosy* transcript has been deduced (Lee et al., 1987; Kieth et al., 1987). The features of a prototypic *rosy* messenger as deduced from cDNA analyses may be summarized as follows: The 3′ end of the transcript is located at base number 3852 of the genomic DNA sequence in Appendix A. There are three introns in the gene. One of these, an intron of 281 base pairs (bp) near the middle of the coding element, has been confirmed by S1 analysis (see Section 5.4). There is also a small intron of 65 bases approximately 140 bases upstream from the polyadenylation site. Finally, there is a large (815-base) intron very near the 5′ end of the gene, splitting off a small exon encoding 14 amino acids. Several mutant derivatives of the ry^{+5} genotype have been determined by genomic DNA sequence analysis to be mutations in splice donor or acceptor sites at the borders of this intron (see Section 7). Of great interest is the observation that the overproducer variant site in ry^{+4} (discussed in Section 6) appears to map within this intron (Clark and Chovnick, unpublished), as have dysgenesis-induced mutations with tissue- and stage-specific effects (see Section 7). Figure 9 compares *rosy* cDNA structure to the molecular map of genomic DNA.

5.4. S1 Analysis

Although cDNA analysis has provided major insight into the transcription of the *rosy* locus, direct analysis of the RNA by S1 protection and other methods is still required. The cDNAs analyzed to date represent a collection of seven

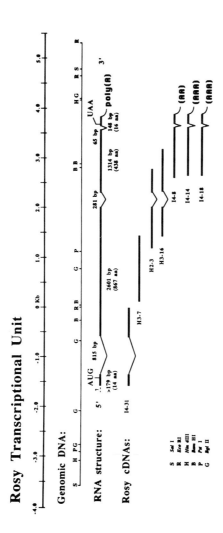

Figure 9. The rosy transcriptional unit. A genomic restriction map for the region of the rosy gene is pictured in the second line, with coordinates in kilobases shown above. The coordinate of 0 kb is placed at the *Eco*RI site near the center of the gene. The RNA structure shown on the third line is a composition deduced from partial cDNAs. The locations of the seven rosy cDNAs that have been isolated are given below the composite picture.

partial cDNAs culled from various libraries. It is thus possible that stage-, strain-, or tissue-specific transcripts have been overlooked in the amalgamated cDNA. S1 analysis using excess single-stranded DNA of the cloned *rosy* sequence permits examination of relatively large quantities of polyadenylated RNA and may therefore permit the future detection of RNA species too rare to be represented in cDNA libraries or detected in northern blot studies (see Section 5.2).

Currently, analysis of the *rosy* transcript(s) by S1 protection is fragmentary. However, several provisional conclusions can be reached from the results obtained so far. We have prepared an M13 clone of the DNA template strand from a 1.68-kb *Pst I–Bam HI* fragment of the structural region of XDH (see Fig. 7). (This fragment will bracket the middle intron defined by cDNA analysis.) Figure 10A,B represents neutral and alkaline agarose gels, respectively, of the fragments protected from S1 digestion by annealing of viral DNA of this clone to *rosy* transcripts from strains ry^{+5} and ry^{+10}. Several important observations are made:

1. Protection of the presumed template strand confirms the orientation of transcription at the locus.
2. There is clearly an intron within this sequence, as shown by the splitting of the single band at approximately 1.5 kb (Fig. 10a) into two smaller fragments in the alkaline analysis (Fig. 10b). (The migration of the RNA–DNA hybrids in the neutral gel reflects the excision of intronic DNA by S1 digestion.)
3. There is a detectable amount of polyadenylated transcript in which the intron remains intact in both strains examined. This presumably represents nuclear precursor present in extracts of unfractionated whole organisms.
4. The relative abundance of this precursor appears to be greater in the underproducer strain ry^{+10}.
5. It does not appear that the difference in mobility between ry^{+5} and ry^{+10} (Fig. 8) can be attributed to different processing near the intron in this region.

In this, as in other systems, the isolation of polyadenylated transcripts containing introns rules out any model in which splicing is coincident with transcription. Sequence analysis (Appendix) shows multiple stops within this intron, which seemingly rules out an alternative messenger RNA that retains the intron (see Chapter 11 for a discussion of alternative processing).

5.5. Questions Raised by Preliminary Transcript Analyses

Clearly, much of the basic molecular biology of *rosy* transcription remains to be investigated. The fragmentary results to date have raised a number of questions of developmental interest, a few of which are highlighted below.

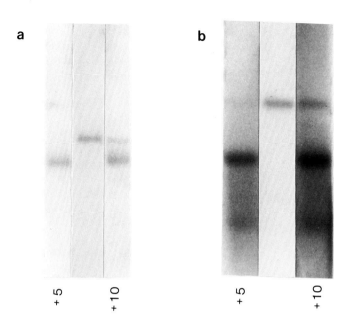

Figure 10. S1 protection analysis of *rosy* transcripts. Polyadenylated transcripts from adults of strains ry^{+5} (30 µg) and ry^{+10} (45 µg) were annealed to viral DNA of an M13 clone containing a 1.7-kb insert of DNA from the XDH structural region of *rosy*. Hybrids were digested with S1 nuclease, electrophoresed through either neutral or alkaline agarose, blotted to nitrocellulose, and probed with a nick-translate of double-stranded DNA representing the same 1.7-kb sequence. (**a**) Neutral gel: both samples contain an abundant species of hybrid that is about 1.5 kb in size. Both samples also contain intact 1.7-kb sequence, although this species is far more abundant in ry^{+10} than in ry^{+5}. (**b**) Alkaline gel: the 1.7-kb sequence is again observed in both samples. However, the 1.5-kb sequence is split into two smaller fragments, consistent with the removal of approximately 200 bases from within the DNA fragment. The marker is a 1.7-kb fragment produced by annealing complementary M13 clones of the sequence and subjecting the resulting hybrids to S1 analysis in parallel with the genomic RNA samples.

5.5.1. Embryonic Transcripts

Developmental northern analyses to date have lumped embryonic stages together with early larval forms, with no transcript being detected in these stages (Covington *et al.*, 1984). However, transcription must begin as early as gastrulation, based on the results of Sayles *et al.* (1973) (see Section 6), which show that production of XDH using transcript synthesized *de novo* can be detected at this early stage.

5.5.2. Larval Transcripts in ry^{+4}

Figure 8B shows a second, higher-molecular-weight transcript in ry^{+4} larvae that has not been seen in young adults (or, for that matter, in earlier analyses of larval RNA). This may represent a tissue-specific transcript or precursor in this strain (which overproduces XDH specifically in the fat body) (see Section 7).

5.5.3. Accumulation of Precursor in ry^{+10}

The strain ry^{+10} has less XDH crm and less mature transcript than ry^{+5} (see Section 7). However, S1 analysis (see Fig. 10) indicates that the ry^{+10} strain may accumulate a higher proportion of precursor than ry^{+5}. It is possible that the ry^{+10} variant involves a site that influences the overall processing of transcripts in this strain.

6. *rosy* Expression in Standard Strains

6.1. Enzymatic Activity

A number of studies have analyzed expression of the *rosy* locus at the product level. These studies provide evidence of (1) stage-specific regulation (i.e., XDH activity is detected only in certain developmental stages of *Drosophila*, and different levels of activity are associated with those stages that do express XDH activity); and (2) tissue-specific regulation (i.e., XDH activity is associated primarily with certain tissues in those stages that do express activity).

6.1.1. Developmental Profile of XDH Activity

The time course of appearance of enzyme during development has been determined for a number of wild-type isoallelic lines (Chovnick *et al.*, 1978). Figure 11 illustrates a normal developmental profile of XDH in the ry^{+0} line. Maximum XDH activity is seen on day 10, and the data are presented as the percent of this activity seen at other times during development at 25°C. A roughly similar result was obtained for the Oregon R wild-type population by Barrett and Davidson (1975). Although the amount of XDH in embryos is expectedly low compared with larvae on a per individual basis, significant activity can be detected quite early in development. By assaying XDH activity in progeny of *ry/ry* females mated to +/+ males, Sayles *et al.* (1973) demonstrated the *de novo* synthesis of enzyme from a paternal gene at the time of gastrulation, or within four hours of fertilization (Fig. 12).

6.1.2. Expression of XDH in Specific Tissues

6.1.2a. Larvae. Ursprung and Hadorn (1961) reported high XDH activity associated with larval fat body and significant levels in malpighian tubules of larvae and adults. Using histochemical staining, J. O'Donnell and A. Chovnick (unpublished data) have demonstrated intense activity in tubules of larvae, with additional activity spread throughout the fat body (Fig. 13). Crm analyses

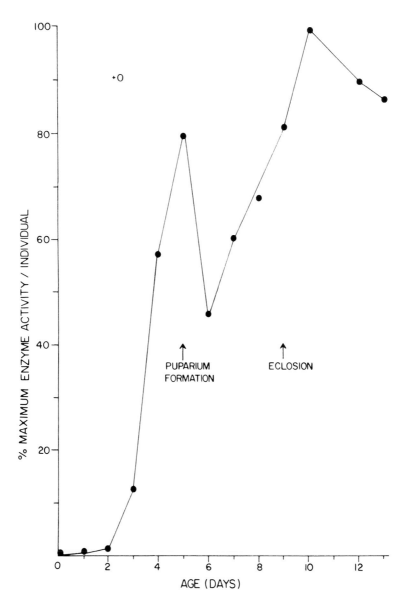

Figure 11. Developmental profile of XDH in ry^{+0} homozygotes.

of these two tissues from several wild-type lines are presented in Fig. 14 (Clark *et al.*, 1984). In all cases, the activity in fat body exceeds that in malpighian tubules. (In the case of ry^{+4}, the fat body activity is disproportionately high, as discussed in Section 7.) In standard strains (e.g., ry^{+5}), fat body generally exhibits roughly three times the activity of tubule, and no other larval tissue has significant XDH activity.

6.1.2b. Adults. *rosy* mutants were initially recognized by adult eye phenotype; indeed, Barrett and Davidson (1975) reported that up to 30% of XDH activity is accumulated in the eye of pharate adults of Oregon R. XDH activity is clearly essential for normal eye color. However, the classic transplantation studies of Hadorn and Schwinck (1956) demonstrate that this activity is non-autonomous. Transplanted pieces of larval tubule and fat body, but not eye discs or other tissues from wild-type individuals, cause mutant larvae to develop into adults with normal eye color. The factor responsible is presumed to be XDH. Enzyme assays conducted on extracts of excised adult tissues indicate that fat body and malpighian tubules are the primary organs of expression of XDH at this stage as well as in larvae (M. McCarron and A. Chovnick, unpublished data).

The autonomy of XDH expression in the intact larval malpighian tubule was dramatically emphasized by the analysis of heterochromatic position effect on the *rosy* gene conducted by Rushlow *et al.* (1984). In this study, malpighian tubules were excised from a strain bearing a rearranged third chromosome, which confers heterochromatic position-effect variegation on *rosy*. When stained for XDH activity, the tissue displays alternating patches of expressing and nonexpressing cells, indicating that staining is dependent upon endogenous (i.e., autonomous) XDH activity (Fig. 15).

That the expression in larval fat body is also autonomous may be inferred from the results presented in Fig. 14. Here, the disproportionate amount of activity in ry^{+4} whole larvae is traced to expression specifically in the fat body

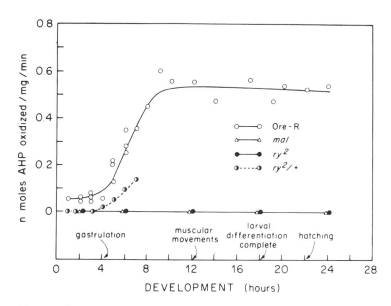

Figure 12. The specific activity of XDH as a function of developmental time (in hr) of Ore-R, *mal*, ry^2, and $ry^2/+$ *Drosophila melanogaster*. Each point on the curve for Ore-R is the average of two independent assays. The *mal*, ry^2, and $ry^2/+$ stocks were assayed once.

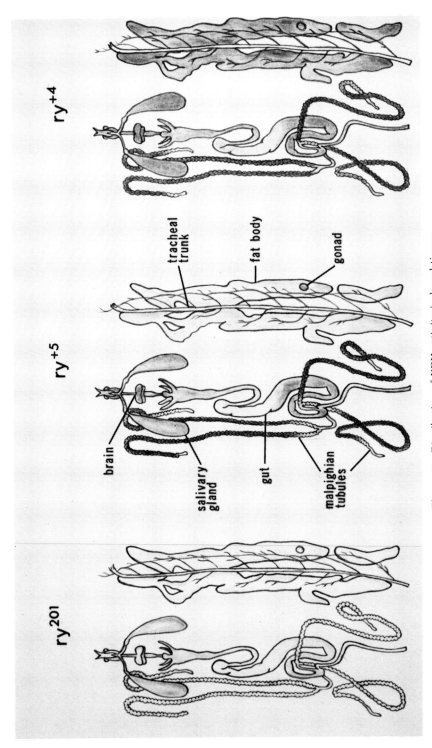

Figure 13. Distribution of XDH activity in larval tissues.

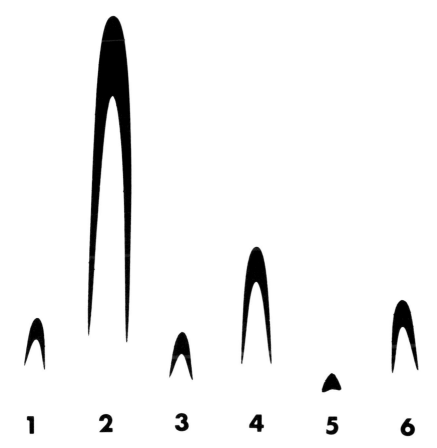

Figure 14. Rocket immunoelectrophoresis of strains ry^{+4}, ry^{+5}, and ry^{+10} (control element variants). Matched extracts came from third instar larval fat bodies (fb) and malpighian tubules (mt). Samples were applied to wells as follows: (1) ry^{+4}: mt; (2) ry^{+4}: fb; (3) ry^{+5}: mt; (4) ry^{+}: fb; (5) ry^{+10}: mt; (6) ry^{+10}: fb. Tissue extracts are matched on a per organ basis. The heights of the rocket peaks are a quantitative reflection of the amount of XDH CRM.

(see Section 7). It is most reasonable to assume that this expression reflects endogenous synthesis of enzyme.

6.1.3. Strain-Specific Differences in XDH Activity

Strain-specific differences in the level of activity of a given enzyme are not uncommon. Such differences might be attributed to (1) *trans*-acting genetic background (including post-translational modifiers); (2) variations in the coding sequence of the structural element of the gene; or (3) noncoding, *cis*-acting control sequences contiguous with the structural element. Two natural variants affecting the level of XDH activity have been identified whose genetic basis maps to the *rosy* locus, but outside of the structural element coding information (see Section 7). The level of activity typical of these types is compared to two other

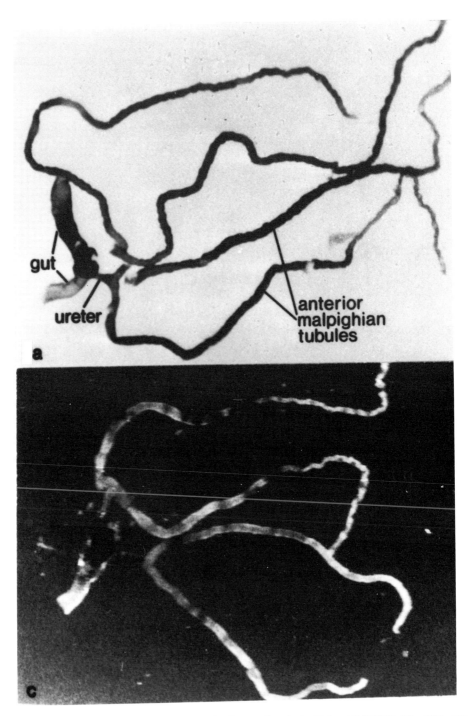

Figure 15. Larval malpighian tubules (late third instar females) of (**a**) ry^{+11}/MKRS; (**b**) ry^{506}/ry^{506}; (**c**) $ry^{ps11136}$/MKRS; and (**d**) ry^{ps1149}/MKRS.

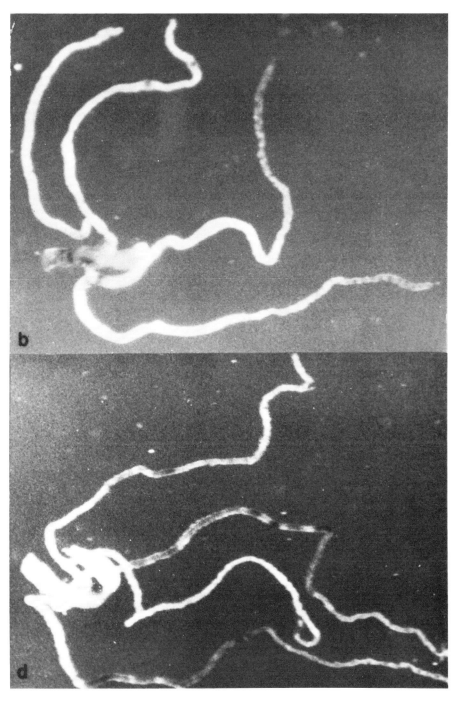

Figure 15. (*continued*)

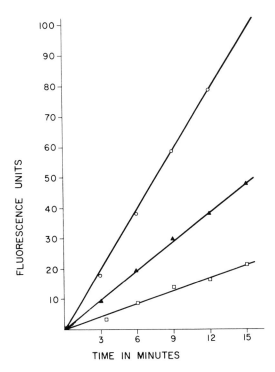

Figure 16. Fluorimetric assay of XDH activities of matched extracts of the indicated homozygous wild-type isoallelic strains.

wild-type alleles in Fig. 16. It can be seen that ry^{+4} displays about twice the activity seen in wild-type controls, whereas ry^{+10} produces about one half the normal amount. These two variants will be considered in detail in Section 7.

6.2. Transcript Analyses

A developmental northern analysis of XDH transcription in strain ry^{+5} was undertaken by Covington *et al.* (1984). These workers found the peaks of *rosy* transcription to coincide with the peaks of XDH enzymatic activity documented in Fig. 11. We can add to this the observation that detectable transcript is lost in aged (1-week-old) adults (F. L. Dutton and A. Chovnick, unpublished data). XDH enzymatic activity also declines with age. It thus appears that the translation of XDH peptide is coincident with transcription of the *rosy* gene. The detection of paternally coded XDH in gastrulae (Sayles *et al.*, 1973) indicates that *rosy* must be transcribed at some level even in the early stages, although this transcript has not yet been detected.

No stage-specific variation in *rosy* transcripts was demonstrated by the analyses of ry^{+5} reported by Covington *et al.* (1984). However, small differences could have been missed due to the large size of the transcript. Indeed, Covington *et al.* (1984) remarked on a possible difference between pupal transcripts and those detected in other stages. The upper signal from strain ry^{+4} shown in Fig. 8b may represent a stage-specific transcript (see Section 5).

7. Regulation of *rosy* Expression

7.1. Analyses of Spontaneous Control Element Variants

McCarron *et al.* (1974) observed several isolate strains of *Drosophila* that not only displayed distinct electrophoretic mobilities of XDH but that also showed stable, characteristically high or low levels of enzyme activity. The level of activity associated with one overproducer variant (ry^{+4}) and an underproducer (ry^{+10}) in whole organisms is illustrated by the fluorimetric assay presented in Fig. 16. In Section 7.1, we present the evidence that these two variants each involve the modification of a 5′ control element(s), whose influence acts in *cis* and pretranslationally to affect accumulation of *rosy* transcripts (and in consequence, the amount of XDH peptide made). Furthermore, one of these variants (the overproducer, ry^{+4}) is shown to exert this influence specifically in the fat body. Finally, we discuss evidence that this tissue-specific site is located within a large intron near the 5′ end of the *rosy* gene. Studies of ry^{+4} (*i409H*) are reviewed in some detail, to illustrate the methods of analysis.

7.1.1. *i409H* (ry^{+4}): A Fat Body-Specific Overproducer Variant of the *rosy* Locus

7.1.1a. Preliminary Mapping of the Overproducer Site.   The strain ry^{+4} is associated with a high level of XDH activity. This high activity level segregates consistently with the electrophoretic variant characteristic of ry^{+4} and thus must belong to the same linkage group as the *rosy* locus. Using standard genetic techniques, Chovnick *et al.* (1976) localized the site responsible for the elevated activity in ry^{+4} (designated *i409H*) to the 0.5-map-unit region between *kar* and *Ace*. (*rosy* represents about 1% of this region.) In fact, the overproducer site is very tightly linked to the *rosy* structural gene, as demonstrated by Chovnick *et al.* (1976). In this analysis, a crossover selector system was used to obtain 123 exchanges within the *kar–Ace* interval. The parental types possessed distinct activity levels associated with characteristic electrophoretic variants. None of the 123 crossovers involved recombination of one parental activity level with the electrophoretic variant of the other parent, strongly suggesting that both phenotypes are under the control of the *rosy* locus itself.

7.1.1b. ry^{+4} Accumulates High Levels of XDH Peptide.   Given an elevated enzymatic activity that maps at or near the structural locus, one must next ask whether the elevated activity reflects an increase in the level of peptide or an increased specific activity of enzyme. Chovnick *et al.* (1976) found no

systematic relationship between the level of XDH enzyme activity and electrophoretic mobility; i.e., elevated activity is not a feature of any particular
class of electrophoretic structural variant. Second, heat inactivation experiments were carried out to compare the thermolability of the ry^{+4} (high activity)
and ry^{+11} (normal activity) enzymes at 60°C. There is no difference between
the two; thus, the increased level of ry^{+4} enzymatic activity does not appear to
be due to greater molecular stability, which could result in increased accumulation of molecules. (These experiments did not address the issue of
relative sensitivity to proteolysis, which could also account for differential
accumulation of peptide in different strains.) Third, Edwards et al. (1977)
purified XDH from ry^{+11} wild-type and from the ry^{+4} strain and found that the
kinetic constants for XDH purified from the two strains differ only slightly.
Finally, Chovnick et al. (1976) conducted antiserum titration experiments,
which revealed roughly twice as much enzyme antigen in extracts of the overproducer strain as in extracts of normal controls. (A similar result is obtained
by comparing peak heights after rocket electrophoresis.)

   7.1.1c. i409H Influence is *Cis*-acting and Tissue-specific.   The crm analysis pictured in Fig. 14 partly anticipates the assertion that ry^{+4} (*i409H*) exerts
its effect specifically in the fat body (Chovnick et al., 1976; Clark et al., 1984).
Referring to this figure, one can see that the ry^{+4} larval fat body exhibits a large
excess of XDH relative to the other wild-types, whereas the rockets due to larval
malpighian tubule in ry^{+4} and ry^{+5} are nearly equal. A priori, the action of the
i409H locus may either be restricted to its *cis*-linked structural element, or it
may also act in *trans* to alter production from the homologous chromosome. If
exclusively *cis*-acting, only production from the linked structural locus should
be high. The fact that the *i409H* site is linked to a specific XDH electromorph
enables one to determine which of these alternatives is the case. Since XDH is a
homodimer, combining two different electromorphs in a heterozygote will result in the production of three electrophoretic forms of enzyme. If both classes
of monomer are produced in equal amounts, the result will be a 1 : 2 : 1 ratio of
enzyme forms (see Fig. 3). Clark et al. (1984) constructed a heterozygote that
can be described as *i409H*, (mobility 1.02)/*i409N*, (mobility 0.90). In Fig. 17, the
XDH enzyme forms produced in such a heterozygote have been examined. In
whole adults, *i409H* results in overproduction of the ry^{+4} electrophoretic variant and an excess of that peptide over the product of the 0.90 allele. Overproduction in the larval fat body results in an excess of fast homodimers. By
contrast, the malpighian tubule shows a 1 : 2 : 1 ratio of abundances of the three
forms, indicating that equal amounts of each type of monomer are available for
random association into dimers (i.e., *i409H* does not result in overproduction
in the tubule).

   7.1.1d. The Influence of i409H is Pretranslational.   Figure 18 presents a
northern blot analysis of third instar larval polyadenylated RNA from three
strains: ry^{+4} (overproducer), ry^{+5} (normal control), and ry^{+10} (underproducer).

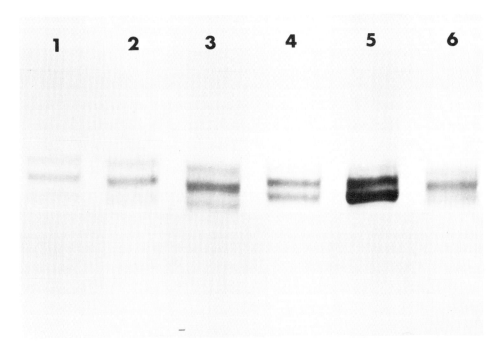

Figure 17. XDH electropherogram indicating the relative amounts of slow homodimers, fast homodimers, and intermediate hybrid dimers present in extracts of adult flies, third instar larval fat bodies, and malpighian tubules from indicated heterozygotes. In lanes 1–3, extracts are from *(i1005N i409N)0.90/(i1005N i409N)1.05* heterozygotes. Lane 1: adults; lane 2: third instar larval fat bodies; lane 3: third instar larval malpighian tubules. In lanes 4–6, extracts are from *(i1005N 1409N)0.90/(i1005N i409H)1.02* heterozygotes. Lane 4: adults; lane 5: third instar larval fat bodies; lane 6: third instar larval malpighian tubules.

ry^{+4} *(i409H)* is associated with an elevated accumulation of transcript. Presumably, this reflects increased initiation of transcription; however, differential efficiency of processing or differential transcript stability are not rigorously excluded.

7.1.1e. Transcript Abundance in ry^{+4} Does Not Reflect Regulation of Template Copy Number.   Both tissues associated with primary expression of XDH (fat body and malpighian tubule) are polytene. In such tissues, it is at least possible that different levels of transcript accumulation are the consequence of differential somatic replication of template DNA. Such a model can be ruled out in the present case by taking advantage of restriction site polymorphisms between the overproducer allele and that of the standard strain, ry^{+5}. Figure 19 presents a Southern blot analysis of DNA from third instar larval fat bodies isolated from the indicated heterozygotes. Template abundance of the sequence linked to *i409H* is not detectably different from that of the normal allele.

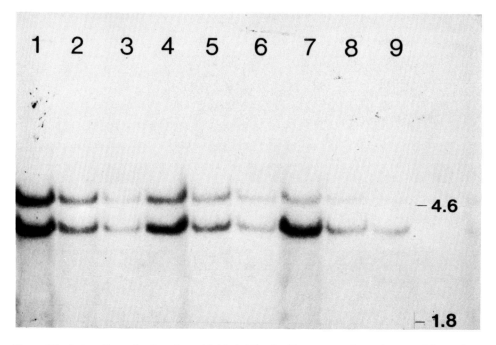

Figure 18. Autoradiograph of northern blot hybridized with a *rosy* probe and exposed for 24 hr, then rehybridized with the P1 probe without washing off the *rosy* probe and exposed for 1.5 hr. Poly-A$^+$ RNA was applied to the lanes as follows: (1) 24-μg *ry*$^{+4}$; (2) 12-μg *ry*$^{+4}$; (3) 6-μg *ry*$^{+4}$; (4) 24-μg *ry*$^{+5}$; (5) 12-μg *ry*$^{+5}$; (6) 6-μg *ry*$^{+5}$; (7) 24-μg *ry*$^{+10}$; (8) 12-μg *ry*$^{+10}$; (9) 6-μg *ry*$^{+10}$. Markers are chicken rRNA: 28S = 4.6 kb and 18S = 1.8 kb.

7.1.1f. Localization of i409H. To this point, it has been demonstrated that the control site, *i409H*, is tightly linked to the *rosy* structural information and exerts its influence in *cis* and pretranslationally to alter the level of XDH peptide accumulated specifically in the larval fat body. The control site has been localized precisely via large-scale fine-structure mapping experiments, and the site has been associated with a defined segment of the molecular (restriction enzyme) map of the cloned DNA. These results are reviewed in Section 7.1.3.

7.1.2. *i1005L* (*ry*$^{+10}$): Underproducer Control Variant

Strain *ry*$^{+10}$ is associated with a low level of XDH activity. This variant was initially detected and subsequently mapped to the *rosy* locus as a purine sensitive site (designated *i1005L*) and has also been characterized as a *cis*-acting regulator of XDH peptide levels. Figures 18 and 19 show, respectively, that the effect of the *i1005L* site is exerted pretranslationally; i.e., it alters accumulation of transcript and does not reflect alterations of template levels in polytene tissue (see Section 7.1.1e.). However, *i1005L* is unlike *i409H* in that it does not appear to be a tissue-specific control element. Both tissues associated with primary expression of XDH show reduced levels of crm (see Fig. 14). The *i1005L* (underproducer) site is separable from the *i409H* (overproducer) site by recombination. *i1005L* appears to lie outside the large 5′ intron in *rosy*, on the centromere-proximal side of the locus.

7.1.3. Fine-Structure and Molecular Mapping of Control Element Variants

The *i409H* control variant site was not separated from *rosy* by any of 123 crossovers in the *kar–Ace* interval (see Section 7.1.1). Chovnick *et al.* (1976) further localized *i409H* by recombination between null mutants induced in the

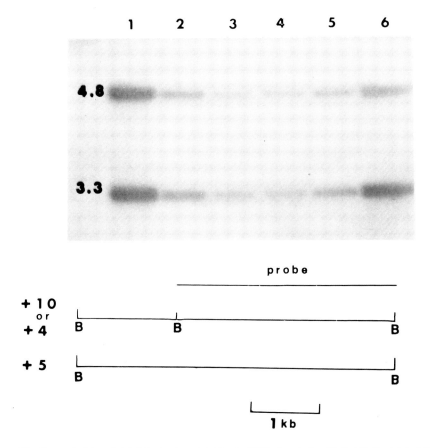

Figure 19. Genomic blot of third instar larval fat body DNA digested with *Bgl II*. Restricted DNA applied to lanes as follows: (1) 0.50-μg ry^{+10}/ry^{+5}; (2) 0.25-μg ry^{+10}/ry^{+5}; (3) 0.125-μg ry^{+10}/ry^{+5}; (4) 0.125-μg ry^{+4}/ry^{+5}; (5) 0.25-μg ry^{+4}/ry^{+5}; (6) 0.50-μg ry^{+4}/ry^{+5}. The blot was probed with a nick-translated ^{32}P-labeled 3.3-kb *Bgl II* fragment.

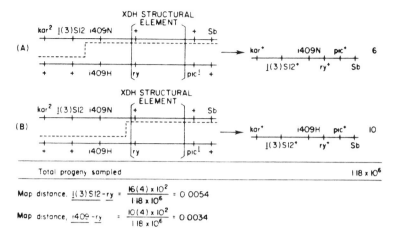

Figure 20. Experimental confirmation of the location of *i409*. Map distances are calculated as 4× percent crossovers since only one-fourth of the crossovers survive and are recoverable.

ry^{+4} structural locus and null sites that had been induced at the left border of an electrophoretically distinct allele. This permitted screening for rare ry$^+$ recombinant progeny on purine-supplemented media and analyses of these recombinants to determine the configuration of flanking markers. Similar crosses mapped *i409H* relative to the *1(3)S12* locus. These analyses are summarized in Fig. 20. The results establish that the *i409H* site lies to the left of all crossover points (i.e., at or beyond the left border of the structural element) and to the right of *1(3)S12*.

Recent analyses (Clark and Chovnick, 1986) have established that both *1(3)S12* and *rosy* DNA are contained in the 8.1-kb *Sal I* DNA fragment pictured in Fig. 7. cDNA and DNA sequence analyses indicate that the coding sequences of XDH extend into the 1.4-kb *Bgl II–Bgl II* fragment highlighted in this figure (see also Section 5.4). Analyses of the transcript produced from the *1(3)S12* locus (Dutton and Chovnick, unpublished) establish that this locus extends into the 0.64 *Bgl II–Bgl II* fragment immediately to the left of the 1.4-kb fragment above. Clearly, *i409H* must lie between these two approximate limits.

By logic and procedures resembling those described for *i409H*, the site responsible for underproduction in ry^{+10} (*i1005L*) was also mapped to the region between *1(3)S12* and the *rosy* structural element. Thus, *i1005L* and *i409H* are closely linked sites with similar effects on the production of XDH (both influencing the level of transcript and peptide accumulated). McCarron *et al.* (1979) established that the two sites were, in fact, separable by recombination. The relative order of *i1005L* and *i409H* sites could not be determined in this analysis because it was impossible to predict with certainty the phenotype of the double variant (and thus it could not be identified among the recombinant progeny). The order of the two sites was ultimately determined by fine structure crosses involving recombination in half-tetrads (i.e., crossing over between arms of a compound 3R chromosome constructed for this purpose) (Clark *et al.*, 1984) (Fig. 21). Twenty recombinant half-tetrads were recovered, detached, and analyzed. Two were found to be reciprocal crossovers between the two control variant sites and established the relative map order of the sites: *kar–i1005L–i409H–ry*. Within this segment, analyses of the exchange of restriction site polymorphisms (occurring in the course of the recombination above to produce the double variant with *i1005L* and *i409H* on the same chromosome) restrict the right-most boundary of *i409H* to a segment of DNA that lies entirely within the large 5′ intron (see Sections 5.4 and 7.4) (S. H. Clark and A. Chovnick, unpublished data). The left-most limit of *i409H* has not been determined with the same precision. However, recombination data establish that this site is sufficiently to the right of *1(3)S12* to lie within the intron as well. Recombination data also establish that the *i1005L* site lies the closest to *1(3)S12*, well to the left of *i409H*, and therefore probably lies proximal to the intron.

7.2. Analyses of P-Element-Mediated Insertions of *rosy*⁺ DNA

P-element-mediated transformation in *Drosophila* was first demonstrated by Rubin and Spradling (1982) using the 8.1-kb *Sal* I DNA fragment (see Fig. 7) to correct the *rosy* defect in mutant flies. A number of lines have since been derived, each carrying the same DNA fragment inserted in a different genomic location (Rubin and Spradling, 1982; Daniels *et al.*, 1985, 1986). This section presents evidence that expression of *rosy* transposons is affected by the genomic site of transposon insertion. Specifically, (1) X-linked transposons of *rosy* place the transposon-borne gene under dosage compensation controls; (2) insertion of the transposon into heterochromatin results in heterochromatic position effect variegation of the gene; and (3) insertion in various euchromatic sites results in either increased or reduced XDH activity and altered levels of *rosy* transcript, revealing site-specific euchromatic position effects.

7.2.1. Isolation and Initial Characterization of Transformants

Daniels *et al.* (1985) generated a number of transformed lines, each of which carries the 8.1-kb transposon in a unique euchromatic location. These lines were obtained by co-injection of embryos of the *rosy* deletion strain ry⁵⁰⁶ with DNA of the 8.1-kb transposon plus helper P element of the clone pπ25.1 (Rubin and Spradling, 1982; Daniels *et al.*, 1985). A variety of stable transposon stocks have been examined as homozygotes to determine their level of expression of the transposed *rosy* gene. Figure 22 presents a rocket analysis of crm levels in various transposon stocks, compared to several wild-type *rosy* strains. One can see that transposon expression ranges from lower to higher than the ry⁺⁵ control level. Table I summarizes the genetic and cytological

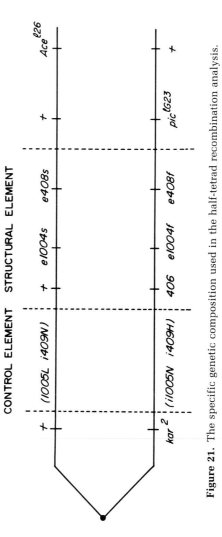

Figure 21. The specific genetic composition used in the half-tetrad recombination analysis.

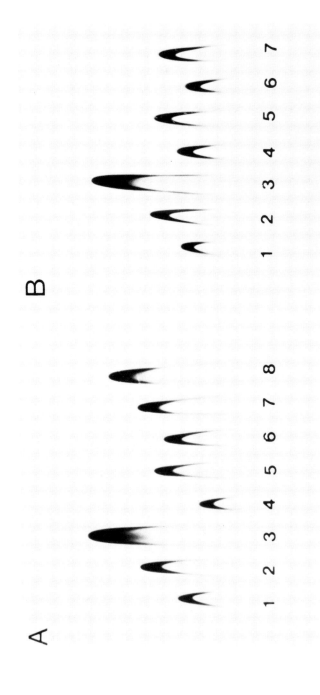

Figure 22. Rocket immunoelectropherograms illustrating the different XDH CRM levels of seven ry$^+$ transformants. Whole-fly extracts containing equivalent amounts of protein were run against anti-XDH serum. In **A** and **B**, wells 1–3 contain, in order, extracts from homozygous ry^{+10}, ry^{+5}, and ry^{+4} males. The ry^{+10} and ry^{+4} samples represent XDH control element variants associated with under- and overproduction of XDH, respectively, wheras the ry^{+5} sample exhibits a normal level of XDH (Clark et al., 1984). (**A**) Autosomal ry$^+$ transposons. Extracts from uniformly aged, homozygous males were applied to wells as follows: (4) [ry$^{+14-1\alpha}$] ry^{506}; (5) ry^{42} [R401.1]; (6) ry^{506} [ry^{+72-1}]; (7) [ry^{+241-8}] ry^{506}, and (8) [ry^{+201-4}] ry^{506}. (**B**) Sex-linked ry$^+$ insertions. Wells 4 and 5 contain extracts of hemizygous males from lines [ry^{+77-1}] ry^{506} and [R403.1] ry^{42}, respectively; wells 6 and 7 contain extracts from the corresponding homozygous females. The ry^{506} and ry^{42} mutant strains are CRM$^-$ (data not shown).

Table I. Genetic and Cytological Features of Seven ry^+ Transformants[a]

Strain	Linkage group	Cytological location	CRM rank[b]	Genotypic designation
$ry^{+i4\text{-}la}$	2	57F	1	$;[ry^{+i4\text{-}1a}]; ry^{506};$
$ry^{+72\text{-}1}$	3	100D	2	$;;ry^{506} [ry^{+72\text{-}1}];$
$ry^{+i77\text{-}1}$	1	16D	2	$[ry^{+i77\text{-}1}]; ;ry^{506};$
$ry^{+201\text{-}4}$	2	43EF	4	$; [ry^{+201\text{-}4}]; ry^{506};$
$ry^{+241\text{-}8}$	3	76F	3	$;;[ry^{+241\text{-}8}] ry^{506};$
R401.1[c]	4	Chromocenter	2	$;;ry^{42}; [R401.1]$
R403.1[c]	1	7D	3	$[R403.1]; ; ry^{42};$

[a]Shown in Fig. 22.
[b]See Table II.
[c]Based on Spradling and Rubin (1983).

features of the seven ry^+ transformants shown in Fig. 22. Table II presents the crm level rankings for a total of 14 stable transformant lines.

Southern blot analysis has shown that the differing levels of expression in transformed lines are not due to transposon copy number or to changes in the DNA organization of the transposon. Furthermore, the orientation of the 8.1-kb sequence within the P element vector does not correlate with high or low expression (Daniels et al., 1986). In no case has the orientation of the transposon with respect to neighboring transcription units been determined.

The electrophoretic mobility of the rosy allele encoded by the transposon DNA has been analyzed (Daniels et al., 1986) and its mobility relative to a standard set of electrophoretic variants determined (illustrated in Fig. 23a). By analogy to arguments presented in earlier sections, one can examine the cis regulation of expression of the transposon gene by construction of heterozygotes between the transformant line and defined wild-types with known (different) electrophoretic mobility. Figure 23b presents the results of such an analysis of two transformant lines, one associated with low crm production and the other with higher than normal levels. The excess of the slow electromorph encoded by the wild-type homologue serves to reveal underexpression of the transposon gene in strain $[ry^{+i4-1a}]$ (lane 5). By contrast, the transposon heterozygote constructed from strain $[ry^{+201-4}]$ produces a ratio of electrophoretic enzyme forms that indicates a slight overproduction of the trans-

Table II. CRM Rankings for the 14 ry^+ Stable Transformants Generated in our Laboratory

Rank	CRM range	No. of transformants
1	ry^{+10} or less	1
2	$>ry^{+10} <ry^{+5}$	2
3	$\simeq ry^{+5}$	8
4	$>ry^{+5} <ry^{+4}$	3

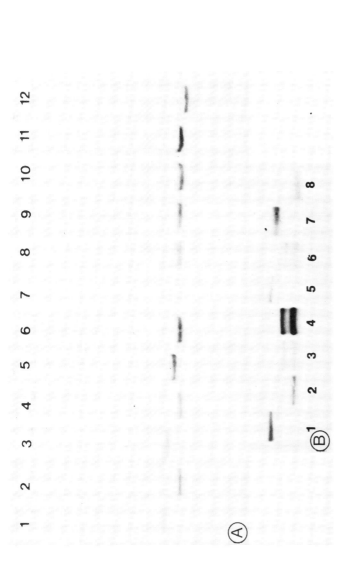

Figure 23. Polyacrylamide gel electropherograms illustrating the *cis*-acting nature of XDH expression associated with ry^+ transposons. (A) Demonstration that a ry^+ transposon derived from the 8.1-kb Canton-S ry^+ fragment (present in pry1 and pry3) encodes an XDH peptide with an electrophoretic mobility of 1.02. Lanes 2, 4, 6, 8 and 11 contain extracts from $[ry^{+4-1}]$. Control extracts were applied to wells as follows: (1) 0.90; (3) 0.94; (5) 0.97; (7) 1.00; (9) 1.02; (10) 1.03; and (12) 1.05. (B) Electropherogram illustrating the relative amounts of slow, intermediate, and fast dimers present in uniform, whole-fly extracts. Homozygous transformant (XDH$^{1.20}$/XDH$^{1.02}$) males were mated to ry^{+13}/MKRS (XDH$^{0.90}$/−) to produce F_1, Sb^+ males that were used to provide the extracts for lanes 5 and 6. Control samples (lanes 1–4, 7, and 8) were obtained from males of the indicated homozygous strains or from heterozygous F_1 males resulting from crosses of homozygous strains. Samples were applied to lanes as follows: (1) and (7) ry^{+13}/ry^{+13} (0.90N/0.90N); (2) and (8) ry^{+11}/ry^{+11} (1.02N/1.02N); (3) ry^{+13}/ry^{+11} (0.90N/1.02N); (4) ry^{+13}/ry^{+4} (0.90N/1.02H); (5) $[ry^{i4-1a}]$/−, ry^{506}/ry^{+13} (1.02H/0.90N); and (6) $[ry^{+201-4}]$/−, ry^{506}/ry^{+13} (1.02L/0.90N).

poson monomer class. These results demonstrate that the expression of the transposon-borne and wild-type alleles in heterozygotes are controlled independently by cis-acting influences. Since both transformants analyzed above carry the same sequence of DNA, the differences in expression between the two transformants must be ascribed to sequences outside the transposon. (Confirmation of this point is presented below.)

In summary, (1) normal levels of expression of transposed rosy$^+$ genes are observed; (2) differences in expression between transformant lines may also be observed; and (3) these differences must be attributable to differences in neighboring genomic sequences at the specific target sites in the various transformed lines.

Northern blot analysis indicates that the different levels of XDH production documented above reflect differing levels of transcript accumulation (F. L. Dutton and A. Chovnick, unpublished data). A number of transformed lines are also known to produce regularly one or more categories of unusual transcripts, some of which show homology to rosy-specific probes (F. L. Dutton and A. Chovnick, unpublished data). However, to date there does not appear to be any correlation between the presence or absence of these unusual transcripts and the level of expression of the transposon gene.

7.2.2. Dosage Compensation

The levels of XDH expression from two transposon-bearing X chromosome lines have been examined by rocket electrophoresis of crm, as shown in Fig. 24 (Daniels et al., 1986). An equal abundance of crm in single dose, hemizygous males and two dose, homozygous females indicates that expression of the transposon is dosage compensated in the male (see also Section 7.3.1).

7.2.3. Heterochromatic Position Effects

Expression of a euchromatic gene is frequently altered by chromosomal rearrangements that bring the locus into proximity with heterochromatin (see Spofford, 1976, for review). Rushlow and Chovnick (1984) established the sensitivity of rosy expression to such heterochromatic position effects. Spradling and Rubin (1983) isolated a transformed line in which the transposon had inserted into centric heterochromatin of the fourth chromosome (designated [ry$^{+R401.1}$]). Expression in this transposon was analyzed at the crm level (Fig. 22) and found to be lower than the ry^{+5} control (although higher than the underproducer variant, ry^{+10}). The effect of added heterochromatin on the level of crm produced from this transposon has been analyzed (Fig. 25). [ry$^{+R401.1}$] responds to the presence of extra heterochromatin, producing more XDH with higher heterochromatin contents. By comparison, the euchromatic transposon [ry^{+i4-1a}] consistently underexpresses and does not respond to added heterochromatin. The response of [ry$^{+R401.1}$] is consistent with a classic heterochromatic position effect.

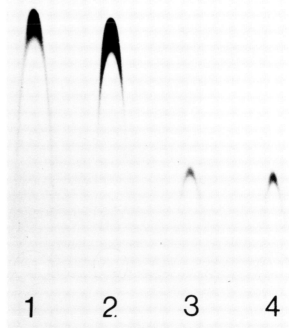

Figure 24. Rocket immunoelectropherogram of whole-fly extracts run against anti-XDH serum. Extracts were prepared from uniformly aged flies and contain equal numbers of males or females. Samples of the following genotypes were applied to wells: (1) $sd[ry^{+2216}]/sd[ry^{+2216}]$; $ry^{506}/ry^{506} ♀♀$; (2) $sd[ry^{+2216}]$; $ry^{506}/ry^{506} ♂♂$; (3) $sd[ry^{2216-547}]/sd[ry^{2216-547}]$; $ry^{506}/ry^{506} ♀♀$; (4) $sd[ry^{2216-547}]$; $ry^{506}/ry^{506} ♂♂$.

7.2.4. Euchromatic Position Effects

It is known that $[ry^{+i4-1a}]$ is an underexpresser and does not respond to added doses of heterochromatin. In this, the strain resembles a number of transformant lines that are associated with stable low production of the transposon-encoded peptide. We have attributed these effects to the *cis* action of neighboring genomic sequences at the target site (above). This assertion has been verified by remobilization of the underexpressing transposon in $[ry^{+i4-1a}]$ and isolation of a new line $[ry^{+i4-1a-4}]$ in which the transposon occupies a new location and expression is restored to normal levels. A comparison of expression in these two strains is documented in Fig. 26. Thus, we can conclude firmly that underexpression of transformant $[ry^{+i4-1a}]$ is due to neighboring DNA sequences.

7.2.5. Tissue-Level Expression

XDH expression in a number of transformed strains has been analyzed at the tissue level. Although quantitative differences are observed between lines, in all cases so far examined the gene appears to be expressed only in those tissues ordinarily associated with *rosy* expression. Although low levels of expression in other tissues or at inappropriate developmental stages have not been rigorously excluded, the evidence gathered thus far is consistent with the hypothesis that the sequences mediating tissue-specific expression of *rosy* re-

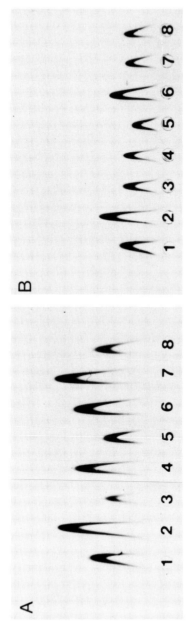

Figure 25. The effects of heterochromatin content on the XDH levels of two transformed lines: (**A**) *R401.1* and (**B**) [ry^{+i4-1q}]. **A** and **B**, Rocket immunoelectropherograms. Extracts were obtained from uniformly aged adults and contain equivalent amounts of protein. Protein synthesis in ovaries contributes substantially to the total protein in extracts from females. Since XDH is not normally produced in ovaries, protein extracts from males will exhibit greater XDH levels than equivalent preparations from females. In this experiment, valid comparisons are limited to preparations from individuals of the same sex. Wells 1 and 2 contain control samples from ry^{+10}/ry^{506} and ry^{+5}/ry^{506} XY flies, respectively, and wells 5 and 6 contain samples from the corresponding XX individuals. Experimental samples from ry^{42}/MKRS; [R401.1]/ or [ry^{+i4-1q}]/−; ry^{506}/MKRS flies were applied to wells as follows: (3) XO; (4) XY; (7) XXY; and (8) XX. The XO and XXY flies were obtained by mating Y^{SX} · Y^{L}/0; ry^{+11}/MKRS males to homozygous ry^{42}; [R401.1] or [ry^{+i4-1q}]; ry^{506} females and selecting ry^{+} Sb progeny. XY and XX flies were obtained in the same fashion, except XY; ry^{+11}/MKRS flies were used in the P$_1$ cross. (**A**) Analysis of [R401.1]. Comparison of XO with XY males and XXY and XX females reveals that XDH CRM levels are dramatically influenced by the addition and subtraction of Y chromosome heterochromatin. (**B**) Analysis of [ry^{+i4-1q}]. Alterations in Y chromosome heterochromatin content do not affect XDH levels.

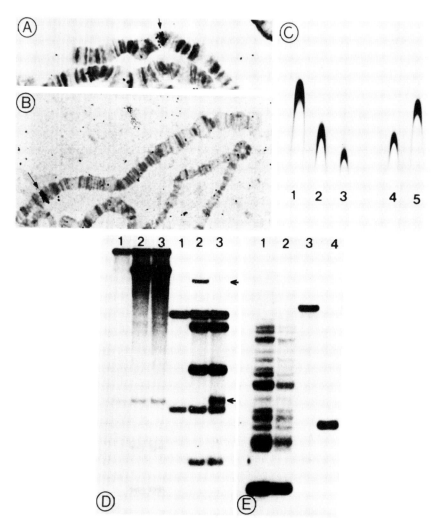

Figure 26. Analysis of [ry$^{+i4-1a-4}$]. The ry$^+$ transposon in this line is a derivative of [ry^{i4-1a}] obtained by transposition. **A** and **B**, *in situ* hybridization of ^{3}H-labeled pry8.1 DNA to salivery gland polytene chromosomes from homozygous larvae of (**A**) [ry^{+i4-1a}];ry^{506} and (**B**) [ry$^{+i4-1a-4}$] ry^{506}. In both cases, two hybridization signals are evident, one at 87D, the location of the *rosy* locus on chromosome 3 (not shown), and the other at the site of insertion of the ry$^+$ transposon. The [ry^{+i4-1a}] transposon is located at 57F (arrow in **A**) on chromosome arm 2R, and its transposition derivative, the [ry$^{+i4-1a-4}$] transposon, is inserted at 68A (arrow in **B**) on chromosome arm 3L. (**C**) Rocket immunoelectropherogram. Extracts were obtained from homozygous adult males and contain equivalent amounts of protein. Well 1 contains an undiluted control sample from ry^{+5}; wells 2 and 3 contain, respectively, dilutions with 50% and 25% of the extract in well 1. Wells 4 and 5 contain samples from [ry^{+i4-1a}]; ry^{506} and [ry$^{+i4-1a-4}$]; ry^{506}, respectively. A nearly twofold increase in XDH CRM is observed in the transposition derivative. (**D**) Verification that the [ry$^{+i4-1a-4}$] transposon DNA was unaltered by the transposition process. The autoradiogram shows 2 sets of 3 lanes. DNA samples are as follows: (1) ry^{506}; (2) [ry^{+i4-1a}] ry^{506}; and (3) [ry$^{+i4-1a-4}$] ry^{506}. The left set of samples was digested with *Hind III*, and the samples in the set on the right were digested with *Pvu II* fragment that extends from the transposon's rightmost site into the flanking genomic DNA. This fragment is unique for each insertion.

side within the 8.1-kb sequence. (Variants in which the control element in the transposon has been altered do show tissue- and stage-specific effects as shown in Section 7.3.)

7.3. *rosy* Expression in Dysgenesis-Induced Control Element Deletions

Daniels *et al.* (1985) described a method for generating small deletions within ry^+ transposon DNA via dysgenic destabilization of stocks previously transformed with the transposon-borne 8.1-kb fragment. In a pilot study, two such deletions were identified within the control element, both associated with tissue- and stage-specific effects in *rosy* expression.

7.3.1. *sd[ry²²¹⁶⁻⁵⁴⁷]*: A Variant Associated with Altered Fat-Body Expression

Southern blot analysis of one derivative, $sd[ry^{2216\text{-}547}]$, is depicted in Fig. 27. The alteration in this strain involves an approximately 400-bp deletion of DNA from the 5′ control element region of the *rosy* gene, resulting in lower XDH activity and increased sensitivity to phenocopy on allopurinol (see Section 2.3). More refined mapping (M. McCarron and A. Chovnick, unpublished data) has shown that this deletion lies entirely within the 5′ intronic sequence defined in cDNA analyses (Lee *et al.*, 1987). Figure 24 presented an analysis of the total level of XDH crm in individuals of the $sd[ry^{2216\text{-}547}]$ strain as compared with the parent type, $sd[ry^{+\,2216}]$, and demonstrates that $sd[ry^{2216\text{-}547}]$ is associated with lower levels of XDH protein, rather than altered specific activity of XDH produced by the derivative line.

M. McCarron and A. Chovnick (unpublished data) found that the reduction in XDH in strain $sd[ry^{2216\text{-}547}]$ is confined largely, if not exclusively, to the fat body of larvae and adults. Malpighian tubule expression is affected only slightly, or not at all. The effects of the alteration in $sd[ry^{2216\text{-}547}]$ thus parallel those of the *i409H* overexpression site in strain ry^{+4}, which also exerts its effects specifically in the fat body. Taken together, analyses of these two variants indicate that sequences essential for proper control of tissue-specific expression of the *rosy* gene lie within the large 5′ intron.

The larval RNA northern blot presented in Fig. 8b includes an extract from strain $sd[ry^{2216\text{-}547}]$ (lane D), in which it appears that there has been an accumulation of large transcripts with homology to the *rosy* probe employed. We are currently mapping DNA sequences into these large transcripts to test the hypothesis that they represent a heterogeneous collection of unprocessed or aberrantly processed *rosy* message.

7.3.2. *sd[ry²²¹⁶⁻⁸⁵³]*: A Variant Associated with Stage-Specific Alteration in *rosy* Expression

A second derivative, designated $sd[ry^{2216\text{-}853}]$, also involves perturbation of the 5′ DNA sequences of the *rosy* gene, including the deletion of at least part

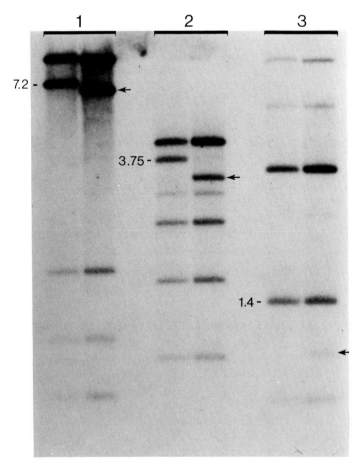

Figure 27. Autoradiogram of whole-genome Southern blot of HPP-sensitive exception. Pairs of lanes contain sd[ry^{+2216}]; ry^{506} parental DNA on the left and DNA from the exception, sd[ry$^{2216-547}$], on the right. Each pair of DNA samples was restriction-enzyme digested as follows: (1) *Hind III*; (2) *Pvu II*; and (3) *Bgl II*. The pry8.1 DNA was used as probe; the sizes of the pertinent sd[ry^{+2216}]; ry^{506} restriction fragments are indicated. The restriction patterns of the HPP-sensitive exception lack the 7.2-kb *Hind III*, the 3.75-kb *Pvu II*, and one of the 1.4-kb *Bgl II* fragments. New fragments appear (arrows), generated as a result of a 0.4-kb deletion in the *rosy* DNA borne by the perturbed transposon.

of the 5′ intron. Tissue-level analyses of this variant (M. McCarron and A. Chovnick, unpublished data) have revealed that larval fat body and malpighian tubule expressions are low. However, unlike the sd[ry$^{2216-547}$] strain, expression in the adult fat body rises to nearly wild-type levels. Tissue- and stage-specific expression in these two strains is compared to that of the parent type sd[ry^{+2216}] in Fig. 28. (It should be noted here that malpighian tubules are retained from larval into adult stages with relatively little change. By contrast, the larval fat body apparently dissociates completely, with none of its cellular constituents incorporated into the adult organ.)

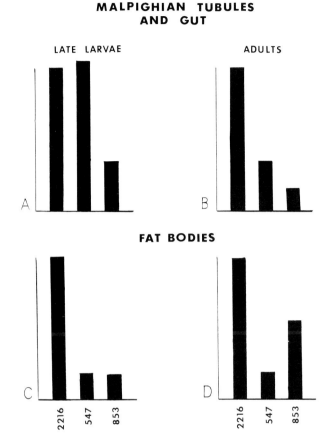

Figure 28. XDH activity in larval and adult tissues of control element deletions. Activity was measured by fluorescence assay for the two deletion-bearing strains sd[ry$^{2216-547}$] and sd[ry$^{2216-853}$], and their parent type sd[ry^{+2216}]. In each panel, the mutants' activities are expressed as percent of control. (**A**) Larval malphigian tubule and gut; (**B**) adult malpighian tubule and gut; (**C**) larval fat body; and (**D**) adult fat body.

7.3.3. Regulatory Sequences within the 5′ Intron of *rosy*

We can now point to analyses of four variants whose effects trace to sequences within the 5′ control element of the *rosy* gene. Two of these (*i1005L* and *i409H*) are natural variants, and two (sd[ry$^{2216-547}$] and sd[ry$^{2216-853}$]) are derivatives of the sd[ry^{+2216}] transposon. Of these four, at least two are associated with effects specific to the fat body, and both of these (*i409H* and sd[ry$^{2216-547}$]) involve sequences that lie within the large 5′ intron of the gene. *i1005L* lies to the left of this intron and results in a general reduction of XDH production. sd[ry$^{2216-853}$] is associated with a tissue-specific effect in the larval stage and also involves sequences that include the 5′ intron. Since the larval

and adult fat bodies have quite distinct developmental origins, the stage specificity of this variant may reflect a corresponding difference in the pattern of *rosy* expression between the two organs.

7.4. Sequence Analyses of Processing Site Mutants

The sequence of genomic DNA at the 5' side of the gene is presented in Appendix A (Lee *et al.*, 1987). Although it is of obvious interest to determine which sequence alterations are responsible for the transcriptional effects in ry^{+10} and ry^{+4}, too many sequence polymorphisms exist between the strains to allow meaningful comparison. (However, see discussion in Section 7.1.3.) As an alternative approach, a large number of mutations induced on the ry^{+5} allele have been screened to identify low or null enzyme activity variants that map at the left end of the locus. Lee *et al.* (1987) analyzed partial sequences for seven such mutants. Figure 29 presents the location of lesions identified in each, which suffice to explain the mutant phenotypes. None of the lesions thus identified are in the intron, which—in the studies reviewed above—has been implicated in the stage- and tissue-specific regulation of transcription. However, several of the mutants have aided in establishing the limits of the intron inferred from cDNA analyses (see Section 5.4). Mutants ry^{5208}, ry^{523}, and ry^{545} all appear to affect splice donor or acceptor sequences at the borders of the intron. ry^{523} and ry^{545} (acceptor site) result in a null enzyme phenotype. ry^{5208} (donor site) reduces XDH activity to less than 5% of wild type. Tissue-level analyses of the residual activity in ry^{5208} have not been completed. (Lee *et al.* (1987) report that this mutant produces a transcript somewhat smaller than the parental strain, ry^{+5}.)

8. Denouement

Until recently, rather limited analyses of the regulation of *rosy* expression have been pursued, primarily to obtain that information needed to elucidate the genetic and molecular organization of the locus. A number of current investigations place new emphasis on developmental questions of stage- and tissue-specific regulation, especially as these may be illuminated by analyses of defined mutations. These studies will exploit the wealth of molecular detail now available, to gain insight into the normal developmental regulation of *rosy* expression. Current goals of analyses include (1) precise (sequence-level) definition of the 5' end of the locus; (2) sequence definition of the *cis*-linked regulatory loci responsible for overproduction in *i409H* and underproduction in *i1005L*; (3) identification of *trans*-acting loci that affect *rosy* expression; and (4) analyses of the differences and similarities of *rosy* expression in fat body and malpighian tubule. Although there are several levels at which control may be exerted, current work focuses on regulation at the transcriptional level.

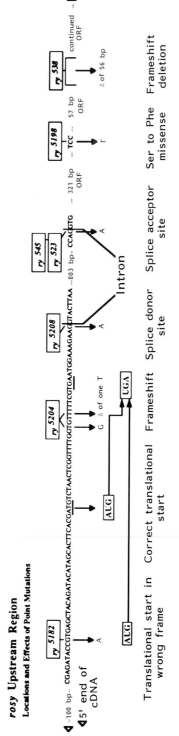

Figure 29. Summary diagram of mutant sequence changes and their proposed effects.

Appendix I

```
                                                          <PstI>              -2901
                                                          CTGCAGGTAA GTGACTGTGA

                                                                              -2801
TTGAAACCGT GCGTCGCCCT TATCACTTAA TTATCAATGT GGCATCCGCA GGTATCAATG GTCCGTCAGG AGATTCGAGC AGCAGCAATT GCGTGAGGCG

                                            <Bgl 2>                           -2701
ATGAGACGAC AAGCCCTTTA TGAGGGCACC CAAAGGGAGA GAGACTTAGA TCTGAAGTCG GCGTAGCATA AGTCAGATCC TTCATTACTA CTTCGATGTT

                                                                              -2601
TGTTTTCACT CTAGGATACC AAGTAATCCT GGCATTGTCA AATCCCTTGT GTTGTAAACT ACCTTTTGTT ACCAAACTTT ATAGTAGCCA ATATATTATA

                                                                              -2501
TATATGTTGT AAGTGAAATC ATAGCTTGGA GTTTGCTTGC CACTCGTAGG GAATTAGTTA TGCCACCATC TTTGATTTGG GATTGACAGA AATCAATCCA

                                                                              -2401
TCATTTTAAT GATTGTGGTT CCTAAAACCT GTTGGGCTTT ATTGTGCTTT CGATTTGATT TAAGCTGATA GAGCTAGTTG TTTACTTTTG AAGAACATTG

                                                                              -2301
AACCCATTTT GACGACCTTT TAACTTTTTA ATAGTTTTTT TTTACCATTT ACATAAAAGC AAAATTTACA GTCTAAAAGT TGTATATTCA ATACGTTCTG

                                                                              -2201
GTACGTAAAA GGGTTATTTA ATGCACCATT TTTACAACTT TAGATAGTAT ACTTTTTACT TGAGTGGGAG CCATATCACA AAAATGAAAT GAAAGGTTAA

                                                                       <Bgl2
AAAAGTATCT AAAAGTAAGT TTCAATTAAT ATTCTGAAAA ATAATTGCAT TTCATGTACG TTTCTCTGAA ATACCCCTAA TTCGATACTA TTTATAGATC

>                                                           C                 -2001
TGGAAGATAT ATAGCTAGGA AATAGAATTA TATATTGCC TAGTTTGAAC ACCTTGTGAA CATTTTGTTT AAAAAAAAAG CGTATGGTGG TACTAGAAAA

                                                                              -1901
CTTTAAGATA CAGATAAATA ACTTCACTAT ACCGAGGTTT TGGTGTTTCT TGAGCTGAA GGACAACTTC CAAAATGGTC TTAATTCTTG AGATAGAAAA

       C          <BclI>                          C                           -1801
GCTGAAGTTC TCATGCTGAA TAGTGATCAG TACGATTTTA TCTGACAGAT TGCTTTCCTA GCTGATCATA CCGATCTCGG ATGGGTATCT GCTCTAATCC
                                                          <BclI>

               G          <BalI>                          G            C
GAGATAGCAG GTAGTAGAAC TGCTGCCTCG TAACCCCTAG ACATGGCCAT TTCCACGCAC ATGTAGGCGA TCCTGCACTC GTCGGTCTCT CGTCTTATCT

                                            <Hinc2>        CG               -1601
CACAAAAATA AAAGTCGGCG TTTGGATGGA TTGCTGGCT GGCTGTTGGC ACATTCATTA CGTTTTCGTT GACGGCGCGA GACGGTTGTG TTGTTTTTTA

                                                              T            -1501
ATCGGCGGGC ACATGAGGTT GTCGGTTCTC TAATTTTGAA ATAAAAAACT TGTTTTCGCT TTACATTCAA GGAACGCACC TCCTCCCCTC CATATATGTA
                                            [<--5' end of I4-31 cDNA clone

                                                A in ry^5182                 -1401
AACGAGCTTC GTAATTGCAG GCTAACACTT ATATAGGCAT TGGTACTGAG ATACAGCGAG ATACCGTGAG CTACAGATAC ATAGCACTTC ACGATGTCTA
                                                                                        MetSerA


               G [-] both in ry^5204    A in ry^5208                    T          <Cla
ACTCGGTTTT GGTGTTTTTC GTGAATGGAA AGAAGGTACT TAAAGATACT CGGATGCCTT GCGTGTCTTT CCTTAAAGTC GGCCCACTCC TACTATATCG
snSerValLe uValPhePhe ValAsnGlyL ysLys<-- intron 1

1>                                                                            -1201
ATTCCCTTTA TTGTTTTTATT GCCCTAATTG CGTTTAACTG GCATTCGATG GAATTTGCCT TCTTGCTCAA TTGAATGTGT CGAATGTATA TAGTATTTGC

                                                                              -1101
TGTGCATATA GTGATAACAA TTGACCCCCA TGTGAAGCCT CGTTGTTTGC TGGTGGCCCC TTATCGCCCG ATCAAGATGG TGCACTTGCC AATCGCGCAT

                             T                        [-]                     -1001
GGGAGCAGTG CTACTCGGAT GTGGGTGAGC TGGGATCGTC ATGGGCATTT GTGTCTGCCA GATAGCTCAG GCCCTCTAAT CCGCACATTC CCACCGGCTG

T          <SstI>                            T                A               -901
GGCGGTGTTT ACAATCCAGT GAGCTCGGCT GGATTATTCA GATAAAGCTG GCCATATTTA GCATCTGGTG TTGGTGGAGT GCAATTGGAG TTCTTTACTT

                                                                              -801
TGTGCTCACA GCTTGCGCAC AAGTGAAGTA AATAAATAATA AATAACCAGT TGCTCGTGCG CCCGGCATTT GATAGTGATA GATAGCAATC TCTGTTTGCT

                   T                                             C            -701
CAGGGCTGGC AGAGCAGATA ACACTTTCGC TGTTTTAAGT TTAAACAAAG TCACAGTTTT GACAAACAAT TGCTTTAAAC CATATAAACT TATAATTAAA

       [----]        <Bgl 2>TA                    C                           -601
GCAATTATAA TGGGTTAAGT AAGCACAGAT CTAGAAGCTA CAAGATACAA GATACAAATT TTCCAATTAA TGGATAAATG GCATTAGATA CTTTTGGGAA
                <Xba 1>

                                      A in ry^523 and ry^545                  -501
ATTGATTTCG TTTTTTTTCAT TTTATTATAT TTATACATTT CTTATTCCAG GTGACCGAAG TGTCGCCTGA TCCGGAGTGC ACGCTCCTCA CATTCCTGCG
                                      intron 1 -> ValThrGluV alSerProAs pProGluCys ThrLeuLeuT hrPheLeuAr

                                                   G                          -401
CGAAAAGCTG CGGCTGTGCG GAACGAAGTT GGGATGTGCG GAAGGCGGAT GCGGCGCCTG CACCGTGATG GTGTCCCGCC TAGACCGTCG GGCCAACAAG
gGluLysLeu ArgLeuCysG lyThrLysLe uGlyCysAla GluGlyGlyC ysGlyAlaCy sThrValMet ValSerArgL euAspArgAr gAlaAsnLys

                                                   G          C    C          -301
ATACGTCACC TGGCCGGTCAA CGCCTGCCTG ACGCCAGTGT GCTCGATGCA CGGATGTGCG GTGACCACTG TGGAGGGTAT AGGGAGCCAC AAAACGCGTC
IleArgHisL euAlaValAs nAlaCysLeu ThrProValC ysSerMetHi sGlyCysAla ValThrThrV alGluGlyIl eGlySerThr LysThrArgL

                   <BamHI>                                   T in ry^5198     -201
TGCATCCGGT GCAGGAGCGA CTGGCCAAGG CATGCTGGGC TTCTGCACGC CGGGCATTGT GATGTCCATG TACGCACTTC TGCGGAACGC
euHisProVa lGlnGluArg LeuAlaLysA laHisGlySe rGlnCysGly PheCysThrP roGlyIleVa lMetSerMet TyrAlaLeuL euArgAsnAl

                   [<-- 56 bp deletion in ry^538                 <Xho 1>  -->]  -101
GGAGCAACCC TCTATGCGAG ACTTGGAGGT GGCATTCCAA GGTAACCTGT GCCGCTGCAC CGGCTATCGA CCCATTCTCG AGGGCTACAA GACGTTCACC
aGluGlnPro SerMetArgA spLeuGluVa lAlaPheGlu GlyAsnLeuC ysArgCysTh rGlyTyrArg ProIleLeuG luGlyTyrLy sThrPheThr

                                                                              -1
                                                                               <
AAGGAGTTTG CCTGCGGAAT GGGCGAGAAG TGCTGCAAAG TTAGTGGGAA AGGATGTGGA ACCGATGCGG AGACCGATGA CAAGCTCTTC GAGCGCAGCG
LysGluPheA laCysGlyMe tGlyGluLys CysCysLysV alSerGlyLy sGlyCysGly SerAspAlaG luThrAspAs pLysLeuPhe GluArgSerG

EcoRI>          <BamHI>                                      <Pvu2>           +100
AATTCCAGCC CTTGGATCCC AGCCAGGAAC CCATCTTCCC ACCGGAACTT CAGCTGAGTG ACGCCTTCGA TTCTCAGAGT TTGATCTTTA GTTCGGATAG
luPheGlnPr oLeuAspPro SerGlnGluP roIlePhePr oProGluLeu GlnLeuSerA spAlaPheAs pSerGlnSer LeuIlePheS erSerAsp
```

The DNA sequence of the upstream region of *rosy* (Lee *et al.*, 1987). The sequence is numbered with the restriction site of *Eco*RI as the 0 point.

Appendix II

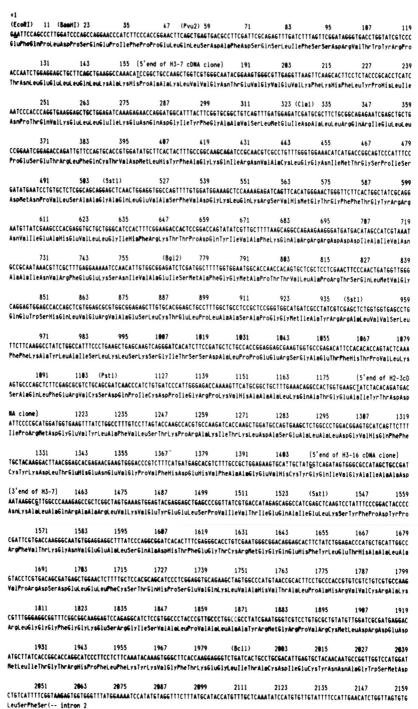

The DNA sequence of the 4.6-kb *EcoRI* fragment containing 70% of the *rosy* locus (Keith *et al.*, 1987). This sequence is numbered as in Lee *et al.* (1987). The predicted XDH protein sequence is shown below the DNA sequence.

Appendix II (*Continued*)

```
          2171      2183      2195      2207      2219      2231      2243      2255      2267      2279
GGAATTTCTAGGACCCCTATATGTTATCATCCTGAGTGCAAATAGTTTGAAACTTTTAAGAATGTTTAGGTCTAATCAGGAAAGCAAGCACCAGGATTATATTGACATAAACAATTATAA

          2291      2303      2315      2327      2339      2351      2363      2375      2387      2399
ATAATAATCTAAAAATTCTTAAAAATGATCTAATATATAAATCCTATGTTTAGGTTCTTGAGCGCGCCATGTTCCACTTTGAGAATTGCTACAGGATTCCCAACGTTCGCGTGGGTGGAT
                                                                intron 2 -->)ValLeuGluArgAlaMetPheHisPheGluAsnCysTyrArgIleProAsnValArgValGlyGlyT

          2411      2423      2435      2447      2459      2471      2483      2495      2507      2519
GGGTCTGCAAGACGAACCTGCCCTCGAATACGGCCTTCCGTGGATTTGGAGGACCACAAGGCATGTAGCCGGTGAGCATATCATCCGGGATGTGGCCCGGATAGTGGGTCGCGATGTGG
          rpValCysLysThrAsnLeuProSerAsnThrAlaPheArgGlyPheGlyGlyProGlnGlyMetTyrAlaGlyGluHisIleIleArgAspValAlaArgIleValGlyArgAspValV

          2531      2543     (5'end of I4-8,14,18 cDNA clones)  2591      2603      2615      2627      2639
TGGATGTGATGCGGCTGAACTTCTACAAGACTGGAGACTACACACACTACCACCAGCAGCTGGAGCACTTCCCCATCGAGCGGTGTCTGGAGGATTGCTTGAAGCAGTCGAGATACGACG
          alAspValMetArgLeuAsnPheTyrLysThrGlyAspTyrThrHisTyrHisGlnGlnLeuGluHisPheProIleGluArgCysLeuGluAspCysLeuLysGlnSerArgTyrAspG

          2651      2663      2675      2687      2699      2711      2723      2735      2747      2759
AGAAGCGGCAGGATATTGCTCGATTCAATCGGGAGAATCGCTGGCGGAAACGCGGCATGGCGGTGGTGCCCACCAAGTATGGAATCGCATTTGGAGTGATGCACTTGAACCAAGCGGGAT
          luLysArgGlnAspIleAlaArgPheAsnArgGluAsnArgTrpArgLysArgGlyMetAlaValValProThrLysTyrGlyIleAlaPheGlyValMetHisLeuAsnGlnAlaGlyS

          (3'end of M2-3)  2783(BamH1)  2795     2807      2819      2831      2843      2855      2867      2879
CGCTGATCAACATCTATGGTGATGGATCCGTGTTGCTTTCGCACGGAGGAGTTGAGATCGGACAAGGTCTGAATACCAAGATGATTCAGTGCGCCGCCAGGGCTCTGGGGATTCCTTCGG
          erLeuIleAsnIleTyrGlyAspGlySerValLeuLeuSerHisGlyGlyValGluIleGlyGlnGlyLeuAsnThrLysMetIleGlnCysAlaAlaArgAlaLeuGlyIleProSerG

          2891      2903      2915      2927      2939     (BamH1)   2963      2975      2987      2999
AACTGATTCACATTTCGAAAACGGCCACGGATAAAGTACCCAACACTTCACCCACGGCGGCGAGTCGGATCTGAACGGAATGGCGTCTGGATGCGTGTGGAAAAGTTGAACA
          luLeuIleHisIleSerGluThrAlaThrAspLysValProAsnThrSerProThrAlaAlaSerValGlySerAspLeuAsnGlyMetAlaValLeuAspAlaCysGluLysLeuAsnL

          3011      3023      3035      3047      3059      3071      3083      3095      3107      3119
AAAGACTGGCGCCCATCAAGGAGGCATTGCCTGGAGGCACCTGGAGGGAGTGGATCAACAAGGCGTATTCGATCGGGTCAGCCTCTCGGCCACAGGATTCTATGCCATGCCCGGGATTG
          ysArgLeuAlaProIleLysGluAlaLeuProGlyGlyThrTrpLysGluTrpIleAsnLysAlaTyrPheAspArgValSerLeuSerAlaThrGlyPheTyrAlaMetProGlyIleG

          3131      3143     (3'end of M3-16)  3167      3179      3191     (ClaI)    3215      3227      3239
GATATCACCCGGAAACGAATCCCAATGCTCGCACCTATAGCTACTACACGAATGGCGTGGGAGTCACTGTGGTAGAGATCGATTGCCTGACTGGCGACCATCAGGTGCTCAGCACAGACA
          lyTyrHisProGluThrAsnProAsnAlaArgThrTyrSerTyrTyrThrAsnGlyValGlyValThrValValGluIleAspCysLeuThrGlyAspHisGlnValLeuSerThrAspI

          3251      3263      3275      3287      3299      3311      3323      3335      3347      3359
TCGTGATGGACATCGGCTCTAGCCTGAATCCGGCTATTGACATTGGTCAGATCGAGGGAGCATTCATGCAGGGCTATGGACTGTTCACTTTGGAGGAACTCATGTACTCACCACAAGGCA
          leValMetAspIleGlySerSerLeuAsnProAlaIleAspIleGlyGlnIleGluGlyAlaPheMetGlnGlyTyrGlyLeuPheThrLeuGluGluLeuMetTyrSerProGlnGlyM

          3371      3383      3395      3407      3419      3431      3443      3455      3467      3479
TGCTTTACTCCAGAGGTCCGGGCATGTACAAGCTGCCCAGGATTTCGCGACATTCCCGGGAGTTCAATGTCAGCCTACTGACCGGTGCCCCCAATCCACGGGCAGTCTACTCTTCCAAGG
          ETLeuTyrSerArgGlyProGlyMetTyrLysLeuProGlyPheAlaAspIleProGlyGluPheAsnValSerLeuLeuThrGlyAlaProAsnProArgAlaValTyrSerSerLysA

          3491      3503      3515      3527      3539      3551      3563      3575      3587      3599
CAGTGGGTGAACCTCCGCTCTTCATTGGATCATCTGCATTCTTTCGCCATTAAGGAGGCCATTGCAGCTGCTCGCGAGGATCAGGGCTTGAGTGGTGACTTCCCACTGGAGGCGCCTTCCA
          laValGlyGluProProLeuPheIleGlySerSerAlaPhePheAlaIleLysGluAlaIleAlaAlaAlaArgGluAspGlnGlyLeuSerGlyAspPheProLeuGluAlaProSerT

          3611      3623      3635      3647      3659      3671      3683      3695      3707      3719
CATCGGCACGCATTCGAATTGCTTGTCAGGATAAGTTCACGGAACTGGTAAGTTACCCTTGGATTAGTTTAGAACATTAAACTAACTTTTATTTATTAATTTATTAATTATAGCTTGAAAT
          hrSerAlaArgIleArgIleAlaCysGlnAspLysPheThrGluLeu(-- intron 3                               intron 3 -->)LeuGluIl

          3731      3743      3755      3767      3779      3791      3803      3815      3827      3839
ACCCGAACCAGGATCATTTACGCCATGGAACATTGTGCCTTAAATTGTTTTTATTGTTTGTACTGCTTAAGCATTTAAAATGACGATTTTATTTGTTAATTTGTTTATATATACATAGT
          eProGluProGlySerPheThrProTrpAsnIleValPro

          (3'end of I4-8,14,18)  3875      3887      3899      3911      3923      3935      3947      3959
GAATTAAATGTTTTAAAAAATAATTGAGTCGTTTAATGTGTAAACTAAGCTGGAGAGCTGTTTGAACAAATTTATATACCACTGATTAATATTAAATATTCCTTTTAGATTAAGAAAAATT

          3971      3983      3995      4007      4019      4031      4043      4055      4067      4079
CAAATGGTTACTTTTTTGTTCTATCTATTATTCAATAGAATCTCGTGGCATGCAATGCATTCTGATTTTCAATTGAAAAGTCTTAGTCACTATTTATTTTGAGTATATTAATGAAAGTTGA

          4091      4103      4115      4127      4139      4151     (Hind3)   4175      4187 (Bgl2) 4199
GCAAGTTTTCCGATGAATTGAAATATGAGTTCCGATTTCGGCGCGACTGCTACCCGAAAATATAAGCTCAATCAAAAGAAGCTTTGTATGGAACTCAGGTTATGAGACGAGATCTTCGCG

          4211      4223     (Sst1)    4247      4259      4271      4283      4295      4307      4319
AGCATCTCAGTCGGCAATCTGAAGTCGTAGAGCTCGGACATCCGAGATGGGAAAGCTATCATTTGGATACAGTTTGCTTTTAGAGTTCGTCTTAATTTCGACGTTATCGTTCGTATCGGC

          4331      4343      4355      4367      4379      4391      4403 (ClaI) 4415      4427      4439
AGAACAATTGTTGCTATCACTACAACCATTATTACAATAGATATGCACAGAATGCGCCTTGGAACGCAGTGGCTCCATCGTATCTATCGATTGACATCTACGGAAATTGCCAAGCCCACG

          4451      4463      4475      4487      4499      4511      4523      4535      4547      4559
ATCGTCCGCTGATTGGTAAGTGTGTGCGGTACGTGGATTGCATTAGTGCCATGCAGGCAGTGCCCCGGGTGACACCACTACTCTGTCCGTCATCGTGGCCCAATCAGTTGGTTTGCTGTC

          4571      4583      4595      4607      4619
CACACGGCGGATACTTGTTGCCGCCGCCCCAGCATCTCGAAGAGCGAACAAGGTGCGTATGAAAGG
```

References

Andres, R. Y., 1976, Aldehyde oxidase and xanthine dehydrogenase from wild-type *Drosophila melanogaster* and immunologically cross-reacting material from *ma-1* mutants, *Eur. J. Biochem.* **62**:591–600.

Baker, B. S., 1973, The maternal and zygotic control of development of *cinnamon*, a new mutant in *Drosophila melanogaster*, *Dev. Biol.* **33**:429–440.

Barrett, D., and Davidson, N. A., 1975, Xanthine dehydrogenase accumulation in developing *Drosophila* eyes, *J. Insect Physiol.* **21**:1447–1452.

Bender, W., Spierer, P., and Hogness, D. S., 1983, Chromosomal walking and jumping to isolate DNA from the *ace* and *rosy* loci and the *bithorax* complex in *Drosophila melanogaster*, *J. Mol. Biol.* **168**:17–33.

Boni, P., DeLerma, B., and Parisi, G., 1967, Effects of the inhibitor of xanthine dehydrogenase, 4-hydroxyprazolo (3,4d) pyrimidine (or HPP) on the red eye pigments of *Drosophila melanogaster*, *Experientia* **23**:186–187.

Chovnick, A., 1966, Genetic organization in higher organisms, *Proc. R. Soc. (Lond.) Ser. B* **164**:198–208.

Chovnick, A., Schalet, A., Kernaghan, R. P., and Talsma, J., 1962, The resolving power of genetic fine structure analysis in higher organisms as exemplified by *Drosophila*, *Am. Nat.* **96**:281–296.

Chovnick, A., Schalet, A., Kernaghan, R. P., and Krauss, M., 1964, The *rosy* cistron in *Drosophila melanogaster*: Genetic fine structure analysis, *Genetics* **50**:1245–1259.

Chovnick, A., Ballantyne, G. H., Baillie, D. L., and Holm, D. G., 1970, Gene conversion in higher organisms: Half-tetrad analysis of recombination within the *rosy* cistron of *Drosophila melanogaster*, *Genetics* **66**:315–329.

Chovnick, A., Gelbart, W., McCarron, M., Osmond, B., Candido, E. P. M., and Baillie, D. L., 1976, Organization of the *rosy* locus in *Drosophila melanogaster*: Evidence for a control element adjacent to the xanthine dehydrogenase structural element, *Genetics* **84**:233–255.

Chovnick, A., McCarron, M., Hilliker, A., O'Connell, J., Gelbart, W., and Clark, S., 1978, Gene organization in *Drosophila*, *Cold Spring Harbor Symp. Quant. Biol.* **42**:1011–1021.

Clark, S. H., and Chovnick, A., 1986, Studies of normal and position-affected expression of *rosy* region genes in *Drosophila melanogaster*, *Genetics* **114**:819–840.

Clark, S. H., Hilliker, A. J., and Chovnick, A., 1979, Genetic organization of the *rosy* locus in *Drosophila melanogaster*: Pilot studies on the induction and analysis of low XDH activity mutants of the *rosy* locus as putative control mutants, in: *Eucaryotic Gene Regulation*, ICN–UCLA Symposia on Molecular and Cellular Biology, Vol. XIV (R. Axel, T. Maniatis, and C. Fred Fox, eds.), pp. 117–121, Academic, New York.

Clark, S. H., Daniels, S., Rushlow, C. A., Hilliker, A. J., and Chovnick, A., 1984, Tissue-specific and pretranslational character of variants of the *rosy* locus control element in *Drosophila melanogaster*, *Genetics* **108**:953–968.

Clark, S. H., Hilliker, A. J., and Chovnick, A., 1986a, Genetic analysis of the right (3′) end of the *rosy* locus in *Drosophila melanogaster*, *Genet. Res. Camb.* **47**:109–116.

Clark, S. H., McCarron, M., Love, C., and Chovnick, A., 1986b, On the identification of the *rosy* locus DNA in *Drosophila melanogaster*: intragenic recombination mapping of mutations associated with insertions and deletions, *Genetics* **112**:755–767.

Collins, J. F., Duke, E. J., and Glassman, E., 1971, Multiple molecular forms of xanthine dehydrogenase and related enzymes. IV. The relationship of aldehyde oxidase to xanthine dehydrogenase, *Biochem. Gen.* **5**:1–13.

Coté, B., Bender, W., Curtis, D., and Chovnick, A., 1986, Molecular mapping of the *rosy* locus in *Drosophila melanogaster*, *Genetics* **112**:769–783.

Covington, M., Fleenor, D., and Devlin, R. B., 1984, Analysis of xanthine dehydrogenase mRNA levels in mutants affecting the expression of the *rosy* locus, *Nucl. Acids Res.* **12**:4559–4573.

Daniels, S. B., McCarron, M., Love, C., Clark, S. H., and Chovnick, A., 1986, The underlying bases of gene expression differences in stable transformants of the *rosy* locus in *Drosophila melanogaster*, *Genetics* **113**:265–285.

Daniels, S. B., McCarron, M., Love, C., Clark, S. H., and Chovnick, A., 1986, The underlying bases of gene expression differences in stable transformants of the *rosy* locus in *Drosophila melanogaster, Genetics* **113**:265–285.

Edwards, T. C. R., Candido, E. P. M., and Chovnick, A., 1977, Xanthine dehydrogenase from *Drosophila melanogaster*. A comparison of the kinetic parameters of the pure enzyme from two wild-type isoalleles differing at a putative regulatory site, *Mol. Gen. Genet.* **154**:1–6.

Forrest, H. S., Glassman, E., and Mitchell, H. K., 1956, Conversion of 2-amino-4-hydroxypteridine to isoxanthopterin in *D. melanogaster, Science* **124**:725–726.

Forrest, H., Hanley, E. W., and Lagowski, J. M., 1961, Biochemical differences between the mutants *rosy-2* and *maroon-like* of *Drosophila melanogaster, Genetics* **40**:1455–1463.

Gelbart, W., McCarron, M., Pandey, J., and Chovnick, A., 1974, Genetic limits of the xanthine dehydrogenase structural element within the *rosy* locus in *Drosophila melanogaster, Genetics* **78**:869–886.

Gelbart, W., McCarron, M., and Chovnick, A., 1976, Extension of the limits of the XDH structural element in *Drosophila melanogaster, Genetics* **84**:211–232.

Glassman, E., 1962, Convenient assay of xanthine dehydrogenase in single *Drosophila melanogaster, Science* **137**:990–991.

Glassman, E., 1965, Genetic regulation of xanthine dehydrogenase in *Drosophila melanogaster, Fed. Proc.* **24**:1243–1251.

Glassman, E., and Mitchell, H. K., 1959, Mutants of *Drosophila melanogaster* deficient in xanthine dehydrogenase, *Genetics* **44**:153–162.

Glassman, E., Karam, J. D., and Keller, E. C., 1962, Differential response to gene dosage experiments involving the two loci which control xanthine dehydrogenase of *Drosophila melanogaster, Z. Vererbungslehre* **93**:399–403.

Glassman, E., Shinoda, I., Moon, H. M., and Karam, J. D., 1966, *In vitro* complementation between non-allelic *Drosophila* mutants deficient in xanthine dehydrogenase. IV. Molecular weights, *J. Mol. Biol.* **20**:419–422.

Grell, E. H., 1962, The dose effect of *ma-1*[+] and *ry*[+] on xanthine dehydrogenase activity in *Drosophila melanogaster, Z. Vererbungslehre* **93**:371–377.

Hadorn, E., and Mitchell, H. K., 1951, Properties of mutants of *Drosophila melanogaster* and changes during development as revealed by paper chromatography, *Proc. Natl. Acad. Sci. USA* **37**:650–665.

Hadorn, E., and Schwinck, I., 1956, A mutant of *Drosophila* without isoxanthopterine which is non-autonomous for the red eye pigments, *Nature (Lond.)* **177**:940–941.

Henikoff, S., Sloan, J. S., and Kelly, J. D., 1983, A Drosophila metabolic gene transcript is alternatively processed, *Cell* **34**:405–414.

Hilliker, A. J., Clark, S. H., Chovnick, A., and Gelbart, W., 1980, Cytogenetic analysis of the chromosomal region immediately adjacent to the *rosy* locus in *Drosophila melanogaster, Genetics* **95**:95–110.

Keller, E. C., and Glassman, E., 1965, Phenocopies of the *ma-1* and *ry* mutants of *Drosophila melanogaster*. Inhibition *in vivo* of xanthine dehydrogenase by 4-hydroxyprazolo (3,4d) pyrimidine, *Nature (Lond.)* **208**:202–203.

Kieth, T. P., Riley, M. A., Kreitman, M., Lewontin, R. C., Curtis, D., and Chambers, G., 1987, Sequence of the structural gene for xanthine dehydrogenase (*rosy* locus) in *Drosophila melanogaster, Genetics* **116**:67–73.

Laurell, C.-B., 1966, Quantitative estimation of proteins by electrophoresis in agarose gel containing antibodies, *Anal. Biochem.* **15**:45–52.

Lee, C. S., Curtis, D., McCarron, M., Love, C., Gray, M., Bender, W., and Chovnick, A., 1987, Mutations affecting expression of the *rosy* locus in *Drosophila melanogaster, Genetics* **116**:55–66.

Maniatis, T., Fritsch, E. F., and Sambrook, J., 1982, *Molecular Cloning. A Laboratory Manual*, Cold Spring Harbor Laboratory, Cold Spring Harbor, New York.

McCarron, M., Gelbart, W., and Chovnick, A., 1974, Intracistronic mapping of electrophoretic sites in *Drosophila melanogaster*: Fidelity of information transfer by gene conversion, *Genetics* **76**:289–299.

McCarron, M., O'Connell, J., Chovnick, A., Bhullar, B. S., Hewitt, J., and Candido, E. P. M., 1979, Organization of the rosy locus in Drosophila melanogaster: Further evidence in support of a cis-acting control element adjacent to the xanthine dehydrogenase structural element, Genetics **91:**275–293.

Mitchell, H. K., Glassman, E., and Hadorn, E., 1959, Hypoxanthine in rosy; and maroon-like mutants of Drosophila melanogaster, Science **129:**268–269.

Morita, T., 1958, Purine catabolism in Drosophila melanogaster, Science **128:**1135.

Rubin, G. M., and Spradling, A. C., 1982, Genetic transformation of Drosophila with transposable element vectors, Science **218:**348–353.

Rushlow, C. A., and Chovnick, A., 1984, Heterochromatic position effect at the rosy locus of Drosophila melanogaster: cytological, genetic and biochemical characterization, Genetics **108:**589–602.

Rushlow, C. A., Bender, W., and Chovnick, A., 1984, Studies on the mechanism of heterochromatic position effect at the rosy locus of Drosophila melanogaster, Genetics **108:**603-615.

Sang, J. H., 1984, in: Genetics and Development, Longman, New York. p. 136.

Sayles, C. D., Browder, L. W., and Williamson, J. H., 1973, Expression of xanthine dehydrogenase activity during embryonic development of Drosophila melanogaster, Dev. Biol. **33:**213–217.

Smith, K. D., Unsprung, H., and Wright, T. R. F., 1963, Xanthine dehydrogenase in Drosophila: Detection of isozymes, Science **142:**226–227.

Spierer, P., Spierer, A., Bender, W., and Hogness, D., 1983, Molecular mapping of genetic and chromomeric units in Drosophila melanogaster, J. Mol. Biol. **168:**35–50.

Spofford, J. B., 1976, Position effect variegation in Drosophila, in: Genetics and Biology of Drosophila, Vol. 1C (M. Ashburner and E. Novitski, eds.), pp. 955–1018, Academic, New York.

Spradling, A. C., and Rubin, G. M., 1983, The effect of chromosomal position on the expression of the Drosophila xanthine dehydrogenase gene, Cell **34:**47–57.

Ursprung, H., and Hadorn, E., 1961, Xanthindehydrogenase in Organen von Drosophila melanogaster, Experientia **17:**230–232.

Yen, T. T., and Glassman, E., 1965, Electrophoretic variants of xanthine dehydrogenase in Drosophila melanogaster, Genetics **52:**977.

Chapter 11

Transcriptional and Post-Transcriptional Strategies in Neuroendocrine Gene Expression

MICHAEL ROSENFELD, E. BRYAN CRENSHAW III,
RONALD EMESON, STUART LEFF, JEFFREY GUISE,
SERGIO LIRA, CHRISTIAN NELSON,
CHARLES NELSON, and ANDREW RUSSO

The developmental and homeostatic control of expression of certain genes in higher eukaryotes is directed by a diverse group of regulatory molecules comprising the neuroendocrine system, which act through specific receptor-mediated events. The precise temporal and spatial expression of genes of this system during development is requisite for the progressively more complex patterns of regulated gene expression that characterize higher eukaryotes. Based on an analysis of the rat and human calcitonin genes, alternative RNA processing represents one developmental strategy used in the neuroendocrine system to direct a tissue-specific pattern of polypeptide production. A second developmental strategy utilized by the neuroendocrine system restricts the expression of genes encoding neuroendocrine peptides to precise groups of cells as a consequence of modulating their transcription. Understanding the molecular mechanisms responsible for generating such patterns of restricted gene expression is likely to provide general insights into the molecular strategies that are critical for development and function of the neuroendocrine system.

1. Regulation of Gene Expression at the Level of Alternative RNA Processing as a Mechanism for Generating Diversity

The expression of eukaryotic genes requires the activities of complex biochemical machinery to transcribe, process, and transport mRNA before it can be translated into a functional product (Darnell, 1982; Breathnach and Cham-

MICHAEL ROSENFELD, E. BRYAN CRENSHAW III, RONALD EMESON, STUART LEFF, JEFFREY GUISE, SERGIO LIRA, CHRISTIAN NELSON, CHARLES NELSON, and ANDREW RUSSO • Howard Hughes Medical Institute, Eukaryotic Regulatory Biology Program, School of Medicine, University of California, San Diego, La Jolla, California 92093.

bon, 1981). Whereas simple transcription units in eukaryotes can be defined as those that are transcribed to generate mature transcripts encoding a simple protein product, complex transcription units, which generate more than one discrete transcript encoding different protein products, have been established to be a common regulatory event in eukaryotic gene expression.

It has been established that multiple messenger RNAs (mRNAs) can be generated from a single transcription unit in several viral and a large number of eukaryotic genes (see Nevins, 1983; Ziff, 1980). In the case of adenovirus and simian virus 40 (SV40), alternative RNA splicing maximizes the functional use of the limited genetic information. The complex arrangement of the gene allows for a diverse group of proteins to be produced from a single transcription matrix 25 kilobases (kb) in length (Ziff, 1980). Alternative use of five separate poly(A) sites, alternative transcriptional termination, and alternative splicing events all serve to generate a series of mature transcripts, many encoding different protein products. Comparable types of RNA processing regulation have been identified in endogenous eukaryotic genes.

A number of mechanistically distinct forms of alternative RNA processing events have been identified. RNA polymorphism can be associated with the use of alternative 3' polyadenylation sites in association with alternative splicing choices, as exemplified by immunoglobulin heavy-chain (Early et al., 1986; Maki et al., 1980; Alt et al., 1980) gene expression. Immunoglobulin heavy-chain gene expression shows multiple changes during B-lymphocyte development (see Blattner and Tucker, 1984). Early immunoglobulin heavy-chain gene expression is in the form of the membrane-associated immunoglobulin M (IGM) monomer before progressive appearance of IgD at the surface. Antigenic or mitogenic stimuli decrease membrane-associated IgM and IgD, with production of secreted IgM. The switch from membrane-bound to secreted immunoglobulin is the consequence of alternative RNA processing, which results in a change of poly(A) site utilization with an accompanying alteration of 3'-exon splicing. The result is a coding change of the C-terminal portion of the molecule, in which the membrane anchoring information, encoded by an exon, permits passage of the immunoglobulin through the membrane. The genomic organization is consistent with the possibility that poly(A) site selection controls the differential processing events. When B cells express immunoglobulins such as IgM and IgD, alternative RNA splicing of nuclear transcripts, which use different poly(A) sites, results in expression of the isotypes. Actions of either a specific or a general *trans*-acting factor, such as an endonuclease that recognizes different poly(A) sites with different efficiencies and whose levels may vary in different tissues (Blattner and Tucker, 1984), has been suggested. Alternative exon splicing in transcripts that utilize a unique poly(A) site can be associated with use of multiple transcription initiation sites or unique CAP sites (see Leff et al., 1986). The stochastic use of alternative splice sites within a single coding exon and mRNAs exhibiting both 5'- and/or 3'-terminal heterogeneity (but with invariant splicing patterns) represent additional mechanisms that can generate further polypeptide product diversity (Leff et al., 1986). Therefore, alternative RNA processing is widespread, and examples of developmental control of such alternative processing events are found in many gene families. The similarity of the alternative RNA-processing events in genes of the neuroendocrine and immune system with those of other families of eukaryotic genes suggest that common underlying biochemical mechanisms may operate in this form of regulated expression of many eukaryotic transcription units. The molecular basis for developmentally regulated alternative RNA processing remains entirely unknown and represents an important issue for understanding normal developmental and potentially pathological events. In addition to immunoglobulin heavy-chain genes, other transcription units produce transcripts subject to alternative processing (see Ziff *et al.*, 1986). Research in our laboratory has been focused on alternative processing of the calcitonin gene transcript to yield calcitonin mRNA and a messenger for calcitonin gene-related peptide (CGRP). Figure 1 compares the processing of the IgM transcript with that of the calcitonin/CGRP transcript.

2. Calcitonin/CGRP Gene Expression: Developmentally Regulated Tissue-Specific Control of Alternative RNA-Processing Events

Evidence that both CGRP and calcitonin mRNAs are generated by differential RNA processing from a single genomic locus is provided by isolation and sequencing of the calcitonin genomic DNA and calcitonin and CGRP cDNAs (Rosenfeld *et al.*, 1981, 1982, 1983; Amara *et al.*, 1982b) (Figs. 1, 2, 3, and 4). Examination of the sequence of the two gene products reveals that CGRP and calcitonin mRNAs share sequence identity through nucleotide 227 of the coding region, predicting that the initial 76 N′-terminal amino acids of each precursor are identical. The nucleotides then diverge entirely, thereby encoding unique C-terminal domains (Amara *et al.*, 1982b). Protein-processing signals within the C-terminal region of CGRP predict the excision of a 37-amino acid polypeptide containing a C-terminal aminated phenylalanine residue (Amara *et al.*, 1982b). Based on the structure of the calcitonin-CGRP gene, production of calcitonin mRNA involves splicing of the first three exons, which is present

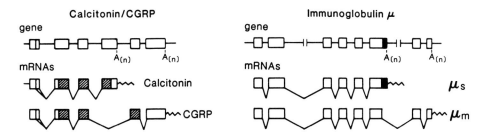

Figure 1. Schematic representation of alternative RNA-processing events in calcitonin/CGRP gene expression and immunoglobulin heavy-chain gene expression.

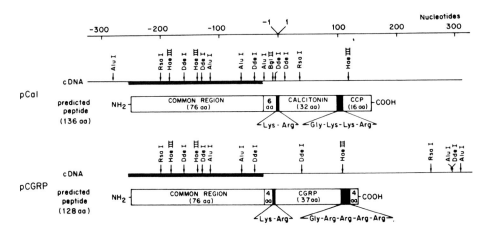

Figure 2. The mRNA structure of calcitonin and CGRP mRNAs were deduced from the sequences of cDNA clones, based on RNAs isolated from rat medullary thyroid C cell tumors, and the open reading frame predicted the primary protein sequence. Production of CGRP has been confirmed by immunological methods and by protein purification. (Adapted from Amara *et al.*, 1983.)

in both mRNAs, to the fourth exon, which encodes the entire calcitonin sequence (Fig. 5). Alternative splicing of the first three exons to the fifth and sixth exons, which contain the entire CGRP coding sequence and the 3′ noncoding sequences, respectively, results in production of CGRP mRNA. In this case, the fourth exon is excised along with the flanking intervening sequences.

Using an antiserum generated against a synthetic peptide corresponding to the fourteen C-terminal amino acids of CGRP and S1 nuclease protection assays, immunoreactive CGRP and CGRP mRNA were identified in a unique distribution in a large number of cell groups and pathways in the central nervous system (CNS) distinct from that of any known neuropeptide (Rosenfeld *et al.*, 1983) (Fig. 6). Gel filtration analysis of brain immunoreactive peptide suggests that this precursor is processed in brain to generate the predicted peptide product (CGRP) (Rosenfeld *et al.*, 1983), and primary cultures of rat

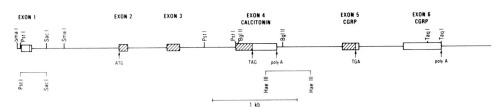

Figure 3. Schematic diagram of the rat calcitonin/CGRP gene, based on the primary sequence. Exon 4 includes calcitonin-coding information; exon 5 includes CGRP coding information. Boxes indicate the component exons. Exons 1, 2, and 3 are present in both calcitonin and CGRP mRNAs. Exon 4 is present exclusively in calcitonin mRNA; exons 5 and 6 are present exclusively in CGRP mRNA.

Exon 1

5'--ccgcggcggg**aataa**gagcagctgcaggcgcttggaaGCACAGGAGCCGCTGCCCAGATCAAGAG

                                    50
TCACCGCCTCGCAACCACCGCCTGGCTCCATCAGGACCCCGCAGTCTCAGCTCCAAGTCATCGCTCACC

          100   **Exon 2**   1200
AGGTGAGCCCTGAGGTTCCTGCTCAGgt--1.06 kb--agGGAGGCATCATGGGCTTTCTGAAGTTCTC
                                                    MetGlyPheLeuLysPheSe

                                    1250
CCCTTTCCTGGTTGTCAGCATCTTGCTCCTGTACCAGGCATGCGGCCTCCAGGCAGTTCCTTTGAGgt
rProPheLeuValValSerIleLeuLeuLeuTyrGlnAlaCysGlyLeuGlnAlaValProLeuArg

          Exon 3   1800
--0.48 kb--ccGTCAACCTTAGAAAGCAGCCCAGGCATGGCCACTCTCAGTGAAGAAGAAGCTCGCCT
              SerThrLeuGluSerSerProGlyMetAlaThrLeuSerGluGluGluAlaArgLe

          1850
ACTGGCTGCACTGGTGCAGAACTATATGCAGATGAAAGTCAGGGAGCTGGAGCAGGAGGAGGAACAGGA
uLeuAlaAlaLeuValGlnAsnTyrMetGlnMetLysValArgGluLeuGluGlnGluGluGluGlnGl

          1900
GGCTGAGGGCTCTAGgtaaggtccccttacccgt--0.64kb--cgcatgtctttccctgcagCTTGG
uAlaGluGlySerSer   LeuA

Exon 4   2600   2650
ACAGCCCCAGATCTAAGCGGTGTGGGAATCTGAGTACCTGCATGCTGGGCACGTACACACAAGACCTCA
spSerProArgSerLysArgCysGlyAsnLeuSerThrCysMetLeuGlyThrTyrThrGlnAspLeuA

                                    2700
ACAAGTTTCACACCTTCCCCCAAACTTCAATTGGGGTTGGAGCACCTGCAAGAAAAGGGATATGGCCA
snLysPheHisThrPheProGlnPheSerIleGlyValGlyAlaProGlyLysLysArgAspMetAlaL

                                    2750
AGGACTTGGAGACAAACCACCACCCCTATTTTGGCAACTAGGTCCCTCCTCTCCTTTCCAGTTTCCATC
ysAspLeuGluThrAsnHisHisProTyrPheGlyAsnStop

          2800   2850
TTGCTTTCTTCCTATAACTTGATGCATGTAGTTCCTCTCTGGCTGTTCTCTGGGCTATTATGGGTTACT

                    2900
TTCATGAGGCAAAGGATGGTATCTGGAATCTCCAATGGGTGAGGAGAAAGGGCCTACAGGCTAAAAGAG

          2950
AATCACCCAGGAAGATGGCAGAGAGCCAGGGCAGTCATCTGGATTCCTAGTAGAGCTTCTCAGTCTAGT

3000   3050
CTTGCTTCATAAAGTGTTGGTTGTTTGGGA**AATAAA**CCTATTTTTCTAAAAGGACttgagccatggtgg

                                    Exon 5
tcattgatgtagcccacatctcaggc--0.71 kb--cactgcatcctgaatatcagTGTCACTGCCCAG
                                                              ValThrAlaGln

          3850   3900
AAGAGATCCTGCAACACTGCCACCTGCGTGACCCATCGGCTGGCAGGCTTGCTGAGCAGGTCGGGAGGT
LysArgSerCysAsnThrAlaThrCysValThrHisArgLeuAlaGlyLeuLeuSerArgSerGlyGly

          3950
GTGGTGAAGGACAACTTTGTGCCCACCAATGTGGGCTCTGAAGCCTTCGGCCGCCGCCGCCAGGGACCTT
ValValLysAspAsnPheValProThrAsnValGlySerGluAlaPheGlyArgArgArgArgAspLeu

          4000
CAGGCTTGAACAGATAATAGCCCCAGAAAGAAGgtgacttccttgtacaactgg--0.51 kb--tattc
GlnAlaStop

          Exon 6   4600
tttttttcctcctagGTTACAC**AATAAA**GATAAACTCTAATTCTTTTCATGTATA**ATTAAA**GTCATGTG

          4650
TCAGAAAGGCTGATGGAAGACACATATATTTGCATCCTTCTTGGTACTGAAACCCTTCTCCCTATGACA

4700   4750
GGA**AATAAA**GCTAAGTGCAGAATAAGCTCACCCATAGTTGGCTATTGTGCATCGTGTTGTATGTGATTCT

          4800
ATATCCATAACATGACAGCCATGGTTCTGGCTTATCTGGTAGCAAATCTAGTCCCCATAAACCACCCTGT

          4850
CGATGTTGATAACTCTGCTAAACCTCAAGCGGATATGAAACCACTGCCTCTTGGTCTTCTGGGGACACA

4900   4950
TGGTAATGTTGTGACTCAATGGAACCATATGCTTAAAGAACTGTTAATGTTGTCACTTGTGAGCTTAAT

CAAA**ATTAAA**AAATATGTATTTTCGATatcctcaagtggagtctttgctgttacttatactgtgttt--3'

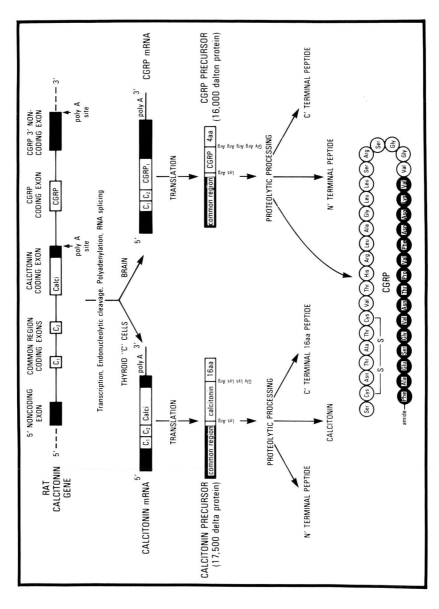

Figure 5. A model of tissue-specific neuropeptide production in calcitonin/CGRP gene expression. The structural organization of the gene and the mature transcripts are based on DNA sequence information. (From Rosenfeld et al., 1983. Reprinted by permission from Nature, Vol. 304, No. 5922, pp. 129–135. Copyright © 1983 Macmillan Journals

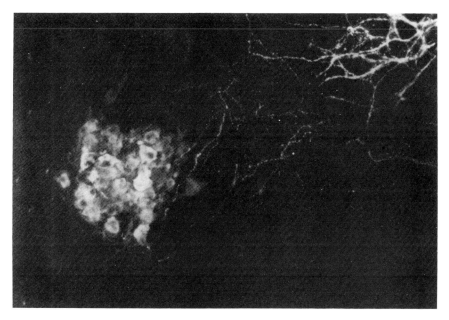

Figure 6. Immunohistochemical demonstration of CGRP in the nucleus ambiguus (cell bodies) and fibers of the spinal tract cranial nerve 5 (trigeminal). Staining was completely and specifically blocked by CGRP (1 μg).

trigeminal ganglia appear to secrete authentic CGRP peptide (Mason *et al.*, 1984). Tissue specificity of the RNA-processing events is suggested because virtually no calcitonin mRNA could be identified in the rat brain (Rosenfeld *et al.*, 1983), whereas in thyroid C cells, calcitonin and CGRP mRNAs (and calcitonin and CGRP) are present in a ratio of approximately 95–98 : 1 (Sabate *et al.*, 1985). Small amounts of CGRP are present histochemically and by radioimmunoassay in thyroid C cells (Sabate *et al.*, 1985 and Tschopp *et al.*, 1984), and both calcitonin and CGRP can be co-produced within the identical cell (Sabate *et al.*, 1985). The distribution of CGRP in pathways and neurons believed to serve specific sensory, integrative, and motor systems (Rosenfeld *et al.*, 1983) (Fig. 7) suggests several possible physiological roles for the peptide, including ingestive behavior (Rosenfeld *et al.*, 1983) at neuromuscular junctions (Gibson *et al.*, 1984), in the relay of thermal and nociceptive information to the brainstem and spinal cord (Rosenfeld *et al.*, 1983), and in the homeostasis of tissues of vascular musculature (Rosenfeld *et al.*, 1982; Gibson *et al.*, 1984).

　　CGRP-containing nerve fibers are widely distributed to most other organ systems, including the genitourinary system (e.g., bladder trigone, vagina, prostate), gastrointestinal (esophagus, duodenum, ileum), pulmonary (bronchiolar cells), and dermal systems (e.g., sensory fibers penetrating epidermis, sweat glands, hair follicles). In fact, it is perhaps the most widely distributed of the known neuropeptides. Consistent with features of its anatomical distribution, administration of synthetic rat CGRP produces a unique pattern of effects on

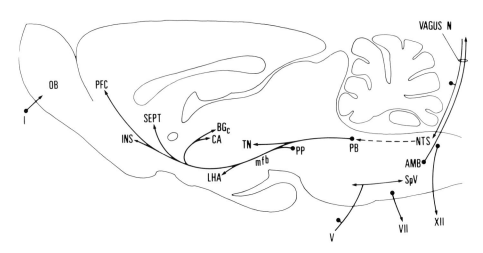

Figure 7. Summary of the major CGRP-stained cell groups (black dots) and pathways (arrows) projected on a sagittal view of the rat brain. This staining was localized in discrete parts of several functional systems. Dense terminal fields were stained throughout the substantia gelatinosa of the spinal cord and caudal part of the trigeminal nucleus; these fibers arise in dorsal root and trigeminal ganglion cells. CGRP is found in most parts of the taste pathways, including sensory endings in taste buds and the central endings of these fibers in the rostral part of the nucleus of the solitary tract (NTS), and in the relay system from the parabrachial nucleus (PB) to the thalamic taste nucleus (TN) and the taste area of the cerebral cortex (posterior agranular insular area, INS). In addition, most motor neurons in the hypoglossal nucleus (XII) are stained. A small group of primary olfactory fibers (I) that end in the glomerular layer of the olfactory bulb (OB) are also stained, suggesting that CGRP is found throughout the caudal part of the NTS, and throughout the PB, suggesting that it plays a part in the relay of visceral sensory information from the vagus (and glossopharyngeal) nerve, by way of an ascending pathway through the medial forebrain bundle (MFB). This pathway appears to arise in the PB and peripeduncular nucleus (PP) and projects to the lateral hypothalamic area (LHA), the central nucleus of the amygdala (CA) to patches in caudal parts of the caudoputamen and globus pallidus (BGc), to the lateral septal nucleus and bed nucleus of the stria terminalis area (PFC), the INS and the perirhinal area. The ascending projections in the MFB are probably modulated by a massive, non-CGRP containing pathway from the NTS to the PB (dashed line). Stained motor neurons in the rostral part of the nucleus ambiguus (AMB) project through the vagus nerve and may innervate the heart and/or branchial muscles in the pharynx.

blood pressure and catecholamine release in dogs and rats and produces gastric hypoacidity (Lenz *et al.*, 1984). CGRP is also widely distributed in the endocrine system, in a subset of adrenal medullary cells, in bronchiolar cells, and in fiber baskets that innervate the pancreatic islets and, interestingly, in thyroid C cells (Sabate *et al.*, 1985 and Tschopp *et al.*, 1984).

2.1. Mechanism of Alternative RNA Processing in Calcitonin Gene Expression

Documentation that calcitonin and CGRP mRNAs share an identical transcriptional start (CAP) site was provided by both S1 nuclease protection and primer extension analyses (Amara *et al.*, 1984) (Fig. 8). Therefore, both RNAs

are products of a single transcription unit. In S1 nuclease protection, the position of the first nucleotide of the mRNA is assessed by hybridizing the RNA to a labeled genomic fragment that spans the putative site, digesting the unhybridized material with S1 nuclease, and determining the size of the protected fragment by gel electrophoresis. Alternatively, by annealing a synthetic oligonucleotide primer complementary to a sequence downstream of the 5′ end of the RNA and transcribing a DNA copy of the RNA template, one can determine the location of the transcription initiation site. These analyses identified a

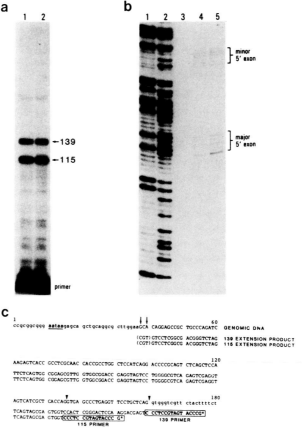

Figure 8. (a) Primer extension analysis of RNA products of calcitonin gene expression. Poly(A)-selected RNA (9 to 13S) from a calcitonin RNA-producing MTC tumor (lane 1) or a CGRP mRNA-producing MTC tumor (lane 2) was used as a template for extension of a 5′-labeled synthetic oligonucleotide complementary to the 5′ end of exon 2, which is common to both calcitonin and CGRP mRNAs; the extension products were separated by denaturing gel electrophoresis. Numbers show 115- and 139-nucleotide sequences. (**b**) S1 nuclease resistance mapping of the 5′ terminus of calcitonin and CGRP mRNAs. A 5′ genomic fragment was used for S1 nuclease resistance analysis with 5 μg of wheat germ tRNA in the absence (lane 3) or presence of 0.1 μg of poly(A)-selected RNA from a CGRP mRNA-producing MTC tumor (lane 4) or a calcitonin mRNA-producing MTC tumor (lane 5) and was displayed on a sequencing gel, using a known and unrelated sequencing ladder (lanes 1 and 2) to provide standards. Because Maxam-Gilbert reactions eliminate the chemically modified nucleotide and leave a 3′ phosphate group, the fragments on the sequencing ladder were assumed to be migrating 1.5 nucleotides faster than the S1 products. (**c**) DNA sequence analysis of regions flanking the first calcitonin/CGRP genomic exon. Sequence analysis of the 139- and 115-nucleotide primer extension products are shown below the genomic sequence. The position of the putative TATA box is indicated; arrows indicate initiation sites predicted by primer extension and S1 mapping analyses. Arrowheads indicate the location of alternative 3′ splice donor sites. 5 flanking and intron sequences are in lower case letters. The sequence of the 139- and 115-nucleotide primer extension products includes the oligonucleotide primer extension products (boxed); the identity of the first 5′ nucleotides was ambiguous and is inferred from genomic data.

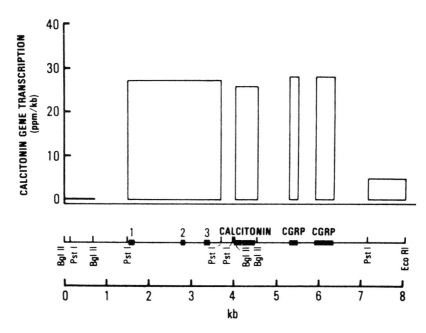

Figure 9. The pattern of transcription across the rat calcitonin gene. Nuclear run-off transcripts were quantitated under conditions of DNA-excess hybridization using the series of indicated genomic fragments. This analysis was performed in a tissue producing predominantly (>92%) calcitonin mRNA (open bars). Transcription continues without attenuation through the CGRP exons, to a point approximately 1 kb 3' of the CGRP poly(A) site.

cryptic splice site generating a 24-nucleotide extension of the first exon in the case of both calcitonin and CGRP mRNAs. The utilization of this site appears to be a stochastic event; it occurs in the untranslated region of the message and thus does not alter the encoded protein products. The polyadenylation site of calcitonin mRNA appears to be 18 or 19 nucleotides 3' to a sequence AATAAA located 226 nucleotides downstream of the calcitonin termination codon (exon 4). CGRP mRNA utilizes an alternative recognition sequence, ATTAAA, and the CGRP poly(A) addition signal and site are situated 1.9 kilobases (kb) downstream from the analogous region for calcitonin mRNA, defining the end of the large 3'-CGRP non-coding exon (the sixth genomic exon). Thus, production of calcitonin and CGRP mRNAs is associated with the selective polyadenylation of transcripts at one of two alternative poly(A) sites.

    Alternative transcriptional termination was evaluated as a possible explanation for the selective poly(A) site utilizing transcriptional runoff assays. In this method, nuclei are isolated, and nascent transcripts are then permitted to elongate by addition of nucleotides, including one radiolabeled to high specific activity. Following purification, the transcripts are annealed to immobilized clonal probes, washed, and treated with ribonuclease to reduce assay background and the specific hybrid quantitated. By using probes distributed over

the length of the calcitonin/CGRP gene, one can quantitate the distribution of transcripts across the gene. The results of these experiments indicate that selective poly(A) site selection, and not alternative RNA transcriptional termination, is the regulated event, because termination occurs >1 kb 3' of the CGRP poly(A) site, regardless of which mature transcript is produced (Amara *et al.*, 1984) (Fig. 9).

A critical question in this system was: Which of the two alternative RNA-processing events represents the regulated choice, and which reflects the potential null, or unregulated, choice? One approach taken to answer this question was to introduce a fusion gene containing the mouse metallothionein I promoter fused to the first exon of the calcitonin/CGRP gene into fertilized mouse eggs to target expression of the gene in tissues not normally expressing the endogenous calcitonin/CGRP gene and to analyze the pattern of calcitonin and CGRP mRNA expression in the resultant transgenic offspring. The metallothionein I promoter was chosen because this gene is widely expressed and would presumably result in widespread expression of fusion gene. In three pedigrees analyzed, calcitonin mRNA was the overwhelmingly predominant mature transcript in all tissues except neuronal tissue, where most—but not all—areas of brain made a clear choice, producing CGRP mRNA (Crenshaw *et al.*, 1987). The simplest interpretation of these data is that production of CGRP mRNA requires a cell-specific factor(s), whereas cells that do not contain this active factor(s) produce calcitonin mRNA as a null choice.

The alternative poly(A) site choice could reflect the primary, regulated event, or it could result from the commitment of nascent transcripts toward alternative splicing pathways involving exons 3, 4 (calcitonin coding), and 5 (CGRP coding). In order to distinguish between these possibilities, an extensive series of mutated calcitonin/CGRP genes and fusion genes containing the regions encompassing the poly(A) cleavage sites were analyzed by DNA-mediated gene transfer in cell lines making unambiguous RNA-processing choices. A chimeric plasmid containing the rat calcitonin/CGRP gene and a dominant selectable marker (Eco gpt) was transferred into several cell lines. Following transfection, cell lines were selected with mycophenolic acid, to permit clonal selection of cells expressing the introduced gene. Transfer of the rat calcitonin gene into terminally differentiated plasma cells or lymphocyte/plasma cell-fusion lines resulted in clear alternative RNA processing regulation, with the selective production of calcitonin mRNA (Leff *et al.*, 1987). By contrast, in several of the cell lines, CGRP mRNA represented >95% of mature transcripts. These experiments suggest that the machinery necessary to choose between the RNA processing choices is operative in specific cell types.

A series of mutant genes were then analyzed in these cell lines. On the basis of these analyses (Leff *et al.*, 1987), we suggest that, in CGRP-producing tissues, there is a factor(s) that is required for exon 3 to exon 5 splice to occur, and that it is the commitment to this splicing event that precedes and dictates the choice of poly(A) site. The exon 3 to exon 4 splice appears to be cryptic in cell types which produce calcitonin mRNA, occurring only when information beyond the calcitonin poly(A) site is removed from the transcript. We therefore

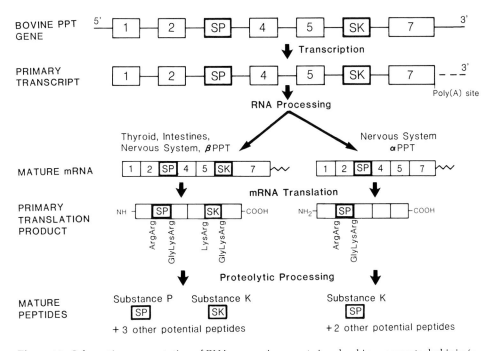

Figure 10. Schematic representation of RNA processing events involved in α-preprotachykinin (α-PPT) and β-preprotachykinin (β-PPT) production. (Based on data of Nawa *et al.*, 1984.)

propose a model in which splice commitment, dictated by a neuron-specific factor(s), precedes and determines the poly(A) cleavage site and the exon 3 to exon 5 splice. Such a factor would likely be similar or identical to that regulating alternative splicing events in other neuroendocrine gene transcripts, such as the substance P gene (see Section 2.2) or in expression of other gene families. The characterization of this neuron-specific factor(s) is the subject of current investigation.

2.2. Alternative RNA Processing in the Neuroendocrine System

Analysis of alternative mRNA products expressed by the calcitonin gene predicted a novel family of neuropeptides. Similar analysis has led to the discovery of an additional member of the tachykinin family as a constituent of the substance P gene. Figure 10 outlines these events. Nucleotide sequence analysis of complementary DNA (cDNA) clones from bovine brain determined that two separate mRNAs can encode substance P (Nawa *et al.*, 1983). These two mRNAs have common 5' and 3' sequences and differ by the inclusion of a region that encodes a peptide sharing considerable homology with substance P and an amphibian peptide kassinin, and was named substance K. The shorter mRNA ecodes a 112-amino acid precursor protein named α-preprotachykinin

(α-PPT). The precursor contains an N-terminal signal sequence and an internal substance P sequence bounded by signals for excision and a carboxy-terminal glycine required for amidation. The second mRNA encodes β-preprotachykinin (β-PPT), a 130-amino acid precursor that differs from α-PPT only by the inclusion of sequences encoding an excisable amidated form of substance K at amino acids 98–107.

The PPT gene is composed of seven exons (Fig. 10); the third and sixth encode substance P and K, respectively (Nawa *et al.*, 1984). Exclusion of the sixth exon during mRNA splicing distinguishes α-PPT mRNA. Although this event parallels the removal of the fourth (calcitonin-coding) exon in calcitonin/CGRP mRNA processing, PPT processing differs in that both α- and β-PPT mRNAs share the same 3′-end poly(A) site. Thus, alternative RNA processing of PPT transcripts may be regulated entirely by splice site selection.

Analysis of PPT and calcitonin/CGRP mRNAs and immunoreactivity indicate that they are expressed in distinct, yet overlapping, tissues and areas of the nervous system. Analysis of the mRNA or peptide products expressed from these two genes in neurons of sensory ganglia and thyroid C cells shows that α-PPT and CGRP mRNAs predominate over β-PPT and calcitonin mRNA levels in sensory neurons, whereas the reverse relationship is found in thyroid C cells. Thus, it is tempting to speculate that common factors or mechanisms may operate on transcripts from these two genes in thyroid C cells to include the substance K and calcitonin-coding exons, whereas in sensory neurons mRNA processing favors the exclusion of these two exons and the greater production of α-PPT and CGRP mRNAs.

The vasoactive peptide encoded by the bradykinin gene is excised from both a high and low molecular weight form of prekininogen (Kitamura *et al.*, 1983). High-molecular-weight prekininogen mRNAs use an upstream poly(A) site and coding sequences to generate a large cofactor for contact activation of coagulation and fibrinolysis. Low-molecular-weight prekininogenin mRNAs are alternatively spliced to a small downstream exon that uses a separate poly(A) site and that does not encode such a function.

These three genes provide examples of the apparently widespread cell-specific developmental utilization of post-transcriptional, RNA processing events to generate an increased diversity of functional peptides in the neuroendocrine system.

3. Developmental Regulation of Neuroendocrine Gene Expression Based on Heritable Patterns of Cell-Specific Gene Transcription

Gene-transfer experiments have provided evidence for *cis*-active regulatory sequences, referred to as enhancers, which can stimulate the transcription of eukaryotic promoters in a position- and orientation-independent fashion. Although protracted sequence homologies are not observed among the various enhancer elements, short sequences of 8–10 core nucleotides have been suggested (Weiber *et al.*, 1983 and Hen *et al.*, 1983) that often exhibit alternation of purine and pyrimidine residues characteristic of left-handed (Z) DNA (Nordheim and Rich, 1983). Based on footprinting and *in vitro* transcriptional analysis, it is suggested that enhancers exert their effects as a consequence of binding *trans*-acting factors (Sassone-Corsi *et al.*, 1983; Sassone-Corsi *et al.*, 1984; Ephrussi *et al.*, 1985; Mercola *et al.*, 1985; and Kriegler and Botchan, 1983). Regions of several eukaryotic genes containing such enhancer sequences have been shown to dictate a cell-specific pattern of expression in transfected cells (for examples, see Walker *et al.*, 1983; Kriegler and Botchan, 1983; DeVilliers *et al.*, 1983; Ott *et al.*, 1984; Spandidos and Paul, 1982; and Chao *et al.*, 1983). Although the precise role of any enhancer element has not yet been unequivocally established, the physiological action of these sequences on development is suggested by the observations that genomic sequences containing the promoter and putative enhancer elements of several genes appear to direct a restricted tissue-specific pattern of gene expression in transgenic animals (Storb *et al.*, 1984; Brinster *et al.*, 1983; Hanahan, 1985; Shani, 1985; Ornitz *et al.*, 1985; Swift *et al.*, 1984; McKnight *et al.*, 1983; Krumlauf *et al.*, 1985; Chada *et al.*, 1985; and Grosschedl *et al.*, 1984). Intact *Drosophila* genes including their flanking chromosome regions are also appropriately expressed when introduced into the germlines of mutant strains of *Drosophila* (see Spradling and Rubin, 1983; Goldberg *et al.*, 1983; and Scholnick *et al.*, 1983).

Because the anterior pituitary gland develops from a common primordium to produce phenotypically distinct cell types, each of which expresses one of at least six discrete trophic hormones, it represents a good model to study the ontogeny and molecular mechanisms that dictate multiple phenotypes. Two of these are the structurally related prolactin and growth hormone genes, which are evolutionarily derived from a single primordial gene (Cooke and Baxter, 1982). These related peptide hormones are produced in discrete cell types, referred to as lactotrophs and somatotrophs, respectively, and their expression is apparently limited to the pituitary gland. Growth hormone and prolactin are temporally the last pituitary hormones expressed during ontogeny (Khorram *et al.*, 1984 and Watanabe and Daikoku, 1979). Rat growth hormone is initially detected at fetal day 17–18, whereas prolactin appears at or after birth, with the percentage of lactotrophs rising dramatically 3–5 days postpartum to constitute approximately 10% of total cells. Because it has been reported that growth hormone (GH) and prolactin are initially co-produced within single cells before the appearance of lactotrophs (Chatelain *et al.*, 1979; Watanabe and Daikoku, 1979; and Hoeffler *et al.*, 1985), it is possible that lactotrophs may arise developmentally from somatotrophs.

To evaluate whether these two closely related neuroendocrine genes are developmentally controlled by different factors interacting with unique structural sequences in each gene, or whether a common *trans*-acting factor regulates their expression, the potential *cis*-active elements dictating cell-specific expression of each gene were examined. Regions of approximately 250 and 150 bp were identified 2.0 and 0.2 kb 5′ upstream of the transcription start site of the rat prolactin and growth hormone genes, respectively. These elements act in a position- and orientation-independent fashion to transfer cell-specific expression to both heterologous genes (Nelson *et al.*, 1986). These sequences permit expression of reporter genes in rat cell lines of pituitary origin that express the endogenous prolactin and growth hormone genes but fail to exert any effects in any other cell lines tested, including those of endocrine origin (thyroid C cells, pancreatic β cells, ovarian cells). The *cis*-active elements appear to act developmentally *in vivo*, because prolactin 5′ sequences target lactotroph-specific expression of SV40 T antigen fusion genes in transgenic mice (Crenshaw III *et al.*, 1987). A lactotroph cell line derived from these transgenic animals contains the putative *trans*-active factors to permit function of the prolactin, but not the growth hormone, gene enhancer. The *trans*-acting factors that bind to each region are discrete, and a lactotroph cell line (which selectively expresses the prolactin gene) fails to express fusion genes containing the growth hormone enhancer. Discrete and separable sequences were identified in the prolactin enhancer, each required but insufficient for biological activity, and which can be separated with reconstruction of enhancer activity (Nelson *et al.*, 1986). The challenge now is to characterize the putative regulations of cell-specific enhancers and the molecular mechanisms by which they developmentally control gene transcription.

ACKNOWLEDGMENTS. Original research supported by grants from the American Cancer Society, the National Institutes of Health, and the Howard Hughes Medical Institute.

References

Amara, S. G., Jonas, V., O'Neil, J. A., Vale, W., Rivier, J., Roos, B. A., Evans, R. M., and Rosenfeld, M. G., 1982a, Calcitonin COOH-terminal cleavage peptide as a model for identification of novel neuropeptides predicted by recombinant DNA analysis, *J. Biol. Chem.* **257:**2129–2132.

Amara, S. G., Jonas, V., Rosenfeld, M. G., Ong, E. S., and Evans, R. M., 1982b, Synthesis of secreted and membrane-bound immunoglobin Mu heavy chains is directed by mRNAs that differ at the 3′ ends. Alternative RNA processing in calcitonin gene expression generates mRNAs encoding different polypeptide products, *Nature (Lond.)* **298:**240–244.

Amara, S. G., Evans, R. M., and Rosenfeld, M. G., 1984, Calcitonin/calcitonin gene related peptide transcription unit: Tissue-specific expression involves selective use of alternative polyadenylation sites, *Mol. Cell. Biol.* **4:**2151–2160.

Amara, S. G., Arrize, J. L., Leff, S. E., Swanson, L. W., Evans, R. M., and Rosenfeld, M. G., 1985, Expression in brain of an mRNA encoding a novel neuropeptide homologous to calcitonin gene-related peptide (CGRP), *Science* **229:**1094–1097.

Banerji, R., Rusconi, S., and Schaffner, W., 1981, Expression of a β-globin gene is enhanced by remote SV40 DNA sequences, *Cell* **27:**299–308.

Banerji, J., Olson, L., and Schaffner, W., 1983, A lymphocyte-specific cellular enhancer is located downstream of the joining region in immunoglobin heavy chain genes, *Cell* **33:**729–740.

Benoist, C., and Chambon, P., 1981, In vivo sequence requirements of the SV40 early promoter region, *Nature (Lond.)* **290:**304–310.

Blattner, F. R., and Tucker, P. W., 1984, The molecular biology of immunoglobulin D, *Nature (Lond.)* **307:**417–422.

Breathnach, R., and Chambon, P., 1981, Organization and expression of eukaryotic split genes coding for proteins, *Annu. Rev. Biochem.*, **50:**349–383.

Brinster, R. L., Ritchie, K. A., Hammer, R. E., O'Brien, R. L., Arp, B., and Storb, U., 1983, Expression of a microinjected immunoglobin gene in the spleen of transgenic mice, *Nature (Lond.)* **306**:332–336.

Chada, K., Magram, J., Raphael, K., Radice, G., Lacy, E., and Costantini, F., 1985, Specific expression of a foreign β-globin gene in erythroid cells of transgenic mice, *Nature (Lond.)* **314**:377–380.

Chao, M. V., Mellon, P., Charnay, P., Maniatis, T., and Axel, R., 1983, The regulated expression of beta-globulin genes introduced into mouse erythroleukemia cells. *Cell* **32**:483–493.

Chatleain, A., Dupuoy, J. P., and Dubois, M. P., 1979, Ontogenesis of cells producing polypeptide hormones (ACTH, MSH, LPH, GH, Prolactin) in the fetal hypophysis of the rat: Influence of the hypothalamus, *Cell Tissue Res* **196**:409–427.

Cooke, N. G., and Baxter, J. D., 1982, Structural analysis of the prolactin gene suggests a separate origin for its 5' end, *Nature (Lond.)* **297**:603–606.

Darnell, J. E., Jr., 1982, Variety in the level of gene control in eukaryotic cells, *Nature (Lond.)* **297**:365–371.

DeVilliers, J. L., Olson, L., Tyndall, C., and Schaffner, W., 1982, Transcriptional 'enhancers' from SV40 and polyoma virus show a cell type preference, *Nucl. Acids Res.* **10**:7965–7976

Dynan, W. S., and Tjian, R., 1985, Control of eukaryotic messenger RNA synthesis by sequence-specific DNA-binding protein, *Nature (Lond.)* **316**:774–778.

Ephrussi, A., Church, G. M., Tonegawa, S., and Gilbert, W., 1985, β-lineage-specific interactions of an immunoglobin enhancer with cellular factors in vivo, *Science* **227**:134–140.

Fromm, M., and Berg, P., 1982, Transcription *in vivo* from SV40 early promoter deletion mutants without repression by larg T antigen, *J. Mol. Appl. Genet.* **1**:457–463.

Gibson, S. T., Polak, J. M., Bloom, S. R., Sabate, I. M., Mulderry, P. M., Ghatei, M. A., McGregory, G. P., Morrison, J. F. B., Kelly, J. S., Evans, R. M., and Rosenfeld, M. G., 1984, Calcitonin gene-related peptide immunoreactivity in the spinal cord of man and of eight other species, *J. Neurosci.* **4**:3101–3111.

Gillies, S. D., Morrison, S. L., Oi, V. T., and Tonegawa, S., A tissue-specific transcription enhancer element is located in the ma'or intron of a rearranged immunoglobin heavy chain gene, *Cell* **33**:717–728.

Gluzman, Y., and Shenk, T. (eds.), 1983, *Enhancers and Eukaryotic Gene Expression*, Cold Spring Harbor Laboratory, Cold Spring Harbor, New York.

Goldberg, P. A., Posakony, J. W., and Maniatis, T., 1983, Correct developmental expression of a cloned alcohol dehydrogenase gene transduced into the Drosophila germ line, *Cell* **34**:59–73.

Grosscehedl, R., Weaver, D., Baltimore, D., and Costantini, F., 1984, Introduction of a u immunoglobin gene into the mouse germ line: Specific expression in lymphoid cells & synthesis of functional antibodv, *Cell* **38**:647–658.

Gruss, P., Dhar, R., and Khoury, G., 1981, Simian virus 40 tandem repeated sequences as an element of the early promoter, *Proc. Natl. Acad. Sci. USA* **78**:943–947.

Hanahan, D., 1985, Heritable formation of pancreatic β-cell tumours in transgenic mice expressing recombinant insulin/simian virus 40 oncogenes, *Nature (Lond.)* **315**:115–122.

Hen, R., Borrelli, E., Sassone-Corsi, P., and Chambon, P., 1983, Far upstream sequences are required for efficient transcription from the adenovirus-2 ElA transcription unit, *Nucl. Acids Res.* **11**:8735–8745.

Hoeffler, J. P., Boockfor, F. R., and Frawley, L. S., 1985, Ontogeny of prolactin cells in neonatal rats: Initial prolactin secretors also release growth hormone, *Endocrinology* **117**:187–195.

Khorram, O., Depalatis, L. R., and McCann, S. M., 1984, Hypothalamic control of prolactin secretion during the perinatal period in the rat, *Endocrinology* **115**:1698–1704.

Kitamura, N., Takagaki, Y., Furuto, S., Tanaka, T., Nawa, H., 1983, A single gene for bovine high molecular weight and low molecular weight kinonogens, *Nature (Lond.)* **305**:545–549.

Kriegler, M., and Botchan, M., 1983, Enhanced transformation by a simian virus containing a Harvey Murine sarcoma virus long terminal repeat, *Mol. Cell Biol.* **3**:325–339.

Krumlauf, R., Hammer, R. E., Tilghman, S. M., and Brinster, R. L., 1985, Developmental regulation of α-fetoprotein genes in transgenic mice, *Mol. Cell Biol.* **5**:1639–1648.

Laimins, L. A., Khoury, G., Gorman, C., Howard, B., and Gruss, P., 1982, Host-specific activation of transcription by tandem repeats from simian virus 40 and Moloney murine sarcoma virus, *Proc. Natl. Acad. Sci. USA* **79**:6453–6457.

Lenz, H. J., Mortrud, M. T., Rivier, J. E., and Brown, M. R., 1984, Calcitonin gene related peptide inhibits basal, pentagastrin, histamine, and bethanecol stimulated gastric acid secretion, *Gut* **26**:550–556.

Luciw, P. A., Bishop, J. M., Varmus, H. E., and Capecchi, M. R., 1983, Location and function of retroviral and SV40 sequences that enhance biochemical transformation after microinjection of DNA, *Cell* **33**:705–716.

Lusky, M., Borg, L., Weiber, H., and Botchan, M., 1983, Bovine papilloma virus contains an activator of gene expression at the distal end of the early transcription unit, *Mol. Cell Biol.* **3**:1108–1122.

McKnight, G. S., Hammer, R. E., Kuenzel, E. A., and Brinster, R. L., 1983, Expression of the chicken transferrin gene in transgenic mice, *Cell* **34**:335–341.

Mason, R. T., Peterfreund, R. A., Sawchenko, P. E., Corrigan, A. Z., Rivier, J. E., and Vale, W. W., 1984, Release of the predicted calcitonin gene-related peptide from cultured rat trigeminal ganglion cells, *Nature (Lond.)* **304**:129–135.

Mercola, M., Goverman, J., Mirell, C., and Calame, K., 1985, Immunoglobin heavy-chain enhancer requires one or more tissue-specific factors, *Science* **227**:266–270.

Nawa, H., Hirose, T., Takashima, H., Inayama, S., Nakanishi, S., 1983, Nucleotide sequences of cloned cDNAs for two types of bovine brain substance P precursor, *Nature (Lond.)* **306**:32–36.

Nawa, H., Kotani, H., Nakanishi, S., 1984, Tissue-specific generation of two prepretachykinin mRNAs from one gene by alternative RNA splicing, *Nature* **313**:729–734.

Nelson, C., Crenshaw, E. B. III, Franco, R., Lira, S. A., Albert, V. R., Evans, R. M., and Rosenfeld, M. G., 1986, Discrete cis-active genomic sequences dictate the pituitary cell type-specific expression of rat prolactin and growth hormone genes, *Nature (Lond.)* **322**:557–562.

Neuberger, M. S., 1983, Expression and regulation of immunoglobin heavy chain gene transfected into lymphoid cells, *EMBO J.* **2**:1373–1378.

Nevins, J. R., 1983, The pathway of eukaryotic mRNA formation, *Annu. Rev. Biochem.* **52**:441–466.

Nordheim, A., and Rich, A., 1983, Negatively supercoiled simian virus 40 DNA contains Z-DNA segments within transcriptional enhancer sequences, *Nature (Lond.)* **303**:674–679.

Ornitz, D. M., Palmiter, R. D., Hammer, R. E., Brinster, R. L., Swift, G. H., and MacDonald, R. J., 1985, Specific expression of an elastase-human growth hormone fusion gene in pancreatic acinar cells of transgenic mice, *Nature (Lond.)* **313**:600–602.

Ott, M. O., Sperling, L., Herbomel, P., Yaniv, M., and Weiss, M. C., 1984, Tissue-specific expression is conferred by a sequence from the 5′ end of the rat albumin gene, *EMBO J.* **3**:2505–2510.

Padgett, R. A., Grabowski, P. J., Konarska, M. M., Seiler, S., and Sharp, P. A., 1986, Splicing of messenger RNA precursors, *Annu. Rev. Biochem.* **55**:1119–1150.

Queen, C., and Baltimore, D., 1983, Immunoglobin gene transcription is activated by downstream sequence elements, *Cell* **33**:741–748.

Rosenfeld, M. G., Amara, S. G., Roos, B. A., Ong, E. S., and Evans, R. M., 1981, Altered expression of the calcitonin gene associated with RNA polymorphism, *Nature (Lond.)* **290**:63–65.

Rosenfeld, M. G., Lin, C. R., Amara, S. G., Stolarsky, L. S., Ong, E. S., and Evans, R. M., 1982, Calcitonin mRNA polymorphism: Peptide switching associated with alternate RNA splicing events, *Proc. Natl. Acad. Sci. USA* **79**:1717–1721.

Rosenfeld, M. G., Mermod, J-J., Amara, S. G., Swanson, L. W., Sawchenko, P. E., Rivier, J., Vale, W. W., and Evans, R. M., 1983, Production of a novel neuropeptide encoded by the calcitonin gene via tissue-specific RNA processing, *Nature (Lond.)* **304**:129–135.

Sebate, M. I., Stolarsky, L. S., Polak, J. M., Bloom, S. R., Verndell, I. M., Ghatei, M. A., Evans, R. M., and Rosenfeld, M. G., 1985, Regulation of neuroendocrine gene expression by alternative RNA processing: Colocalization of calcitonin and calcitonin gene-related peptide (CGRP) in thyroid C-cells, *J. Biol. Chem.* **260**:2589–2592.

Sassone-Corsi, P., Wildeman, A., and Chambon, P., 1983, A *trans*-acting factor is responsible for the simian virus 40 enhancer activity *in vitro*, *Nature (Lond.)* **313**:458–463.

Sassone-Corsi, P., Dougherty, J. P., Wasylyk, B., and Chambon, P., 1984, Stimulation of *in vitro* transcription from heterologous promoters by the simian virus 40 enhancer, *Proc. Natl. Acad. Sci USA* **81**:308–312.

Scholnick, S. B., Morgan, B. A., and Hirsh, J., 1983, The cloned dopa decarboxylase gene is developmentally regulated when reintegrated into the Drosophila genome, *Cell* **34**:37–45.

Shani, M., 1985, Tissue-specific expression of rat myosin light-chain 2 gene in transgenic mice, *Nature (Lond.)* **314**:283–286.

Spandidos, D. A., and Paul, J., 1982, Transfer of human globin genes to erythroleukemic mouse cells, *EMBO J.* **1**:15–20.

Spradling, A. C., and Rubin, G. M., 1983, The effect of chromosomal position on the expression of the Drosophila xanthine dehydrogenase gene, *Cell* **34**:47–57.

Storb, U., O'Brien, R. L., McMullen, M. D., Gollahon, K. A., and Brinster, R. L., 1984, High expression of cloned immunoglobin K gene in transgenic mice is restricted to β-lymphocytes, *Nature (Lond.)* **310**:238–241.

Swift, G. H., Hammer, R. E., MacDonald, R. J., and Brinster, R. L., 1984, Tissue-specific expression of the rat pancreatic elastase I gene in transgenic mice, *Cell* **38**:639–646.

Tschopp, F. A., Tobler, P. H., Fischer, J. A., 1984, Calcitonin gene-related peptide in the human thyroid, pituitary and brain, *Mol. Cell. Endocrinol.* **36**:53–57.

Walker, M. D., Edlund, T., Boulet, A. M., and Rutter, W. J., 1983, Cell-specific expression controlled by the 5'-flanking region of insulin & chymotrypsin genes, *Nature (Lond.)* **306**:557–561.

Watanabe, Y. G., and Daikoku, S., 1979, An immunohistochemical study on the cytogenesis of adenohypophysial cells in fetal rats, *Dev. Biol.* **68**:557–567.

Weiber, H., König, M., and Gruss, P., 1983, Multiple point mutations affecting the simian virus 40 enhancer, *Science* **219**:626–631.

Yaniv, M., 1982, Enhancing elements for activation of eukaryotic promoters, *Nature (Lond.)* **297**:17–18.

Ziff, E. B., 1980, Transcription and RNA processing, *Nature (Lond.)* **287**:491–499.

Chapter 12

Hypomethylation of DNA in the Regulation of Gene Expression

LOIS A. CHANDLER and PETER A. JONES

1. Introduction

Cellular differentiation involves the activation of a specific set of genes as well as the suppression of those genes associated with the differentiation of other cell types. Differentiated cells also possess tissue-specific patterns of DNA methylation that are somatically heritable. The tissue-specific nature of methylation patterns implies that changes in these patterns must take place during normal cellular development. These observations have generated much speculation as to the potential role of this postreplication modification of DNA in eukaryotic gene regulation. The experimental evidence accumulated to date strongly supports the hypothesis that methylation of critical sites in DNA serves to lock some (but not all) genes in an inactive state. Therefore, cellular differentiation, which involves the ordered switching of genes, must also involve tightly regulated switches in DNA methylation patterns.

2. Nonrandom Distribution of 5-Methylcytosine in Eukaryotic Genomes

Approximately 3–4% of all cytosine residues in mammalian DNA are modified enzymatically in a postreplication process to give **5-methylcytosine (5-mCyt;** Fig. 1) (Riggs and Jones, 1983). 5-Methylcytosine is the only naturally occurring modified base yet detected in mammalian DNA; 90% or more of 5-mCyt occurs in the dinucleotide CpG (Ehrlich and Wang, 1981). The sequence CpG is statistically underrepresented in vertebrate DNA, being present at a frequency only one fourth of that predicted by random distribution (Ehrlich and Wang, 1981). This underrepresentation has been hypothesized to result from the fact that 50–70% of all cytosine residues within CpG sequences are

LOIS A. CHANDLER and PETER A. JONES • Urological Cancer Research Laboratory, Comprehensive Cancer Center and Department of Biochemistry, University of Southern California, Los Angeles, California 90033.

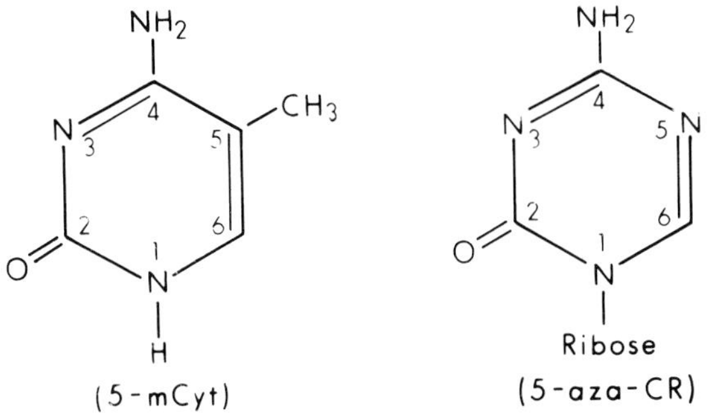

Figure 1. Structures of 5-methyl-cytosine (5-mCyt) and 5-azacytidine (5-aza-CR).

methylated (Gruenbaum *et al.,* 1981) and because 5-mCyt has a propensity to deaminate to thymidine, which cannot be recognized by repair enzymes (Salser, 1977). This explanation has gained support from findings that a deficiency of CpG sequences is generally accompanied by an overabundance of TpG and CpA dinucleotides (Bird, 1980). Furthermore, a high frequency of sequence polymorphisms has been demonstrated at restriction endonuclease sites containing the sequence CpG (Barker *et al.,* 1984).

Methylated cytosine residues are nonrandomly distributed within eukaryotic DNA. For example, 5-mCyt occurs in repetitive sequences in a severalfold higher abundance than in middle repetitive or unique sequences (Solage and Cedar, 1978; Ehrlich *et al.,* 1982). 5-Methylcytosine is also nonrandomly distributed with respect to the nucleosomal structure of chromatin; DNA associated with nucleosomes is generally more methylated than DNA within linker regions (Razin and Cedar, 1977; Solage and Cedar, 1978). Insight into gene-specific patterns of eukaryotic DNA methylation has been obtained through the use of restriction endonucleases that recognize CpG-containing sequences (see Section 4.1). One striking finding has been that potential methylation sites often appear in highly CpG-enriched domains or clusters. Bird *et al.* (1985) estimated that about 30,000 CpG-rich clusters are present per haploid genome in the mouse. These clusters appear to have been protected from methylation and are often associated with specific genes, including numerous housekeeping genes (Jones, 1986). The clusters may have escaped the usual depletion of CpG sequences, which are otherwise underrepresented in the genome of mammals, because they have escaped methylation. The one exception to the rule that housekeeping gene clusters are unmethylated occurs when they are located on inactive X chromosomes (Wolf *et al.,* 1984). This finding has provided support for the hypothesis that DNA methylation plays a role in regulating eukaryotic gene expression.

3. Clonal Inheritance of Methylation Patterns

3.1. Semiconservative Replication of Methylation Patterns

Levels and patterns of cytosine methylation are tissue-specific (Ehrlich *et al.,* 1982; Razin and Szyf, 1984); DNA-mediated gene-transfer experiments

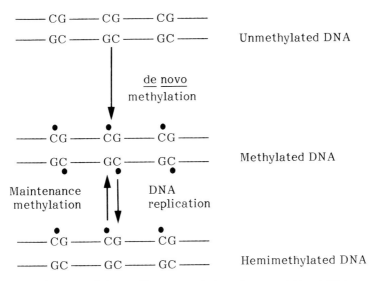

Figure 2. States of DNA methylation. *De novo* methyltransferase activity establishes methylation of CpG sites in DNA (closed circles represent methyl groups). A maintenance methyltransferase acts on hemimethylated sites generated by DNA replication, thus ensuring heritability of the methylation pattern.

have provided evidence demonstrating the somatic heritability of these patterns (Wigler *et al.*, 1981; Stein *et al.*, 1982). A methylation site (CpG) can exist in three states: unmethylated, methylated in both strands, or methylated in one strand (hemimethylated) (Fig. 2). Hemimethylated sites are generated from methylated sites by semiconservative DNA replication (Riggs and Jones, 1983).

3.2. Methylases and Their Substrate Specificities

Two types of methyltransferase activities have been postulated to exist in eukaryotic cells (Riggs, 1975; Holliday and Pugh, 1975). *De novo* methylation establishes the methylation of previously unmethylated sites, whereas maintenance methyltransferase activity methylates newly replicated, hemimethylated DNA, thus ensuring the faithful inheritance of methylation patterns (Fig. 2). Central to the theory of maintenance methylation is the fact that the dinucleotide CpG is a simple palindrome. Therefore, the two hemimethylated sites produced by replication of a fully methylated site are identical with regard to recognition by the maintenance methyltransferase. The enzyme thus uses the methyl group on the template strand to direct the accurate methylation of the newly synthesized strand.

Numerous studies have established that the predominant DNA methyltransferase activity in mammalian cells is a maintenance methyltransferase, which uses S-adenosylmethionine as a methyl group donor (Riggs and Jones, 1983). Mammalian DNA methyltransferase has been partially purified from several sources (Simon *et al.*, 1978; Pfeifer *et al.*, 1985; Zucker *et al.*, 1985).

However, the enzyme has not been purified to homogeneity, and no firm evidence exists for the presence of two distinct methyltransferase enzymes in eukaryotic cells. Indeed, in many instances, *de novo* and maintenance methyltransferase activities copurify, with the ratio of their activities remaining unchanged during the purification process (Razin, 1984; Zucker *et al.*, 1985). In most studies, however, *de novo* methyltransferase activity is very inefficient relative to maintenance methyltransferase activiy (Razin, 1984; Zucker *et al.*, 1985). Furthermore, the sequence specificities of the eukaryotic DNA methyltransferase enzymes vary from a high level of specificity for CpG sequences exerted by the maintenance methyltransferase (Gruenbaum *et al.*, 1982; Stein *et al.*, 1982) to a low level of specificity exerted by the *de novo* methyltransferase (Simon *et al.*, 1983). The activity of *de novo* methyltransferase at CpG sites appears to be highly modulated by flanking sequences. Using a partially purified HeLa cell methyltransferase, Bolden *et al.* (1985) demonstrated that the most efficient methylation of synthetic oligodeoxynucleotides occurs with multiple, separated CpG sites in a region of high G + C content (>65%).

4. Evidence for Gene Regulation by DNA Methylation

A potential role for DNA methylation in controlling eukaryotic gene expression was initially proposed in 1975 (Holliday and Pugh, 1975; Riggs, 1975). A large body of experimental evidence has accumulated since that time that supports the hypothesis that hypomethylation of specific cytosine residues in DNA is a necessary, but not sufficient, condition for gene expression in higher eukaryotes (Razin and Riggs, 1980; Doerfler, 1983; Riggs and Jones, 1983).

4.1. Methylation-Sensitive Restriction Endonucleases

The availability of bacterial restriction endonucleases sensitive to the methylation of cytosine residues in CpG sequences opened up the field of DNA methylation by allowing for determination of the methylation status of specific CpG sites within DNA. Enzymes such as Hpa II (recognizes CCGG) and Hha I (recognizes GCGC), which are inhibited by methylation at their restriction site CpG residues, are frequently used together with the Southern blotting technique (Southern, 1975) to assay the methylation state of CpG sites in specific genes. This approach has been made even more powerful with the discovery of pairs of enzymes with identical sequence specificities (**isoschizomers**) but different sensitivities to methylation. For example, both Hpa II and Msp I cleave at CCGG sequences, but only Msp I can cleave DNA when the internal cytosine of this sequence is methylated (McClelland, 1981). Therefore, the existence of the sequence C^mCGG in DNA can be assessed by comparing the size distribution of DNA fragments generated by digestion with these two enzymes.

The use of methylation-sensitive restriction enzymes coupled with probes for specific genes has shown that certain CpG sites in the vicinity of a wide variety of genes are less methylated in tissues in which the gene is expressed than in nonexpressing tissue (Razin and Riggs, 1980; Riggs and Jones, 1983; Cooper, 1983). For example, McGhee and Ginder (1979) found that certain Hpa II sites within the adult β-globin gene of chicken were undermethylated in cells where the gene was being expressed or had been expressed (erythrocytes and reticulocytes) as compared with cells where the gene was never expressed (oviduct).

Together with the demonstration of tissue-specific methylation patterns, these observations strongly suggest a role for methylation in the regulation of gene expression. Consistent with this hypothesis is the fact that correlations between gene activity and DNA hypomethylation are best seen in the 5′-regulatory regions of genes (Riggs and Jones, 1983; Doerfler *et al.*, 1984). Several experiments, however, show an apparent discordance with this theory (Gerber-Huber *et al.*, 1983; Macleod and Bird, 1983). For example, the α2(I) collagen gene of chicken has a pattern of Hpa II site methylation that is invariant whether the gene is expressed or not (McKeon *et al.*, 1982).

A number of problems are associated with the interpretation of results obtained using restriction enzymes to probe methylation patterns. Only about 10% of all CpG sites in DNA can be examined using available restriction enzymes (Razin and Riggs, 1980). Therefore, CpG sites important for regulating a gene would be undetected by this method if they do not reside within a recognition sequence for one of the methylation-sensitive enzymes. In addition, it is not possible to analyze very closely spaced restriction sites by the Southern blotting method because of the inefficient transfer of small DNA fragments to nitrocellulose. Therefore, it is possible that experiments in which no correlation between gene expression and DNA methylation was found failed to detect the biologically relevant sites.

4.2. Hypomethylation and Gene Expression

In spite of the striking correlation between gene undermethylation and expression revealed by restriction endonuclease analyses, studies of this kind cannot distinguish between undermethylation being a cause or consequence of gene activity. Therefore, alternative experimental approaches have been used to test the effect of DNA methylation on gene activity. One way to resolve the question of cause or effect is to examine the expression of methylated and unmethylated genes after insertion into animal cells. To date, some of the most convincing evidence relating DNA methylation to transcriptional inactivity has come from such experiments (Cooper, 1983). Furthermore, the data again suggest that methylation in the 5′ region of a gene may play a direct role in regulating gene expression. For example, *in vitro* methylation of the 5′-flanking region of the human δ-globin gene prevents expression upon introduction into mouse L cells (Busslinger *et al.*, 1983). Similarly, methylation of three specific Hpa II sites in the 5′ region of the E2a gene of adenovirus type 2 leads to transcriptional inactivation after microinjection into *Xenopus laevis* oocytes (Langner *et al.*, 1984).

As with restriction endonuclease analysis, the correlations in these experiments have not been absolute. For example, when cloned globin genes are introduced into mouse thymidine kinase⁻ (tk⁻) cells, together with the herpesvirus tk gene, the globin genes are not expressed despite the fact that they are not methylated (Chen and Nienhuis, 1981). Therefore, undermethylation cannot, by itself, ensure transcription in these assays. Nonetheless, these transfection and microinjection experiments are significant, because they provide direct evidence that DNA methylation can indeed cause transcriptional inactivation. The mechanism by which DNA methylation represses genes is unknown. However, it has been postulated that cytosine methylation alters the specific binding of regulatory proteins to DNA. The experiments demonstrating that genes are influenced by methylation in their 5′-regulatory regions are consistent with this hypothesis.

Keshet *et al.* (1986) inserted methylated and unmethylated M13 gene constructs into mouse L cells in order to study the mechanism of gene repression by DNA methylation. They found that methylation decreased overall DNase I sensitivity and inhibited the formation of DNase I- and restriction endonuclease-hypersensitive sites within the transfected sequences. In other words, upon integration into the genome, methylated sequences assume a chromatin structure that is characteristic of inactive genes (Conklin and Groudine, 1984). These findings suggest that one mechanism of action of DNA methylation may involve a generalized alteration in DNA–protein interactions, leading to the formation of an inactive chromatin conformation.

5. 5-Azacytidine and Gene Expression

5.1. Inhibition of Methylation by 5-Azacytidine

Further evidence supporting a causative role for methylation in suppressing gene activity comes from experiments using the nucleoside analogue **5-azacytidine (5-aza-CR)** (see Fig. 1). Originally developed as a cancer chemotherapeutic agent (Vesely and Cihak, 1978), 5-aza-CR is a cytosine analogue with a nitrogen atom instead of a carbon atom at the 5-position of the pyrimidine ring. 5-Azacytidine has remarkable effects on the differentiated state of cultured cells and on the activation of specific genes within cells. Although the drug can be incorporated into replicating DNA (Jones, 1984), it cannot be methylated because of its chemical structure (nitrogen in the 5-position). Hence, the biological effects of 5-aza-CR were proposed to stem from its ability to inhibit DNA methylation (Jones and Taylor, 1980).

Friedman (1979) was the first to demonstrate the inhibition of DNA methylation by 5-aza-CR in *Escherichia coli* K12 cells. 5-Aza-CR was subsequently shown to inhibit the methylation of newly synthesized DNA in mammalian cells (Jones and Taylor, 1980; Tanaka *et al.*, 1980; Christman *et al.*, 1983; Wilson *et al.*, 1983). Jones and Taylor (1980) showed that 5% substitution of cytosine residues by 5-aza-CR resulted in greater than 80% inhibition of DNA methylation in 10T ½ cells. These results suggested that incorporated 5-aza-CR inhibits the methylation of cytosine residues at unsubstituted CpG sites.

Early studies by Drahovsky and Morris (1971) suggested that DNA methyltransferase is a processive enzyme in that it remains associated with the DNA substrate after each catalytic event and "walks" along the helix, scanning for available methylation sites. Given this model, it has been proposed that the occurrence of a 5-aza-CR residue at a potential methylation site might impede the progress of the enzyme along the DNA helix. Thus, low levels of the fraudulent base could cause considerable inhibition of DNA methylation such as that seen in treated 10T ½ cells (Jones and Taylor, 1980).

Tanaka et al. (1980) demonstrated that treatment of Ehrlich's ascites tumor cells with 5-aza-CR led to a marked reduction in extractable DNA methyltransferase activity. Similarly, Creusot et al. (1982) found treatment of Friend erythroleukemia cells with 5-aza-CR to lead to a rapid, time- and dose-dependent decrease of methyltransferase activity in addition to the synthesis of markedly undermethylated DNA. The DNA methyltransferase apparently forms a high salt-resistant, tight-binding complex with hemimethylated DNA containing 5-aza-CR (Creusot et al., 1982; Taylor and Jones, 1982), explaining the loss of extractable enzyme activity from 5-aza-CR treated cells.

Santi et al. (1983) proposed a mechanism by which these tight-binding complexes form. In their proposed mechanism, the methyltransferase forms a covalent bond with the 6 position of the 5-aza-CR pyrimidine ring, leading to complete inactivation of the enzyme. This mechanism-based inhibition explains how incorporation of small amounts of 5-aza-CR into DNA results in extensive hypomethylation due to loss of enzyme activity.

5.2. Induction of Gene Expression by 5-Azacytidine

In addition to inducing generalized hypomethylation of DNA in treated cells, 5-aza-CR can induce the expression of previously dormant genes (Jones, 1984, 1985). For example, expression of the tk gene is increased as much as 10^5–10^6-fold following exposure of tk$^-$ Chinese hamster ovary (CHO) cells to 5-aza-CR (Harris, 1982). Particular excitement was generated by the finding that 5-aza-CR could induce the expression of genes located on inactive X chromosomes. For example, the human hypoxanthine–guanine phosphoribosyltransferase (HPRT) gene, encoded on an inactive X chromosome in mouse–human somatic cell hybrids, can be reactivated with 5-aza-CR (Mohandas et al., 1981). Induction of human HPRT gene expression is cell cycle dependent, with reactivation occurring maximally in cells treated in late S phase, when the inactive X chromosome is replicating (Jones et al., 1982). Evidence supporting the idea that inhibition of DNA methylation is the mechanism of drug action was provided by experiments in which DNA extracted from reactivated hybrids was used successfully to transform HPRT$^-$ recipient cells (Venolia et al., 1982). Taken together, the accumulated data strongly support the hypothesis initially put forth by Riggs (1975) that X chromosome inactivation is mediated by DNA methylation.

One feature of 5-aza-CR that has emerged is its ability to activate genes in a selective, rather than global, manner (Doerfler et al., 1984; Jones, 1985). This finding, together with results from restriction endonuclease and gene-transfer studies, leads to the inevitable conclusion that methylation is not the only factor governing gene expression in eukaryotic cells (Bird, 1984).

6. DNA Methylation and Cellular Differentiation

6.1. Induction of New Cellular Phenotypes by 5-Azacytidine

In addition to inducing the expression of certain selectable genes, 5-aza-CR alters the differentiated state of certain eukaryotic cells (Jones, 1984, 1985). These wide-ranging phenotypic changes induced by 5-aza-CR presumably reflect the concerted switching of many cellular genes. Perhaps the most remarkable example is the formation of twitching muscle cells from nonmuscle mouse embryo (10T ½) cells after exposure to 5-aza-CR (Constantinides et al., 1977). The induced muscle cells were found to be biochemically and functionally identical to bona fide myocytes. In subsequent experiments, differentiated adipocytes and chondrocytes were also observed in 5-aza-CR-treated cell cultures (Taylor and Jones, 1979). Such phenotypic changes are not restricted to cells of the 10T ½ line, nor are they restricted to phenotypes of the same developmental lineage as the cells treated with the drug (Jones, 1984, 1985). Thus, the ability of 5-aza-CR to change the differentiated state of cells is quite general.

There is an absolute requirement of cell division for new phenotypes to develop after 5-aza-CR treatment, and in many instances the induced phenotypes remain stable in the absence of further drug treatment (Jones, 1984). Thus, the effects of 5-aza-CR on cellular phenotypes are heritable. Concentrations of 5-aza-CR that induce the muscle phenotype in 10T ½ cells are highly effective inhibitors of methylation of newly-synthesized DNA (Jones and Taylor, 1980). The effects of 5-aza-CR on differentiation are specific for the 5-position of the cytosine ring. Furthermore, the deoxyanalog of 5-aza-CR (**5-aza-CdR;** which—in contrast to 5-aza-CR—is not incorporated into RNA) was found to be a more potent inducer of cellular differentiation than 5-aza-CR (Jones and Taylor, 1980). Taken together, these observations strongly suggest that the effects of 5-aza-CR on cellular phenotypes are linked to its ability to incorporate into DNA and cause heritable alterations in methylation patterns. The availability of undetermined cell lines (e.g., 10T ½) and determined clones isolated after 5-aza-CR treatment (Chapman et al., 1984; Konieczny and Emerson, 1984; Liu et al., 1986) provides a valuable system for studying regulatory gene switching during the differentiation process.

6.2. Demethylation Model for Differentiation

If cellular differentiation is indeed linked to alterations in methylation patterns, how might these patterns be established during development? One

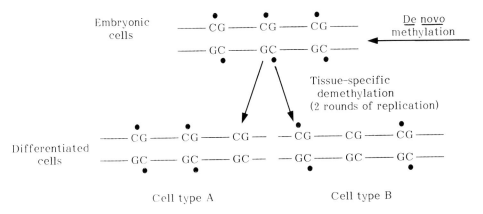

Figure 3. A demethylation model for the establishment of tissue-specific methylation patterns. Inhibition of methylation by sequence-specific DNA binding proteins during DNA replication leads to the establishment of specific methylation patterns in various cell types (closed circles represent methyl groups). The maintenance methyltransferase ensures the faithful inheritance of these patterns, as illustrated in Fig. 2.

model that has been proposed suggests that the ground state in the early embryo is one of full methylation, established by *de novo* methyltransferase activity (Razin and Riggs, 1980; Riggs and Jones, 1983). Recently, another model was proposed, in which *de novo* methylation during spermatogenesis plays a role in the templating of genetic information in sperm DNA (Groudine and Conklin, 1985). This *de novo* methylation proceeds so that regions within and around genes become methylated, excluding those sequences within the hypersensitive sites that are associated with constitutively expressed genes. These sites of undermethylation may serve to mark genes in the paternal genome for selective expression during early development. There is also considerable evidence for the *de novo* methylation of viral genomes and cellular sequences during embryonic development (Jahner and Jaenisch, 1984).

In more fully differentiated cells, the predominant methyltransferase is the maintenance type (Riggs and Jones, 1983). Assuming that *de novo* methylation occurs infrequently during the normal process of differentiation, tissue-specific methylation patterns must be the result of demethylation events that have occurred during development (Fig. 3). The mechanisms governing the selective loss of preexisting methyl groups are not well understood, and earlier reports providing evidence of demethylating enzymes (Gjerset and Martin, 1982) have not been confirmed. Instead, most experimental evidence is compatible with a passive mechanism, which is based on DNA replication in the absence of methylation.

Demethylation events at critical regulatory sites during development could be directed by sequence-specific proteins that sterically inhibit the maintenance methyltransferase. During the next round of replication, a totally unmethylated site would be generated. The methyl group in the 5-position of the cytosine ring is exposed in the major groove of the DNA helix (Razin and Riggs,

1980) and is indeed known to have profound effects on the interaction of DNA-binding proteins with DNA. For example, changing a specific thymine residue in the Lac operator to uracil or to cytosine greatly reduces the affinity of the repressor for the operator, whereas changing to 5-mCyt restores the repressor-binding affinity (Fisher and Caruthers, 1979). Therefore, binding of the Lac repressor is strongly influenced by the presence of a methyl group in its recognition sequence. The interferences by cytosine methylation of the activity of restriction endonucleases constitutes another clearly documented example of DNA methylation affecting DNA–protein interactions.

Recently, a mammalian protein(s) was identified for which DNA binding affinity is modulated by cytosine methylation. This protein(s) was isolated from human placental nuclei and was found to bind preferentially to 5-mCyt-enriched DNA (Huang *et al.*, 1984). More recently, by comparing its ability to bind to various *in vitro*-methylated restriction fragments of pBR322, it was determined that this protein(s) is sequence specific (Wang *et al.*, 1986). The findings of Keshet *et al.* (1986) (see Section 4.2) suggest that DNA methylation may exert its effects on gene transcription by altering both specific and non-specific protein–DNA interactions. Future experiments will no doubt attempt to determine whether the placental protein functions as a regulatory protein that directly controls transcriptional activity or as a structural protein that alters the local conformation of chromatin. Another important goal will be to identify other DNA-binding proteins that may modulate DNA methylation during differentiation, thus creating the patterns of methylation and gene expression that characterize fully differentiated cells.

7. DNA Methylation and Cancer

There is an emerging consensus that many cancers are caused by the aberrant expression of normal cellular genes (Comings, 1973; Alitalo *et al.*, 1984). If DNA methylation plays an important role in regulating gene expression, it follows that aberrations within this controlling mechanism may be implicated in the abnormal gene expression seen in cancer. Holliday (1979) was the first to propose a model of carcinogenesis that involved losses of 5-mCyt residues in DNA. This model was based on the finding that damage to DNA, followed by repair, could lead to the loss of methyl groups resulting in heritable, but potentially reversible, switches in gene activity.

One line of evidence supporting Holliday's theory of carcinogenesis comes from experiments showing that chemical carcinogens inhibit DNA methylation (Riggs and Jones, 1983; Jones *et al.*, 1986). The advantage of the hypothesis that deranged methylation is implicated in transformation is that it does not require the continued presence of a carcinogen in order for effects on gene expression to be manifest. Furthermore, an epigenetic mechanism is favored because the frequency of oncogenic transformation in culture is often greater than that predicted for a purely mutagenic mechanism (Holliday, 1979).

The hypothesis that oncogenic changes may develop as a result of abnormal gene regulation has also prompted investigators to study levels of methylation in transformed cells. The results of these studies have varied, with increased (Gunthert *et al.*, 1976), decreased (Gama-Sosa *et al.*, 1983), and unchanged (Flatau *et al.*, 1983) levels of methylation being reported. Recent experiments have also demonstrated hypomethylation of specific genes in tumor cells (Jones, 1986). These studies reveal considerable heterogeneity of methylation patterns in tumor cells, suggesting that these patterns are not as rigorously maintained in these cells as they are in their normal counterparts.

More direct evidence that methylation and transformation are linked comes from studies in which methylation patterns within tumor cells have been altered with 5-aza-CR. Numerous reports have demonstrated that 5-aza-CR has marked effects on the stability of the malignant phenotype (Jones, 1986). For example, Frost *et al.* (1984) found that 5-aza-CR influenced the expression of tumor antigens in highly tumorigenic mouse cell lines, resulting in conversion to a nontumorigenic phenotype.

One means by which methylation patterns may be altered *in vivo* is variation in the levels of DNA methyltransferase. Kautiainen and Jones (1986) have found that tumorigenic cell lines contain higher levels of methyltransferase activity than nontumorigenic cell lines. Since methylation patterns are thought to be determined by methylase levels (Razin and Szyf, 1984), the increased level of enzyme activity in tumor cells may allow for greater flexibility in the establishment of new methylation patterns, which are in turn responsible for the plasticity of the transformed phenotype.

8. Summary

Although numerous possible functions of eukaryotic DNA methylation have been proposed (Ehrlich and Wang, 1981), current evidence suggests an association between hypomethylation of specific gene regions and transcriptional activity (Doerfler, 1983; Riggs and Jones, 1983). Considerable excitement has been generated by the finding that potential methylation sites are often clustered at the 5' regions of genes (Bird *et al.*, 1985). Furthermore, these clusters appear to be protected from methylation when associated with actively expressed genes. These findings have generated a new line of thinking among researchers, namely that genes may be regulated by domains of methylation sites (Jones, 1986).

Although the evidence supporting a role for methylation in eukaryotic gene regulation is convincing, it is by no means unequivocal (Bird, 1984). To date, however, determination of the methylation status of specific cytosine residues in DNA has been limited by the experimental tools at hand. Therefore, development of a method that allows for evaluation of the methylation status of all CpG sites within, and in the vicinity of, genes may be necessary to elucidate the exact relationship between DNA methylation and gene expression.

The emerging consensus is that DNA methylation is one component of a hierarchy of control mechanisms that regulate eukaryotic gene expression and cellular differentiation. Furthermore, the realization that methylation plays an important role in regulating gene expression during normal cellular differentiation suggests that aberrations in this potential controlling mechanism may be implicated in the abnormal gene expression seen in cancer. Indeed, a large body of experimental evidence demonstrates altered levels and patterns of methylation in tumor cells (Riggs and Jones, 1983; Jones, 1986).

The regulatory function of 5-mCyt is most likely exerted via DNA–protein interactions. Two key questions remain unanswered: How does 5-mCyt affect these interactions? More importantly, what are the proteins that modulate DNA methylation during differentiation?

References

Alitalo, K., Saksela, K., Winqvist, R., Schwab, M., and Bishop, J. M., 1984, Amplification and aberrant expression of cellular oncogenes in human colon cancer cells, in: *Genes and Cancer* (J. M. Bishop, J. D. Rowley, and M. Greaves, eds.), pp. 383–397, Alan R. Liss, New York.

Barker, D., Schafer, M., and White, R., 1984, Restriction sites containing CpG show a higher frequency of polymorphism in human DNA, *Cell* **36**:131–138.

Bird, A. P., 1980, DNA methylation and the frequency of CpG in animal DNA, *Nucleic Acids Res.* **8**:1499–1504.

Bird, A. P., 1984, DNA methylation-how important in gene control?, *Nature (Lond.)* **307**:503–504.

Bird, A., Taggart, M., Frommer, M., Miller, O. J., and Macleod, D., 1985, A fraction of the mouse genome that is derived from islands of nonmethylated, CpG-rich DNA, *Cell* **40**:91–99.

Bolden, A. H., Nalin, C. M., Ward, C. A., Poonian, M. S., McComas, W. W., and Weissbach, A., 1985, DNA methylation: Sequences flanking C-G pairs modulate the specificity of the human DNA methylase, *Nucleic Acids Res.* **13**:3479–3494.

Busslinger, M., Hurst, J., and Flavell, R. A., 1983, DNA methylation and the regulation of globin gene expression, *Cell* **34**:197–206.

Chapman, A. B., Knight, D. M., Dieckmann, B. S., and Ringold, G. M., 1984, Analysis of gene expression during differentiation of adipogenic cells in culture and hormonal control of the developmental program, *J. Biol. Chem.* **259**:15548–15555.

Chen, M.-J., and Nienhuis, A. W., 1981, Structure and expression of human globin genes introduced into mouse fibroblasts, *J. Biol. Chem.* **256**:9680–9683.

Christman, J. K., Mendelsohn, N., Herzog, D., and Schneiderman, N., 1983, Effect of 5-azacytidine on differentiation and DNA methylation in human promyelocytic leukemia cells (HL-60), *Cancer Res.* **43**:763–769.

Comings, D. E., 1973, A general theory of carcinogenesis, *Proc. Natl. Acad. Sci. USA.* **70**:3324–3328.

Conklin, K. F., and Groudine, M., 1984, Chromatin structure and gene expression, in: *DNA Methylation. Biochemistry and Biological Significance* (A. Razin, H. Cedar, and A. D. Riggs, eds.), pp. 293–351, Springer-Verlag, New York.

Constantinides, P. G., Jones, P. A., and Gevers, W., 1977, Functional striated muscle cells from nonmyoblast precursors following 5-azacytidine treatment, *Nature (Lond.)* **267**:364–366.

Cooper, D. N., 1983, Eukaryotic DNA methylation, *Hum. Genet.* **64**:315–333.

Creusot, F., Acs, G., and Christman, J. K., 1982, Inhibition of DNA methyltransferase and induction of Friend erythroleukemia cell differentiation by 5-azacytidine and 5-aza-2'-deoxycytidine, *J. Biol. Chem.* **257**:2041–2048.

Doerfler, W., 1983, DNA methylation and gene activity, *Annu. Rev. Biochem.* **52**:93–124.

Doerfler, W., Langner, K.-D., Kruczek, I., Vardimon, L., and Renz, D., 1984, Specific promoter methylations cause gene inactivation, in: *DNA Methylation. Biochemistry and Biological Significance* (A. Razin, H. Cedar, and A. D. Riggs, eds.), pp. 221–247, Springer-Verlag, New York.

Drahovsky, D., and Morris, N. R., 1971, Mechanism of action of rat liver DNA methylase I. Interaction with double-stranded methyl-acceptor DNA, *J. Mol. Biol.* **57**:475–489.

Ehrlich, M., and Wang, R. Y.-H., 1981, 5-Methylcytosine in eukaryotic DNA, *Science* **212**:1350–1357.

Ehrlich, M., Gama-Sosa, M. A., Huang, L.-H., Midgett, R. M., Kuo, K. C., McCune, R. A., and Gehrke, C., 1982, Amount and distribution of 5-methylcytosine in human DNA from different types of tissues or cells, *Nucleic Acids Res.* **10**:2709–2721.

Fisher, E. F., and Caruthers, M. H., 1979, Studies on gene control regions. XII. The functional significance of a Lac operator constitutive mutation, *Nucleic Acids Res* **7**:401–416.

Flatau, E., Bogenmann, E., and Jones, P. A., 1983, Variable 5-methylcytosine levels in human tumor cell lines and fresh pediatric tumor explants, *Cancer Res.* **43**:4901–4905.

Friedman, S., 1979, The effect of 5-azacytidine on *E. coli* DNA methylase, *Biochem. Biophys. Res. Commun.* **89**:1328–1333.

Frost, P., Liteplo, R. G., Donaghue, T. P., and Kerbel, R. S., 1984, Selection of strongly immunogenic "Tum⁻" variants from tumors at high frequency using 5-azacytidine, *J. Exp. Med.* **159**:1491–1501.

Gama-Sosa, M. A., Slagel, V. A., Trewyn, R. W., Oxenhandler, R., Kuo, K. C., Gehrke, C. W., and Ehrlich, M., 1983, The 5-methylcytosine content of DNA from human tumors, *Nucleic Acids Res.* **11**:6883–6894.

Gerber-Huber, S., May, F. E. B., Westley, B. R., Felber, B. K., Hosbach, H. A., Andres, A. C., and Ryffel, G. U., 1983, In contrast to other Xenopus genes the estrogen-inducible vitellogenin genes are expressed when totally methylated, *Cell* **33**:43–51.

Gjerset, R. A., and Martin, D. W., 1982, Presence of a DNA demethylating activity in the nucleus of murine erythroleukemic cells, *J. Biol. Chem.* **257**:8581–8583.

Groudine, M., and Conklin, K. F., 1985, Chromatin structure and de novo methylation of sperm DNA: Implications for activation of the paternal genome, *Science* **228**:1061–1068.

Gruenbaum, Y., Stein, R., Cedar, H., and Razin, A., 1981, Methylation of CpG sequences in eukaryotic DNA, *FEBS Lett.* **124**:67–71.

Gruenbaum, Y., Cedar, H., and Razin, A., 1982, Substrate and sequence specificity of eukaryotic DNA methylase, *Nature (Lond.)* **295**:620–622.

Gunthert, U., Schweiger, M., Stupp, M., and Doerfler, W., 1976, DNA methylation in adenovirus, adenovirus-transformed cells, and host cells, *Proc. Natl. Acad. Sci. USA* **73**:3923–3927.

Harris, M., 1982, Induction of thymidine kinase in enzyme-deficient Chinese hamster cells, *Cell* **29**:483–492.

Holliday, R., 1979, A new theory of carcinogenesis, *Br. J. Cancer* **40**:513–522.

Holliday, R., and Pugh, J. E., 1975, DNA modification mechanisms and gene activity during development, *Science* **187**:226–232.

Huang, L.-H., Wang, R., Gama-Sosa, M. A., Shenoy, S., and Ehrlich, M., 1984, A protein from human placental nuclei binds preferentially to 5-methylcytosine-rich DNA, *Nature (Lond.)* **308**:293–295.

Jahner, D., and Jaenisch, R., 1984, DNA methylation in early mammalian development, in: *DNA Methylation. Biochemistry and Biological Significance* (A. Razin, H. Cedar, and A. D. Riggs, eds.), pp. 189–219, Springer-Verlag, New York.

Jones, P. A., 1984, Gene activation by 5-azacytidine, in: *DNA Methylation. Biochemistry and Biological Significance* (A. Razin, H. Cedar, and A. D. Riggs, eds.), pp. 164–187, Springer-Verlag, New York.

Jones, P. A., 1985, Altering gene expression with 5-azacytidine, *Cell* **40**:485–486.

Jones, P. A., 1986, DNA methylation and cancer, *Cancer Res.* **46**:461–466.

Jones, P. A., and Taylor, S. M., 1980, Cellular differentiation, cytidine analogs and DNA methylation, *Cell* **12**:85–93.

Jones, P. A., Taylor, S. M., Mohandas, T., and Shapiro, L. J., 1982, Cell cycle-specific reactivation of an inactive X-chromosome locus by 5-azadeoxycytidine, *Proc. Natl. Acad. Sci. USA* **79**:1215–1219.

Jones, P. A., Chandler, L., Kautiainen, T., Wilson, V., and Flatau, E., 1986, DNA methylation, in: *Biochemical and Molecular Epidemiology of Cancer*, pp. 149–154.

Kautiainen, T. L., and Jones, P. A., 1986, DNA methyltransferase levels in tumorigenic and non-tumorigenic cells in culture, *J. Biol. Chem.* **261:**1594–1598.

Keshet, I., Leiman-Hurwitz, J., and Cedar, H., 1986, DNA methylation affects the formation of active chromatin, *Cell* **44:**535–543.

Konieczny, S. F., and Emerson, C. P., 1984, 5-Azacytidine induction of stable mesodermal stem cell lineages from 10T ½ cells: Evidence for regulatory genes controlling determination, *Cell* **38:**791–800.

Lanner, K.-D., Vardimon, L., Renz, D., and Doerfler, W., 1984, DNA methylation of three 5'CCGG3' sites in the promoter and 5' region inactivate the E2a gene of adenovirus type 2, *Proc. Natl. Acad. Sci. USA* **81:**2950–2954.

Liu, L., Harrington, M., and Jones, P. A., 1986, Characterization of myogenic cell lines derived by 5-azacytidine treatment, *Dev. Biol.* **117:**331–336.

Macleod, D., and Bird, A., 1983, Transcription in oocytes of highly methylated rDNA from *Xenopus laevis* sperm, *Nature (Lond.)* **306:**200–203.

McClelland, M., 1981, The effect of sequence specific DNA methylation on restriction endonuclease cleavage, *Nucleic Acids Res.* **9:**5859–5866.

McGhee, J. D., and Ginder, G. D., 1979, Specific DNA methylation sites in the vicinity of the chicken β-globin genes, *Nature (Lond.)* **280:**419–420.

McKeon, C., Ohkubo, H., Pastan, I., and deCrombrugghe, B., 1982, Unusual methylation pattern of the α2(I) collagen gene, *Cell* **29:**203–210.

Mohandas, T., Sparkes, R. S., and Shapiro, L. J., 1981, Reactivation of an inactive human X chromosome: Evidence for X inactivation by DNA methylation, *Science* **211:**393–396.

Pfeifer, G. P., Grunwald, S., Palitti, F., Kaul, S., Boehm, T. L. J., Hirth, H-P., and Drahovsky, D., 1985, Purification and characterization of mammalian DNA methyltransferases by use of monoclonal antibodies, *J. Biol. Chem.* **260:**13787–13793.

Razin, A., 1984, DNA methylation patterns: Formation and biological function, in: *DNA Methylation. Biochemistry and Biological Significance* (A. Razin, H. Cedar, and A. D. Riggs, eds.), pp. 127–146, Springer-Verlag, New York.

Razin, A., and Cedar, H., 1977, Distribution of 5-methylcytosine in chromatin, *Proc. Natl. Acad. Sci. USA* **74:**2725–2728.

Razin, A., and Riggs, A. D., 1980, DNA methylation and gene function, *Science* **210:**604–610.

Razin, A., and Szyf, M., 1984, DNA methylation patterns. Formation and function, *Biochim. Biophys. Acta* **782:**331–342.

Riggs, A. D., 1975, X inactivation, differentiation, and DNA methylation, *Cytogenet. Cell Genet.* **14:**9–25.

Riggs, A. D., and Jones, P. A., 1983, 5-Methylcytosine, gene regulation, and cancer, *Adv. Cancer Res.* **40:**1–30.

Salser, W., 1977, Globin mRNA sequences: Analysis of base pairing and evolutionary implications, *Cold Spring Harbor Symp. Quant. Biol.* **42:**985–1002.

Santi, D. V., Garrett, C. E., and Barr, P. J., 1983, On the mechanism of inhibition of DNA-cytosine methyltransferases by cytosine analogs, *Cell* **33:**9–10.

Simon, D., Grunert, F., vonAcken, U., Doring, H. P., and Kroger, H., 1978, DNA methylase from regenerating rat liver: Purification and characterisation, *Nucleic Acids Res.* **5:**2153–2167.

Simon, D., Stuhlmann, H., Jahner, D., Wagner, H., Werner, E., and Jaenisch, R., 1983, Retrovirus genomes methylated by mammalian but not bacterial methylase are non-infectious, *Nature (Lond.)* **304:**275–277.

Solage, A., and Cedar, H., 1978, Organization of 5-methylcytosine in chromosomal DNA, *Biochemistry* **17:**2934–2938.

Southern, E. M., 1975, Detection of specific sequences among DNA fragments separated by gel electrophoresis, *J. Mol. Biol.* **98:**503–517.

Stein, R., Gruenbaum, Y., Pollack, Y., Razin, A., and Cedar, H., 1982, Clonal inheritance of the pattern of DNA methylation in mouse cells, *Proc. Natl. Acad. Sci. USA* **79:**61–65.

Tanaka, M., Hibasami, H., Nagai, J., and Ikeda, T., 1980, Effect of 5-azacytidine on DNA methylation in Ehrlich's ascites tumor cells, *Aust. J. Exp. Biol. Med. Sci.* **58:**391–396.

Taylor, S. M., and Jones, P. A., 1979, Multiple new phenotypes induced in 10T ½ and 3T3 cells treated with 5-azacytidine, *Cell* **17**:771–779.

Taylor, S. M., and Jones, P. A., 1982, Mechanism of action of eukaryotic DNA methyltransferase. Use of 5-azacytosine-containing DNA, *J. Mol. Biol.* **162**:679–692.

Venolia, L., Gartler, S. M., Wassman, E. R., Yen, P., Mohandas, T., and Shapiro, L. J., 1982, Transformation with DNA from 5-azacytidine-reactivated X chromosomes, *Proc. Natl. Acad. Sci. USA* **79**:2352–2354.

Vesely, J., and Cihak, A., 1978, 5-Azacytidine: Mechanism of action and biological effects in mammalian cells, *Pharm. Ther.* **2**:813–840.

Wang, R. Y.-H., Zhang, X.-Y., and Ehrlich, M., 1986, A human DNA-binding protein is methylation-specific and sequence-specific, *Nucleic Acids Res.* **14**:1599–1614.

Wigler, M., Levy, D., and Perucho, M., 1981, The somatic replication of DNA methylation, *Cell* **24**:33–40.

Wilson, V. L., Jones, P. A., and Momparler, R. L., 1983, Inhibition of DNA methylation in L1210 leukemic cells by 5-aza-2'-deoxycytidine as a possible mechanism of chemotherapeutic action, *Cancer Res.* **43**:3493–3496.

Wolf, S. F., Dintzis, S., Toniolo, D., Persico, G., Lunnen, K. D., Axelman, J., and Migeon, B. R., 1984, Complete concordance between glucose-6-phosphate dehydrogenase activity and hypo-methylation of 3' CpG clusters: Implications for X chromosome dosage compensation, *Nucleic Acids Res.* **12**:9333–9348.

Zucker, K. E., Riggs, A. D., and Smith, S. S., 1985, Purification of human DNA (cytosine-5-)-methyltransferase, *J. Cell. Biochem.* **29**:337–349.

III

The Molecular Biology of Plant Growth and Development

Chapter 13

Molecular Biology of Plant Growth and Development
Arabidopsis thaliana as an Experimental System

ELLIOT M. MEYEROWITZ and CAREN CHANG

1. Introduction

Plant development differs from animal development in several fundamental respects. Since plant cells are immobilized in a rigid cell wall, morphogenesis is dependent upon control of cell growth and plane of cell division, rather than cell migration as occurs in animal development. The meristematic cells of plants display a degree of plasticity not found in animal cells; they remain embryonic throughout the life of the plant and produce both adult organs and germ cells. Furthermore, individual differentiated cells from vegetative plant parts are able to dedifferentiate and regenerate into new, fertile plants.

Arabidopsis thaliana is a flowering plant that has many characteristics that recommend it for genetic and molecular studies of development, much as certain attributes of Drosphila recommend it for such studies. In fact, A. thaliana was recognized as a botanical Drosphila during the 1940s (Whyte, 1946) because of its low chromosome number, short generation time, ease of culturing, and high fecundity. Over the past 40 years an extensive literature has accumulated on the genetics and ecology of this member of the mustard family (reviewed by Rédei, 1970, 1975a,b). Recently, molecular characterization of the Arabidopsis genome has shown that the plant has an extraordinarily small and simple genome, properties that facilitate the molecular analysis of development (Leutwiler et al., 1984; Pruitt and Meyerowitz, 1986). This chapter provides an overview of A. thaliana as an experimental system by describing the classical and molecular genetics of the plant and discussing the attributes that make it a valuable model system for genetic and molecular studies of plant development.

ELLIOT M. MEYEROWITZ and CAREN CHANG • Division of Biology, California Institute of Technology, Pasadena, California 91125.

2. General Description

Arabidopsis thaliana is a typical member of the mustard family (Cruciferae). Although a number of other crucifers, such as cabbage and turnips, are crops, *Arabidopsis* has no known food or other economic value. The mature plant is small but is as morphologically complex as other flowering plants (Figs. 1 and 2). Typically about 30 cm tall, the mature plant consists of a rosette of small leaves surrounding a main stem. Flowers in racemes are at the apex of the stem. Lateral branches, arising from the main stem as the plant ages, also develop apical inflorescences. The flowers are bisexual and are approximately 3 mm long × 1 mm wide. They consist of four sepals surrounding four white petals, alternating with the sepals; within the petals are six stamens (two short

Figure 1. The *erecta* mutant of the *Arabidopsis thaliana* Landsberg ecotype. Landsberg *erecta* is commonly used in genetic and molecular genetic experiments because of its rapid life cycle and reduced stature.

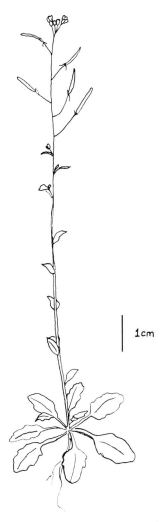

Figure 2. *Arabidopsis thaliana* plant, Bensheim ecotype.

and four long) and a pistil. The anthers of the longer stamens rise above the stigma surface and release pollen, whereas the two shorter stamens are not believed to contribute to fertilization (Meinke and Sussex, 1979a). The pistil consists of two fused carpels with parietal placentation (i.e., the ovules are borne on the ovary wall), and the ovary is divided into two chambers by a vertical partition of complex origin. After fertilization, the ovary matures to a fruit called a silique, 12–16 mm in length, containing 30–60 seeds arranged in two rows. A single plant can produce several hundred siliques and may therefore yield more than 10,000 seeds. The seeds are very small, measuring approximately 0.5 × 0.3 × 0.2 mm and weighing 16–20 μg. Detailed descriptions of the wild-type plant are given by Röbbelen (1957a), Müller (1961), and Napp-Zinn (1963).

The length of the *A. thaliana* life cycle depends on growth conditions and genetic background. A typical time from germination to mature silique is 5–6 weeks for the most rapidly flowering ecotypes in long day conditions. Seed germination occurs within a few days of the start of imbibition, followed by vegetative growth of rosette leaves. The primary stem bolts in 2–3 weeks, and the primary inflorescence begins its development shortly thereafter. As the plant ages, lateral branches develop from the leaf axils of the main stem and occasionally also develop from the base of the plant. Although the flowers normally self-fertilize, cross-pollination is easily effected under a dissecting microscope. In nature, up to several percent of the flowers may be cross-pollinated, evidently by small insects and perhaps wind; under laboratory conditions, in the absence of insect infestation, spontaneous cross-fertilization is virtually absent. The siliques reach maturity about 2 weeks after pollination. The most apical fruits on each stem are the youngest, so that developmental sequences of seeds may be obtained simply by picking the fruits in the order in which they appear on each stem. If left on the plant, the siliques dry out, their valves open, and vibration causes the seeds to drop. For experimental work, individual fruits may be harvested before they open, or the seeds of dried siliques can be harvested by inverting the plants and shaking off the seeds. An *Arabidopsis* plant will continue to grow and produce seed for months. Mature seeds remain germinable for years, and probably decades, if stored in dry conditions. A detailed description of the life cycle is given by Meinke and Sussex (1979a).

The precise taxonomic placement of *A. thaliana* is unclear; different taxonomic treatments of the mustards regard the plant differently. The precise geographical origin of *Arabidopsis* is also unclear. Most collections have been made in Europe and Asia, and the plant is thought to be native to the Old World. However, *Arabidopsis* has been found on all continents (except the Antarctic) in a wide variety of environments, ranging from the temperate Himalayas to semidesert regions to high elevations of the tropics. A great deal of genetic variation is found in wild populations even within small geographical areas; different populations vary in response to day length, requirements for vernalization, times to flowering, and leaf morphology and color. Many ecotypes that are adapted to a wide range of ecological niches have been collected and described (Röbbelen, 1965); seeds from these collections are among those available to researchers from the seed bank maintained at the Goethe University in Frankfurt, Federal Republic of Germany, by Professor Dr. A. R. Kranz (Kranz, 1978; Kirchheim and Kranz, 1981).

Growing *Arabidopsis* in the laboratory is not difficult. Several dozen plants can be grown to maturity in a 5–6-cm diameter pot. The plants grow well in soil with occasional watering and will also grow in a variety of defined simple media (Langridge, 1957). Rapid growth occurs in a broad range of temperatures (14–26°C); above 30°C, the number of seeds per silique is reduced, and at colder temperatures growth and maturation are slow. Continuous lighting produces the most rapid growth to maturity; on short days, the rosette stage lasts several times longer than in long days, resulting in much larger rosettes. Growth in short day conditions is thus a method for producing large amounts of

plant tissue from individual plants. Plants can be grown aseptically in either solid or liquid media. Whole plants can be regenerated from callus at high frequency (Feldmann and Marks, 1986). Haploid callus and plants have been generated by anther culture (Gresshof and Doy, 1972).

3. Classical Genetics

Arabidopsis thaliana has a haploid chromosome number of 5 (Laibach, 1907). The small chromosomes, ranging in length from 1.1 to 3.7 μm at meiotic telophase I, can be distinguished by their different lengths and arm ratios (Steinitz-Sears, 1963).

Classical genetic analysis of the plant began about 40 years ago with the isolation of the first induced mutations by Reinholz (1947), who used X-rays as the mutagenic agent. It has since been shown that a large number of chemical mutagens are effective in the induction of new mutations (Rédei, 1970; Ehrenberg, 1971). Since *Arabidopsis* is self-fertilizing and matures rapidly, it has been a convenient plant for evaluating the effects of mutagenic agents. This is exemplified by the embryo-lethal assay of Müller (1963), in which plants grown from mutagenized seeds (M_1 seeds) are screened for siliques that contain approximately 25% aborted and 75% normal M_2 seed. Such siliques result from the induction of recessive embryo-lethal mutations in the cells of the M_1 seed that give rise to flowers. When the flowers self-fertilize, one fourth of the seeds that develop are homozygous for the newly induced mutation, whereas two thirds of the phenotypically normal seeds maintain the mutation in a heterozygous condition. Since each seed contains several cells that will develop into reproductive structures, new mutations are normally limited to one sector of the mature plant; only the first five siliques need to be scored to assay all of the sectors. The embryo-lethal screen is simple and reliable: aborted seeds are distinct in color and size in both immature and mature siliques, and the frequency of spontaneous abortion in *Arabidopsis* is low and unaffected by fluctuations in temperature, water, and nutrient supply (Meinke and Sussex, 1979a).

Rédei (1970) reviewed the genetic effects of about 80 mutagens on *A. thaliana*, measured by various techniques. Ethyl methanesulfonate (EMS) is an effective mutagen, giving a high ratio of mutants to nonsurvivors, in addition to producing a large number of visible mutations relative to induced sterility. Soaking dry or imbibed seeds in solutions of EMS for several hours is sufficient for mutagenesis. Other methanesulfonic esters (hydroxyethyl, methoxy, and methoxyethyl) are comparable to EMS. Certain nitrosoguanidines, nitrosoamines, and nitrosoureas also serve as effective mutagens. Although it is not as efficient as chemical mutagenesis, ionizing radiation applied to seeds has also been used successfully to induce mutations. *A. thaliana* is highly resistant to ionizing radiation, presumably because its extremely small nuclear volume (chromosome volume) presents such a tiny target (Sparrow, 1965). This resistance was initially observed by Reinholz (1947). As a result of these and

other studies, a wide spectrum of morphological and physiological mutations have been induced (Reinholz, 1947; Röbbelen, 1957b; McKelvie, 1962, 1963; Bürger, 1971; Koornneef et al., 1983), a few of which are discussed here to provide an introduction to the range of mutations available for developmental, physiological, and molecular genetic research.

The easiest mutations to identify are those that have a visible or lethal phenotype. Visible mutations affecting every part of the plant have been identified. There are mutations affecting plant and embryo color (*albina, chlorina, sulfurata*), leaf morphology (*serrate, angustifolia*), trichome morphology (*distorted trichomes, glabra*), and plant size (*erecta, compacta, miniature*), to name only a few of the mapped mutations. Another mapped mutation, *immutans*, interferes with plastid differentiation such that some cells are normal (green), whereas other cells lack both chlorophylls and carotenoids (white) (Rédei, 1967). This mutation is suppressed by 6-azauracil or low light intensity. *Chloroplast mutator* also causes variegated leaves; mutations in this nuclear gene induce plastid mutations that are then cytoplasmically inherited (Rédei, 1973).

Many mutations are known to affect floral development in ways that are of particular interest to developmental biologists because these mutations cause homeotic transformations of floral organs. For example, *apetala-2* is a recessive mutation that shows a conversion of petals to stamens that ranges from partial to complete and an apparent conversion of sepals to leaves. The flowers thus have ten stamens, no petals or sepals, and are surrounded by four leaves. Another striking homeotic transformation is shown by *apetala-3* homozygotes, in which the petals are converted to sepals, and each of the six stamens is changed to a single unpaired carpel having what appear to be ovules attached to its margin. In *pistillata*, petals are replaced by sepals and anthers are absent. There are other floral mutations that cause fundamental changes in floral structure that cannot be properly considered homeotic, such as *agamous*, which causes flowers to develop that have a large and indefinite number of sepals and petals, but no stamens or carpels. Another such mutant is *apetala-1*, in which flowers have rudimentary or absent petals and frequently have flower buds arising from within the whorl of sepals. All of the floral mutations described here, and others as well, have been mapped to their chromosomal locations by Koornneef et al. (1983) and J. Bowman and E. Meyerowitz (unpublished).

Another group of developmental mutations has been isolated using Müller's embryo-lethal screen. Isolation and characterization of embryo-lethal mutants is the first step in identifying essential functions at the various stages of plant embryogenesis. In maize, defective kernel mutants and viviparous mutants have been studied in detail; in carrot, somatic embryogenesis *in vitro* has been examined in mutant cell lines. In *Arabidopsis*, about 40 EMS-induced embryo-lethal mutants have been analyzed, representing a wide range of phenotypes (Meinke and Sussex, 1979b; Meinke, 1985). These aborted seeds and embryos differ with respect to color, size, stage of developmental arrest, and extent of abnormal development. Evidence for the action of essential genes in the gametophyte, before fertilization, is provided by the nonrandom distribution of aborted seeds within the siliques of 10 of the mutants (Meinke, 1982;

Meinke, 1985). The low proportion of aborted embryos in some of the mutants indicates an effect on fertilization. Other embryo-lethal mutants have been used for ultrastructural analysis (Marsden and Meinke, 1985) and for embryo culture. From the culture of arrested embryos, Meinke *et al.* (1985) were able to produce homozygous plants exhibiting mutant phenotypes. By culturing arrested embryos on a nutrient medium, it should be possible to identify nutritional lethals and to rescue them, whereas mutants with a cellular basis would not be rescued in this manner; developmental lethals should not be able to differentiate in culture but should be able to produce callus. In addition to their use in understanding essential developmental processes, embryo-lethals may be useful for studying chlorophyll synthesis during embryogenesis, since defects in chlorophyll and carotenoid pigments (i.e., white or pale green arrested seeds and embryos) are commonly seen, but mutants for pigmentation apparently undergo normal embryo development.

In addition to mutations causing morphological abnormalities and arrested embryonic development, a number of mutations have been obtained that affect synthesis of, and response to, phytohormones. Mutations at five different genetic loci in *A. thaliana* are known to result in plants deficient in gibberellins. The phenotype of some of the mutations is failure of germination; those that allow germination cause dwarfing (Koornneef and van der Veen, 1980). When exogenous gibberellins are sprayed on the mutant plants, they are restored to normal or near-normal phenotype. Abscisic acid-deficient mutants have been obtained by selecting for second site revertants of the nongerminating gibberellin mutations, based on the rationale that seed germination is stimulated by gibberellins but inhibited by abscisic acid (Koornneef *et al.*, 1982). The phenotype of many of the abscisic acid-deficient mutants is precocious seed germination and wilting of maturing plants. Phytohormone-insensitive mutants have also been isolated. Mutations in at least three loci confer resistance to high levels of abscisic acid; these mutants resemble the abscisin-deficient mutants in phenotype, except that endogenous abscisin levels are normal or greater than normal (Koornneef *et al.*, 1984). Mutants resistant to the auxin 2,4-dichlorophenoxyacetic acid, including a dominant mutation, have been mapped to two loci (Maher and Martindale, 1980). Alleles at both loci display increased rates of root growth and altered geotropism (Mirza *et al.*, 1984).

Mutations affecting the levels of activity of a number of specific enzymes are also known. Mutants with reduced nitrate reductase activity have been selected by isolating plants resistant to chlorate, which is converted to toxic chlorite by nitrate reductase (Braaksma and Feenstra, 1982). Some of these mutations affect chlorate uptake, others reduce nitrate reductase activity by affecting its regulation or the molybdenum cofactor necessary for its function. Mutants lacking alcohol dehydrogenase have been selected by isolating plants resistant to allyl alcohol, which is converted to a toxic aldehyde by alcohol dehydrogenase. These mutants appear to have genetic lesions in the structural gene for the enzyme (Negrutiu *et al.*, 1984). A number of mutations affect the levels of enzymes in the photorespiration pathway; among the affected enzymes are glutamate synthase, glycine decarboxylase, phosphoglycolate phos-

phatase, serine–glyoxylate aminotransferase, and serine transhydroxy-methylase (Somerville and Ogren, 1982). It is not known whether any of these mutations is in the structural genes for the enzymes.

Other biochemical mutations are known in *Arabidopsis*, including muta-tions in several loci that cause the plant to require exogenous thiamine or thiamine precursors (Langridge, 1955; Li and Rédei, 1969). Some of these con-ditional lethals are temperature sensitive. It is likely that temperature-sensitive mutations and other conditional mutations can be isolated for different genes.

Many of the known *Arabidopsis* mutations have been mapped to their positions on the five *Arabidopsis* linkage groups, which correspond to the five chromosomes. Koornneef *et al.* (1983) published a map showing the genetic locations of 76 loci; this map is a compilation of all previously published linkage information and a great deal of new linkage data. The correspondence between linkage groups and chromosomes has been established by analysis of trisomic lines (Koornneef and van der Veen, 1983); centromeres have been mapped by study of complementation in telotrisomic lines. The entire genetic map currently measures approximately 430 centimorgans (cM).

4. The *Arabidopsis* Genome

Although *Arabidopsis* has a large number of useful and developmentally interesting mutations, it is not distingushed from other experimentally useful angiosperms such as maize and tomato in this respect. While it is easier and quicker to isolate new mutations in *Arabidopsis* than in these other plants, the properties that truly set *Arabidopsis* apart are those that are important for experiments in molecular genetics, primarily an extremely small genome and paucity of dispersed repetitive DNA in the nuclear genome.

Arabidopsis has the smallest known angiosperm genome. The haploid nuclear genome of *Arabidopsis* is 70,000 kilobase pairs (kb), as measured both by DNA reassociation kinetic analysis (Leutwiler *et al.*, 1984) and by quan-titative genome blot hybridization (Pruitt and Meyerowitz, 1986). This is only five times the genome size of the yeast, *Saccharomyces cerevisiae*, and only 20 times the size of the *Escherichia coli* genome. Plants frequently used in mo-lecular genetic experiments have much larger genomes; tobacco and wheat, for example, have genomes of 1,600,000 and 5,900,000 kb, respectively (Zimmer-man and Goldberg, 1977; Flavell and Smith, 1976).

The importance of the tiny *Arabidopsis* genome is that it simplifies the molecular cloning of any specific gene. First, only 16,000 random λ clones need be screened with any probe to have a 99% chance of finding the desired DNA in the library. By contrast, 370,000 clones would be required for tobacco and 1,400,000 clones for wheat (Meyerowitz and Pruitt, 1985). Second, in genome blot experiments with heterologous probes, the signal-to-noise ratio is much higher than in other plants because of the low complexity of the *Arabidopsis* genome. This allows very diverged probes to be tested for hybridization to an angiosperm genome. An example in which a heterologous probe was success-

fully used in the isolation of an *Arabidopsis* gene is the use of the *Saccharomyces cerevisiae* acetolactate synthase gene to isolate the acetolactate synthase gene of *Arabidopsis* (Mazur *et al.*, personal communication). This gene codes for an enzyme that is inhibited by an important class of herbicides; mutations in this gene confer herbicide resistance in plants (Chaleff 1984).

The 70,000 kb of *Arabidopsis* haploid nuclear DNA has been shown to be largely single copy by solution hybridization kinetic measurements (Leutwiler *et al.*, 1984). The organization of the genomic DNA has been examined by molecular characterization of 50 λ clones chosen randomly from an *Arabidopsis* genomic library made from whole-plant DNA (Pruitt and Meyerowitz, 1986). These clones represent 0.8% of the nuclear genome; they were analyzed by a variety of restriction endonclease digestions and gel-blotting procedures designed to reveal the nature and interspersion of repetitive sequence elements in the genome. In agreement with the results obtained in reassociation experiments, the majority of the clones (34 of 50) contained low-copy-number sequences; 32 of these were entirely unique, whereas two contained restriction fragments that may be present twice in the genome. The 16 remaining random clones contained DNA belonging to the middle repetitive sequence class. Eight of these clones contained portions of the ribosomal DNA repeat unit, a 9.9-kb unit repeated 570 times per haploid genome, largely or entirely in tandem arrays, and making up 7.5% of the nuclear DNA. Four of the clones were derived from the chloroplast genome, one was likely derived from the mitochondrion, and only three were nuclear sequences containing dispersed middle repetitive elements. From the frequency and length of the unique sequence and middle repetitive sequence-containing clones, the average amount of single-copy DNA separating adjacent dispersed repeats was estimated to be 120 kb. This is very different from the DNA of the other angiosperms that have been analyzed: The average size in the predominant class of single-copy sequences in tobacco, for instance, is only 1.4 kb; in wheat it is 1 kb, and in pea only 300 base pairs (bp) (summarized in Meyerowitz and Pruitt, 1985). The evolutionary significance of the small genome and long period interspersion of repetitive DNA in the *Arabidopsis* genome is unclear. The experimental significance of the extraordinarily long sequence interspersion pattern in *Arabidopsis* is that it permits chromosome walking experiments; i.e., experiments in which successive isolations of overlapping cloned segments lead from an initial starting clone to any nearby genetic location. Only *Arabidopsis*, among all known flowering plants, has a genome with a low enough frequency of interspersed repeats to allow such a procedure to be performed practically.

5. Molecular Genetics

Several genes have already been cloned in *A. thaliana*, making it possible to begin using *Arabidopsis* for examining the developmental and environmental control of plant gene expression. Characterization of these genes provides another look at the *Arabidopsis* genome and, additionally, demonstrates its utility for certain molecular approaches. These cloning experiments show (1) that genes having many copies in other plants exist in a low number of copies, or in only a single copy, in *Arabidopsis*; and (2) when multiple gene copies have been found in *Arabidopsis*, the copies are usually clustered. Among the genes that demonstrate these principles are the light harvesting chlorophyll a/b binding protein genes, the alcohol dehydrogenase gene, the 12S-seed-storage protein gene, and the 70,000 M_r heat-shock protein (hsp 70) genes.

The light-harvesting chlorophyll a/b binding protein (LHCP) is a light-inducible nuclear-encoded component of the light-harvesting antenna complex of the chloroplast thylakoid membranes. The *Arabidopsis* LHCP genes were cloned from a genomic recombinant library by cross-hybridization with a LHCP gene probe from the aquatic monocot *Lemna gibba* (Leutwiler *et al.*, 1986). Although other plants are known to contain LHCP gene families of 7–16 genes, the *Arabidopsis* family contains only three genes clustered within a 6.5-kb region. The DNA sequences of the three open reading frames differ at 4% of the nucleotides. The deduced amino acid sequences encoded by the three clustered genes, excluding the transit peptides, are identical; thus, the three copies of the LHCP genes do not provide protein heterogeneity.

The alcohol dehydrogenase (ADH) enzymes in higher plants are under environmental and developmental control; ADH activity has been localized in dry seeds, pollen, germinating roots, and anaerobically treated seedling roots. In maize and *Arabidopsis*, ADH is induced by anaerobic treatment and by the synthetic auxin 2,4-dichlorophenoxyacetic acid (Dennis *et al.*, 1985; Freeling, 1973; Dolferus *et al.*, 1985). The *Arabidopsis Adh* gene was cloned from a genomic recombinant library by cross-hybridization with a maize *Adh1* gene probe (Chang and Meyerowitz, 1986). In agreement with previous genetic data, genome blots probed with the *Adh* clone indicate that only a single *Adh* gene exists in *Arabidopsis*. By contrast, most plants have two or more *Adh* genes. The single *Arabidopsis Adh* gene is similar to the two maize *Adh* genes in both sequence and structure, although it is smaller: The *Arabidopsis* ADH protein sequence (deduced from the DNA sequence) and the maize ADH protein sequences are the same length and are approximately 80% conserved, and the six intervening sequence positions of the *Arabidopsis Adh* gene coincide with six of the nine intervening sequences in both of the maize genes.

Developing plant embryos store nitrogen in the form of various seed storage proteins, such as the 12S globulin storage protein in *Arabidopsis*, which accumulate to high levels in the latter half of embryogenesis and are rapidly broken down during germination. Genes that code for abundant seed RNAs have been isolated from a genomic recombinant library by cross-hybridization with *A. thaliana* seed pod cDNA probes (Pang *et al.*, 1988). These clones fall into four homology classes based on cross-hybridization and restriction fragment patterns. One class contains the *Arabidopsis* 12S storage protein gene, which was identified by homology with a 12S storage-protein complementary DNA (cDNA) clone of the crucifer *Brassica napus*. Although there appears to be only one copy in this class of 12S genes in the Columbia and Bensheim ecotypes of *Arabidopsis*, there are two tandemly arranged 12S genes in the Landsberg *erecta* ecotype. The 5'-upstream DNA sequences of the two tightly clustered genes are not conserved except in the immediate TATA box region. Such close clustering of genes has not been observed for the 12S protein gene families of other plant species, nor has a gene copy number as low as one or two. The amino acid sequence (deduced from the DNA sequence of one of the two genes) is homologous with both the *B. napus* and the pea 12S storage protein sequences. The intron positions of the *Arabidopsis* gene are coincident with those of the pea gene.

Using a *Drosophila* hsp70 gene clone to probe an *Arabidopsis* genomic recombinant library, two tandemly arranged genes in *Arabidopsis* have been cloned that resemble both the maize and *Drosophila* hsp 70 genes (C. Somerville, personal communication). One of the genes is induced by exposure to elevated temperatures for 2 hr, whereas the other gene is constitutively expressed at low levels.

Another approach for identifying developmentally important genes is to examine the differential expression of genes in specific tissues. In *Arabidopsis*, such work has been initiated by Mesnard *et al.* (1985), who constructed a cDNA library from leaf mRNA. The abundance and relative expression of random genes in two different organs were examined by using individual leaf cDNA clones to probe blots containing seed pod and leaf mRNA. In this manner, two clones were found that have mRNA levels 5–10 times greater in pods than in leaves.

A number of additional approaches to the isolation of environmentally and developmentally regulated *Arabidopsis* genes are currently being taken in different laboratories. Although no deep understanding of developmental or environmental regulation of transcription has yet been achieved, the ease with which molecular techniques can be applied to analysis of *Arabidopsis* genes ensures that progress will continue to be made at a rapid pace.

6. The Future

The unique combination of molecular and genetic properties of *Arabidopsis* will soon make possible the cloning of *Arabidopsis* genes about which no more is known than their mutant phenotype and genetic map location. The importance of this for plant developmental biology is that this cloning will be the first step in understanding the wild-type functions of the products of genetic loci such as *apetala-2*, *apetala-3*, and *pistillata*, and thus a first step in understanding organogenesis and pattern formation in plants. One way such cloning will be attempted is to start with a clone shown to map near one of these developmental mutations in a meiotic recombination map, then to walk to the developmentally active gene by successive isolations of overlapping clones.

There are two requirements for such a walk to succeed. One is that an appropriate starting clone must be available, the other is that there must be a way to ascertain when the goal of the walk has been achieved. A method of providing starting clones is to produce a restriction fragment length polymorphism map of *Arabidopsis*. Different ecotypes of the plant differ in the detailed restriction endonuclease map of their genomic DNAs; these restriction map polymorphisms can be detected by hybridizing genome blot filters with appropriate labeled recombinant clone inserts. Standard crosses permit genetic mapping of such polymorphisms. Because of the small size of the *Arabidopsis* genome, a genetic map consisting of only 100 such polymorphisms will provide molecular probes on average only a few hundred kilobase pairs from any location in the DNA of the plant. The distance between any polymorphism and any visible or biochemical mutation can be determined by the appropriate crosses. Mapping of polymorphisms and their relation to known developmental mutations is currently in progress in our laboratory (C. Chang, J. L. Bowman, A. W. DeJohn and E. M. Meyerowitz, work in progress). The near absence of dispersed repetitive DNA will make isolation of overlapping clones possible and rapid; the final problem will be recognizing when to end the walk. This can be done by transforming the cloned DNA segments into the plant, and assaying their ability to complement mutations in the target gene. Because only a small number of clones near the gene of interest need to be tested for their ability to complement, a high frequency of transformation will not be required. Transformation of *Arabidopsis* has already been demonstrated using Ti-plasmid constructs (Lloyd *et al.*, 1986). Leaf explants were infected with an *Agrobacterium tumefaciens* strain containing an altered Ti plasmid that carried a chimeric bacterial hygromycin-resistance gene. This gene was transferred into the plant genome by the bacteria; after regeneration of fertile plants from transformed cells, the resistance gene was inherited as part of one of the plant chromosomes.

The ability to transform *Arabidopsis* with cloned DNA fragments will permit not only the molecular cloning of genes of developmental interest but the type of detailed analysis of gene structure and function currently in progress with the genes of yeast, *Drosophila*, and mice. Thus, *A. thaliana* may soon become the first plant to enter the group of organisms widely used as model systems for the molecular and genetic analysis of development.

ACKNOWLEDGMENTS. Our *Arabidopsis* work is supported by grant PCM-8408504 from the National Science Foundation to E.M.M.

References

Braaksma, F. J., and Feenstra, W. J., 1982, Isolation and characterization of nitrate reductase-deficient mutants of *Arabidopsis thaliana*, *Theoret. Appl. Genet.* **64**:83–90.

Bürger, D., 1971, Die morphologischen Mutanten des Göttinger *Arabidopsis*-Sortiments, einschliesslich der mutanten mit abweichender Samenfarbe, *Arab. Inform. Serv.* **8**:36–42.

Chaleff, R. S., and Mauvais, C. J., 1984, Acetolactate synthase is the site of action of two sulfonylurea herbicides in higher plants, *Science* **224**:1443–1445.

Chang, C., and Meyerowitz, E. M., 1986, Molecular cloning and DNA sequence of the *Arabidopsis thaliana* alcohol dehydrogenase gene, *Proc. Natl. Acad. Sci. USA* **83**:1208–1212.

Dennis, E. S., Sachs, M. M., Gerlach, W. L., Finnegan, E. J., and Peacock, W. J., 1985, Molecular analysis of the alcohol dehydrogenase 2 (Adh2) gene of maize, *Nucleic Acids Res.* **13**:727–743.

Dolferus, R., Marbaix, G., and Jacobs, M., 1985, Alcohol dehydrogenase in *Arabidopsis:* Analysis of the induction phenomenon in plantlets and tissue cultures, *Mol. Gen. Genet.* **199:**256–264.

Ehrenberg, L., 1971, Higher plants, in: *Chemical Mutagens* (A. Hollaender, ed.), pp. 365–386, Plenum, New York.

Feldman, K. A., and Marks, M. D., 1986, Rapid and efficient regeneration of plants from explants of *Arabidopsis thaliana, Plant Sci. Lett.* **47:**63–69.

Flavell, R., and Smith, D., 1976, Nucleotide sequence organization in the wheat genome, *Heredity* **37:**231–252.

Freeling, M., 1973, Simultaneous induction by anaerobiosis or 2,4-D of multiple enzymes specified by two unlinked genes: Differential *Adh1-Adh2* expression in maize, *Mol. Gen. Genet.* **127:**215–227.

Gresshoff, P. M., and Doy, C. H., 1972, Haploid *Arabidopsis thaliana* callus and plants from another culture, *Aust. J. Biol. Sci.* **25:**259–264.

Kirchheim, B., and Kranz, A. R., 1981, New population samples of the AIS-Seed Bank, *Arab. Inform. Serv.* **18:**173–176.

Koornneef, M., and van der Veen, J. H., 1980, Induction and analysis of gibberellin sensitive mutants in *Arabidopsis thaliana* (L.) Heynh, *Theoret. Appl. Genet.* **58:**257–263.

Koornneef, M., and van der Veen, J. H., 1983, The trisomics of *Arabidopsis* and the location of linkage groups, *Genetica* **61:**41–46.

Koornneef, M., Jorna, M. L., Brinkhorst-van der Swan, D. L. C., and Karssen, C. M., 1982, The isolation of abscisic acid (ABA) deficient mutants by selection of induced revertants in non-germinating gibberellin sensitive lines of *Arabidopsis thaliana* (L.) Heynh., *Theoret. Appl. Genet.* **61:**385–393.

Koornneef, M., van Eden, J., Hanhart, C. J., Stam, P., Braaksma, F. J., Feenstra, W. J., 1983, Linkage map of *Arabidopsis thaliana, J. Hered.* **74:**265–272.

Koornneef, M., Reuling, G., and Karssen, C. M., 1984, The isolation and characterization of abscisic acid-insensitive mutants of *Arabidopsis thaliana, Physiol. Plant* **61:**377–383.

Kranz, A. R., 1978, Demonstration of new and additional population samples and mutant lines of the AIS-Seed Bank, *Arab. Inform. Serv.* **15:**118–139.

Laibach, F., 1907, Zur Frage nach der Individualität der Chromosomen in Pflanzenreich, *Beih. Bot. Cbl. 1 Abt.* **22:**191–210.

Langridge, J., 1955, Biochemical mutations in the crucifer *Arabidopsis thaliana* (L.) Heynh, *Nature (Lond.)* **176:**260–261.

Langridge, J., 1957, The aseptic culture of *Arabidopsis thaliana* (L.) Heynh., *Aust. J. Biol. Sci.* **10:**243–252.

Leutwiler, L. S., Hough-Evans, B. R., and Meyerowitz, E. M., 1984, The DNA of *Arabidopsis thaliana, Mol. Gen. Genet.* **194:**15–23.

Leutwiler, L. S., Meyerowitz, E. M., and Tobin, E. M., 1986, Structure and expression of three light-harvesting chlorophyll a/b binding proteins in *Arabidopsis thaliana, Nucl. Acids Res.* **14:**4051–4064.

Li, S. L., and Redei, G. P., 1969, Thiamine mutants of the crucifer, *Arabidopsis, Biochem. Genet.* **3:**163–170.

Lloyd, A. M., Barnason, A. R., Rogers, S. G., Byrne, M. C., Fraley, R. T., and Horsch, R. B., 1986, Transformation of *Arabidopsis thaliana* with *Agrobacterium tumefaciens, Science* **234:**464–466.

Maher, E. P., and Martindale, S. J., 1980, Mutants of *Arabidopsis thaliana* with altered responses to auxins and gravity, *Biochem. Genet.* **18:**1041–1053.

Marsden, M. P. F., and Meinke, D. W., 1985, Abnormal development of the suspensor in an embryo-lethal mutant of *Arabidopsis thaliana. Am. J. Bot.* **72:**1801–1812.

McKelvie, A. D., 1962, A list of mutant genes in *Arabidopsis thaliana* (L.) Heynh, *Radiat. Bot.* **1:**233–241.

McKelvie, A. D., 1963, Studies in the induction of mutations in *Arabidopsis thaliana* (L.) Heynh, *Radiat. Bot.* **3:**105–123.

Meinke, D. W., 1982, Embryo-lethal mutants of *Arabidopsis thaliana:* Evidence for gametophytic expression of the mutant genes. *Theoret. Appl. Genet.* **63:**381–386.

Meinke, D. W., 1985, Embryo-lethal mutants of *Arabidopsis thaliana*: Analysis of mutants with a wide range of lethal phases. *Theoret. Appl. Genet.* **69**:543–552.

Meinke, D. W., and Sussex, I. M., 1979a, Embryo-lethal mutants of *Arabidopsis thaliana*: A model system for genetic analysis of plant embryo development. *Dev. Biol.* **72**:50–61.

Meinke, D. W., and Sussex, I. M., 1979b, Isolation and characterization of six embryo-lethal mutants of *Arabidopsis thaliana*. *Dev. Biol.* **72**:62–72.

Meinke, D. W., Franzmann, L., Baus, A., Patton, D., Weldon, R., Heath, J. D., and Monnot, C., 1985, Embryo-lethal mutants of *Arabidopsis thaliana*, *Plant Genetics*, U.C.L.A. Symposia on Molecular Cellular Biology, vol. 35, pp. 129–146.

Mesnard, J.-M., Lebeurier, G., Lacroute, F., and Hirth, L., 1985, Use of labelled cDNA as probe to study differences in messenger RNAs abundance in the leaves and in the pods of *Arabidopsis thaliana*, *Plant Sci.* **40**:185–191.

Meyerowitz, E. M., and Pruitt, R. E., 1985, *Arabidopsis thaliana* and plant molecular genetics, *Science* **229**:1214–1218.

Mirza, J. I., Olsen, G. M., Iversen, T.-H., and Maher, E. P., 1984, The growth and gravitropic responses of wild-type and auxin resistant mutants of *Arabidopsis thaliana*, *Physiol. Plant.* **60**:516–522.

Müller, A., 1961, Zur Charakterisierung der Blüten und Infloreszenzen von *Arabidopsis thaliana* (L.) Heynh, *Kulturpflanze* **9**:364–393.

Müller, A. J., 1963, Embryonenstest zum Nachweis rezessiver Letalfaktoren bei *Arabidopsis thaliana*, *Biol. Zentralbl.* **83**:133–163.

Napp-Zinn, K., 1963, Zur Genetik der Wuchsformen, *Beitr. Biol. Pflanz*, **38**:161–177.

Negrutiu, I., Jacobs, M., and Cattoir-Reynaerts, A., 1984, Progress in cellular engineering of plants: Biochemical and genetic assessment of selectable markers from cultured cells, *Plant Mol. Biol.* **3**:289–302.

Pang, P., Pruitt, R. E., and Meyerowitz, E. M., 1988, Molecular cloning and nucleotide sequence of two genes which are abundantly expressed in the seed of *Arabidopsis*, in preparation.

Pruitt, R. E., and Meyerowitz, E. M., 1986, Characterization of the genome of *Arabidopsis thaliana*, *J. Mol. Biol.* **187**:169–183.

Rédei, G. P., 1967, Biochemical aspects of a genetically determined variegation in *Arabidopsis*, *Genetics* **56**:431–443.

Rédei, G. P., 1970, *Arabidopsis thaliana* (L.) Heynh. A review of the genetics and biology, *Bibl. Genet.* **20**:1–151.

Rédei, G. P., 1973, Extrachromosomal mutability determined by a nuclear gene locus in *Arabidopsis*, *Mutat. Res.* **18**:149–162.

Rédei, G. P., 1975a, *Arabidopsis* as a genetic tool. *Annu. Rev. Genet.* **9**:111–127.

Rédei, G. P., 1975b, *Arabidopsis thaliana*, in: *Handbook of Genetics*, Vol. 2, pp. 151–180, Plenum, New York.

Reinholz, E., 1947, Auslösung von Röntgenmutationen bei *Arabidopsis thaliana* (L.) Heynh. und ihre Bedeutung für die Pflanzenzüchtung und Evolutionstheorie, *Field Inform. Agency Tech. Rep.* **1006**:1–70.

Röbbelen, G., 1957a, Über heterophyllie bei *Arabidopsis thaliana* (L.) Heynh, *Ber. Dtsch. Bot. Ges.* **70**:39–44.

Röbbelen, G., 1957b, Untersuchungen an strahleninduzierten Blattforbmutanten von *Arabidopsis thaliana* (L.) Heynh, *Z. Indukt. Abst. Vererb.-Lehre* **88**:189–252.

Röbbelen, G., 1965, The Laibach standard collection of natural races, *Arab. Inform. Serv.* **2**:36–47.

Somerville, C. R., and Ogren, W. L., 1982, Isolation of photorespiration mutants in *Arabidopsis thaliana*, in: *Methods in Chloroplast Molecular Biology* (M. Edelman, R. B. Hallick, and N. H. Chua, eds.), pp. 129–138, Elsevier Biomedical Press, New York.

Sparrow, A. H., 1965, Comparisons of the tolerances of higher plant species to acute and chronic expsoures of ionizing radiation, *Jpn. J. Genet.* **40**(Suppl.):12–37.

Steinitz-Sears, L. M., 1963, Chromosome studies in *Arabidopsis thaliana*, *Genetics* **48**:483–490.

Whyte, R. O., 1946, *Crop Production and Environment*, Faber and Faber, London.

Zimmerman, J. L., and Goldberg, R. B., 1977, DNA sequence organizations in the genome of *Nicotiana tabacum*, *Chromosoma* **59**:227–252.

Chapter 14

Regulation of Gene Expression during Seed Development in Flowering Plants

MARTHA L. CROUCH

1. Introduction

Seeds are complex structures, made up of genetically and physiologically distinct components whose functions change during development (Bewley and Black, 1978, 1982; Rubenstein *et al.*, 1978; Johri, 1984; Murray, 1984). The regulation of gene expression can thus be expected to be equally complex, as each part of the seed differentiates according to its own program. Most of the research effort to date has been directed toward describing the kinds of genes expressed in seeds and their temporal and spatial patterns of expression. Our knowledge of factors controlling the observed patterns is limited and comes from genetic analyses, studies of the effects of environmental and hormonal perturbations, and *in vitro* culture of isolated seed parts. In no case has the product of a regulatory gene been isolated and characterized, and the mechanisms by which hormones and environmental stimuli exert their effects on specific genes remain obscure. This chapter describes the systems and approaches being used to study regulation of expression of genes in each part of the seed, with special emphasis on evidence for interactions between tissues that are likely to involve *trans*-acting regulatory molecules, even though the identities of such molecules are unknown. My hope is that by pointing to our areas of ignorance, more systematic research will be undertaken, resulting in some clear answers about gene regulation in seeds.

2. The Significance of the Seed

The seed habit evolved in plants some 350 million years ago as life on land became drier and the conditions during embryo development became less pre-

MARTHA L. CROUCH • Department of Biology, Indiana University, Bloomington, Indiana 47405.

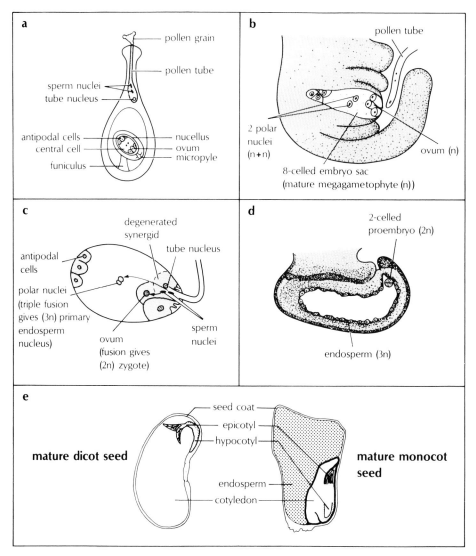

Figure 1. Fertilization and relationship of embryo and seed. (**a**) The male gametophyte (pollen grain) must germinate and deliver the sperm cells to the egg via a pollen tube for fertilization to occur. The ovule (enlarged to show detail) consists of several layers of tissue called the integuments and nucellus. These layers completely surround the megagametophyte (embryo sac) except for a gap called the micropyle. The megagametophyte is composed of seven cells: three antipodal cells, the binucleate central cell, two synergids, and the ovum. It is connected to the ovary wall by the funiculus. (**b**) The pollen tubes grow through the style to enter the ovule through the micropyle. (**c**) A pollen tube enters the megagametophyte, growing through a degenerated synergid. The two sperm nuclei are released from the tube to undergo fusion: One sperm will fuse with the egg to form the diploid zygote, and the other sperm will fuse with the polar nuclei of the central cell to form the triploid endosperm. (**d**) The zygote divides to form a 2-celled proembryo that will develop within the endosperm. (**e**) Mature seeds of monocots and dicots. The legume seed on the left is a representative dicot, although the second cotyledon cannot be seen from this view. The monocot grass seed on the right contains the same structural components of the embryo (i.e., epicotyl, hypocotyl) but

dictable (Steeves, 1983). In seed plants, fertilization is not dependent on extracellular water. The developing embryo is usually retained on the mother plant until the basic tissues and organ systems are formed, and nutrients from the mother are transferred to the heterotrophic embryo and seed storage organs during its development. The seed then becomes specialized for dispersal in time and space: Cells accumulate storage reserves, then desiccate and exist in suspended animation until the detached and dispersed seed encounters conditions favorable for reactivation of growth. When reactivation does occur, growth is directed toward the establishment of an autotrophic organism. Root and shoot meristems in the embryo become active and form new organs, storage reserves in the embryo and other seed parts are metabolized, and photosynthetic machinery is assembled and put into use. Thus, the prolonged association of the new sporophyte generation with the mother during seed development allows for a more predictable environment in embryogeny and a better chance of survival during subsequent independent establishment. The same qualities that make seeds good propagules for sedentary, land-dwelling plants also make them good food sources for humans. Seed-storage reserves provide concentrated nutrition, the metabolic inactivity of seeds permits them to be stored relatively unchanged for long periods of time, and the predictable ability of seeds to propagate the species when activated at a particular time and place is thought to be a prerequisite of complex civilizations. Thus, seed development is of intrinsic importance to us.

3. Fertilization: The Trigger of Seed Development

The conversion of an ovule into a seed is triggered by fertilization (for review, see Van Went and Willemse, 1984). In flowering plants, details of fertilization are hidden from view inside the ovule, which itself is enclosed in an ovary (Fig. 1). Before fertilization, the ovule consists of one or two outer folds of tissue (integuments), inner tissue (nucellus) surrounding the megagametophyte (embryo sac), and a stalk connecting the ovule to the placenta of the ovary (raphe along the ovary wall, funiculus between the placenta and ovule). Compared with more primitive plants, the megagametophyte in flowering plants is greatly reduced, generally consisting of just seven cells. However, two of the seven cells are destined to participate directly in fertilization.

Each male gametophyte (pollen) produces two nonflagellated sperm cells. These sperm cells are carried to the egg through the style by the pollen tube, which grows by tip growth. By a combination of physical constraint in the transmitting tissue of the style and chemotaxis, the pollen tube finds its way to the ovule and enters through a gap in the integuments (micropyle), where it bursts into one of the synergids. This synergid actually starts degenerating

has only one cotyledon. The monocot's endosperm is quite developed, whereas many dicots, including bean, contain little or no endosperm at maturity. The seed of grasses is actually an entire fruit, with the fruit wall tightly fused to the coat of the single seed enclosed. (**a** and **c** adapted from Harris *et al.*, 1984; **b** and **d** adapted from Raven *et al.*, 1986; **e** adapted from Keeton, 1972.)

before the tube gets there, presumably in response to a chemical signal from the tube. The two genetically identical sperm cells are released, and one fuses with the egg to form the zygote. The other sperm cell fuses with the two polar nuclei of the central cell to form the triploid endosperm.

Fertilization thus simultaneously triggers the formation of the next sporophytic generation and the development of the nutritive endosperm tissue. If fertilization does not occur, very little energy has been invested. This whole process of fertilization is circumvented in species with apomictic seeds (Nogler, 1984). In these species, seed development proceeds autonomously from either unreduced gametes or maternal cells and results in an embryo with the maternal genotype. Thus, all the benefits of seed development are attained without the genetic risk of sex.

Although development of the seed coat, endosperm, and embryo are usually all triggered by the same event, their development proceeds at different rates. Development of each part will now be considered in more detail.

4. Development of Maternal Tissues

The integuments of the ovule develop into the seed coat (testa); in some cases, the ovary wall (pericarp) remains associated with the seed as well. For example, in the one-seeded fruit of grasses the pericarp and the testa are fused (Fig. 1e). Also, the maternal nucellus may persist in the seed as a nutrient storage tissue (perisperm). There is no direct vascular connection between the endosperm and embryo and the mother, so the first role of the maternal seed tissues is to transfer nutrients from the rest of the plant to the rapidly growing products of fertilization. The kinds of nutrients transported, the specific sources, and the nutrient conversions that take place in the seed are discussed by Pate (1984). Shortly after fertilization, the integuments undergo cell division and expansion, sometimes increasing in diameter 100-fold, and vascular tissue differentiates (Bouman, 1984). The extent of vascularization is generally proportional to the size of the seed. The inner cells of the integument may have wall ingrowths characteristic of transfer cells (Mogensen, 1985).

Very little is known about the role of seed coat gene expression in the regulation of endosperm and embryo development, but the study of maternal-effect mutations resulting in defective seeds may provide some information. In barley there are eight nonallelic maternal-effect shrunken endosperm mutants called seg 1–8 (Felker et al., 1985). Histological analysis shows that four of these mutants have necrotic maternal cells close to the area in which vascular tissue enters the grain. Thus, the early cessation of development of the endosperm is assumed to be the result of an interruption of nutrient flow from the mother. This conclusion is further supported by the ability of cultured mutant endosperm cells to take up sucrose and synthesize starch normally. In addition, cultured mutant embryos develop into normal plants. The other four seg mutants have normal-appearing seed coats and maternal tissue, but abnormal or absent endosperm tissue. Even in seeds with virtually no endosperm, the embryos are capable of giving rise to healthy plants germinated on a sucrose solution. Since these abnormal endosperm mutations are maternally inherited, the phenotype must be due either to an alteration in maternal metabolism that is not manifest in the cytology of the maternal cells or to a defect that is carried over from the meiocyte to the central cell and subsequent endosperm. Although maternal effects in animals can sometimes be traced to defective egg cytoplasms, in plants the alternation of generations makes direct carryover from the mother seem unlikely, since several cell divisions take place between meiosis and central cell/egg formation. Unraveling the basis of the defects in these *seg* mutants will be interesting, because each mutation affects endosperm development at a particular time.

After nutrient transfer from the maternal plant to the embryo is complete, the seed coat can become specialized for its functions in the mature seed and in subsequent germination. The nature and timing of specialization are species specific and may involve the synthesis of pigments, secondary metabolites, and structural elements (reviewed in Bewley and Black, 1978, 1982). The seed coat may be involved in the inhibition of germination due to permeability restriction, physical constraint of the embryo, the presence of chemical inhibitors, or modification of the light reaching the embryo. Special substances may accumulate in the seed coat to help regulate water entry during germination, or enzymes for storage reserve degradation may be localized in the inner seed coat wall. Thus, the mature seed coat is an assemblage of specialized cell types that finally become desiccated and compressed when the seed is ready to be shed.

Many seed-coat characteristics can be environmentally regulated. For example, seed coats of *Ononis sicula* are quite different, depending on the photoperiod (Gutterman and Heydecker, 1973). On long days, the seeds are large with thick yellow coats and exhibit slow imbibition and a low germination rate. On short days, the seeds are smaller, with thin green or brown coats, are readily imbibed, and have a high germination rate. Seed coats of many species exhibit differential development, depending on their position in the inflorescence (Bewley and Black, 1982). Also, seed-coat characteristics can be influenced by the specific genotype of the underlying endosperm and embryo, as shown by variations in cotton fiber length (Brink and Cooper, 1947).

There are few studies of the biochemical events responsible for the differentiated functions of the maternal seed tissues. One such study outlines the development of impermeability in the seed coat of the weed *Sida spinosa* (Egley *et al.*, 1983). These seeds take about 24 days to mature, but impermebability of the seed coat to water begins 16 days after pollination, when the coats are still soft and before the seed has dehydrated. Impermeability is correlated with coat browning and high levels of peroxidase activity localized by immunocytochemistry to the palisade layer of the seed coat. It appears that peroxidase is involved in the polymerization of soluble phenolics to insoluble lignin polymers and that lignin is responsible for barring entry of water. By contrast, in domestic peas, the seed coat remains green and permeable to water and exhibits low peroxidase activity in the seed coat throughout development (Marbach and Mayer, 1975).

Studying the regulation of gene expression in maternal seed tissues is inherently difficult because of the number of cell types and the changing function of each cell type during development. Analysis of mutants, coupled with *in situ* localization techniques, might result in a description of the patterns of gene expression and the isolation of seed-coat specific genes. Also, regulatory influences of other seed parts are undoubtedly important, since the seed coat does not normally begin development without concomitant growth of the embryo and endosperm, and there is usually a high degree of coordination, as evidenced by proportional growth of seed parts. Perhaps seed coats could be cultured to determine the effects of growth regulators and other factors on expression of specific genes. For example, cotton fiber differentiation will occur in culture.

5. Endosperm

The endosperm is a tissue unique to flowering plants; its normal development is usually required for seed formation (reviewed by Brink and Cooper, 1947; Vijayraghavan and Prabhakar, 1984). Although fertilization of egg and central cell by brother sperm cells occurs roughly at the same time, the endosperm develops more rapidly and surrounds the embryo everywhere except where the suspensor attaches to the embryo sac wall. Thus, the endosperm is the milieu in which the embryo develops and may be the intermediary between the maternal tissues and the embryo. The extent to which endosperm cells are retained in the mature seed varies among species. After hydration, endosperm storage products are digested and transferred to the growing seedling. Virtually nothing is known about expression of specific genes during the early stages of endosperm development, when it is presumed to function as a nurse tissue for the developing embryo. However, there is a wealth of knowledge about genes expressed during differentiation of storage tissues and aleurone cells of persistent endosperms, particularly in cereals.

5.1. Early Endosperm Development

Activation of cell division in the endosperm is usually a result of fertilization, although in some apomictic seeds the process will begin autonomously (Nogler, 1984). In most species, development of the endosperm is coenocytic at first but usually becomes cellular with a slowing of the rate of nuclear division. In general, it is necessary to have a particular ratio of maternal to paternal genomes in the endosperm itself in order for endosperm cells to proliferate normally. This has been studied in interploidy-intraspecific and -interspecific crosses in a wide variety of species; usually the maternal : paternal genome ratio must be 2 : 1. Johnston *et al.* (1980) proposed the notion of **endosperm balance number (EBN)** to explain this phenomenon. By comparing crosses between different tetraploid and diploid potatoes, they concluded that all interploidy, intraspecific crosses conform to the 2 : 1 hypothesis but that some of the interspecific crosses behave as though the species involved have different effective ploidies. For example, *Solanum acaule* is a tetraploid, but it will not form viable hybrid seeds when crossed with other South American tetraploids, although plump seeds are formed with diploids. *S. acaule* behaves consistently as though it has a diploid EBN. How the ratio of male- to female-derived chromosomes is detected and translated into endosperm abortion is unclear. However, there is evidence that in *Datura stramonium*, genes on at most 2 of its 12 chromosomes are required to determine its EBN (Johnston and Hanneman, 1982). It is easy to imagine the EBN being uncoupled from the ploidy level by gene duplication or deletion, or by changes in expression at a few loci. A result of the requirement for a particular EBN is that autotetraploids are immediately reproductively isolated from their progenitor diploids, even though the triploid embryos formed from such a union are capable of normal sporophytic development if rescued from the seed and cultured *in vitro*. Thus, an early role of the endosperm may be as a barrier to hybridization when prefertilization barriers fail.

Some genes expressed in the endosperm are altered in expression when they are contributed by the male, rather than the female, gamete and thus have been imprinted somehow by their passage through the male gametophyte (Kermicle, 1970). This is an epigenetic phenomenon, since it is erased in the next generation. It has been best described for anthocyanin formation controlled by the R locus in maize (Kermicle, 1970). Endosperms are mottled when heterozygous rr/R, where R is from the pollen, but the reciprocal cross (RR/r) generates a solid-colored endosperm. Dosage effects have been ruled out by translocation studies, and thus it appears that the level of action of the gene is strongly affected in a persistent way by the male gametophyte. Other genes have also been described that show this phenomenon (Schwartz, 1965), but its significance in regulation of endosperm development is unknown.

Since normal endosperm development is usually required for embryo growth, it is surprising that the specific role of the endosperm during development is so poorly understood. It is assumed that the endosperm is a source of nutrition and hormones required for early embryo development. However, peaks in endogenous hormones, such as cytokinins, correspond more closely with rapid cell divisions in the endosperm itself than with changes in the embryo (Quatrano, 1986a,b). Furthermore, although the route of nutrient transfer to the embryo presumably traverses endosperm cells, at least at the apical and cotyledon end of the embryo (Marinos, 1970), the contribution of endosperm cells themselves to embryo nutrition has not been demonstrated. In fact, it could be argued that the endosperm and embryo are competitors for maternal nutrients. Chemical analysis of the liquid in the embryo sac during early seed development does not distinguish between endosperm and seed coat origin of the constituents (Smith, 1973; Murray, 1979). Since the extent of endosperm development varies so widely among species, generalizations may be unfounded in any case.

Evidence that endosperm is necessary for embryo growth is based primarily on the appearance of abnormalities in interspecific hybrid seeds

(Williams and De Lautour, 1980). Abnormalities could arise because the unrelated endosperms are toxic to the embryo, because they develop at an incongruous rate and thus fail to synthesize specific chemical or physical factors at the right times, or because they outcompete the embryos for nutrients and water. Hybrid embryos often can be successfully converted into plantlets if removed from their own abnormal endosperms, but the degree to which they progress through the normal stages of embryogeny is not documented. Thus, these experiments with hybrid seeds provide few clear answers about the early role of the endosperm.

A more direct analysis of embryo–endosperm interactions is afforded by mutations that specifically alter seed development, such as the *defective kernel* (*dek*) mutants of maize (Neuffer and Sheridan, 1980) and the embryo lethal mutants in *Arabidopsis* (Meinke, 1985). These mutations can be classified according to the part of the seed affected as well as the time during development when the defect can first be recognized. No temperature-sensitive mutants are available to help define the time of gene action. The *dek* mutants are grouped into four types (Sheridan and Neuffer, 1986; Clark and Sheridan, 1986): (1) endosperm and embryo growth both affected and embryo nonviable; (2) endosperm and embryo affected, but embryo able to germinate into a seedling with a mutant phenotype; (3) only the endosperm exhibiting a mutant phenotype; and (4) only the embryo showing abnormalities, thus forming a germless seed. In *Arabidopsis*, the endosperm is only present during early seed development and is not easily monitored. Therefore, *Arabidopsis* seed mutations have been described on the basis of embryo morphology (Meinke, 1985). Embryo development arrests at a specific stages of development in each line, ranging from zygotic to linear and curled cotyledon stages.

A major problem in interpreting the origin of defects in development in defective seed mutants is that the endosperm and embryo have qualitatively identical genetic constitutions and thus both contain the mutation. It is possible, for example, that an embryo is arrested at a specific stage because the endosperm is not providing some essential product. Even if the mutant embryo cannot be rescued in culture, the phenotype may still be ascribed to a previous interaction with mutant endosperm rather than expression of the mutant gene in the embryo itself. In order to be able to distinguish autonomous expression from interactions, Neuffer and Sheridan (1980) have taken advantage of maize translocations to construct seeds with different combinations of mutant and normal genotypes, so that either the embryo or endosperm can carry the mutant allele. When the appropriate crosses are performed, some of the kernels on an ear will have hypoploid mutant (mm$^-$) endosperms associated with hyperploid normal (m^{++}) embryos. There should be an approximately equal number of kernels with hyperploid normal (mm^{++}) endosperms associated with hypoploid mutant (m$^-$) embryos. Most of the kernels will have normal endosperm and embryo. These classes can be compared with homozygous mutant material, in which both embryo and endosperm are carrying the mutant allele. Nineteen different *dek* mutants were examined in this way for interactions between endosperm and embryo. When the embryo was hyperploid nonmutant, it affected the mutant endosperm in various ways: In three cases, the phenotype of the mutant endosperm was less extreme than if both were mutant, in seven cases the mutant endosperm had a more extreme phenotype, and in nine cases there was no effect of a normal embryo on the defective endosperm. Conversely, the mutant endosperm had no effect on development of the normal embryo in 17 of 19 cases. Thus, even severely defective endosperms do not generally result in abnormal embryos. However, the two exceptions are interesting in light of the role of the endosperm: In one instance, the mutant endosperm resulted in a dead embryo, and in the other mutation the embryo developed to term, but the resultant seedling was weak. Analysis of these mutants may lead to a clearer understanding of the relationship between embryo and endosperm in the young seed.

The ability to transfer mutations independently to the endosperm or embryo in a seed with a transient endosperm would be an extremely useful tool. The reproductive process in angiosperms is remarkably variable, so it should be possible to use species with particular characteristics for these studies. For example, in some apomictic species the endosperm is fertilized, but the embryo develops parthenogenically (Nogler, 1984). Also, fertilization of endosperm and embryo by different pollen grains has been reported in maize (Sarkar and Coe, 1971) and may also occur in other plants, if searched with appropriate markers.

5.2. Differentiation of Persistent Endosperms

Cellularization of an originally coenocytic endosperm is usually accompanied by an increase in the cell cycle time and often DNA amplification. For example, in some strains of maize individual nuclei reach DNA levels of 690C (Kowles and Phillips, 1985). The amount of DNA per nucleus increases as development proceeds, followed by a decrease in most strains. The significance of DNA amplification and loss during endosperm development is unknown.

After cellularization, endosperm cells differentiate according to their position in the seed to accumulate storage products, pigments, defense compounds, and, in some cases, the enzymatic and cellular machinery to break down their storage products after reactivation of metabolism in germination. Among the most extensively studied systems are castor bean (Greenwood and Bewley, 1982), fenugreek (Reid and Bewley, 1979), and the economically important grasses such as wheat, barley, and maize (Soave and Salamini, 1984).

The amount of information on the genes expressed during endosperm differentiation is relatively extensive, and generalizations are difficult to make. Two examples where there is knowledge about the regulation of differentiation will be discussed. One example is the regulation of storage protein synthesis in maize endosperm, in which mutations in *trans*-acting genes are being analyzed. The other example is the acquisition of sensitivity to hormones by barley aleurone cells during grain maturation, where the normal drying of the seed is likely to be involved in acquisition of competence to respond to signals that regulate α-amylase expression.

5.2.1. Regulation of Zein Synthesis in Maize

Endosperm cells in maize differentiate to accumulate large amounts of starch and storage protein, which are sequestered in plastids and membrane-enclosed protein bodies, respectively (Soave and Salamini, 1984). The major storage protein is a family of related alcohol-soluble polypeptides called **zein,** which accounts for 50–60% of total seed protein. These polypeptides are encoded by a large family of genes, with estimates of up to 150 genes, and the members are dispersed on different linkage groups among three chromosomes. Zein has never been detected in any part of the plant other than the seed (Boston *et al.*, 1986), and within the seed it is much more abundant in the endosperm than in the embryo (Tsai, 1979). In fact, the study by Tsai (1979) is unique in reporting the occurrence of zein in embryo cells; this protein study has not been followed by an analysis at the molecular level. Thus, the entire family is restricted in its expression to the seed and is differentially expressed within the seed. The level of tissue-specific control is transcriptional, since run-on transcripts are not detected in isolated nuclei from other plant parts (Boston *et al.*, 1986). The degree to which family members are coordinately expressed and the action of mutations that affect the expression of whole classes of zeins have recently been investigated using molecular techniques.

Zein gene expression during development has been described by Marks *et al.* (1985*a,b*), using complementary DNA (cDNA) clones for distinct family members to measure mRNA levels in differentiating endosperms of the Inbred W64A. By restriction mapping and cross hybridization analysis, they were able to group zein cDNA clones into nine classes. Three of the classes encoded proteins of M_r 22,000, five encoded proteins of M_r 19,000, and one encoded a protein of M_r 15,000. Using RNA dot blots with a limit of detection of 0.5% of the total poly(A)$^+$ RNA, zein mRNA representing eight of the nine classes could first be detected at 12 days after pollination (DAP) (Fig. 2). This corresponds roughly to the period when cell division in the central endosperm is terminating and endosperm cells have completely filled the seed cavity (Kowles and Phillips, 1985). Nuclear volume and DNA content are increasing dramatically at this time. The levels of all nine groups increased at 18 DAP and remained at these higher levels until at least 28 DAP. Older stages were not examined for family-specific expression. At the earliest stage, the mRNAs encoding the M_r 19,000 polypeptides were already at 25–30% of their highest levels, whereas the messenger RNAs (mRNAs) for the 15,000- and 22,000-M_r proteins were only at 10% of their highest levels. Thus, it appears that some of the family members are expressed earlier than others.

In order to examine transcription more directly, Boston *et al.* (1986) measured the levels of some of the different classes of zein mRNAs in run-on transcription assays from endosperm nuclei isolated at different times in development between 12 and 18 DAP. Each zein subclass was transcribed at approximately the same rate over this period of development. The transcription rates for all classes were higher at 12 DAP than later, in contrast to the mRNA levels, which were highest after 16 DAP. However, the number of transcripts representing the 22,000-M_r zeins was greater than for the other subfamilies, which would not be predicted by the levels of mRNA that accumulate on polysomes. Thus, although the rate of transcription of the 22,000-M_r zeins was twice the rate for the 19,000-M_r zeins, the relative level of mRNA in the cytoplasm was threefold lower. This indicates that there is post-transcriptional, as well as transcriptional, regulation of the different zein genes.

Not all the zein genes contribute equally to the mRNA population, although the large size of the family makes it difficult to analyze in detail. However, by examining two-dimensional protein gels and Southern blots, it can be determined that there are only one or two 15,000-M_r zein genes, compared with several 22,000- and 19,000-M_r genes. Marks *et al.* (1985a) calculated that the 15,000-M_r zein genes are expressed at a higher level per gene than the other classes, although the transcription data indicate that this may be due to post-transcriptional regulation of mRNA levels.

Zein synthesis continues until the seed begins to desiccate. Whether zein gene transcription is actively terminated is unknown.

This description of zein gene expression points to two major questions concerning gene regulation: What is involved in tissue-specific regulation of the entire family? How is the expression of individual zein genes modulated during development? Unfortunately, it is not yet possible to transform and regenerate corn plants to analyze developmentally regulated promoters, although progress is being made toward this technology (Fromm *et al.*, 1986). However, zein genes can be transferred into heterologous hosts, such as tobacco or petunia, in which their promoters may be recognized in a tissue-specific way. Results in heterologous systems have so far been limited to analysis of callus tissue (Boston *et al.*, 1985). There are no mutants with altered tissue specificity.

Some interesting *trans*-acting mutations affect the level of zein expression as a whole or the relative expression of zein subfamilies during seed development (Soave and Salamini, 1984). For example, *opaque-2* (*o2*) and *defective endosperm-B30* (*De*-B30*) result in a preferential reduction of the 22,000-M_r zeins, and *opaque-7* (*o7*) preferentially reduces the 19,000-M_r zeins. *Opaque-6* (*o6*), *floury-2* (*fl2*), and *mucronate* (*Mc*) decrease the synthesis of all zeins equally. Some mutations also delay the timing of zein deposition. Because none of these genes maps at zein loci, they are thought to exert their influence by diffusible factors.

To determine how these mutations cause changes in zein levels, the nature of the defects is being studied in more detail, and the protein products of the loci are being sought. *Opaque-2* and *-o6* are perhaps the best characterized (Langridge *et al.*, 1982; Soave *et al.*, 1981; Marks *et al.*, 1985a; Di Fonzo *et al.*, 1986). Although all classes of zein are substantially reduced by the *o2* mutation, the levels of the 22,000-M_r protein are particularly low. *Opaque-2* reduces

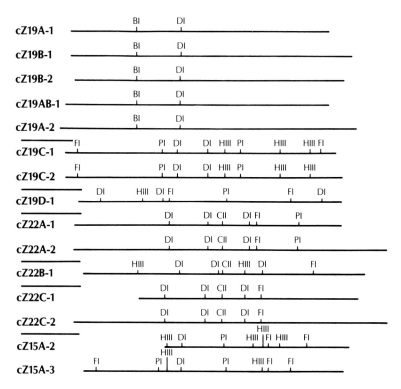

Figure 2. Restriction endonuclease maps of zein clones and expression of zein mRNAs. (**a**) Restriction endonuclease maps of zein clones. Classification of these clones was based on their homology to zein mRNAs. The clones are named such that the first two letters designate whether the clone was isolated from a cDNA (cZ) or genomic (gZ) library, and the numbers indicate the apparent weight (in kilodaltons) of the protein that the clone encodes. (**b**, next page) Levels of zein mRNAs in developing endosperm. Solid bars represent mRNA levels at 12 days after pollination (DAP), striped bars represent mRNA levels at 18 DAP, and open bars represent mRNA levels at 28 DAP. (Upper panel) Zein levels in W64, a normal inbred line of maize. (Lower panel) Zein expression in W64 opaque, a line mutant in zein synthesis. (Adapted from Marks *et al.*, 1985.)

total zein by 50%, and the seeds are viable, whereas *o6* further reduces the levels and is lethal when homozygous. Levels of mRNA for the various classes of zeins have been determined during development of *o2* endosperm, and none of the zein mRNAs can be detected at 12 DAP, as they would be in normal endosperm (Marks *et al.*, 1985a). The mRNAs for the 19,000-M_r zein attain relatively normal levels by 28 DAP, although they arrive at that level slowly, being about three times lower than normal at 18 DAP. Thus, it appears that expression of the 19,000-M_r zeins is delayed. The appearance of the 15,000-M_r

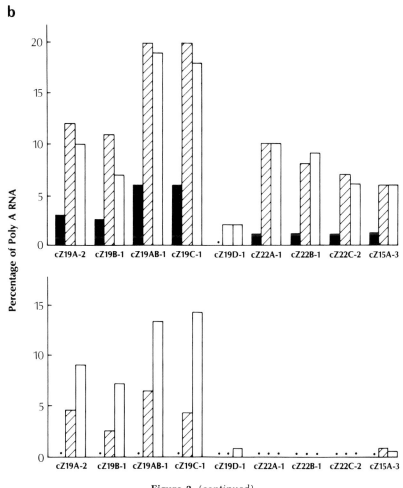

Figure 2. (continued)

zein mRNA is also delayed and is greatly reduced even at the later developmental stage. As predicted, the 22,000-M_r classes of mRNAs were not detectable at any of the time periods. Transcription studies with $o2$ nuclei have not been reported. Although the effects of $o2$ on zein protein levels can be explained by mRNA levels, this does not preclude additional levels of regulation (Soave and Salamini, 1984). It should be mentioned that endosperm mutants such as $o2$ do not interfere with the increase in DNA content per endosperm nucleus and, if anything, result in higher DNA content (Kowles and Phillips, 1985).

Both $o2$ and $o6$ lack a soluble protein of 32,000-M_r (b-32 protein) (Soave et al., 1981), which has been mapped to the $o6$ locus. This protein is coordinately expressed with zein during development of wild-type endosperm but is cytoplasmic (rather than in protein bodies), has a distinct amino acid composition, and is present as a monomer. It is abundant, comprising 8% of total salt-soluble protein in maturing endosperm cells. Soave *et al.* (1981) proposed that *o2* regulates the synthesis of *o6*, which encodes b-32 protein, in turn regulating zein synthesis. However, the abundance and location of b-32 protein would argue against a direct interaction with zein or zein genes. The product of the *o2* gene has not been identified.

The only other candidate for a regulatory protein that has been described is a 70,000-M_r polypeptide (b-70) associated with the outside of protein body membranes in endosperm cells. This protein is overproduced in the mutants *f12, Mc* and *De*-B30* (Galante *et al.*, 1983). The b-70 protein is abundant, and the level of its regulation of zein is thought to be post-translational, perhaps by interfering with zein deposition.

It is unlikely that proteins involved in the direct regulation of zein genes will be abundant enough to detect differences in their levels in mutant endosperms (Croy and Gatehouse, 1985). The ability to clone genes by transposon tagging will make it possible to analyze directly the products of regulatory genes, however. Being able to combine the genetic identification of regulatory genes with direct analyses of their products should lead to real progress in understanding how coordination of multigene families occurs in endosperm development.

5.2.2. Regulation of α-Amylase during Grain Maturation

Most events of differentiation in endosperm cells are aimed at accumulation of storage products, which by definition must be protected from degradation during seed development and yet must be rapidly broken down after germination. Storage product turnover involves several levels of control and is a good example of the complexity of regulation in the seed. It has been studied most extensively in barley, in which the breakdown of storage products during the malting process has economic significance, and also in wheat and triticale, in which premature grain sprouting is a problem. Most of the research has focused on hormonal and environmental regulation of the enzymes involved in reserve hydrolysis after germination. However, the factors responsible for preventing reserve breakdown before germination are also an important part of grain maturation.

The major storage reserve in barley endosperm is starch, which is stored in amyloplasts in dead cells (MacGregor, 1983). Therefore, during seed germination various amylases must be secreted into the dead endosperm from adjacent living tissues, such as the scutellum of the embryo and the aleurone cells at the periphery of the endosperm. α-Amylases degrade intact granules of cereal starches to form a complex mixture of saccharides, by attacking $\beta(1 \rightarrow 4)$ bonds. The α-amylases in barley are encoded by a gene family that can be divided into two subfamilies on the basis of extent of nucleotide homology (Rogers, 1985), isoelectric variants of the encoded enzymes (Jacobsen and Higgins, 1982), and chromosomal location (Brown and Jacobsen, 1982). Group A isozymes have isoelectric variants of the encoded enzymes (Jacobsen and Higgins, 1982), and chromosomal location (Brown and Jacobsen, 1982). The low pI isozymes have isoelectric points of 4.3–5.2 (group A), and the high pI isozymes have iso-electric points of 5.9–6.6 (group B). The precise number of α-amylase genes in barley is unknown, but genomic Southern blots indicate that there are approx-imately seven genes, some of which may be pseudogenes (Rogers and Milliman, 1984). At any rate, the differences between the group A and B genes isolated so far are great enough to permit separate analysis of the expression of these two families, and there are interesting differences in their patterns of expression.

The first expression of α-amylase is early in seed development, when the low pI isozymes are localized in maternal pericarp cells (MacGregor, 1983). Disappearance of starch in wheat pericarp has been correlated with the begin-ning of cellularization of the endosperm (Bennett et al., 1975). Thus, pericarp starch is synthesized and degraded before starch begins to accumulate in the endosperm. The next period of α-amylase synthesis does not occur until after germination, when both pI forms are secreted into the endosperm from the scutellum of the embryo and then become the major component of protein synthesis and secretion in aleurone cells (MacGregor et al., 1984). Later, during sporophytic growth, starch is also synthesized and degraded in photo-synthetically active leaves, but only one α-amylase isozyme is found in barley leaves (Jacobsen et al., 1986). Genetic analyses show that the low pI isozyme of leaves is probably the same as one of those expressed in the aleurone. Thus, α-amylase genes are expressed progressively in different tissues and organs dur-ing development. Current evidence suggests that the low pI forms are expressed in maternal pericarp and leaf tissue, whereas both the low and high pI forms are expressed in aleurone and embryo.

There is good evidence that diffusible factors emanating from the embryo during germination are involved in stimulating α-amylase transcription in al-eurone cells. Exogenous gibberellic acid can substitute for the embryo in em-bryoless half-grains (Varner and Chandra, 1964), aleurone layers (Chrispeels and Varner, 1967), and even in protoplasts isolated from aleurone layers (Fig. 3). Exogenous gibberellic acid effectively stimulates the level of both pI forms. In isolated aleurone layers, the low pI mRNAs increase 5–20-fold soon after treatment with gibberellic acid, and the high pI mRNAs increase about 100-fold, but more slowly (Rogers, 1985). The embryonic origin and role of gib-berellic acid in vivo are points of controversy, however.

Although the level of regulation of the α-amylase gene expression by gib-berellic acid can be explained by changes in transcription, as measured by run-on transcription from nuclei isolated from aleurone protoplasts (Jacobsen and Beach, 1985), other levels of control are also implicated (Higgins et al., 1982). For example, formation of endoplasmic reticulum increases within hours of treatment with gibberellic acid, which may facilitate translation of the higher mRNA levels or may be involved in mRNA stability (Belanger et al., 1986).

All the effects of gibberellic acid in aleurone cells can be reversed by exogenous abscisic acid added at the same time (Ho et al., 1985). Studies with inhibitors indicate that the effects of abscisic acid require both protein and RNA synthesis, and many new proteins are seen on gels of in vitro translation products after the addition of abscisic acid. One of the abscisic acid-stimulated

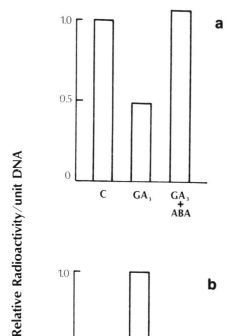

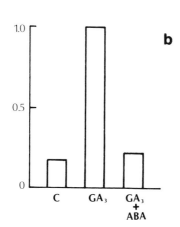

Figure 3. Comparison of the effects of gibberellin A$_3$ (GA$_3$) and abscisic acid (ABA) on RNA accumulation in isolated nuclei. Barley aleurone protoplasts were treated with Ga$_3$, GA$_3$ plus ABA, or neither (C) for 24 hr. The nuclei were isolated and permitted to undergo transcription *in vitro* in the presence of [α-^{32}P]-UTP for 1 hr. To measure total RNA accumulation, aliquots were taken at the start and finish of the reaction and precipitated with trichloroacetic acid. To measure α-amylase mRNA, radiolabeled RNA isolated from the nuclei was hybridized to filters containing cloned α-amylase sequences. The relative values for each RNA species were calculated from cpm per 100 μg DNA. (a) Total RNA. (b) α-Amylase mRNA. (Adapted from Jacobsen and Beach, 1985. Reprinted by permission from *Nature*, Vol. 316, No. 6025, pp. 275–277. Copyright © 1985 Macmillan Journals Limited.)

proteins in the aleurone is an inhibitor of endogenous barley α-amylase (Mundy, 1984). Whether abscisic acid plays a role in regulation of germination of mature seeds is unknown, although there are some interesting possibilities. Endogenous abscisic acid has been measured during steeping of barley to make malt (Yamada, 1985). During aerobic steeping, the endogenous abscisic acid, which is primarily in the embryo, greatly decreases before germination, whereas anaerobic steeping does not result in a decrease in abscisic acid levels, and germination is inhibited. Thus, destruction of endogenous abscisic acid in the mature grain may be required before α-amylase can be secreted by aleurone cells. Alternatively, anaerobiosis may have a direct effect on α-amylase response. It is also possible that abscisic acid increases if a seed becomes water stressed during germination, reversibly halting α-amylase production, although this has not been tested.

A more pertinent question concerns what keeps the aleurone cells from expressing α-amylase genes during the period in which starch is being synthesized. There are several possibilities, based on what has been learned about inhibition of α-amylase in germinating grains. For example, levels of inducers such as gibberellic acid could be too low, or the aleurone cells may be insensitive to inducers during development, either because they lack receptors or because inhibitor concentrations are high. In fact, aleurone cells from immature grain, although morphologically differentiated, are unable to respond to exogenous gibberellic acid by synthesizing α-amylase mRNA and only acquire that ability late in maturation (Evans et al., 1975).

The basis for the insensitivity of immature aleurone cells to gibberellic acid is unknown. However, sensitivity can be induced prematurely by drying the seed or isolated aleurone layers (Evans et al., 1975). For example, in wheat the aleurone of a normal grain is fully responsive to gibberellic acid (optimal concentration of 10^{-6}M) at about 50 DAP when water content is 25% (Armstrong et al., 1982). If seeds are removed from the plant between 11 and 38 DAP, they do not respond to gibberellic acid (even at 10^{-4} M) until they are dried to a critical water content of 25%. Also, pretreatment of grains at about 30°C for several hours promotes sensitivity to gibberellic acid in immature seeds and even protoplasts (Norman et al., 1983). These drying and heating treatments may cause changes in receptor levels or orientation or may destroy inhibitors such as abscisic acid. Levels of abscisic acid are high during the period in which the grain is insensitive to gibberellic acid (King, 1976), but the effects of drying and heat treatments on abscisic acid levels have not been measured directly in these systems.

If some α-amylase were produced during grain development, it might be inhibited at the functional level by the endogenous α-amylase inhibitor synthesized in the endosperm during starch accumulation (Mundy, 1984). This protein has an interesting pattern of expression, since it is apparently specific to the starchy endosperm during seed development, but inducible by ABA in the aleurone after germination (Mundy et al., 1986).

As cereal grains approach maturity, abscisic acid levels decrease, water content falls, and they enter developmental arrest or dormancy, ready to respond to gibberellic acid upon rehydration. Apparently, perturbations in the coordination of these events can lead to premature starch hydrolysis. Hybrid seeds such as Triticale (wheat × rye) are particularly susceptible (King et al., 1979), presumably due to imbalances in the rates of development of different parts of the seed.

Several principles emerge from these studies of α-amylase expression in seeds. First, different gene family members have distinct patterns of expression, both in terms of tissue and response to exogenous factors. Even the same family member appears to be capable of different responses, depending on where it is being expressed. For example, one of the low-pI α-amylase genes is expressed in the leaf of barley, as well as the aleurone (Jacobsen *et al.*, 1986). However, the level of its mRNA is about 5000-fold lower in the leaf than in gibberellic acid-treated aleurone cells. Also, wilting causes an increase in α-amylase RNA in the leaf, along with increased endogenous abscisic acid levels. The effects of gibberellic acid and ABA added exogenously to the leaf have not been tested, but at face value it appears that the same gene that is inhibited by abscisic acid in aleurone cells can be upregulated in the presence of abscisic acid in the leaf. In the pericarp and scuttelum of barley, there is some evidence from inhibitor studies that gibberellic acid promotes α-amylase synthesis in those tissues (MacGregor, 1983). Another point from these studies is that there is an antagonism between the two naturally occurring growth regulators, gibberellic acid and abscisic acid, although their roles and mechanisms of action *in vivo* are still quite obscure. Both hormones cause dozens of changes in the aleurone cells over a period of hours and probably affect α-amylase expression at other levels besides the transcription rate. Their interaction with α-amylase genes may be indirect via a chain of intermediate steps. No *bona fide* receptors have been found for either growth regulator, nor are their sites of synthesis known. Thus, this best described system of regulation of gene expression by exogenous growth regulators in plants is still a long way from being understood. Furthermore, the induction of sensitivity to gibberellic acid during development is a clear example of the importance of taking both the levels of inducers and tissue responsiveness into account.

6. Embryo Development

The embryo is the next sporophyte generation. Thus, it is the only part of the seed destined to proliferate after germination. The embryo in most flowering plants passes through three major phases of development (Steeves and Sussex, 1972; Walbot, 1978): (1) the formation of the bipolar root–shoot axis with epidermal, vascular, and ground tissue; (2) synthesis of storage reserves in the cotyledons and/or axis, especially in seeds with transient endosperms; and (3) desiccation and developmental arrest or dormancy (Fig. 4). These major processes often occur sequentially, and each phase presumably requires activation of different sets of genes.

6.1. The Suspensor

The first division of the zygote partitions it into two cells with different fates: One cell will give rise to the tissues and organs of the embryo, and the other cell will form the suspensor. The suspensor is a group of similar cells, usually in a single file or column, that attaches the embryo proper to the wall of the ovule and—by its growth—positions the embryo deeper into the embryo sac. The egg has an asymmetric distribution of organelles and has a specific orientation relative to the other cells of the megagametophyte, so there are

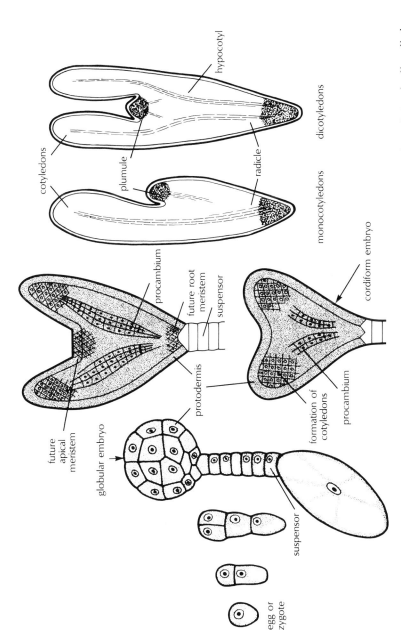

Figure 4. Embryo development. The zygote undergoes a number of mitotic divisions to form a short chain of cells called the proembryo. From its beginning, the proembryo has polarity that will become further developed later in embryogeny. The basal cells of the proembryo form a separate structure called the suspensor, which will not become part of the plant. The more apical cells divide to form a globular mass, which is the embryo proper. The globular embryo will develop into the heart-stage embryo. At this stage, the cotyledons emerge at the apical pole. The shoot apical meristem develops between the cotyledons; the root meristem develops at the suspensor pole. Besides the cotyledons, the organs of the fully developed embryos consist of the plumule or primordial shoot apex, the hypocotyl or stemlike axis, and the radicle or primordial root.

many possible cues for determining which daughter cell will form the suspensor. It is interesting that a suspensor-like structure will sometimes form when embryogenesis is initiated from isolated microspores in liquid culture, even when there is no obvious environmental polarity.

Although there are many specific patterns of cell divisions during embryogenesis in flowering plants, in most cases the suspensor cells become highly differentiated during the period in which the cells of the embryo proper remain small and undifferentiated. The suspensor cells become large, vacuolated, and have the cytological characteristics of a high metabolic rate. Suspensor cells often develop wall ingrowths characteristic of transfer cells (Yeung and Clutter, 1979) and increased nuclear DNA content, sometimes in the form of polytene chromosomes (Brady and Clutter, 1974). Unfortunately, the banding patterns of polytene chromosomes examined so far are not distinct enough to make them useful for cytological analyses of gene expression (Nagl, 1974), although certain regions of polytene chromosomes in *Phaseolus coccineus* suspensor cells appear to puff (Tagliasacchi *et al.*, 1983, 1984). The suspensor usually reaches its peak of differentiation while the embryo is a globular mass of cells; as the embryo differentiates into tissues and organs, the suspensor declines in activity. The anatomy of the suspensor suggests that it may have a role in biosynthesis and transfer of nutritional and regulatory molecules to the very young embryo. There is some experimental evidence to support this view. When early heart stage embryos of *Phaseolus coccineus* are cultured with their suspensors intact, they develop more normally than if they are cultured after their suspensors have been surgically removed (Yeung and Sussex, 1979). In addition, embryos with suspensors synthesize characteristic proteins at higher rates than those without suspensors (Walthall and Brady, 1986). The addition of the severed suspensor or exogenous gibberellic acid to the medium will partially restore development and protein synthesis in early heart stage embryos. Older heart stage embryos seem to be more independent. Since suspensors from young embryos have high levels of endogenous gibberellic acid (Alpi *et al.*, 1975), which can substitute for the suspensor in promoting growth of isolated embryos at that stage, it has been suggested that one role of the suspensor is to supply gibberellic acid to the embryo. Other hormones have been implicated as well, but conclusions are difficult to make on correlative evidence alone. Some of the embryo-lethal mutations in *Arabidopsis* alter development of the suspensor (Marsden and Meinke, 1984).

6.2. Tissue Differentiation and Establishment of a Bipolar Axis

The three basic tissue systems (epidermal, vascular, and ground) are established early in embryo development. Nothing is known about molecular events associated with tissue formation in embryos, since the cells cannot be easily separated and sorted into types for analysis. The basic pattern of development in the embryo proper varies among species (Maheshwari, 1950). In a typical dicot, the epidermis is the first tissue to be recognized, as division in the outer cells of the globular embryo becomes restricted to the plane perpendicular to the surface of the embryo and cuticle polymerizes on the outer walls. This results in a distinct epidermal cell layer that is usually maintained in an unbroken lineage throughout sporophyte development. Next, localized cell divisions in two cotyledon ridges opposite the suspensor result in a heart-shaped embryo with bilateral symmetry. The shoot apical meristem will form between the two cotyledons, and the root apical meristem will form from cells near the suspensor, sometimes including some of the suspensor cells in the process. Thus, the root–shoot axis is established. At the heart stage, cells destined to become vascular tissue can be recognized as a cylinder of cells extending from beneath the cotyledons through the hypocotyl, which is continuous with a central core of provascular tissue at the root pole. Thus, the shoot generally has ground tissue to the inside of the vascular tissue (pith) as well as to the outside (cortex), whereas the root has ground tissue to the outside only. These three basic tissue systems will continue to differentiate in relationship to the pattern established in the embryo throughout subsequent organ formation. Presumably, the cells in the globular and heart-shaped embryo differentiate according to their positions and thus must sense where they are. Preliminary experimental evidence suggests that diffusible substances from epidermal cells in heart-shaped embryos may regulate differentiation of vascular tissue (Walker and Bruck, 1985). If a strip of epidermis is removed before vascular tissue patterning in *Citrus* embryos, the vascular tissue does not form. However, if the incision is made but the flap of epidermal tissue is allowed to remain attached to the embryo, vascular tissue formation does occur. The molecular basis of communication between cells in plants is of great interest and may be important throughout development (Sussex *et al.*, 1985).

The whole process of axis and tissue formation from a single cell is remarkably autonomous in plants. Cells derived from many tissue types are capable of initiating and completing this sequence of events under a wide variety of environmental conditions (Williams and Maheswaran, 1986). Because the egg of flowering plants is so inaccessible, these somatic embryo systems may be the best way to study early molecular events in embryogenesis, although early division patterns are usually abnormal. The most extensively studied somatic embryo system is wild carrot (*Daucus carota*), in which suspension cultures initiated from leaf petioles, root parenchyma, flower peduncles, or epidermal strips from the hypocotyls of germinating seedlings will all form embryos under certain culture conditions (Street, 1976). The usual procedure is to put the initial explant in culture on a medium containing high levels of the synthetic auxin 2,4-D to initiate callus and then to transfer the callus to a similar liquid culture medium to form a proliferating suspension culture. Embryo differentiation is obtained by transferring small cell clumps to medium without 2,4-D, in which under optimal conditions up to 400 embryos per milliliter of suspension can be obtained (Sung *et al.*, 1979). Although the embryos do not develop synchronously and may show characteristic abnormalities, they can be separated on the basis of size to permit biochemical analysis of changes during development. Evidence from analyses of proteins on two-dimensional gels

(Sung and Okimoto, 1983) and randomly picked clones from cDNA libraries (Thomas and Wilde, 1986) indicates that only a few percent of the detectable proteins and mRNAs are different between cultures proliferating in the presence of 2,4-D and differentiating into embryos in the absence of 2,4-D. However, the protein and mRNA profiles of the original explant are very different from the cultures. This supports earlier cytological descriptions (Halperin and Jensen, 1967) that led to the conclusion that the initiation of embryo-specific gene expression occurs in some cells of the original explant shortly after it is placed in culture. These proembryogenic cells subsequently proliferate while exhibiting some of the cellular aspects of embryogeny, even though they do not form organs and tissues. Thus, carrot somatic embryos may be good for studying morphogenesis, but not for the initiation of embryo-specific events from nonembryo cells, since the initiation step occurs early and is ill defined. Temperature-sensitive mutants blocked at various steps in somatic embryogenesis are also being analyzed (Sung et al., 1984).

If somatic embryogenesis is so easily initiated, what keeps it from occurring during normal development, particularly in the ovule, in which the environmental conditions are obviously favorable? Studies with *Citrus* species indicate that there may be specific inhibitors of somatic embryogenesis in ovules (Tisserat and Murashige, 1977). Many varieties of *Citrus* are prone to polyembryony originating from maternal nucellar cells; this trait is genetically controlled. Extracts of monoembryogenic citrus ovules have been shown to inhibit carrot somatic embryogenesis more than extracts of polyembryonic varieties, for example.

Some of the embryo-lethal mutations in maize and *Arabidopsis* are specifically blocked at very early stages of development and may be defective in genes required for tissue and organ formation (Neuffer and Sheridan, 1980; Meinke, 1985). However, direct biochemical analysis of the mutants is hindered by the size and location of the embryos, just as in normally developing seeds.

6.3. Inhibition of Precocious Germination and Synthesis of Storage Reserves

After the shoot and root meristems and basic tissue systems have been formed, there presumably is no intrinsic barrier to continued development of organs at the meristems. In some species, this does occur during seed development resulting in leaf primordia, lateral or adventious roots, and sometimes even flower primordia. In most species, however, organ formation from meristems is restricted to postgerminative growth. In both cases, cell elongation must be inhibited during the last two thirds of seed development in order for storage reserve deposition and subsequent developmental arrest to occur and for the embryo to remain confined within the seed. When immature embryos are removed from the seed and placed in culture on a simple nutrient medium, they will exhibit cell elongation (Norstog, 1979), indicating that some factor(s) in the seed must normally inhibit this precocious germination while permitting the rest of embryogeny to proceed. Because it appears that inhibition of precocious germination is controlled by factors external to the embryo, the process has been particularly attractive for biochemical and molecular studies. After removing embryos from the seed, putative regulatory molecules are added to the culture medium, and the expression of marker genes is monitored and compared to expression during normal development.

6.3.1. Gene Expression during Embryogeny

The first step in the study of gene regulation is to describe the expression of the gene or genes in question during normal embryo development. The approach has been to identify coordinately regulated sets of genes expressed with different patterns. Because of the extremely small size of embryos before the period of histogenesis, these experiments have been performed with embryos older than heart-shaped stage and presumably capable of germination in culture.

At the global level, Goldberg *et al.* (1981) and Galau and Dure (1981) have described the complexity and abundance of transcripts during soybean and cotton embryogeny, respectively, using reassociation kinetics. The total complexity of mRNA sequences on polysomes is similar in both species and increases from 15,000 to 30,000 average-sized mRNAs during the late stages of embryo development. Most of these sequences are rare, being present at an average of only a few copies per cell, and the proteins they encode are unidentified. A few thousand low to moderately abundant sequences can be analyzed using cloned probes hybridized to embryo RNA on blots. Only a few hundred are abundant enough to be detected on two-dimensional gels of *in vitro* translation products. The examination of *in vitro* translation products from cotton embryo RNA isolated at different stages of development has led to the identification of sets of genes expressed with characteristic patterns (Dure *et al.*, 1981) (Fig. 5). The functions of several of these gene products have been determined. For example, the seed proteins are known to fall into a set of genes that is expressed during mid-embryogenesis, whereas other genes that code for proteins such as proteinase inhibitors and lectins are expressed in late embryogenesis and during early germination (Dure, 1985).

The best characterized genes expressed in embryos code for storage proteins, since the transcripts are extremely abundant and the products are of agronomic significance. The details of gene organization and expression (Higgins, 1984; Croy and Gatehouse, 1985) vary. However, several generalizations can be made. Storage proteins synthesized in embryos are encoded by multigene families. From nucleotide sequence data, it is now apparent that many of these families are evolutionarily related between species. In addition, the embryo-specific storage proteins are related to the storage proteins present in endosperm cells, although within a species the endosperm and embryo usually store different proteins. The embryo storage protein genes are only transcribed in embryos, but within the embryo different genes may show specific temporal and spatial patterns of expression. In most species, storage protein mRNAs are

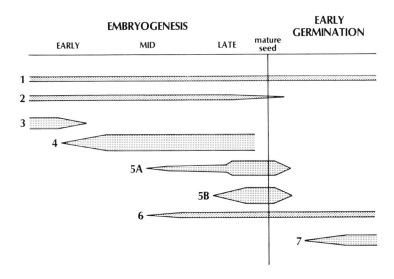

Figure 5. Groups of genes expressed during cotton seed embryogenesis. Diagrammed are the periods of expression of a number of sets of genes abundantly expressed during cotton embryogenesis. Group 1 mRNAs are derived from genes such as actin and tubulin, which are expressed throughout embryogenesis and early germination. Group 2 genes are on throughout embryogeny but appear to be turned off upon germination. Group 3 genes are expressed during early embryogenesis but are off once cell division ceases in mid-embryogenesis. Group 4 contains mRNAs for the seed storage proteins. Their levels are high during mid-embryogenesis but decline in embryo maturation. Group 5 is composed of genes whose mRNA levels are high in mature embryos but decline at germination. In cotton, group 5 can be divided into two subgroups, 5A and 5B. Group 6 contains mRNA at levels that become high late in embryogeny and remain high in the first few days of germination. Group 7 genes appear to be turned on during the process of germination. (Adapted from Dure, 1985.)

first detected while cell division is still going on, but mRNAs peak at superabundant levels after cell division stops. Often cotyledon cells become polyploid, but there is no evidence for specific amplification of expressed genes. Since storage protein synthesis spans a long time period, usually weeks, the need for specific gene amplification would not be predicted. At the end of seed development, storage protein mRNA levels crash to become part of the low or rare abundance class before disappearing completely at germination. This general pattern of gene expression points to three questions about regulation: What is involved in the initial expression of storage proteins genes? How is the level of expression modulated? What is responsible for turning off storage protein gene expression at the end of embryo development? Similar questions can be posed for other coordinately regulated sets of genes.

A genetic analysis of embryo gene expression that might bear on these questions of regulation has not been reported. No mutations have been described that alter the developmental specificity of genes expressed in embryos, although selection procedures for finding such mutations have probably not been employed. Also, I am unaware of mutations analogous to opaques in maize endosperm that specifically alter expression of classes of embryo proteins. Environmental extremes, such as nutritional deficiencies, can preferentially alter translation of classes of proteins, and are being investigated (Beach et al., 1985).

Progress in understanding the control of embryo specificity is most likely to come from analysis of transgenic plants, in which embryo-specific genes are altered in various ways and integrated into the genomes of homologous or heterologous host plants. For example, a gene for the major storage protein from french bean (*Phaseolus vulgaris*) has been cloned and introduced into tobacco via *Agrobacterium tumefaciens* mediated transformation (Sengupta-Gopalan et al., 1985). Expression of the bean gene in transformed, regenerated tobacco plants is seed specific, although it is found in cells of both the endosperm and the embryo instead of the embryo alone. Other gene–host species combinations have been successfully engineered and also show tissue specificity. The DNA sequences responsible are being analyzed now. The plan is to work backward from a *cis*-acting sequence to *trans*-acting factors involved in coordinate gene regulation.

The analysis of quantitative regulation of gene expression in transgenic plants is tricky, because the level of expression of an introduced gene can vary between individual transformants. Also, a complete understanding of the expression of individual family members (Nagy et al., 1985) is required to interpret the data (Dean et al., 1985). Another caveat is that the cues that regulate the level of expression may be different in heterologous hosts. Co-introduction of genes as internal standards may circumvent some of these difficulties.

6.3.2. Gene Expression in Culture on Basal Medium

Since embryos are often capable of precocious germination before the midpoint of development, most of the gene-expression pattern just described is apparently dispensable and can be manipulated experimentally without resulting in a dead plant. This permits hypotheses about the cues for coordinate gene control to be tested. The approach was pioneered by Dure and associates with immature cotton embryos (Ihle and Dure, 1972) and has since been extended to other species, including both dicots and monocots (reviewed by Quatrano, 1986a,b).

What happens if an immature embryo is removed from its normal seed environment and is challenged prematurely with environmental conditions for germination? Possible responses are (1) the embryo continues expression of the genes that were on when it was excised, because the cues that would permit it to progress in sequence are absent; (2) the embryo has internal regulatory cues that permit it to go through the rest of embryogeny, and then it responds to the germination signals; (3) the embryo proceeds directly to germination, shutting off embryo-specific gene expression, and skipping the interposed stages; or (4) the embryo proceeds to germination but does not shut off embryogeny because specific regulators are required to end programs as well as initiate new ones. There is evidence in the literature for each of the postulated patterns, although there are so many limitations of these studies (discussed at the end of this section) that only tentative conclusions can be drawn. Also, the pattern reported for a species may vary with stage of the embryo at culture.

Pattern 1. Embryos of some species continue expression of genes that were on when they were excised from the seed, without continuing to the next pattern of embryonic expression or skipping to germination. They appear to be stuck at that stage. Examples are (1) *Vicia faba* cotyledons, which show a 10-fold increase in storage protein content during four days of culture on a basal medium (Barratt, 1986); and (2) young (<30 DAP) *Brassica napus* embryos, which continue to make storage proteins at a low level for months on a basal medium (Crouch *et al.,* 1985). A variation of this pattern may occur in soybean, in which the embryonic pattern of expression at the time of embryo excision from the seed is either down-modulated or turned off during precocious germination, but germination characteristics are apparently not acquired (Eisenburg and Mascarenhas, 1985; Bray and Beachy, 1985).

Pattern 2. In some cases, embryos appear to complete their own program of expression and then switch into germination. Axes of *Phaseolus vulgaris* cultured on a basal medium will not germinate immediately (Long *et al.,* 1981). During the lag period they make storage proteins and become desiccation tolerant, as they would during maturation in the seed, before switching to germination, after which storage protein synthesis ceases.

Pattern 3. Embryos of grasses seem to skip the rest of embryo development and proceed directly into germination. For example, when 12 DAP rice embryos are cultured, the embryonic lectin expression switches off, as it would during normal germination (Stinissen *et al.,* 1984). In wheat, lectin synthesis persists at a low level, but only in the organs and cell types that were already expressing those genes at the time of embryo excision (Raikhel and Quatrano, 1986). They do not progress through the pattern of lectin expression characteristic of maturation, nor do they make other maturation proteins. They do commence synthesis of germination proteins, such as ribulose-bisphosphate carboxylase (Quatrano, 1986a,b).

Pattern 4. The apparent simultaneous expression of both embryonic and germinative genes during precocious germination has been described in cotton. The studies of cotton have involved the most sets of genes and paint the most complex picture. When mid-maturation cotton embryos are germinated on water, the abundant proteins that normally disappear in late embryogeny (such as storage proteins) cease to be synthesized, most of the proteins characteristic of late embryogenesis do not appear, and germination-associated proteins do appear (Dure *et al.,* 1981). It looks as though late stages of embryogeny are being bypassed, and germination ensues. However, when the experiments are repeated using a nutrient medium optimized for most rapid radicle elongation, instead of water alone, synthesis of the late embryogenesis-associated proteins does continue, albeit at a lower level than in the seed (Galau *et al.,* 1986). Choinski *et al.* (1981) examined the activity of germination-associated enzymes during precocious germination in cotton. They germinated embryos at 40 DAP when catalase and hydroxyacyl CoA dehydrogenase were already being expressed at low levels, malate synthase activity was not yet detectable (although it would be during late embryogeny), and isocitrate lyase activity was also undetectable (isocitrate lyase normally does not appear until after germination). Precociously germinating embryos maintained the expression of the enzymes that were already on at the time of excision, began synthesizing malate synthase, and turned on isocitrate lyase. Also, studies by Ihle and Dure (1972) demonstrated precocious appearance of carboxypeptidase C, a day after germination of young embryos. Thus, cotton embryos are expressing genes characteristic of both germination (isocitrate lyase) and late embryo development (malate synthase) during precocious germination.

Our studies of B. napus also provide evidence of overlap between embryo and germination programs (Finkelstein et al., 1986). For example, a 33-DAP embryo exhibits the following characteristics a week after culture. Embryo-specific storage protein genes are expressed at low levels in the preexisting organs (cotyledons and hypocotyl) and in the newly formed elongating root. By contrast, during normal germination, storage protein genes are not expressed after late embryogeny in the original embryonic organs, and expression is never initiated in any new growth. In precocious germination, storage proteins accumulate rather than being degraded, indicating a failure to induce significant levels of the proteases involved. Two glyoxylate cycle enzymes, malate synthase and isocitrate lyase, have detectable activities. Malate synthase activity would normally not appear until 45 DAP, and isocitrate lyase activity is undetectable until after germination. Therefore, genes associated with embryogeny and germination are expressed concurrently. Also, morphological characteristics indicate mixed programs. For example, the shoot meristem becomes active and makes primordia, but they develop into cotyledons instead of leaves. Sometimes appendages arise that have areas resembling leaves intermixed with areas resembling cotyledons, as if they were developmental chimaeras.

All the studies cited above, including our own, suffer from one or more of the following limitations. Very few markers specific to individual stages have been examined; culture periods have been too short to follow a progression of events, especially in species in which there is a lag period before precocious germination; the levels of marker mRNAs have not been compared to levels in situ after similar time periods; only one or two stages of embryos have been examined; the results from young embryos germinated on water may be complicated by the inability of such embryos to carry on normal metabolism without nutrients; different investigators have used isolated axes, cotyledons, whole embryos, or entire seeds, with no comparisons made among results of culturing different parts; and the assay for gene expression (steady-state levels of mRNA, in vitro or in vivo protein synthesis patterns, steady-state levels of protein, enzyme activities) varies between experiments.

Another limitation of these studies is that expression of genes has by necessity been analyzed in whole embryos and organs. Levels of RNA thus represent an average expression in the different cell types. This can lead to erroneous conclusions about regulation. For example, a gradual increase in the level of transcription over developmental time resulting from a changing rate of transcription initiation could not be distinguished from a gradual recruitment of new cells into the population of expressors.

The availability of techniques for measuring enzyme activities, antigen concentrations, and mRNA levels in histological sections has made it possible to examine gene expression in individual cells. It is not surprising that the expression patterns are much more complex than the averages would indicate. A pertinent example is lectin accumulation in embryos of wheat. Raikhel and Quatrano (1986) used immunolocalization to show that wheat germ agglutinin does not appear uniformly throughout the embryo during development but accumulates first in the radicle and coleorhiza, then in the epiblast, and finally in the coleoptile. There also seems to be tissue-specific expression: In the radicle and coleoptile, wheat germ agglutinin is confined to the epidermal layer, but in the coleorhiza and epiblast, it is also found in the ground tissue. In an earlier study (Triplett and Quatrano, 1982), wheat germ agglutinin could not be detected in extracts of whole embryos early in development, presumably because expression is in only a small fraction of the cells at that time. Thus, the two methods of measurement result in different conclusions about when expression of wheat germ agglutinin begins. During precocious germination of embryos removed from the seeds at 10 DAP (when wheat germ agglutinin is present in the radicle and coleorhiza, but not in the epiblast or coleoptile), wheat germ agglutinin is still present in the organs that were already expressing it but is not induced in the latter organs. Thus, it appears that maturation-characteristic expression is being skipped, while early expression is being maintained. It would have been impossible to make this distinction by measuring average WGA levels.

6.3.3. Role of Exogenous Factors in Regulating Precocious Germination

It is assumed that some factor(s) in the seed must normally inhibit precocious germination, while allowing the rest of embryo development to occur. Two factors have been implicated in the nontoxic inhibition of precocious germination: high levels of abscisic acid and low levels of water. The evidence for inhibition by abscisic acid is that (1) mutations resulting in low levels or insensitivity to abscisic acid often cause vivipary (reviewed in Koornneef 1986); (2) inhibitors of abscisic acid biosynthesis may result in vivipary (Fong et al., 1983); (3) exogenous abscisic acid at physiological concentrations will reversibly inhibit the germination of isolated immature embryos (Quatrano, 1986b); and (4) abscisic acid levels are high during seed development (Black, 1983) and are correlated with lag time for precocious germination (Ackerson, 1984; Prevost and LePage-Degivry, 1985; Finkelstein et al., 1985). However, restricting the rate of water uptake directly by including high concentrations of osmotica in the culture medium can substitute for the effects of exogenous abscisic acid on isolated embryos (Norstog, 1979). The osmotic potential of developing embryos is quite low (Yeung and Brown, 1982); yet, water uptake in the seed is much slower during embryo development than it is during germination and seedling growth. In fact, late in embryogeny there is a net loss of water.

Therefore, both abscisic acid and restricted water uptake may play roles in inhibiting precocious germination *in situ*, and their ability to regulate gene expression in cultured embryos has been assessed.

6.3.3a. Abscisic Acid.   In cultured embryos of several species of monocots and dicots, abscisic acid at 10^{-6}–10^{-5} M will prevent precocious germination, and mRNAs and proteins associated with germination are not synthesized (Dure, 1985; Quatrano, 1986*b*). The concentration of abscisic acid required to prevent germination generally increases as the embryos mature. In most cases, embryos inhibited from germinating by abscisic acid continue to make mRNAs and proteins characteristic of mid- or late embryogeny; the particular response depends on the age of the embryo at culture. Cotton, wheat, and soybean have been studied in the most detail. For example, when young embryos of cotton are cultured for 4 days on 5×10^{-6} M abscisic acid, most of the constitutive genes are still expressed, the mRNAs that disappear during maturation are decreased prematurely, and the mRNAs characteristic of late embryogeny (*lea* mRNAs) appear early (Dure *et al.*, 1981). In fact, the *lea* mRNAs are the only sequences induced by abscisic acid in excised embryos (Galau *et al.*, 1986). Thus, the cotton embryos are precociously "maturing" in the presence of abscisic acid. Recently, Galau *et al.* (1986) cloned 18 different *lea* cDNAs and found 13 of them inducible at the mRNA level by abscisic acid in cultured embryos, although with different patterns of accumulation. At the same time, enzymes that are normally not synthesized until germination, such as isocitrate lyase (Choinski *et al.*, 1981), are inhibited by abscisic acid. A particularly interesting observation is that immature embryos will synthesize several enzymes involved in postgerminative reserve catabolism, such as malate synthase, in response to abscisic acid. Malate synthase is first synthesized at low levels during embryo maturation and then resynthesized at higher levels during germination. However, mature seeds inhibited from germinating by abscisic acid do not synthesize malate synthase. Thus, synthesis of the same enzyme can be either stimulated or inhibited by abscisic acid, depending on whether the synthesis is associated with embryogeny or germination (Choinski *et al.*, 1981).

Studies of the effects of exogenous regulators on embryos in culture can provide information about what is possible, but the role of the regulators in the seed should be studied *in situ*, because what an embryo can do and what it actually does may differ. Seeds can be made deficient in abscisic acid using the chemical biosynthetic inhibitor fluridone, or by mutations in the biosynthetic pathway. Alternatively, mutants can be isolated that do not respond to abscisic acid.

6.3.3b. Fluridone.   Fluridone   (1-methyl-3-phenyl-5-(3(trifluromethyl)-phenyl)-4(1H)-pyridone) is a herbicide that acts by inhibiting the desaturation of phytoene to phytofluene and thus prevents the accumulation of carotenoids (Vaisberg and Schiff, 1976). Lack of carotenoids results in inhibition of photosynthesis, and other biosynthetic pathways that branch off of the carotenoid pathway are disrupted. This results in a decrease in the synthesis of abscisic acid. Fluridone has been used to investigate the effects of lowering abscisic acid levels; if exogenous abscisic acid counteracts the fluridone effects, then the observed phenotype is presumed to be the result of interrupting abscisic acid biosynthesis rather than other processes. For example, Bray and Beachy (1985) applied fluridone directly to cultured soybean cotyledons and showed that the mRNA for the β-subunit of β-conglycinin was decreased along with the endogenous abscisic acid levels. In corn, fluridone has been used to determine the time in development when abscisic acid is important for inhibiting precocious germination in the seed. Fong *et al.* (1983) applied a single fluridone treatment to kernels at different times during development. Although carotenoid synthesis was always disrupted, only treatments between 9 and 13 DAP resulted in subsequent vivipary.

6.3.3c. ABA-Deficient Mutants.   Several mutants of corn look just like fluridone-treated plants: They are carotenoid deficient, and the seeds are viviparous (Robertson, 1955). These *vp* mutants are blocked at different steps in the carotenoid pathway and therefore have reduced abscisic acid levels. Unlike fluridone treatment, it is not possible to manipulate the time of expression in these mutants (e.g., they are not temperature sensitive). They are not useful for studying the window of abscisic acid vulnerability and have pleitropic effects. Nevertheless, most of the abscisic acid-deficient mutants in different species have reduced seed dormancy and/or vivipary, indicating a role for abscisic acid *in situ* (Koornneef, 1986). Also, in the abscisic acid-deficient mutants, reduced vegetative ABA levels are always correlated with reduced seed ABA, indicating that seeds use the same biosynthetic pathways, or perhaps obtain abscisic acid from the maternal plant. Using various heterozygous combinations of abscisic acid-deficient mutants of *Arabidopsis*, Karssen *et al.* (1983) were able to demonstrate that both the maternal and the embryo genotypes contributed to the abscisic acid in seeds and that embryonic abscisic acid was necessary for seed dormancy. (The role of abscisic acid in quiescence or prevention of precocious germination rather than true dormancy was not addressed.) A comparison of gene expression in abscisic acid-deficient and normal embryos throughout development has not been done with any of the available mutants.

6.3.3d. ABA-Insensitive Mutants.   These mutants are easy to select by simply looking for plants that grow in inhibiting concentrations of abscisic acid. For seeds, this type of selection has been done in *Arabidopsis* (Koornneef *et al.*, 1984). Mutants were selected on 10 μM abscisic acid that were 5–20 times less sensitive to abscisic acid during germination than wild-type, but endogeneous abscisic acid levels are comparable. The mutants mapped to three different loci. Mutations at two of the loci resulted in both leaf wilting and reduced seed dormancy, as in the deficient mutants. However, the third locus only affected seed dormancy and thus separated seed from plant responses. These insensitive mutants are attractive for studies of the role of abscisic acid *in situ* because they have the potential of being specific: If the abscisic acid receptor is the site of the mutation, only ABA effects will be reflected in the phenotype.

It cannot simply be assumed that insensitive mutants are receptor mutants, even if abscisic acid uptake and turnover changes have been ruled out. For example, *vpl* is a mutant of corn that has normal carotenoid and abscisic acid levels and requires higher levels of exogenous ABA to inhibit germination, indicating insensitivity to abscisic acid in seeds (Robichaud *et al.*, 1980). When the endosperm is mutant, *vpl* reduces the expression of several genes normally expressed in late aleurone development (Dooner 1985). The connection, if any, between abscisic acid sensitivity and the pleitotropic endosperm effects is unknown. It is possible that *vpl* is a regulatory gene that results in expression of many structural genes, including the abscisic acid receptor. Alternatively, these genes may all be under abscisic acid control. Consequently, lack of the receptor may decrease their expression. More characterization is required to determine the basis of this mutation.

6.3.3e. Restricted Water Uptake and Desiccation. In cases in which it has been tested, high osmotic conditions will substitute for abscisic acid in suppressing precocious germination and allowing embryo maturation to occur (Norstog 1979). In *Brassica*, osmotic enhancement of embryo development is effective over a wider range of stages than abscisic acid (Finkelstein and Crouch, 1986). Throughout the later stages of embryogeny, embryos cultured on high osmoticum accumulate levels of storage protein and storage protein mRNA similar to levels in embryos developing *in situ*. Comparison of the kinetics of induction of storage protein mRNA by abscisic acid or osmotic treatment showed that the osmotic effect was more rapid and therefore possibly more direct. Endogenous abscisic acid in osmotically treated embryos remained low, indicating that high abscisic acid levels are not required for the observed effects on gene expression. These results are consistent with the hypothesis that abscisic acid is important in regulating embryogeny before the desiccation phase, possibly by restricting water uptake, but that desiccation itself is required to complete embryo maturation.

Desiccation is operationally different from osmotic treatment, in that embryos are usually subjected to desiccation in air over graded salt solutions, and they lose water (reviewed by Kermode and Bewley, 1986). During osmotic treatments, however, the water potential gradient may still favor movement of water into the embryo, although at a lower rate than would occur on a basal medium. Desiccation occurs naturally late in embryogeny. Most living cells cannot withstand desiccation, and neither can embryo cells until a specific time in their development. Thus, desiccation tolerance is a stage-specific event.

Desiccation has been proposed as the environmental trigger that irreversibly switches the pattern of gene expression from embryogeny to germination (Misra *et al.*, 1985). Results with *Phaseolus* seeds show that premature drying results in a redirection of *in vitro* protein synthesis from an embryonic to a germinative pattern after rehydration. This finding is reminiscent of the induction of sensitivity to gibberellic acid in aleurone cells by desiccation. Regulation of the timing and rate of water loss in seeds is thus essential for completion of seed development but is not well understood.

7. Concluding Remarks

It is obvious that seed development is a complex process involving sequential changes in the role and function of many different components of the seed. Tens of thousands of genes are expressed in dozens of cell types. Thus, in order to understand the regulation of expression of genes during seed development, more sophisticated tools are required. For example, *in situ* localization of transcripts using cloned probes will permit cell-level analysis of relatively abundant RNAs and will result in a more precise description of gene expression, particularly during early stages of development. The ability to create transgenic plants will result in rapid accumulation of information about the nucleotide sequences within gene loci that are required for coordinate regulation, developmentally regulated expression, and the function of specific proteins. However, the identification of *trans*-acting regulatory proteins and higher-level cues, such as hormones and environmental signals, will be greatly facilitated by genetic analyses. Products of regulatory loci are likely to be rare, and thus direct cloning of regulatory genes with methods such as transposon tagging will be essential to study the protein product. Also, the remarkable diversity in reproductive processes among flowering plants can be used to design experiments to answer specific questions.

ACKNOWLEDGMENTS.   I would like to acknowledge the assistance of Karen Muskavitch and Mark Estelle in editing, Sherri Brown for helping with illustrations, Glenn Galau for showing unpublished results, and Ian Sussex, Brian Larkins, and Derek Bewley for valuable discussions, although they certainly are not responsible for the opinions expressed in this chapter.

References

Ackerson, R. C., 1984, Abscisic acid and precocious germination in soybeans, *J. Exp. Bot.* **35**:414–421.

Alpi, A., Tognoni, F., and D'Amato, F., 1975, Growth regulator levels in embryo and suspensor of *Phaseolus coccineus* at two stages of development, *Plant* **127**:153–162.

Armstrong, C., Black, M., Chapman, J. M., Norman, H. A., and Angold, R., 1982, The induction of sensitivity to gibberellin in aleurone tissue of developing wheat grains. I. Effect of dehydration, *Planta* **154**:573–577.

Barratt, D. H. P., 1986, Regulation of storage protein accumulation by abscisic acid in *Vicia faba* L. cotyledons cultured *in vitro*, *Ann. Bot.* **57**:245–256.

Baulcombe, D. C., and Buffard, D., 1983, Gibberellic-acid-regulated expression of α-amylase and six other genes in wheat aleurone layers, *Planta* **157**:493–501.

Beach, L. R., Spencer, D., Randall, P. J., and Higgins, T. J. V., 1985, Transcriptional and post-transcriptional regulation of storage protein gene expression in sulfur deficient pea seeds, *Nucleic Acids Res.* **13**:999–1013.

Belanger, F. C., Brodl, M. R., and Ho, T. D., 1986, Heat shock causes destabilization of specific mRNAs and destruction of endoplasmic reticulum in barley aleurone cells, *Proc. Natl. Acad. Sci. USA* **83**:1–5.

Bennett, M. D., Smith, J. B., and Barclay, I., 1975, Early seed development in the Triticeae, *Philos. Trans. R. Soc. Lond. Ser. B* **272**:199–227.

Bewley, J. D., and Black, M., 1978, *Physiology and Biochemistry of Seeds in Relation to Germination*, Vol. 1, Springer-Verlag, New York.

Bewley, J. D., and Black, M., 1982, *Physiology and Biochemistry of Seeds in Relation to Germination*, Vol. 2, Springer-Verlag, New York.

Black, M., 1983, Abscisic acid in seed germination and dormancy, in: *Abscisic Acid* (F. T. Addicott, ed.), pp. 331–363, Praeger, New York.

Boston, R. S., Goldsbrough, P. B., and Larkins, B. A., 1985, Transcription of a zein gene in heterologous plant and animal systems, in: *Plant Genetics* (M. Freeling, ed.), pp. 629–639, Alan R. Liss, New York.

Boston, R. S., Kodrzycki, R., and Larkins, B. A., 1986, Transcriptional and post-transcriptional regulation of maize zein genes, in: *Molecular Biology of Seed Storage Proteins and Lectins* (L. M. Shannon, and M. J. Chrispeels, eds.), pp. 117–126, American Society of Plant Physiologists, Rockville, Maryland.

Bouman, F., 1984, The ovule, in: *Embryology of Angiosperms* (B. M. Johri, ed.), pp. 123–158, Springer-Verlag, New York.

Brady, T., and Clutter, M. E., 1974, Structure and replication of Phaseolus polytene chromosomes, *Chromosoma* **45**:63–79.

Bray, E. A., and Beachy, R. W., 1985, Regulation by ABA of β-conglycinin expression in cultured developing soybean cotyledons, *Plant Physiol.* **79**:746–750.

Brink, R. A., and Cooper, D. C., 1947, The endosperm in seed development, *Bot. Rev.* **13**:423–541.

Brown, A. H. D., and Jacobsen, J. V., 1982, Genetic basis and natural variation of α-amylase isozymes in barley, *Genet. Res.* **40**:315–324.

Choinski, J. S., Jr., Trelease, R. N., and Doman, D. C., 1981, Control of enzyme activities in cotton cotyledons during maturation and germination. III. *In vitro* embryo development in the presence of abscisic acid, *Planta* **152**:428–435.

Chrispeels, M. J., and Varner, J. E., 1967, Gibberellic acid-enhanced synthesis and release of α-amylase and ribonuclease by isolated barley aleurone layers, *Plant Physiol.* **42**:398–406.

Clark, J. K., and Sheridan, W. F., 1986, Developmental profiles of the maize embryo-lethal mutants *dek* 22 and *dek* 23, *J. Hered.* **77**:83–92.

Crouch, M. L., Tenbarge, K., Simon, A., Finkelstein, R., Scofield, S., and Solberg, L., 1985, Storage protein mRNA levels can be regulated by abscisic acid in *Brassica* embryos, in: *Molecular Form and Function of the Plant Genome* (L. van Vloten-Doting, G. S. P. Groot, and T. C. Hall, eds.), pp. 555–566, Plenum, New York.

Croy, R. R. D., and Gatehouse, J. A., 1985, Genetic engineering of seed proteins: Current and potential applications, in: *Plant Genetic Engineering* (J. H. Dodds, ed.), pp. 143–268, Cambridge University Press, Cambridge.

Dean, C., van den Elzen, P., Tamaski, S., Dunsmuir, P., and Bedbrook, J., 1985, Differential expression of the eight genes of *Petunia* ribulose bisphosphate carboxylase small subunit multigene family, *EMBO J.* **4**:3055–3061.

Di Fonzo, N., Manzocchi, L., Salamini, F., and Soave, C., 1986, Purification and properties of an endospermic protein of maize associated with the Opaque-2 and Opaque-6 genes, *Planta* **167**:587–594.

Dooner, H. K., 1985, Viviparous-1 mutation in maize conditions pleitropic enzyme deficiencies in the aleurone, *Plant Physiol.* **77**:486–488.

Dure, L. III, 1985, Embryogenesis and gene expression during seed formation, *Oxf. Surv. Plant Mol. Cell Biol.* **2**:179–197.

Dure, L. III, Greenway, S. C., and Galau, G. A., 1981, Developmental biochemistry of cottonseed embryogenesis and germination: Changing messenger ribonucleic acid populations as shown by *in vitro* and *in vivo* protein synthesis, *Biochemistry* **20**:4162–4168.

Egley, G. H., Paul, R. N., Jr., Vaughn, K. C., and Duke, S. O., 1983, Role of peroxidase in the development of water-impermeable seeds coats in *Sida spinosa L, Planta* **157**:224–232.

Eisenberg, A. J., and Mascarenhas, J. P., 1985, Abscisic acid and the regulation of synthesis of specific seed proteins and their messenger RNAs during culture of soybean embryos, *Planta* **166**:505–514.

Evans, M., Black, M., and Chapman, T., 1975, Induction of hormone sensitivity by dehydration is one positive role for drying in cereal seed, *Nature (Lond.)* **258**:144–145.

Felker, F. C., Peterson, D. M., and Nelson, O. E., 1985, Anatomy of immature grains of eight maternal effect shrunken endosperm barley mutants, *Am. J. Bot.* **72**:248–256.

Finkelstein, R. R., Tenbarge, K. M., Shumway, J. E., and Crouch, M. L., 1985, Role of ABA in maturation of rapeseed embryos, *Plant Physiol.* **78**:630–636.

Finkelstein, R. R., DeLisle, A. J., Simon, A. E., and Crouch, M. L., 1986, Role of abscisic acid and restricted water uptake during embryogeny in Brassica, *Mol. Biol. Plant Growth Control* **44**:73–84.

Finkelstein, R. R., and Crouch, M. L., 1986, Rapeseed embryo development in culture on high osmoticum is similar to that in seeds, *Plant Physiol.* **81**:907–912.

Fong, F., Koehler, D. E., and Smith, J. D., 1983, Fluridone induction of vivipary during maize seed development, in: *Third International Symposium on Preharvest Sprouting in Cereals* (J. E. Krueger, and D. E. LaBerge, eds.), pp. 188–196, Westview Press, Boulder, Colorado.

Fromm, M. E., Taylor, L. P., and Walbot, V., 1986, Stable transformation of maize after gene transfer by electroporation, *Nature (Lond.)* **319**:791–793.

Galante, E., Vitale, A., Manzocchi, L. A., Soave, C., and Salamini, F., 1983, Genetic control of a membrane component and zein deposition in maize endosperm, *Mol. Gen. Genet.* **192**:316–321.

Galau, G. A., and Dure, L. III, 1981, Developmental biochemistry of cottonseed embryogenesis and germination: Changing messenger ribonucleic acid populations as shown by reciprocal heterologous complementary deoxyribonucleic acid–messenger ribonucleic acid hybridization, *Biochemistry* **20**:4169–4178.

Galau, G. A., Hughes, D. W., and Dure, L. III, 1986, Abscisic acid induction of cloned cotton late embryogenesis-abundant (Lea) mRNAs, *Plant Mol. Biol.* **7**:155–170.

Goldberg, R. B., Hoschek, G., Tan, S. H., Ditta, G. S., and Breidenbach, R. W., 1981, Abundance, diversity and regulation of mRNA sequence sets in soybean embryogenesis, *Dev. Biol.* **83**:201–217.

Gosling, P. G., Butler, R. A., Black, M., and Chapman, J. M., 1981, The onset of germination ability in developing wheat, *J. Exp. Bot.* **32**:621–627.

Greenwood, J. S., and Bewley, J. D., 1982, Castor bean (*Ricinus communis* L. cv. Hale) seed development. I. Descriptive morphology, *Can. J. Bot.* **60**:1751–1760.

Gutterman, Y., and Heydecker, W., 1973, Studies of the surfaces of desert plant seed. I. Effect of day length, *Ann. Bot.* **37**:1049–1050.

Halperin, W., and Jensen, W. A., 1967, Ultrastructural changes during growth and embryogenesis in carrot cell culture, *J. Ultrastruct. Res.* **18**:428–443.

Harris, P. J., Anderson, M. A., Bacic, A., and Clarke, A. E., 1984, Cell–cell recognition in plants with special references to the pollen–stigma interaction, *Oxford Surveys of Plant Molecular and Cell Biology*, Vol. 1 (P. J. Davies, ed.), Clarendon Press, Oxford.

Higgins, T. J. V., 1984, Synthesis and regulation of major proteins in seeds, *Annu. Rev. Plant Physiol.* **35**:191–221.

Higgins, T. J. V., Jacobsen, J. V., and Zwar, J. A., 1982, Gibberellic acid and abscisic acid modulate protein synthesis and mRNA levels in barley aleurone layers, *Plant Mol. Biol.* **1**:191–215.

Ho, D. T., Nolan, R. C., and Uknes, S. J., 1985, On the mode of action of abscisic acid in barley aleurone layers, *Current Topics Plant Biochem. Phys.* **4**:118–125.

Ihle, J. N., and Dure, L. S. III, 1972, The developmental biochemistry of cottonseed embryogenesis and germination. III. Regulation of the biosynthesis of enzymes utilized in germination, *J. Biol. Chem.* **247**:5048–5055.

Jacobsen, J. V., and Higgins, T. J. V., 1982, Characterization of the α-amylases synthesized by aleurone layers of Himalaya barley in response to gibberellic acid, *Plant Physiol.* **70**:1647–1653.

Jacobsen, J. V., and Beach, L. R., 1985, Control of transcription of α-amylase and rRNA genes in barley aleurone protoplasts by gibberellin and abscisic acid, *Nature (Lond.)* **316**:275–277.

Jacobsen, J. V., Zwar, J. A., and Chandler, P. M., 1985, Gibberellic-acid-responsive protoplasts from mature aleurone of Himalaya barley, *Planta* **163**:430–439.

Jacobsen, J. V., Hanson, A. D., and Chandler, P. C., 1986, Water stress enhances expression of an α-amylase gene in barley leaves, *Plant Physiol.* **80**:350–359.

Johnston, S. A., and Hanneman, R. E., Jr., 1982, Manipulations of endosperm balance number overcome crossing barriers between diploid *Solanum* species, *Science* **217**:446–448.

Johnston, S. A., den Nijs, T. P. M., Peloquin, S. J., and Hanneman, R. E., Jr., 1980, The significance of genic balance to endosperm development in interspecific crosses, *Theoret. Appl. Genet.* **57**:5–9.

Johri, B. M., 1984, *Embryology of Angiosperms*, Springer-Verlag, New York.

Karssen, C. M., Brinkhorst-vanderSwan, D. L. C., Breekland, A. E., and Koornneef, M., 1983, Induction of dormancy during seed development by endogenous abscisic acid: studies on abscisic acid deficient genotypes of *Arabidopsis thaliana* (L.) Heynh, *Planta* **157**:158–165.

Keeton, W. T., 1972, *Biological Science*, 2nd ed., Norton, New York.

Kermicle, J. L., 1970, Dependence on the R-mottled aleurone phenotype in maize on mode of sexual transmission, *Genetics* **66**:69–85.

Kermode, A. R., and Bewley, J. D., 1986, Alteration of genetically regulated synthesis in developing seeds by desiccation, in: *Membranes, Metabolism, and Dry Organisms* (A. C. Leopold, ed.), pp. 59–84, Cornell University Press, Ithaca, New York.

King, R. W., 1976, Abscisic acid in developing wheat grains and its relationship to grain growth and maturation, *Planta* **132**:43–57.

King, R. W., Salminen, S. O., Hill, R. D., and Higgins, T. J. V., 1979, Abscisic-acid and gibberellin action in developing kernels of Triticale (CV.6A190), *Planta* **146**:249–255.

Koornneef, M., 1986, Genetic aspects of abscisic acid, in: *A Genetic Approach to Plant Biochemistry*, (A. D. Blonstein and P. J. King, eds.), pp. 35–65, Springer-Verlag, New York.

Koornneef, M., Reuling, G., and Karssen, C. M., 1984, The isolation and characterization of abscisic acid insensitive mutant of *Arabidopsis thaliana*, *Physiol. Plant.* **61**:377–383.

Kowles, R. V., and Phillips, R. L., 1985, DNA amplification patterns in maize endosperm nuclei during kernel development, *Proc. Natl. Acad. Sci. USA* **82**:7010–7014.

Langridge, P., Pintor-Toro, J. A., and Feix, G., 1982, Transcription effects of the opaque-2 mutation of *Zea mays* L., *Planta* **156**:166–170.

Long, S. R., Dale, R. M. K., and Sussex, I. M., 1981, Maturation and germination of *Phaseolus vulgaris* embryonic axes in culture, *Planta* **153**:405–415.

MacGregor, A. W., 1983, Cereal α-amylases: Synthesis and action pattern, in: *Seed Proteins* (J. Daussant, J. Mossé, and J. Vaughan, eds.), pp. 1–34, Academic, London.

MacGregor, A. W., MacDougall, F. H., Mayer, C., and Daussant, J., 1984, Changes in levels of α-amaylase components in barley tissues during germination and early seedling growth, *Plant Physiol.* **75**:203–206.

Maheshwari, P., 1950, *An Introduction to the Embryology of Angiosperms*, McGraw-Hill, New York.

Marbach, I., and Mayer, A. M., 1975, Changes in catecholoxidase and permeability to water in seed coats of *Pisum elatuis* during seed development and maturation, *Plant Physiol.* **56**:93–96.

Marinos, N. G., 1970, Embryogenesis of the Pea (*Pisum sativum*). I. The cytological environment of the developing embryo, *Protoplasma* **70**:261–279.

Marks, M. D., Lindell, J. S., and Larkins, B. A., 1985a, Quantitative analysis of the accumulation of zein mRNA during maize endosperm development, *J. Biol. Chem.* **260**:16445–16450.

Marks, M. D., Lindell, J. S., and Larkins, B. A., 1985b, Nucleotide sequence analysis of zein mRNAs from maize endosperm, *J. Biol. Chem.* **260**:16451–16459.

Marsden, M. P. F., and Meinke, D. W., 1984, Abnormal development of the suspensor in an embryo-lethal mutant of *Arabidopsis thaliana*, *Am. J. Bot.* 71:(No. 5, Part 2), 15 (abst).

Meinke, D. W., 1985, Embryo-lethal mutants of *Arabidopsis thaliana*: Analysis of mutants with a wide range of lethal phases, *Theoret. Appl. Genet.* **69**:543–552.

Misra, S., Kermode, A., and Bewley, J. D., 1985, Maturation drying as the "switch" that terminates

seed development and promotes germination, in: *Molecular Form and Function of the Plant Genome* (L. van Vloten-Doting, G. S. P. Groot, and T. C. Hall, eds.), pp. 113–128, Plenum, New York.

Mogensen, H. L., 1985, Ultracytochemical localization of plasma membrane-associated phosphatase activity in developing tobacco seeds, *Am. J. Bot.* **72:**741–754.

Mundy, J., 1984, Hormonal regulation of α-amylase inhibitor synthesis in germinating barley, *Carlsberg Res. Commun.* **49:**439–444.

Mundy, J., Hejgaard, J., Hansen, A., Hallgren, L., Jorgensen, K. G., and Munck, L., 1986, Differential synthesis *in vitro* of barley aleurone and starchy endosperm proteins, *Plant Physiol.* **81:**630–636.

Mundy, J., Svendsen, I., and Hejgaard, J., 1983, Barley α-amylase/subtilisin inhibitor. I. Isolation and characterization, *Carlsberg Res. Commun.* **48:**81–90.

Murray, D. R., 1979, Nutritive role of the seedcoats during embryo development in *Pisum sativum* L., *Plant Physiol.* **64:**753–769.

Murray, D. R., 1984, *Seed Physiology*, Vol. 2: *Germination and Reserve Mobilization*, Academic, Orlando, Florida.

Nagl, W., 1974, The *Phaseolus* suspensor and its polytene chromosomes, *Z. Pflanzenphysiol.* **73:**1–44.

Nagy, F., Morelli, G., Fraley, R. T., Rogers, S. G., and Chua, N. H., 1985, Photoregulated expression of a pea rbc S gene in leaves of transgenic plants, *EMBO J.* **4:**3063–3068.

Nesling, F. A. V., and Morris, D. A., 1979, Cytokinin levels and embryo abortion in interspecific *Phaseolus* crosses, *Z. Pflanzenphysiol.* **91:**345–358.

Neuffer, M. G., and Sheridan, W. F., 1980, Defective kernel mutants of maize. I. Genetic and lethality studies, *Genetics* **95:**929–944.

Nogler, G. A., 1984, Gametophytic apomixis, in: *Embryology of Angiosperms* (B. M. Johri, ed.), pp. 475–518, Springer-Verlag, New York.

Norman, H. A., Black, M., and Chapman, J. M., 1983, The induction of sensitivity to gibberellin in aleurone tissue of developing wheat grains. III. Sensitisation of isolated protoplasts, *Planta* **158:**264–271.

Norstog, K., 1979, Embryo culture as a tool in the study of comparative and developmental morphology, in: *Plant Cell and Tissue Culture, Principles, and Applications* (W. R. Sharp, P. O. Larsen, F. F. Paddock, and V. Raghaven, eds.), pp. 179–202, Ohio State University Press.

Pate, J. S., 1984, The carbon and nitrogen nutrition of fruit and seed—Case studies of selected grain legumes, in: *Seed Physiology*, Vol. 1 (D. R. Murray, ed.), pp. 41–82, Academic, Sydney.

Prevost, I., and LePage-Degivry, M. T., 1985, Inverse correlation between ABA content and germinability throughout the maturation and the *in vitro* culture of the embryo of *Phaseolus vulgaris*, *J. Exp. Bot.* **36:**1457–1464.

Quatrano, R. S., 1986, Regulation of gene expression by abscisic acid during angiosperm development, in: *Oxford Surveys of Plant Molecular and Cell Biology* (B. Mivlin, ed.), pp. 467–477, Oxford University Press.

Quatrano, R. S., 1987, The role of hormones during seed development, in: *Plant Hormones and Their Role in Plant Growth and Development* (P. J. Davies, ed.), pp. 494–514, Martinus Niijhoff, The Netherlands.

Raghavan, V., 1986, *Embryogenesis in Angiosperms. A Developmental and Experimental Study*, Cambridge University Press, Cambridge.

Raikhel, N. V., and Quatrano, R. S., 1986, Location of wheat germ agglutinin in developing wheat embryos and those cultured in abscisic acid, *Planta* **168:**433–440.

Raven, P., Evert, R., and Eichorn, S., 1986, *Biology of Plants*, 4th ed., Worth, New York.

Reid, J. S. G., and Bewley, J. D., 1979, A dual role for the endosperm and its galactomannan reserves in the germinative physiology of fenugreek (*Trigonella foenum-graecum* L.), an endospermic leguminous seed, *Planta* **147:**145–150.

Robertson, D. S., 1955, The genetics of vivipary in maize, *Genetics* **40:**745–760.

Robichaud, C. S., Wong, J., and Sussex, I. M., 1980, Control of *in vitro* growth of viviparous embryo mutants of maize in abscisic acid, *Dev. Genet.* **1:**325–330.

Rogers, J. C., 1985, Two barley α-amylase gene families are regulated differently in aleurone cells, *J. Biol. Chem.* **260:**3731–3738.

Rogers, J. C., and Milliman, C., 1984, Coordinate increase in major transcripts from the high pI α-amylase multigene family in barley aleurone cells stimulated with gibberellic acid, *J. Biol. Chem.* **259:**12234–12240.

Rubenstein, I., Phillips, R. L., Green, C. E., and Gengenbac, B. G., 1979, *The Plant Seed: Development, Preservation, and Germination*, Academic, Orlando, Florida.

Sarkar, K. R., and Coe, E. H., 1971, Analysis of events leading to heterofertilization in maize, *J. Hered.* **62:**118–120.

Schwartz, D., 1965, Regulation of gene action in maize, in: *Genetics Today*, pp. 131–136, Pergamon, Oxford.

Sengupta-Gopalan, C., Reichert, W. A., Barker, R. F., Hall, T. C., and Kemp, J. D., 1985, Developmentally regulated expression of the bean β-phaseolin gene in tobacco seed, *Proc. Natl. Acad. Sci. USA* **82:**3320–3324.

Sheridan, W. F., and Neuffer, M. G., 1986, Genetic control of embryo and endosperm development in maize, in: *Gene Structure and Function in Higher Plants* (G. M. Reddy, ed.), pp. 105–122, Oxford–IBH, Calcutta.

Smith, J. G., 1973, Embryo development in *Phaseolus vulgaris*. II. Analysis of selected inorganic ions, ammonia, organic acids, amino acids, and sugar in the endosperm liquid, *Plant Physiol.* **51:**454–458.

Soave, C., and Salamini, F., 1984, The role of structural and regulatory genes in the development of maize endosperm, *Dev. Genet.* **5:**1–25.

Soave, C., Tardani, L., Di Fonzo, N., and Salamini, F., 1981, Zein level in maize endosperm depends on a protein under control of the opaque-2 and opaque-6 loci, *Cell* **27:**403–410.

Steeves, T. A., 1983, The evolution and biological significance of seeds, *Can. J. Bot.* **61:**3550–3560.

Steeves, T. A., and Sussex, I. M., 1972, *Patterns in Plant Development*, Prentice-Hall, Englewood Cliffs, New Jersey.

Stinissen, H. M., Peumans, W. J., and De Langhe, E., 1984, Abscisic acid promotes lectin biosynthesis in developing and germinating rice embryos, *Plant Cell Rep.* **3:**55–59.

Street, H. E., 1976, Experimental embryogenesis: The totipotency of cultured plant cells, in: *The Developmental Biology of Plants and Animals* (C. F. Graham and P. F. Wareing, eds.), pp. 73–91, W. B. Saunders, Philadelphia.

Sung, Z. R., and Okimoto, R., 1983, Coordinate gene expression during somatic embryogenesis in carrot, *Proc Natl. Acad. Sci. USA* **80:**2661–2665.

Sung, Z. R., Smith, R., and Horowitz, J., 1979, Quantitative studies of embryogenesis in normal and 5-methyltryptophan resistant cell lines of wild carrot, *Planta* **147:**236–240.

Sung, Z. R., Fienberg, A., Chorneau, R., Borkird, C., Furner, I., Smith, J., Terzi, M., Lo-Shiavo, F., Giuliano, G., Pitto, L., and Nuti-Ronchi, V., 1984, Developmental biology of embryogenesis from carrot culture, *Plant Mol. Biol. Rep.* **2:**3–14.

Sussex, I., Ellingboe, A., Crouch, M., and Malmberg, R., 1985, *Plant Cell/Cell Interations*, Cold Spring Harbor Laboratory, Cold Spring Harbor, New York.

Tagliasacchi, A. M., Forino, L. M. C., Frediani, and Avanzi, S., 1983, Different structure of polytene chromosomes of *Phaseolus coccineus* suspensors during early embryogenesis. 2. Chromosome pair VII. *Protoplasma* **115:**95–103.

Tagliasacchi, A. M., Forino, L. M. C., Cionini, P. G., Cavallini, A., Durante, M., Cremonini, R., and Avanzi, S., 1984, Different structure of polytene chromosome of *Phaseolus coccineus* suspensors during early embryogenesis. 3. Chromosomes pair VI, *Protoplasma* **122:**98–107.

Tasi, C. Y., 1979, Tissue-specific zein synthesis in maize kernel, *Biochem. Genet.* **17:**1109–1119.

Thomas, T., and Wilde, D., 1986, Analysis of gene expression in carrot somatic embryos, in: *Somatic Embryogenesis* (M. Terzi, L. Pitto, and Z. R. Sung, eds.), pp. 77–85, IPRA, Rome.

Tisserat, B., and Marashige, T., 1977, Probable identity of substances in *Citrus* that repress asexual embryogenesis, *In Vitro* **13:**785–789.

Triplett, B. A., and Quatrano, R. S., 1982, Timing, localization, and control of wheatgerm agglutinin synthesis in developing wheat embryos, *Dev. Biol.* **91:**491–496.

Vaisberg, A. J., and Schiff, J. A., 1976, Events surrounding the early development of *Euglena* chloroplasts, *Plant Physiol.* **57**:260–269.

Van Went, J. L., and Willemse, M. T. M., 1984, Fertilization, in: *Embryology of Angiosperms* (B. M. Johri, ed.), pp. 273–317, Springer-Verlag, New York.

Varner, J. E., and Chandra, G. R., 1964, Hormonal control of enzyme synthesis in barley endosperm, *Proc. Natl. Acad. Sci. USA* **52**:100–106.

Vijayraghavan, R. M., and Probhakar, K., 1984, The endosperm, in: *Embryology of Angiosperms* (B. M. Johri, ed.), pp. 319–376, Springer-Verlag, New York.

Walbot, V., 1978, Control mechanisms for plant embryogeny, in: *Dormancy and Developmental Arrest. Experimental Analysis in Plants and Animals*, (M. E. Clutter, ed.), pp. 113–166, Academic, New York.

Walker, D. B., and Bruck, D. K., 1985, The control of positional cell differentiation in plants, in: *Plant Cell/Cell Interactions* (I. Sussex, A. Ellingboe, M. Crouch, and R. Malmberg, eds.), pp. 53–56, Cold Spring Harbor Laboratory, Cold Sring Harbor, New York.

Walthall, E. E., and Brady, T., 1986, The effect of the suspensor and gibberellic acid on *Phaseolus vulgaris* embryo protein synthesis, *Cell Diff.* **18**:37–44.

Williams, E. G., and de Lautour, G., 1980, The use of embryo culture with transplanted nurse endosperm for the production of interspecific hybrids in pasture legumes, *Bot. Gaz.* **141**:252–257.

Williams, E. G., and Maheswaran, G., 1986, Somatic embryogenesis: factors influencing coordinated behavior of cells as an embryogenic group, *Ann. Bot.* **57**:443–462.

Yamada, K., 1985, Endogenous abscisic acid in barley and use of abscisic acid in malting, *Agric. Biol. Chem.* **49**:429–434.

Yeung, E. C., and Brown, D. C. W., 1982, The osmotic environment of developing embryos of *Phaseolus vulgaris*, *Z. Pflanzenphysiol.* **106**:149–156.

Yeung, E. C., and Clutter, M. E., 1979, Embryogeny of *Phaseolus coccineus:* The ultrastructure and development of the suspensor, *Can. J. Bot.* **57**:120–136.

Yeung, E. C., and Sussex, I. M., 1979, Embryogeny of *Phaseolus coccineus:* The suspensor and the growth of the embryo-proper *in vitro. Z. Pflanzenphysiol.* **91**:423–433.

Chapter 15

Development and Differentiation of the Root Nodule
Involvement of Plant and Bacterial Genes

N. A. MORRISON, T. BISSELING, and D. P. S. VERMA

1. Introduction

The *Rhizobium*–legume symbiosis is an example of an interaction between a prokaryote (*Rhizobium*) and an eukaryote (legume) that brings about the development of an entirely new organ on the plant, the root nodule. The initial signals for induction of this organ must come from the bacteria, as these organs do not develop in the absence of an appropriate *Rhizobium* strain. The cytology of early infection events leading to nodule formation in a number of legumes is well known (Callaham and Torrey, 1981; Bhuvaneswari *et al.*, 1980, 1981; Turgeon and Bauer, 1982), and the ontogeny of nodule development has been described (Dart, 1974, 1977; Bauer, 1981). This chapter describes the salient features of this process as they pertain to the development and differentiation of this structure leading to symbiotic nitrogen fixation.

Two distinct genera of bacteria share genes necessary for forming nitrogen-fixing nodules on legume plants: *Rhizobium* (fast-growing) and *Bradyrhizobium* (slow-growing) (Jordan, 1982; Stanley *et al.*, 1985). These two types of bacteria behave similarly during the development of endosymbiosis and generally can be considered as one. Once in contact with the root, the *Rhizobium* induces morphological deformations known as **curling** in the specialized epidermal root hair cells (see reviews by Bauer, 1981; Dazzo and Gardiol, 1984). Bacteria that have adhered to the surface of the root hair apparently reorient root hair growth causing the growing tip to curl through 180–360°, entrapping *Rhizobium* between the opposing cell walls of the root hair. In this microenvironment, a small colony of bacteria develops, and the plant cell wall material is loosened in the proximity of the colony (Callaham and Torrey, 1981;

N. A. MORRISON and D. P. S. VERMA • Department of Biology, Centre for Plant Molecular Biology, Montreal, Quebec, H3A 1B1 Canada.   T. BISSELING • Department of Molecular Biology, Agricultural University, De Dreijen 11/6703 BC, Wageningen, The Netherlands.

Turgeon and Bauer, 1985; Calvert *et al.*, 1984; Ridge and Rolfe, 1985). After about 24 hr, the plant cell redirects wall synthesis, depositing cellulose microfibrils at the site of penetration and forming a tubelike structure (the **infection thread**), which encloses the invading bacteria. Concomitant with root hair curling and infection thread initiation, cortical cells below the site of incipient infection are induced to divide, forming a **nodule primordium.** As this focus of division enlarges, the infection thread passes from cell to cell, ramifying throughout the growing nodule (Dart, 1977). Before the commencement of nitrogen-fixation activity, bacteria are released from the infection thread into the host cell. An unknown cue causes a switch from infection thread growth to dissolution when the thread reaches the target cell. The end of the infection thread becomes disordered, and the fibrillar matrix of the thread wall is apparently broken down and internalized in small vesicles into the plant cell (Bassett *et al.*, 1979). As thread material is removed, the end of the infection thread develops an unwalled droplet or release vesicle in which small groups of bacteria are presented to the plasmalemma, free of an intervening cell wall. The bacteria are then taken up or internalized by a process resembling endocytosis, in which individual bacteria are packaged within membrane vesicles, topologically outside the cell cytoplasm (Verma *et al.*, 1978).

Once internalized, bacteria proliferate and differentiate into the nitrogen-fixing bacteroid, producing considerable populations of membrane-enclosed bacteroids in the infected cells. Development of nitrogen-fixing bacteroids requires the induction of genes essential for nitrogen fixation (*nif* and *fix*) and other symbiotic processes and represents a substantial differentiation from a free-living state (Sutton *et al.*, 1981; Verma *et al.*, 1986). Against this overview of the key features of nodule ontogeny, we address certain topics in which recent cytological and molecular evidence has revealed further details and has resolved some of the questions concerning organogenesis and function of this organ system.

2. Specificity of Infection

2.1. Host/*Rhizobium* Specificity

The induction and development of a successful nitrogen-fixing symbiosis are dependent on genetic factors in both partners, and specificity exists at different stages during infection and development of the nodule (Verma and Long, 1983; Verma and Nadler, 1986). A direct manifestation of specificity is reflected in the cross-inoculation group, whereby *Rhizobium* species are defined by their host range of nodulation. Generally, a *Rhizobium* strain isolated from one plant will normally infect that plant again, as well as a few other legume plants, which may or may not be related (see Allen and Allen, 1981). For instance, *R. trifolii* will only nodulate clovers (*Trifolium* species) while *R. leguminosarum* nodulates various garden and field peas (*Pisum, Lens, Vicia,* and *Lathyrus* species), whereas *R. meliloti* nodulates only *Medicago, Trigonella,* and *Melilotus* species and will not nodulate other legumes. The soybean has a cross-inoculation group of its own with *Bradyrhizobium japonicum* being the principal symbiont. Cross-inoculation specificity is presumably the result of complex evolutionary pressures that select for specialized or highly tuned interactions between host and endosymbiont, which maximize nitrogen fixation under specific ecological conditions (Verma and Stanley, 1986). Host-specific lectins have been implicated in controlling specificity in the attachment of *Rhizobium* to root hair (Dazzo and Gardiol, 1984). However, the compatibility of a particular host-endosymbiont pair depends on more than attachment; a communication between the plant and *Rhizobium* must be established before contact, activating genes responsible for the initiation of the symbiotic process.

2.2. Symbiotic Plasmid and Nodulation Genes

It is now well established in fast-growing *Rhizobium* that genes controlling host specificity of nodulation (*hsn*), nodulation (*nod*), and nitrogen fixation functions (*nif* and *fix*) reside on large (Sym)biosis plasmids (for review, see Rolfe and Shine, 1984). The symbiotic host range of certain strains can be altered by exchanging the Sym plasmids; however, this does not apply to every combination of species and Sym plasmid. In a broad host range, *Rhizobium* strain, a Sym plasmid is involved in the ability to nodulate legumes and the nonlegume *Parasponia* sp. (Ulmaceae), suggesting that nodulation in nonlegumes shares at least some processes in common with legumes (Morrison *et al.*, 1983).

Rhizobium nodulation genes are best understood in fast-growing rhizobia (Djordjevic *et al.*, 1985; Downie *et al.*, 1985; Kondorosi *et al.*, 1984; Jacobs *et al.*, 1985). On the Sym plasmid of each particular strain, there is a cluster of common *nod* genes that have functional and sequence conservation in a range of rhizobia; genes controlling host-specific nodulation are closely linked (Kondorosi *et al.*, 1984; Schofield *et al.*, 1984; Downie *et al.*, 1985). The common *nod* genes are essential for root hair curling events, since mutations in these genes are Hac$^-$ (no root hair curling). In the common *nod* region are three genes: *nod*ABC, which constitute an operon (Rossen *et al.*, 1984; Torok *et al.*, 1984). Close to the start of *nod*A is the *nod*D gene, which is transcribed divergently from the nodABC operon (Egelhoff *et al.*, 1985). Transposon-induced nodD mutants have a Nod$^-$ phenotype in R. *trifolii* and R. *leguminosarum* but have a delayed nodulation phenotype in R. *meliloti*.

Fusion experiments within the *Escherichia coli* lac operon have shown that only the *nod*D gene is constitutively expressed in free-living *Rhizobium*; however, the *nod*ABC operon is induced by low-molecular-weight organic compounds found in root exudates. The *nod*D gene is now recognized as a regulatory gene, since the plant exudate-mediated induction of *nod*ABC requires the product of *nod*D (Mulligan and Long, 1985). The structure of the inducing compound from alfalfa roots has been elucidated (Peters *et al.*, 1986); it is a tricyclic flavone, luteolin, found in many plant tissues. It is now known that the inducer molecule interacts with the *nod*D gene product to bring about nodABC gene induction, following the general operon model, since the nodD gene product resembles DNA-binding regulatory proteins and can bind to the promoter region of nodABC (Hong et al., 1987).

Downstream from the nodABC operon is a region (nodIJ) in which mutations cause a delayed nodulation phenotype in R. leguminosarum (Downie et al., 1985) and an exaggerated root hair curling response coupled with a Nod⁻ phenotype in R. trifolii (Djordjevic et al., 1985). Infection threads in curled root hairs are aborted in the case of R. trifolii, suggesting that genes in this region are responsible for infection thread growth.

Downstream of nodD and transcribed on the same strand is an additional nod gene region (nodFE) in R. trifolii, which controls the host range of nodulation. Random lac fusions show that these genes are also induced by factors from root exudates (Innes et al., 1986). Mutants in this region are able to nodulate peas, including a narrow-specificity cultivar (Afghanistan), and beans (Phaseolus vulgaris), which are not normally nodulated by R. trifolii. This suggests that Rhizobium host specificity of nodulation genes exert a restrictive or repressive effect, causing a reduction in host range ability. Apparently, R. trifolii has the ability to nodulate peas and beans, but this characteristic is demonstrated only by mutation.

Host specificity in symbiosis is seen at stages other than initial infection. Sym plasmid transfer experiments have usually resulted in ineffective responses on the new host plants (Rolfe and Shine, 1984). In most cases, a failure of bacterial release, bacteroid development, or nodule development has been claimed as the cause of poor symbiosis. These data suggest that host-specific phenomena also control late symbiotic development of both partners before the activation of nitrogen fixation genes. In one case, the host genotype has been implicated in the control of late symbiotic incompatibility. Certain fast-growing Rhizobium japonicum (also called fredii) strains from China form normal nitrogen-fixing symbiosis with a wild soybean cultivar (Peking) but form ineffective (Fix⁻) nodules on commercial soybean cultivars (Keyser et al., 1982). Genetic studies (Devine, 1984) show that this trait is controlled by a single Mendelian recessive allele in the plant, which conditions an ineffective response. It is interesting to note that whereas this allele has a profound effect on symbiosis with fast-growing R. japonicum (fredii) strains, it has no apparent effect on normal (slow-growing) B. japonicum strains, which nodulate these plants effectively. Other plant genes influencing nodulation in strain-specific manners have been identified (Verma and Nadler, 1984) in natural populations and by chemical mutagenesis (LaRue et al., 1985).

3. Induction of Host Cell Division

3.1. Diffusible Factors from *Rhizobium*

The sites of incipient infection in soybean roots inoculated with B. japonicum have recently been mapped by Calvert et al. (1984). They observed that subepidermal cortical cell divisions that give rise to nodule primordia can occur in the absence of local root hair curling. In addition, infection events were found where the thread had penetrated to the cortex but no cortical cell divisions had occurred. Thus, the previously held view that infection thread formation in the root hairs results in a signal for cell division below the site of infection is now questioned. The induction of cortical divisions and infection of the root hair cell are, therefore, independent processes, which have to coincide for successful infection.

The fact that cortical cell division leading to nodule formation can be uncoupled from infection thread formation and bacteroid differentiation has been confirmed by a number of studies with the R. meliloti–alfalfa system (Hirsch et al., 1984; Truchet et al., 1984; Hirsch et al., 1985) using cloned nod genes or entire Sym plasmids in Agrobacterium tumefaciens. Normally, A. tumefaciens does not nodulate alfalfa, but after introduction of the R. meliloti nod genes it induces pseudonodules, which are devoid of infection threads and do not contain endosymbiotic bacteria. In addition, transposon-induced mutants have been isolated from R. japonicum (Fig. 1) and R. meliloti (Finan et al., 1985) that produce bacteria-free pseudonodules lacking infection threads. Moreover, subcortical cell divisions seen in normal infections could be mimicked by cocultivating soybean seedlings separated from Rhizobium by ultrafiltration membranes (Bauer et al., 1985). Since Rhizobium nod genes are activated by diffusible small molecules from root exudates, this present finding raises the implication that a return signal is generated by the bacterial cells and that the signal induces cortical cell division. This may explain how infection thread formation can be uncoupled from cortical cell division.

The existence of a return signal was demonstrated by van Brussel et al. (1986), who used a combination of R. leguminosarum and vetch (Vicia sativa nigra) that results in a reaction known as thick short-root (Tsr), in which the plant develops a calluslike swollen root surface. The Tsr reaction is mediated by root exudate factors and the Rhizobium nod genes. Tsr can be produced by growing plants in concentrated root exudate in which bacteria had been cultured previously, showing that the physical presence of the Rhizobium is unnecessary. Mutations in the common nod region of R. leguminosarum abolish the Tsr effect, demonstrating a direct involvement of the nod gene products with the formation of a plant growth-regulating substance. These results suggest a ping-pong nature of plant–bacterial signaling in the initiation of symbiosis.

It is likely that at least two signals (hormonelike activities) are produced by the invading bacteria, one for root hair curling phenomena and one for cortical divisions, and that the nod genes are required for both those signals. Although phytohormones have been extensively implicated in nodule development, Rhizobium nod gene sequences do not show homologies to phytohormone gene sequences involved in phytohormone production.

3.2. Host Plant Control of Nodule Initiation

The host plant exerts control over the timing, frequency, and ultimate success of infection. In soybean, the region of highest susceptibility of the root

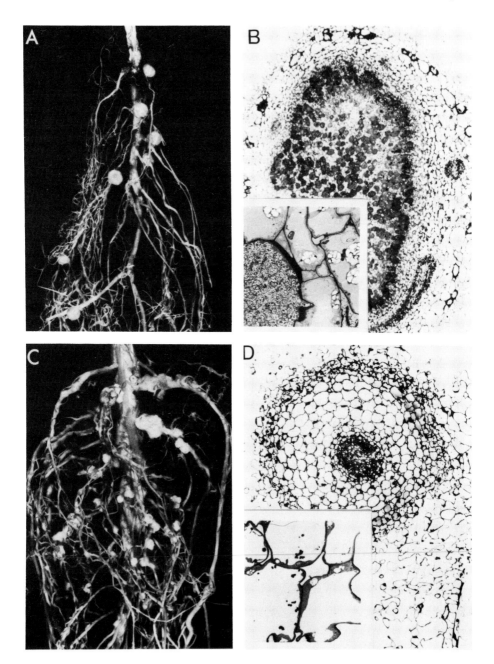

Figure 1. *Rhizobium japonicum (fredii)* mutant that induces uncoupled bacteria-free nodules on soybean. (**A**) Normal nodulation induced by the wild-type parent strain USDA 191. (**B**) Micrograph showing normal nodule morphology with bacteroid-packed infected cells and uninfected cells (inset) containing starch granules. (**C**) Uncoupled nodules of abnormal morphology resembling tumors on the roots of a soybean plant inoculated with a transposon-induced mutant, MU042. (**D**) Microscopy exhibits a central meristematic region and a lack of infection threads in the

coincides with a region near the root tip, where root hair growth initiates from otherwise undifferentiated epidermal cells (Bhuvaneswari et al., 1981). The highly susceptible zone moves down with root growth and any one point on the root is only transiently susceptible to infection. Regions of the root containing mature root hairs are not susceptible to infection in soybean, although they are in other legumes (clover, pea, alfalfa).

The development of an infection requires rhizobia inoculated near the root tip (the so-called zone of no root hairs) in which auxin-dependent cell extension is occurring. Maximal susceptibility coincides roughly with the cessation of elongation and the appearance of the growing root hair. In soybean, susceptibility declines as root hairs mature and cease growing, indicating that infection requires a young, actively growing root hair cell. It is unclear why the root hair cell is the predominant site of entry of invading rhizobia, since rare cases of infection thread formation in an unspecialized epidermal cell have been reported (Dart, 1977) and several other legumes, notably peanut (Arachis hypogea) and Sesbania rostrata, have different modes of infection (Chandler, 1978; Tsien et al., 1983) involving penetration through root cracks and aerial root primordia, respectively.

The number, size, and position of effective nodules are strictly regulated by the host plant. Many more infection foci are formed than the final nodule number, so a process must exist that terminates the development of legitimate nodule primordia. This control mechanism takes the form of feedback inhibition of new nodule development, and it apparently originates from nodules that have reached a certain size. The inhibition is fast acting, since prior inoculation of the root substantially inhibits nodulation by a second inoculum applied only a few hours afterwards (Pierce and Bauer, 1983). Subsequent studies (Calvert et al., 1984) showed that the secondary infections were blocked at an early stage after a cortical meristem had been established but before emergence (visible nodules) of the nodule from the root. If a nodule has reached the emergence stage, it continues to develop. It may be significant that vascular connections to the root are formed at this stage. A possible explanation is that once a particular nodule reaches this stage of development, it has formed a sink, which may perturb nutritional/hormonal balance and lead to inhibition of new nodule emergence.

Combined nitrogen (e.g., nitrate, urea) has a dramatic inhibitory effect on nodulation (see Dart, 1977) with complete inhibition of most legume nodulation being attained at 15 mM nitrate. Recently, soybean mutants have been isolated that nodulate in the presence of inhibitory levels of nitrate (Carroll et al., 1985). The nodules of these mutants are effective in nitrogen fixation, showing that the host rather than the bacteria does indeed control repression of nodulation by combined nitrogen. One class of mutant nodulates superabundantly (up to five times the number and density of nodules per root), even with

pseudonodule. Analysis of nodulin gene expression by immunoblotting (Stanley et al., 1986) and Northern hybridization using specific nodulin cDNA clones (Morrison and Verma, 1987) show that nodulin genes are not detectably expressed in this tissue.

added nitrate. In this mutant, nodulation occurs almost simultaneously as the root matures, indicating that self-inhibitory effects controlling nodule number are not operative. Grafting experiments between mutant root and wild-type shoot showed that the supernodulation phenotype is controlled by the shoot (P. Gresshoff, personal communication). Taken together, these data suggest that nodule number is controlled by an interaction between nodule primordia at a certain stage and the shoot of the plant.

4. Intercellular and Intracellular Compartmentalization

4.1. Nodule as an Organ

Two major types of nodule organization exist, the so-called **determinate** and **indeterminate nodules** (described in detail by Rolfe and Shine, 1984). Determinate nodules have a globose structure with meristematic activity for a limited period during early development (see Fig. 1). The cortex is punctuated by lenticel structures, which serve for gas exchange to the body of the nodule. The symbiotic region of the nodule has a dark red color attributable to leghemoglobin (Lb), a well-known plant-encoded hemoprotein with oxygen buffering characteristics (Appleby, 1984). Within the symbiotic region of the nodule, two plant cell types exist. Larger cells are packed with membrane enclosed bacteroids, the site of nitrogen fixation. Generally, in equal numbers and interspersed between the infected cells are smaller uninfected cells, which are specialized for the assimilation of fixed nitrogen to ureides and allantoin, the transport form of fixed nitrogen in these plants.

Indeterminate nodules have a persistent meristem, which forms the growing tip of cigar-shaped nodules. Behind the meristem, infection threads continue to release bacteria into the newly formed cells, which expand and become bacteroid filled; this symbiotic region comprises the bulk of the nodule. In this region, two cell types are again found; however, the function of the uninfected cell is unknown. The transport form of combined nitrogen in these plants is composed of amides (glutamine and asparagine), which can be made in the infected cells (Verma *et al.*, 1986). The region closest to the root contains cells that are senescing. The organ therefore has a spatial separation of developmental phases (meristematic activity, symbiosis, senescence), rather than the temporal separation seen in determinate nodules. Senescence in a nodule tissue can be triggered by physiological conditions affecting carbon–nitrogen metabolism.

4.2. Nodule Cell Types Express Different Genes

Nodule-specific gene products (**nodulins**) have been identified in both nodule types (Legocki and Verma, 1979, 1980; Bisseling *et al.*, 1983; Govers *et al.*, 1985; Lang-Unnash and Ausubel, 1985). Complementary DNAs (cDNAs) coding for these proteins have been prepared from RNA isolated from soybean (Fuller *et al.*, 1983). Analysis of genomic clones and subcellular localization of gene products (Nguyen *et al.*, 1985) have shown that leghemoglobin and nodule-specific uricase (nodulin-35) can serve as markers for the differentiation of two cell types in determinate nodules, since leghemoglobin is only found in the infected cells and nodulin-35 is found in the peroxisomes of the uninfected cell (Fig. 2). Studies with ineffective *Rhizobium* mutants (Fuller and Verma, 1984; Morrison and Verma, 1987) have shown that the differentiation of these two cell types with their characteristic expression of marker genes is independent

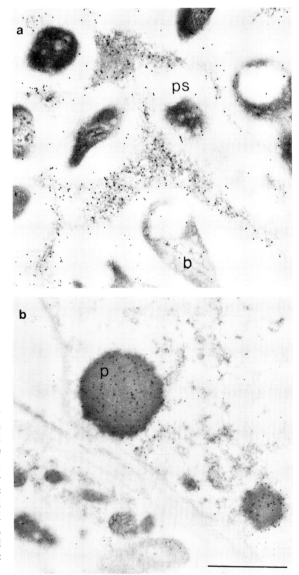

Figure 2. Immunocytochemical localization of (**a**) leghemoglobin in the cytoplasm of the infected cell and (**b**) nodulin-35 (uricase II) in the peroxisomes of the uninfected cell. Epon-embedded tissues were sectioned and reacted with specific antibodies to Lb and N-35 followed by Protein A-conjugated colloidal gold as described in Nguyen *et al.*, 1985. b, bacteroid; ps, peribacteroid space; p, peroxisomes. Scale bars: 1 μm.

of nitrogen fixation activity. Furthermore, this induction occurs before nitrogen fixation is detectable in the wild-type situation, indicating that cellular differentiation is dependent on the infection process rather than nitrogen fixation. These marker genes are still active (at a reduced level) in a mutant that could form infection threads but could not produce bacteroids (Morrison and Verma, 1987), but these genes were not expressed in pseudonodules induced by an uncoupled mutant of R. japonicum (fredii) (Stanley et al., 1986). These data support the hypothesis that differentiation of the two cell types requires the physical presence of bacteria in the infection thread.

Although cell commitment has not been studied in indeterminate nodules, the ease of genetic manipulation of the fast-growing Rhizobium had led to strain constructs with defined symbiotic genes. It is apparent that the bacterial nod genes themselves, although essential for nodule induction, are not sufficient for the persistence of the nodule meristem. A Sym plasmid-cured R. trifolii strain carrying only the essential nodulation genes produced nodules with only a limited meristematic activity (Schofield et al., 1984). The lack of meristem persistence in these nodules is not a result of ineffectivity per se, since in both clover and pea, ineffective Rhizobium mutants (Fix$^-$) are known that produce quite normal-looking nodules having a persistent meristem and bacteroid-packed symbiotic regions. Thus, it is likely that bacterial genes apart from the clustered nod genes are required for nodule development processes involving meristem persistence, differentiation, and bacteroid development (Stanley et al., 1986).

4.3. Intracellular Specialization

As invading bacteria are released from infection threads, they are taken up and packaged in a novel membrane system, the **peribacteroid membrane** **(PBM)**. This membrane is derived from the host plasma membrane, and a considerable increase in membrane biogenesis at this stage of nodulation occurs (Verma et al., 1978). The process of bacterial release and uptake of PBM is poorly understood biochemically. Electron microscopy (Bassett et al., 1979) shows that the infection thread fibrillar material is internalized in vesicles and presumably broken down before, during, and after uptake of bacteria in PBM. Coated pits characteristically associated with endocytotic events are also seen blebbing off the plant membrane surrounding the unwalled droplet (Robertson and Lyttleton, 1982). It is not known whether these coated pits are involved in internalizing infection thread wall or matrix material or are responsible for the uptake of the (larger) bacteria by a concerted endocytosis. However, attachment sites between bacterial outer membranes and the PBM during uptake and bacteroid development (Robertson and Lyttleton, 1984) are suggestive of ligand–receptor-mediated events that are usually necessary for endocytosis. The large increase in membrane biogenesis required for the synthesis of PBM is provided directly from the endoplasmic reticulum through the Golgi apparatus, producing smooth vesicles that fuse with PBM (Fig. 3) (Kijne and Planque, 1979; Robertson and Lyttleton, 1982; Brewin et al., 1985). A new choline kinase

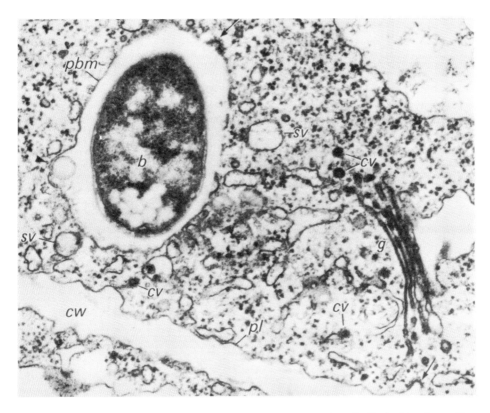

Figure 3. Biogenesis of the peribacteroid membrane in a young infected cell of white clover. Note the flow of smooth membrane vesicles (sv) from the Golgi apparatus (g) to the peribacteroid membrane (pbm), as well as the fusion of smooth vesicles with pbm and the presence of a coated pit (arrowed) and coated vesicles (cv) of an unknown function in the cytoplasm. b, bacteroid; cw, cell wall; pl, plasmalemma. (From Robertson and Lyttleton, 1982.)

activity has been associated with the synthesis of phosphatidylcholine for PBM (Mellor *et al.*, 1986), and the newly synthesized membrane incorporates several nodulins (Fig. 4) targeted specifically to PBM (Fortin *et al.*, 1985), presumably after synthesis on the rough endoplasmic reticulum and processing through the Golgi.

Nodulin-23 (Mauro *et al.*, 1984), nodulin-24 (Katinakis *et al.*, 1985), and nodulin-27 (Fuller *et al.*, 1983) of soybean are now known to be nodulins of PBM. This was demonstrated by immunoprecipitation (using PBM-specific antisera) of translation products from mRNA hybrid-selected using specific cDNA clones (M. Fortin, F. Jacobs, and D. P. S. Verma, unpublished data). The mechanism of targeting of these nodulins to PBM is presently unknown. However, cDNA sequences should help answer whether specific protein signals are involved. These nodulins should serve a necessary function in symbiosis, since the central importance of PBM to the success and stability of symbiosis is well established (Dilworth and Glenn, 1984; Werner *et al.*, 1984).

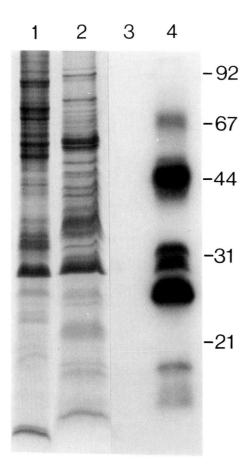

Figure 4. Nodule-specific host proteins of the peribacteroid membrane detected by specific antiserum in an immunoblot. Proteins of the uninfected root plasma membrane (lane 1) and the peribacteroid membrane (lane 2) were separated by gel electrophoresis and transferred to a nitrocellulose filter. The filter was reacted with an antiserum prepared against peribacteroid membranes and adsorbed with total root plasma membrane to remove cross-reacting antibodies. The resultant blot (visualized using radioactively labeled protein A) shows no reactivity with root plasma membrane (lane 3), whereas at least nine nodule-specific proteins of PBM are identified in lane 4.

Amino acid hydropathy analysis of PBM nodulins suggests that at least some of them are integral membrane proteins with characteristic membrane spanning helices. Although PBM nodulins may have a structural role in preserving the integrity of a membrane surrounding a foreign organism, they may have metabolic functions as well. PBM is the site of transit of many molecules to and from the cytoplasm and the bacteroid. Dicarboxylic acids, the energy source of bacteroids (Ronson *et al.*, 1981), must pass through this membrane, and the ammonium ion (the product of nitrogen fixation) must be imported to the plant cell. If PBM nodulins are involved in these processes, the characteristics associated with ion binding, counter-ion exchange, or facilitated diffusion should be found in the tertiary structure of these proteins.

5. Induction of Host Genes during Nodulation

5.1. Early Nodulins

Nodulins were initially identified by preparing antisera (Legocki and Verma, 1980; Bisseling *et al.*, 1983; Lang-Unnash and Ausubel, 1985) or cDNA clones (Fuller *et al.*, 1983) from mature nodule tissues. Time courses of nodulin induction were then followed by immunoblotting or Northern blotting during development. These approaches have not permitted identification of transiently expressed nodulins that may be involved in early infection processes or the establishment of a nodule meristem. In pea as well as in soybean, nodulin genes are expressed differentially during nodule development (Govers *et al.*, 1985, 1986). Most of the nodulins are expressed concomitantly with the leghemoglobin genes shortly before the onset of nitrogen fixation. However, a small number (two or three) of nodulin genes have been identified that are expressed at least 1 week before the leghemoglobin genes and are termed early nodulin genes. In soybean, the expression of these genes is first detectable in young nodules that just distend the epidermis (4 days after infection). If the induced meristems reach this stage of development, they will continue the differentiation process, whereas before this stage they can be aborted (Calvert *et al.*, 1984). However, whether there is a causative relation between the expression of early nodulin genes and the fact that the nodule meristems pass this critical developmental stage cannot be concluded yet.

The mechanism by which the expression of nodulin genes is elicited is still unclear. Studies on nodules formed by *Agrobacterium* transconjugants and rhizobia that carry only small parts of the Sym plasmid suggest that the nodulation genes of *Rhizobium* are involved in the induction of the early nodulin genes. In pea nodules formed by an *Agrobacterium* transconjugant carrying a Sym plasmid from *R. leguminosarum*, an early nodulin gene (ENOD2) is expressed, while the other nodulin mRNAs (represented by leghemoglobin in Fig. 5) are not detectable. This result suggests that at least two different signals are involved in the induction of these genes. *Agrobacterium tumefaciens* itself is not capable of inducing the expression of the early nodulin gene, suggesting that the expression of this gene is regulated by *Rhizobium* genes located on the Sym plasmid. The region of the Sym plasmid that is involved in the induction of the early nodulin gene was identified by using a *Rhizobium* strain cured of its own Sym plasmid. When *nod* gene clones are transferred to this *Rhizobium* strain, the recipients regain the ability to form nodules in which the early nodulin gene is expressed. This result suggests that the *nod* genes of *Rhizobium* are involved in the induction of the early nodulin gene. Since most nodulin genes are not expressed in pea nodules formed by *Agrobacterium* carrying the Sym plasmid, it seems plausible that *R. leguminosarum* genes not located on this plasmid are required to elicit the expression of these nodulin genes (Govers *et al.*, 1986).

5.2. Structural and Metabolic Nodulins

Some nodulins have been assigned enzymatic activities and can be considered metabolic nodulins. Other nodulins may account for increases in enzyme activities seen in mature nodules (Verma and Nadler, 1984). A clear example of a metabolic nodulin, nodulin-35 of soybean, is the subunit of a nodule-specific uricase (Nguyen *et al.*, 1985). Apart from nodulin-35, a nodule-specific form of glutamine synthetase has been identified in *Phaseolus vulgaris* (Cullimore *et*

a

b

1　　**2**　　**3**　　　**1**　　**2**　　**3**

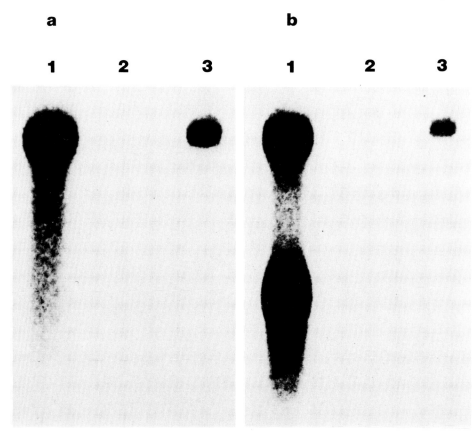

Figure 5. Northern blot of pea nodule messenger RNA (mRNA) showing the expression of an early nodulin gene in empty (bacteroid-free) nodules induced by an *Agrobacterium tumefaciens* strain carrying the *Rhizobium leguminosarum* Sym plasmid. (From Hooykaas *et al.*, 1981.) (**a**) Lane 1, wild-type nodule; lane 2, root; lane 3, "empty" nodule hybridized with a specific cDNA clone ENOD2. (**b**) Same blot hybridized with a leghemoglobin cDNA probe, showing that leghemoglobin genes are not induced, whereas the early nodulin gene is induced by the *A. tumefaciens* pSym hybrid.

al., 1983). The nodule glutamine synthetase is an isoenzyme that shares immunological and sequence (based on mRNA cross-hybridization) homologies with root and leaf forms of the enzyme. Since ammonium is assimilated by glutamine synthetase and the intracellular environment of the infected cell in the nodule is different from that in root cells, it is possible that a new form of the enzyme is involved in processing the large quantity of ammonium fixed by the bacteroids. Assimilating ammonium requires a high supply of carbon skeletons from the tricarboxylic acid (TCA) cycle to provide substrates for glutamate and glutamine synthetases, and the complexities of balancing the carbon–nitrogen metabolism in nodules has been a prime concern of physiologists (Dilworth and Glenn, 1984). Less evidence is available on other metabolic nodulins, although enzymes required for ureide metabolism, such as xanthine dehydrogenase (Triplett, 1985), in determinate nodule plants are probably nodulins. In addition, several other nodule-specific isoenzymes may exist that could not be detected by the approaches used so far in identifying nodulins.

5.3. Control of Nodulin Gene Expression

Tissue-specific expression of nodulin genes suggests that these genes share some regulatory sequences. Most nodulin messenger RNAs (mRNAs) appear in a roughly coordinated fashion during normal nodule development (Fuller and Verma, 1984; Govers et al., 1985), although full accumulation of individual messages may vary in time and extent of induction. However, the regulation of nodulin gene induction is not a simple case of coordinate expression governed by a single regulatory sequence motif. We have seen, for example, that nodulin-35 and leghemoglobin are differentially expressed in two cell types of the nodule. Sequences in the 5′ region of the nodulin-35 gene contain no significant homology to those in the 5′ region of the leghemoglobin, nodulin-23, and nodulin-24 genes. The latter three nodulin genes do have a significantly conserved and repeated sequence motif (Fig. 6), which may be involved in regulating a coordinate induction of at least these three genes by putative *trans*-acting nuclear factors (Mauro et al., 1985). Nodulins-23 and -24 are PBM nodulins, and since these are only found in the infected cells (as is leghemoglobin), it is assumed (but not proven) that the genes are only expressed in this cell type. Nodulin-35 is currently the only nodulin known from the uninfected cell; we must await elaboration of the sequence of another such nodulin gene before concluding that a different sequence motif controls nodulin gene expression in the uninfected cell.

It is clear from studies using ineffective mutants that nitrogen fixation activity is not necessary for the induction of any nodulin gene, and indeed most nodulin genes are maximally expressed before nitrogen fixation in the normal situation. In pea, nodulin-21 was originally identified as being late induced and absent from ineffective nodules (Govers et al., 1985). It now appears that this nodulin is present in Fix⁻ nodules (Govers et al., 1986). However, *Rhizobium* mutants do influence the level of expression of nodulin genes, resulting in depressed mRNA and protein levels. Another level of control of expression may be indicated by the fact that a Fix⁻ mutant that produces bacteroids causes maximal induction of PBM nodulin genes, whereas a Tn5-induced bacteroid development mutant (Bad⁻), which cannot be packaged in PBM, results in low expression of these plant genes (Morrison and Verma, 1987).

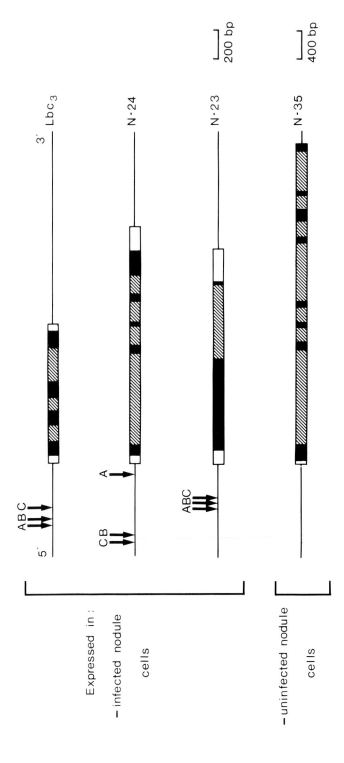

Figure 6. General structure of four soybean nodulin genes. (■) Coding regions. (□) Untranslated regions of the messenger RNA (mRNA). Sequences A, B, and C are sequence motifs found in the 5′ regions of nodulin genes expressed only in the infected cells and are potential candidates for tissue-specific regulatory sequences. The consensus sequence for A is TAAAAAAATTGTTTCCCTATT; B is TAT-AAGGTAGTGACAAATTAATA; and C is TCTGGGAAA. For further information, see Mauro et al. (1985).

6. Perspectives

The nodule is a simple organ composed of a symbiotic region, with two differentiated cell types, surrounded by a cortex and vascular connections to the root. This organ is derived through complex interactions between prokaryotic and eukaryotic partners, involving genetically and environmentally determined susceptibility and compatibility at various stages of development. The crucial importance of molecular signaling in the interaction is obvious from the induction of bacterial *nod* genes by root exudate substances. The *nod* genes, in turn, are responsible for eliciting the nodule and, with a contribution from genes on the *Rhizobium* chromosome, cause (directly or indirectly) the induction of a number of nodulin genes in the host. Thus, a ping-pong mechanism is involved in *Rhizobium*–plant communication, and some of these signals may be common to the phytopathogenic bacteria (e.g., *A. tumefaciens*).

A program of cellular specialization in the nodule leads to two major cell types, each expressing a different set of nodulins. It is not known how early in development this commitment is made or how each cell type is controlled. However, the problem can now be addressed at a subcellular level by *in situ* hybridization of nodulin sequences or by immunocytochemical approaches.

Since nodulin gene induction is not dependent on nitrogen fixation, some events during infection must be the triggers for cell commitment and nodulin induction. If the nodulin inducer is a diffusible factor, perhaps emanating from the infection thread, the uninfected cell type may have some means of ignoring the signal or switching on another developmental program to the same signal. Nucleotide sequence motifs with potential regulatory activity may explain the induction of nodulins in the infected cell type. Other regulatory mechanisms may account for expression of nodulins in the uninfected cells of nodules. So far, *Rhizobium* mutants have not been found that can alter cell commitment and induce only leghemoglobin or nodulin-35 (in soybean) in the nodule. The possibility of multiple signals for nodulin gene induction still remains.

Another unsolved problem of nodule biology is how the infection thread grows through the developing nodule and finally chooses a cell in which to release bacteria. In indeterminate nodules, release occurs in a region just behind the nodule meristem. In the root, cytokinin-induced cell division gives way to an auxin-stimulated expansion in equivalent zones behind the meristem. The timing of release therefore may follow the ratio of putative nodule-specific cytokinin- and auxinlike activities. Following release, cell expansion, rather than division, accounts for the growth of the symbiotic zone, just as it does in the main part of the root. By contrast, the determinate nodule has temporal, rather than spatial, distribution of meristematic activity and bacteroid development; in soybean, release events and subsequent bacteroid development occur just after the cessation of meristematic activity.

The development of a specific structure, the nodule, on the plant meets the physiological needs for the process of nitrogen fixation by separating the endosymbiont from the rest of the plant in a favorable microenvironment. Intracellular lodging of the bacteroids in a distinct subcellular compartment separates the microaerophilic requirement for nitrogen fixation from oxidative processes of the host cell. PBM protects the host cytoplasm from the bacteria and serves as an interface for symbiosis, regulating exchanges between the two organisms. A specific set of (nodulin) genes has evolved in legumes to fulfill the physiological and structural requirements to maintain the large population of bacteria endosymbiotically. The discovery of hemoglobins in nonlegume symbiotic nodules suggests that the nodulin concept also applies to other symbiotic nitrogen-fixing systems.

References

Allen, O. N., and Allen, E. K., 1981, The Leguminosae, A Source Book of Characteristics, Uses and Nodulation, University of Wisconsin Press, Madison, Wisconsin.

Appleby, C. A., 1984, Leghemoglobin and Rhizobium respiration, Annu. Rev. Plant Physiol. **35**:443–478.

Bassett, B., Goodman, R. N., and Novacky, A., 1979, Ultrastructure of soybean nodules. I. Release of rhizobia from the infection thread, Can. J. Microbiol. **23**:873–883.

Bauer, W. D., 1981, Infection of legumes by rhizobia, Annu. Rev. Plant Physiol. **32**:407–449.

Bauer, W. D., Bhuvaneswari, T. V., Calvert, H. E., Malik, N. S. A., and Vesper, S. J., 1985, Recognition and infection by slow growing rhizobia, in: Nitrogen Fixation Research Progress (H. J. Evans, P. J. Bottomley, and W. E. Newton, eds.), pp. 247–253, Niihoff, Dordrecht.

Bhuvaneswari, T. V., Turgeon, B. G., and Bauer, W. D., 1980, Early events in the infection of soybean Glycine max (L.) Merr. by Rhizobium japonicum. I. Localization of infective root cells, Plant Physiol. **66**:1027–1031.

Bhuvaneswari, T. V., Bhagwat, A. A., and Bauer, W. D., 1981, Transient susceptibility of root cells in four common legumes to nodulation by rhizobia, Plant Physiol. **68**:1144–1149.

Bisseling, T., Been, C., Klugkist, J., van Kammen, A., and Nadler, K., 1983, Nodule-specific host proteins in effective and ineffective root nodules of Pisum sativum, EMBO J. **2**:961–996.

Brewin, N. J., Robertson, J. G., Wood, E. A., Wells, B., Larkins, A. P., Galfre, G., and Butcher, G. W., 1985, Monoclonal antibodies to antigens in the peribacteroid membrane from Rhizobium-induced root nodules of pea cross-react with plasma membranes and Golgi bodies, EMBO J. **4**:605–611.

Callaham, D. A., and Torrey, J. G., 1981, The structural basis for infection of root hairs of Trifolium repens by Rhizobium, Can. J. Bot. **59**:1647–1664.

Calvert, H. E., Pence, M., Pierce, M., Malik, N. S. A., and Bauer, W. D., 1984, Anatomical analysis of the development and distribution of Rhizobium infections in soybean roots, Can. J. Bot. **62**:2375–2384.

Carroll, B. J., McNeil, D. L., and Gresshoff, P. M., 1985, Isolation and properties of novel soybean (Glycine max (L.) Merr.) mutants that nodulate in the presence of high nitrate concentrations, Proc. Natl. Acad. Sci. USA **82**:4162–4166.

Chandler, M. R., 1978, Some observations of infection of Arachis hypogaea L. by Rhizobium, J. Exp. Bot. **29**:749–755.

Cullimore, J. V., Lara, M., Lea, P. J., and Miflin, B. J., 1983, Purification and properties of two forms of glutamine synthetase from the plant fraction of Phaseolus root nodules, Planta **157**:245–253.

Dart, P. J., 1974, The development of root nodule symbioses. The infection process, in: Biology of Nitrogen Fixation (A. Quispel, ed.), pp. 381–429, North-Holland, Amsterdam.

Dart, P. J., 1977, Infection and development of leguminous nodules, in: A Treatise on Dinitrogen Fixation (R. W. F. Hardy, ed.), pp. 367–472, Wiley, New York.

Dazzo, F. B., and Gardiol, A. E., 1984, Host-specificity in Rhizobium–legume interactions, in: Genes Involved in Microbe-Plant Interactions (D. P. S. Verma and T. Hohn, eds.), pp. 3–31, Springer-Verlag, Vienna.

Devine, T. E., 1984, Inheritance of soybean nodulation response with a fast-growing strain of Rhizobium, Heredity **75**:359–361.

Dilworth, M., and Glenn, A., 1984, How does a legume nodule work?, Trends Biochem. Sci. **9**:519–523.

Djordjevic, M. A., Schofield, P. R., and Rolfe, B. G., 1985, Tn5 mutagenesis of Rhizobium trifolii host-specific nodulation genes result in mutants with altered host range ability, Mol. Gen. Genet. **200**:463–471.

Downie, J. A., Knight, C. D., Johnston, A. W. B., and Rossen, L., 1985, Identification of genes and gene products involved in the nodulation of peas by Rhizobium leguminosarum, Mol. Gen. Genet. **198**:255–262.

Egelhoff, T. T., Fisher, R. F., Jacobs, T. W., Mulligan, J. T., and Long, S. R., 1985, Nucleotide sequence of Rhizobium meliloti 1021 nodulation genes: nodD is read divergent from nodABC, DNA **4**:241–248.

Finan, T. M., Hirsch, A. M., Leigh, J. A., Johansen, E., Kuldau, G. A., Deegan, S., Walker, G. C., and Signer, E. R., 1985, Symbiotic mutants of Rhizobium meliloti that uncouple plant from bacterial differentiation, Cell **40**:869–877.

Fortin, M. G., Zelechowska, M., and Verma, D. P. S., 1985, Specific targetting of membrane nodulins to the bacteroid-enclosing compartment in soybean nodules, EMBO J. **4**:3041–3046.

Fuller, F., Kunstner, P. W., Nguyen, T., and Verma, D. P. S., 1983, Soybean nodulin genes: Construction and analysis of cDNA clones reveals several tissue-specific sequences expressed in nitrogen-fixing root nodules, Proc. Natl. Acad. Sci. USA **80**:2594–2598.

Fuller, F., and Verma, D. P. S., 1984, Accumulation of nodulin mRNAs during the development of effective root nodules of soybean, Plant Mol. Biol. **3**:21–28.

Govers, S., Gloudemans, T., Moerman, M., van Kammen, A., and Bisseling, T., 1985, Expression of plant genes during the development of pea root nodules, EMBO J. **4**:861–867.

Govers, S., Moerman, M., Downie, A. J., Hooykaas, P., Franssen, H., Louwerse, J., van Kammen, A., and Bisseling, T., 1986, Nodulation genes of Rhizobium are involved in inducing the expression of an early nodulin gene, Nature (Lond.) submitted.

Hirsch, A. M., Wilson, K. J., Gones, J. D. G., Bang, M.. Walker, V. V., and Ausubel, F. M., 1984, Rhizobium meliloti nodulation genes allow Agrobacterium tumefaciens and Escherichia coli to form pseudonodules on alfalfa, J. Bacteriol. **158**:1133–1143.

Hirsch, A. M., Drake, D., Jacobs, T. W., and Long, S. R., 1985, Nodules are induced on alfalfa roots by Agrobacterium tumefaciens and Rhizobium trifolii containing small segments of the Rhizobium meliloti nodulation region, J. Bacteriol. **161**:223–230.

Hong, G. F., Burn, J. E., and Johnston, A. W. B., 1987, Evidence that DNA involved in the expression of nodulation (nod) genes in Rhizobium binds to the product of the regulatory gene nod D, Nucl. Acid. Res. **15**:9677–9690.

Hooykaas, P. J. J., van Brussel, A. A. N., den Dulk-Ras, H., van Slogteren, G. M. S., and Schilperoort, R. A., 1981, Sym plasmid of Rhizobium trifolii expressed in different rhizobial species and Agrobacterium tumefaciens, Nature (Lond.) **291**:351–353.

Innes, R. W., Kuempel, P. L., Plazinski, J., Canter-Cremers, H., Rolfe, B. G., and Djordjevic, M. A., 1985, Plant factors induce expression of nodulation and host-range genes in Rhizobium trifolii, Mol. Gen. Genet. **201**:426–432.

Jacobs, T. W., Egelhoff, T. T., and Long, S. L., 1985, Physical and genetic map of a Rhizobium meliloti nodulation gene region and nucleotide sequence of nodC, J. Bacteriol. **162**:469–476.

Jordan, D. C., 1982, Transfer of Rhizobium japonicum Buchanan 1980 to Bradyrhizobium gen. nov., a genus of slow-growing root nodule bacteria from leguminous plants, Int. J. Syst. Bacteriol. **32**:136–139.

Katinakis, P., and Verma, D. P. S., 1985, Nodulin-24 gene of soybean codes for a polypeptide of the peribacteroid membrane and was generated by tandem duplication of a sequence resembling an insertion element, Proc. Natl. Acad. Sci. USA **82**:4157–4161.

Keyser, H. H., Bohlool, B. B., Hu, T. S., and Weber, D. F., 1982, Fast-growing rhizobia isolated from root nodules of soybean, Science **215**:1631–1632.

Kijne, J. W., and Planque, K.. 1979, Ultrastructural study of the endomembrane system in infected cells of pea and soybean root nodules, Physiol. Plant Pathol. **14**:339–345.

Kondorosi, E., Banfalvi, Z., and Kondorosi, A., 1984, Physical and genetic analysis of a symbiotic region of *Rhizobium meliloti:* Identification of nodulation genes, *Mol. Gen. Genet.* **193:**445–452.

Lang-Unnasch, N., and Ausubel, F. M., 1985, Nodule-specific polypeptides from effective alfalfa root nodules and from ineffective nodules lacking nitrogenase, *Plant Physiol.* **77:**833–839.

LaRue, T. A., Kneen, B. E., and Gartside, E., 1985, Plant mutants defective in symbiotic nitrogen fixation, in: *Analysis of the Plant Genes Involved in the Legume-Rhizobium Symbiosis,* pp. 39–48, Organization for Economic and Cultural Development, Paris.

Legocki, R. P., and Verma, D. P. S., 1979, A nodule-specific plant protein (Nodulin-35) from soybean, *Science* **205:**190–193.

Legocki, R. P., and Verma, D. P. S., 1980, Identification of "nodule-specific" host proteins (nodulins) involved in the development of *Rhizobium*–legume symbiosis, *Cell* **20:**153–163.

Mauro, V. P., Nguyen, T., Katinakis, P., and Verma, D. P. S., 1985, Primary structure of nodulin-23 gene and potential regulatory elements in the 5′ flanking regions of nodulin and leghemoglobin genes, *Nucl. Acids Res.* **13:**239–249.

Mellor, R. B., Christensen, T. M. I. E., and Werner, D., 1986, Choline kinase II is present only in nodules that synthesize stable peribacteroid membranes, *Proc. Natl. Acad. Sci. USA* **83:**659–663.

Morrison, N. A., Hau, C. Y., Trinick, M. J., Shine, J., and Rolfe, B. G., 1983, Heat curing of a sym plasmid in a fast-growing *Rhizobium* sp. that is able to nodulate legumes and the non-legume *Parasponia* sp., *J. Bacteriol.* **153:**527–531.

Morrison, N. A., and Verma, D. P. S., 1987, A block in the endocytosis of *Rhizobium* allows cellular differentiation in nodules but affects the expression of some peribacteroid membrane nodulins, *Plant Molecular Biology* **9:**185–196.

Mulligan, J. T., and Long, S. R., 1985, Induction of the *nodC* gene by root exudate requires the *nodD* product in *Rhizobium meliloti, Proc. Natl. Acad. Sci. USA* **82:**6609–6613.

Nguyen, T., Zelechowska, M. G., Foster, V., Bergmann, H., and Verma, D. P. S., 1985, Primary structure of the soybean nodulin-35 gene encoding nodule-specific uricase localized in peroxisomes of uninfected cells of soybean, *Proc. Natl. Acad. Sci. USA* **82:**5040–5044.

Peters, K. N., Frost, J. W., and Long, S. R., 1986, A plant flavone, luteolin, induces expression of *Rhizobium meliloti* nodulation genes, *Science* **223:**977–979.

Pierce, M., and Bauer, W. D., 1983, A rapid regulatory response governing nodulation in soybean, *Plant Physiol.* **73:**286–290.

Ridge, R. W., and Rolfe, B. G., 1985, *Rhizobium* sp. degradation of legume root hair cell wall at the site of infection thread origin, *Appl. Environ. Microbiol.* **50:**717–721.

Robertson, J. G., and Lyttleton, P., 1982, Coated and smooth vesicles in the biogenesis of cell walls, plasma membranes, infection threads and peribacteroid membranes in root hairs and nodules of white clover, *J. Cell Sci.* **58:**63–78.

Robertson, J. G., and Lyttleton, P., 1984, Division of peribacteroid membranes in root nodules of white clover, *J. Cell Sci.* **69:**147–157.

Rolfe, B. G., and Shine, J., 1984, *Rhizobium*–leguminosae symbiosis: The bacterial point of view, in: *Genes Involved in Microbe–Plant Interactions* (D. P. S. Verma and T. Hohn, eds.), pp. 95–128, Springer-Verlag, Vienna.

Ronson, C. W., Lyttleton, P., and Robertson, J. G., 1981, C_4-dicarboxylate transport mutants of *Rhizobium trifolii* from ineffective nodules of *Trifolium repens, Proc. Natl. Acad. Sci. USA* **78:**4284–4288.

Rossen, L., Johnston, A. W. B., and Downie, J. A., 1984, DNA sequence of the *Rhizobium leguminosarum* nodulation genes *nodAB* and C required for root hair curling, *Nucl. Acids Res.* **12:**9497–9508.

Schofield, P. R., Ridge, R. W., Rolfe, B. G., Shine, J., and Watson, J. M., 1984, Host-specific nodulation is encoded on a 14 kb DNA fragment in *Rhizobium trifolii, Plant Mol. Biol.* **3:**3–11.

Stanley, J., Brown, G. G., and Verma, D. P. S., 1985, Slow-growing *Rhizobium japonicum* comprises two highly divergent symbiotic types, *J. Bacteriol.* **163:**148–154.

Stanley, J., Longtin, D., and Verma, D. P. S., 1986, A genetic locus in *Rhizobium japonicum (fredii)* affecting soybean root nodule differentiation, *J. Bacteriol.* **166:**628–634.

Sutton, W. D., Pankhurst, C. E., and Craig, A. S., 1982, The Rhizobium bacteroid state, in: *International Review of Cytology, Suppl. 13, Biology of Rhizobiaceae* (K. L. Giles, A. G. Atherly, eds.), pp. 149–177, Academic Press, New York.

Torok, I., Kondorosi, E., Stepkowski, T., Posfai, J., and Kondorosi, J., 1984, Nucleotide sequence of *Rhizobium meliloti* nodulation genes, *Nucl. Acids Res.* **12:**9509–9524.

Triplett, E., 1985, Intercellular nodule localization and nodule specificity of xanthine dehydrogenase in soybean, *Plant Physiol.* **77:**1004–1009.

Truchet, G., Rosenberg, C., Vasse, J., Julliot, J. S., Camut, S., and Denarie, J., 1984, Transfer of *Rhizobium meliloti* pSym genes into *Agrobacterium tumefaciens:* Host-specific nodulation by atypical infection, *J. Bacteriol.* **157:**134–142.

Tsien, H. C., Dreyfus, B. L., and Schmidt, E. L., 1983, Initial stages in the morphogenesis of nitrogen-fixing stem nodules of *Sesbania rostrata, J. Bacteriol.* **156:**888–897.

Turgeon, B. G., and Bauer, W. D., 1982, Early events in the infection of soybean by *Rhizobium japonicum.* Time course and cytology of the initial infection process, *Can. J. Bot.* **60:**152–161.

Turgeon, B. G., and Bauer, W. D., 1985, Ultrastructure of infection thread development during the infection of soybean by *Rhizobium japonicum, Planta* **163:**328–349.

van Brussel, A. A. N., Zaat, S. A. J., Canter Cremers, H. C. J., Wijffelman, C. A., Pees, E., Tak, T., and Lugtenberg, B. J. J., 1986, Role of plant root exudate and Sym plasmid localized nodulation genes in the synthesis by *Rhizobium leguminosarum* of Tsr factor, which causes thick and short roots on common vetch, *J. Bacteriol.* **165:**517–522.

Verma, D. P. S., and Long, S., 1983, The molecular biology of *Rhizobium*–legume symbiosis, *Int. Rev. Cytol.* **14:**211–245.

Verma, D. P. S., and Nadler, K., 1984, Legume–*Rhizobium* symbiosis: Host's point of view, in: *Genes Involved in Microbe–Plant Interactions,* pp. 57–93, Springer-Verlag, Vienna.

Verma, D. P. S., and Stanley, J., 1986, The *Rhizobium*–legume equation: A co-evolution of two genomes, in: *Advances in Legume Systematics* (C. Stirton, ed.), pp. 50–70.

Verma, D. P. S., Kazazian, V., Zogbi, V., and Bal, A. K., 1978, Isolation and characterization of the membrane envelope enclosing the bacteroids in soybean root nodules, *J. Cell. Biol.* **78:**919–936.

Verma, D. P. S., Fortin, M. G., Stanley, J., Mauro, V. P., Purohit, S., and Morrison, N., 1986, Nodulins and nodulin genes of *Glycine max:* A perspective, *Plant Mol. Biol.* **7:**51–61.

Werner, D., Morschel, E., Kort, R., Mellor, R. B., and Bassarab, S., 1984, Lysis of bacteroids in the vicinity of the host cell nucleus in an ineffective (Fix⁻) root nodule of soybean (*Glycine max*), *Planta* **162:**8–16.

Index